Boundary Element Topics

Springer
Berlin
Heidelberg
New York
Barcelona
Budapest
Hong Kong
London
Milan
Paris
Santa Clara
Singapur
Tokyo

Wolfgang L. Wendland (Editor)

Boundary Element Topics

Proceedings of the Final Conference
of the Priority Research Programme
Boundary Element Methods 1989-1995
of the German Research Foundation
October 2-4, 1995 in Stuttgart

With 151 Figures

Springer

Prof. Dr.-Ing. Wolfgang L. Wendland

Universität Stuttgart
Mathematisches Institut A
Pfaffenwaldring 57
D-70569 Stuttgart / Germany

ISBN 3-540-62850-9 Springer-Verlag Berlin Heidelberg New York

CIP data applied for

Die Deutsche Bibliothek - CIP-Einheitsaufnahme

Boundary element topics : proceedings of the Final Conference of the Priority Research Programme Boundary Element Methods 1989 - 1995 of the German Research Foundation, October 2 - 4, 1995 in Stuttgart / Wolfgang L. Wendland (ed.). - Berlin ; Heidelberg ; New York ; Barcelona ; Budapest ; Hong Kong ; London ; Milan ; Paris ; Santa Clara ; Singapur ; Tokyo : Springer, 1998
ISBN 3-540-62850-9

Printed in Germany

Typesetting: camera ready by author
Cover: Design & Production, Heidelberg
SPIN:1051767 61/3020-5 4 3 2 1 0 - Printed on acid-free paper

List of Contributors

Dr. H. Andrä
Karlsruhe University, Institute of Solid Mechanics, Kaiserstraße 12, Box 6980, 76131 Karlsruhe, Germany.

Prof. Dr. H. Antes
Technical University of Brunswick, Institute for Applied Mechanics, Spielmannstr. 11, 38106 Brunswick, Germany.

Dr. D. Berthold
Technical University of Chemnitz–Zwickau, Faculty of Mathematics, 09107 Chemnitz, Germany.

Dipl.-Math. U. Brink
University of Hanover, Institute for Structural and Computational Mechanics, Appelstr. 9A, 30167 Hanover, Germany.

Prof. Dr. W. Dahmen
RWTH Aachen, Institute of Geometry and Practical Mathematics, Templergraben 55, 52062 Aachen, Germany.

Prof. Dr.-Ing. L. Gaul
University of Stuttgart, Institute A of Mechanics, Pfaffenwaldring 9, 70550 Stuttgart, Germany.

Dr.-Ing. C. Haack
Technical University Hamburg–Harburg, Ocean Engineering Section II, Eißendorfer Str. 42, 21073 Hamburg, Germany.

Dr. G. Haase
Johannes Kepler University Linz, Institute of Mathematics, Altenberger Str. 69, A 4040 Linz, Austria.

Prof. Dr. W. Hackbusch
Christian-Albrechts-University of Kiel, Lehrstuhl für Praktische Mathematik, Mathematisches Seminar, 24098 Kiel, Germany.

Dr. T. Hartmann
University of Hanover, Institute for Applied Mathematics, Welfengarten 1, 30167 Hanover, Germany.

Dr. B. Heise
Johannes Kepler University Linz, Institute of Mathematics, Altenberger Str. 69, A 4040 Linz, Austria.

Dr. R. Hinder
TH Darmstadt, Department of Mathematics, Schloßgartenstr. 7,
64289 Darmstadt, Germany.

Dr. M. Jäger
Technical University of Brunswick, Institute for Applied Mechanics,
Spielmannstr. 11, 38106 Brunswick, Germany.

Prof. Dr. L. Jentsch
Technical University of Chemnitz–Zwickau, Faculty of Mathematics,
09107 Chemnitz, Germany.

Prof. Dr. K. Kalik
University of Stuttgart, Mathematical Institute A, Pfaffenwaldring 57,
70569 Stuttgart, Germany.

Dipl.-Math. B. Kleemann
Berlin Institute of Optics (BIFO), Rudower Chaussee 5,
12484 Berlin, Germany.

Dr. M. Kreienmeyer
University of Hanover, Institute for Structural and Computational
Mechanics, Appelstr. 9A, 30167 Hanover, Germany.

Prof. Dr.-Ing. G. Kuhn
University of Erlangen–Nürnberg, Lehrstuhl für Technische Mechanik,
Box 3429, 91022 Erlangen, Germany.

Dipl.-Math. M. Kuhn
Johannes Kepler University Linz, Institute of Mathematics,
Altenberger Str. 69, A 4040 Linz, Austria.

Dr. C. Lage
Christian-Albrechts-University of Kiel, Lehrstuhl für Praktische Mathematik,
Mathematisches Seminar, 24098 Kiel, Germany.

Prof. Dr. U. Langer
Johannes Kepler University Linz, Institute of Mathematics,
Altenberger Str. 69, A 4040 Linz, Austria.

Prof. Dr.-Ing. O. Mahrenholtz
Technical University Hamburg–Harburg, Ocean Engineering Section II,
Eißendorfer Str. 42, 21073 Hamburg, Germany.

Dr. M. Maischak
University of Hanover, Institute for Applied Mathematics, Welfengarten 1,
30167 Hanover, Germany.

Prof. Dr. E. Meister
TH Darmstadt, Department of Mathematics, Schloßgartenstr. 7,
64289 Darmstadt, Germany.

Dr. D. Mirschinka
Technical University of Chemnitz–Zwickau, Faculty of Mathematics,
09107 Chemnitz, Germany.

Dipl.-Ing. P. Partheymüller
University of Erlangen–Nürnberg, Lehrstuhl für Technische Mechanik,
Box 3429, 91022 Erlangen, Germany.

Prof. Dr. S. Prößdorf
Weierstrass Institute of Applied Analysis and Stochastics (WIAS),
Mohrenstr. 39, 10117 Berlin, Germany.

Dipl.-Math. R. Quatember
University of Stuttgart, Mathematical Institute A, Pfaffenwaldring 57,
70569 Stuttgart, Germany.

Dr. A. Röttgermann
University of Stuttgart, Institut of Aerodynamics and Gasdynamics,
Pfaffenwaldring 21, 70569 Stuttgart, Germany.

Prof. Dr. A.–M. Sändig
University of Stuttgart, Mathematical Institute A, Pfaffenwaldring 57,
70569 Stuttgart, Germany.

Dr. S. Sauter
Christian-Albrechts-University of Kiel, Lehrstuhl für Praktische Mathematik,
Mathematisches Seminar, 24098 Kiel, Germany.

Dr.-Ing. M. Schanz
Technical University of Brunswick, Institute for Applied Mechanics,
Spielmannstr. 11, 38106 Brunswick, Germany.

Dr.-Ing. V. Schlegel
Technical University Hamburg–Harburg, Ocean Engineering Section II,
Eißendorfer Str. 42, 21073 Hamburg, Germany.

Prof. Dr. E. Schnack
Karlsruhe University, Institute of Solid Mechanics, Kaiserstraße 12, Box 6980,
76131 Karlsruhe, Germany.

Dr. R. Schneider
TH Darmstadt, Department of Mathematics, Schloßgartenstr. 7,
64289 Darmstadt, Germany.

Dr. H. Schulz
University of Stuttgart, Mathematical Institute A, Pfaffenwaldring 57,
70569 Stuttgart, Germany.

Prof. Dr. B. Silbermann
Technical University of Chemnitz–Zwickau, Faculty of Mathematics,
09107 Chemnitz, Germany.

Prof. Dr. E. Stein
University of Hanover, Institute for Structural and Computational Mechanics, Appelstr. 9A, 30167 Hanover, Germany.

Dr. O. Steinbach
University of Stuttgart, Mathematical Institute A, Pfaffenwaldring 57, 70569 Stuttgart, Germany.

Prof. Dr. E. P. Stephan
University of Hanover, Institute for Applied Mathematics, Welfengarten 1, 30167 Hanover, Germany.

Dr. Sz. Szikrai
Karlsruhe University, Institute of Solid Mechanics, Kaiserstraße 12, Box 6980, 76131 Karlsruhe, Germany.

Dr. G. Tröndle
Technical University of Brunswick, Institute for Applied Mechanics, Spielmannstr. 11, 38106 Brunswick, Germany.

Dr. K. Türke
Karlsruhe University, Institute of Solid Mechanics, Kaiserstraße 12, Box 6980, 76131 Karlsruhe, Germany.

Prof. Dr. S. Wagner
University of Stuttgart, Institut of Aerodynamics and Gasdynamics, Pfaffenwaldring 21, 70569 Stuttgart, Germany.

Prof. Dr. W. L. Wendland
University of Stuttgart, Mathematical Institute A, Pfaffenwaldring 57, 70569 Stuttgart, Germany.

Dipl.-Ing. W. Wenzel
University of Stuttgart, Institute A of Mechanics, Pfaffenwaldring 9, 70550 Stuttgart, Germany.

Prof. Dr. B. Zastrau
University of Wuppertal, Institute of Mechanics, Pauluskirchstr. 7, 42285 Wuppertal, Germany.

Table of Contents

Introduction

The so-called boundary element methods BEM, i.e. finite element approximations of boundary integral equations have been improved recently even more vividly then ever before and found some remarkable support by the German Research Foundation DFG in the just finished Priority Research Program "boundary element methods".

When this program began, we could start from several already existing particular activities which then during the six years initiated many new results and decisive new developments in theory and algorithms. The program was started due to encouragement by E. Stein, when most of the later participants met in Stuttgart at a Boundary Element Conference 1987. Then W. Hackbusch, G. Kuhn, S. Wagner and W. Wendland were entrusted with writing the proposal which was 1988 presented at the German Research Foundation and started in 1989 with 14 projects at 11 different universities. After German unification, the program was heavily extended by six more projects, four of which located in Eastern Germany.

When we started, we were longing for the following goals:

1. Mathematicians and engineers should do joint research.
2. Methods and computational algorithms should be streamlined with respect to the new computer architectures of vector and parallel computers.
3. The asymptotic error analysis of boundary element methods should be further developed.
4. Non-linear material laws should be taken care of by boundary element methods for crack-mechanics.
5. The coupling of finite boundary elements should be improved.
6. The traditional aerodynamical panel methods should be extended to transonic flow computations.
7. Boundary element methods should be used for solving free boundary value problems.
8. Adaptivity criterions and error estimators should be found and adaptive methods should be created.
9. Boundary element methods for transient and general time-dependent problems should be developed.

During the six years of support, several of these key-issues could successfully be accomplished and, clearly, many new ideas, techniques and open problems arose. Particular enhancement is due to the unification of Germany since several excellent research teams from East Germany contributed significantly to new boundary element analyses and algorithms. Also the exploitation of the new generation of computers, in particular, of parallel machines has triggered an overwhelming improvement of BEM as well as of their applicability. In spite of the enormous progress, there is much more potential in the BEM for future analysis and simulation of complex problems in physics and engineering. Looking at the applications in solid mechanics, one can say that here the BEM have left the experimental state; the second generation of BEM software for engineering applications was produced. On the other hand, we can now see lots of new questions and problems; and alsø the number of fields for the application of BEM has increased significantly.

Research of the following specific topics was supported:

1. Efficient methods for three-dimensional acoustics. (Antes in Brunswick (Braunschweig))
2. BEM-FEM discretizations of solids with generalized visco-elastic material laws. (Gaul and Zastrau in Hamburg, Stuttgart and Wuppertal)
3. Extension of cluster-techniques and numerical quadrature. (Hackbusch in Kiel)
4. Hypersingular integrals and adaptive methods for the Kirchhoff-plate. (Hartmann in Kassel)
5. Bimetal problems. (Jentsch in Chemnitz)
6. The treatment of three-dimensional elasto-plastic notch- and crack-problems with boundary elements. (Kuhn in Erlangen)
7. Adaptive domain decomposition with finite and boundary elements. (Langer in Chemnitz and Linz/Austria)
8. Boundary element methods for nonlinear surface waves and the coupling of rigid elastic structures; with and without viscosity. (Mahrenholtz and Schlegel in Hamburg)
9. Integral equations for instationary flows around profiles. (Meister in Darmstadt)
10. Approximation methods for one and multi-dimensional pseudodifferential and Mellin equations/Algorithms for the wavelet approximation of integral equations. (Prößdorf in Berlin)
11. The computation of three-dimensional singularities in elastic structures. (Schnack in Karlsruhe)
12. The coupling of FEM and BEM for three-dimensional elastic structures. (Schnack in Karlsruhe)
13. The numerical treatment of problems in mechanics on heterogeneous computer architectures. (Schnack in Karlsruhe)
14. Collocation methods for singular equations with conjugation/Stability of approximation methods for pseudo-differential operators. (Silbermann in Chemnitz)
15. The continuous coupling of BEM and FEM in the Trefftz sense/FEM-BEM coupling in view of parallelization. (Stein in Hanover (Hannover))

16. Error analysis for BEM. (Stephan in Hanover (Hannover))
17. The extension of BEM to transonic flow problems by coupling with field methods. (Wagner in Munich (München) / Stuttgart)
18. Boundary elements of higher order. (Wendland in Stuttgart)
19. Adaptive BEM. (Wendland in Stuttgart)
20. Parallel computing and domain decomposition. (Wendland in Stuttgart)
21. The coupling of nonlinear FEM with BEM in foundation structural engineering. (Werner in Munich (München))

Based on the original goals, the following main topics came up in our priority research group:

- The coupling of FEM and BEM which initiated several new developments:
  - Traditional coupling by using finite elements in domains of nonlinear phenomena with applications to hyperelastic and elasto-plastic stress analyses and the extension of boundary element methods to contact problems by using variational inequalities.
  - The coupling of field methods in the form of non-linear Newton potentials with applications to elasto-plastic notch- and crack-problems and also to transonic flows by using generalized panel-field methods.
  - Coupling of far fields in form of nonlocal boundary conditions on fictitious boundaries by using appropriate boundary integral equations for the approximation of generalized Dirichlet-Neumann maps with applications in acoustics and the computation of stationary transonic flows around profiles.
  - The modelling of hybrid Trefftz FEM by using local boundary element formulations and the application to problems in solid mechanics involving complex geometries as in notch- and crack-problems.
- The domain decomposition methods are closely intertwined with coupling methods which arose from the previous substructural techniques.
  - One variant of the domain decomposition methods are the hybrid Trefftz models, the so called macro-elements. Here the decisive improvement is based on the weak formulation of the continuity properties on the interfaces and the introduction of mortar-elements in combination with the use of the Neumann series for the realization of the local Poincaré-Steklov operators.
  - Domain decompositions are particularly well suited for parallel algorithms which can also be formulated as multigrid and multilevel methods. Parallel algorithms used to be the main aim in several of the projects and became more and more useful for the numerical solution of the discrete equations also in several more applied projects. These algorithms created a new quality of BEM and brought about new opportunities for the simulation of much more complicated complex engineering problems. In order to use CG or GMRES iteration for the large systems of equations

efficiently, one needs appropriate and effective preconditionings which have been developed also in several of the projects.

- In traditional BEM, the computational effort for generating the influence matrix, needed about 90% of the overall computing time since numerical integration of the singular and almost singular integrals was performed by non-appropriate standard numerical integration. Thanks to the close collaboration between engineers and mathematicians, we have the impression that this problem can now pretty well be controlled:
  - The systematic collection of panels to clusters generalizes classical multipole expansions and yields fast matrix times vector operations. Here, for the regular integrals, one only needs integration of polynomials on the finite elements on the boundary. The combination of clustering with multigrid methods provided very efficient solvers for rather complex boundary element applications. In several of the engineering projects the cluster methods are meanwhile used as solution algorithms.
  - For the singular, hypersingular and weakly singular integrals, special numerical integration procedures and quadrature formulae were set up with only a few integration knots, in combination with local regularizations. Appropriate coordinate transformations together with the parity properties of the kernels lead to powerful integration routines which now allow the efficient implementation of Galerkin methods which for given accuracy became as efficient as traditional collocation methods but which are much more robust. For the almost singular integrals, adaptive bisection methods have been created for the numerical integration where the asymptotic order of the error and relative accuracy are controlled automatically.
  - For the evaluation of the boundary potentials near to the boundary, new efficient asymptotic methods have been developed, first for smooth boundaries, which seem to be extendible also to boundaries with corners and edges. These asymptotic methods are closely related to new algorithms for the superconvergent postprocessing and computation of arbitrary derivatives of the solution, for example of the complete stress-field, whose realization only needs additional solution of the discrete equations involving only the original influence matrix.
- The extension of efficient BEM algorithms for the treatment of nonlinear problems becomes more and more important.
  - For nonlinear problems of solid mechanics now exist boundary element methods for visco-elastic and hyperelastic materials and also for contact problems and three-dimensional crack propagation.
  - The panel methods of fluid mechanics have been extended to transonic flows.
  - The free boundary value problems of fluid mechanics can now be treated with BEM.

  - Coupling methods for the combination of different physical models, e.g. exterior problems with artificial boundary conditions, finite elements in the plastic zones and boundary elements in the remaining domain as well as finite volumes or finite differences for vorticity densities and boundary elements for velocities become more and more important for future applications of BEM multifield problems.
- The adaptive methods for BEM have been developed in two rather different directions. Both concepts require appropriate indicators. Because of the nonlocality and different, possibly negative orders of the boundary integral operators as pseudodifferential operators it turned out that finding appropriate and reliable error estimators and adaptivity criteria is a nontrivial and not yet completed task for boundary element analysis. However, in our priority research groups we could accomplish several decisive results.
  - For unstructured grids, error indicators, grid adaption and adaption in the cluster technique could be based on a posteriori estimates and the strategy by C. Johnson.
  - For hp-methods, decisive accomplishments could be obtained with Johnson's strategy.
  - For h-methods, asymptotic local estimators and indicators of Babuška-Rheinboldt type were created.
  - On structured grids, spectral methods on one hand and wavelets on the other hand have led us to methods where adaptivity is efficiently used as data-compression.
- The approximation of boundary integral equations with wavelets became particularly important for some of the mathematicians involved in the program, who have developed the analysis of these methods far beyond our original expectations. These results clearly became one of the outstanding accomplishments and they will surely generate new BEM of a yet unknown quality.
- Of course, one of the main topics of the program was error and stability analysis of BEM associated with the whole variety of discretizations as collocation, Galerkin methods, qualocation and the combination and fully discrete realization of these discretizations.
  - For most of the algorithms, grid refinements with h-methods were used, in particular adaptive methods on unstructured grids, the current variant of domain decomposition methods and efficient solution procedures of the discrete equations like cluster and multigrid methods. Also wavelets belong to the class of h-methods which converge of algebraic order if h tends to zero. During the six years the activities moved more and more from the simpler two-dimensional boundary value problems with one-dimensional boundary integral equations to the three-dimensional problems of boundary integral equations on two-dimensional surfaces. For special equations and regularly graded meshes on special boundary surfaces such as plane screens, also for the collocation methods one has

obtained many new results concerning stability and convergence. However, the breakthrough for the analysis of collocation methods on general triangulations was still not possible.
  - However, desicive new results could be found for Galerkin procedures including wavelet analysis. Here, the classification and analysis of the operators involved including algebraic methods play a central rôle. Correspondingly, new error indicators and error estimators could be created.
  - In the contrary to the h-methods, spectral methods and h-p-methods converge exponentially. For the h-p-methods, pioneering work was done within the program and one can be sure that together with the availability of fundamental boundary element software these h-p-methods are going to be very important in the future due to their high efficiency.
  - Clearly, a reliable error analysis is always based on fine results in solvability and regularity theory of the basic boundary integral equations which partly were also investigated within the Priority Research Program. The asymptotic behavior and expansion near to corners, edges and collision points of different boundary conditions belong to this theoretical background and new results have been obtained in weighted spaces applied to anisotropic elasticity including assembled structures with different materials.
- BEM for transient time depending problems based on retarded potentials or on time dependent boundary elements, for example with the Faedo-Galerkin methods, were treated in some of the projects. Although the problems and concepts for time dependent problems became much clearer, in this field still a lot of work needs to be done. Here, integral transforms for the the transition from the time to the frequency domain and backwards play a more and more important rôle. Examples are the Fourier and Laplace transformation as well as wavelets. The numerical realization of these applications are demanding and rather difficult; hence the further development of corresponding algorithms will be one of the most important tasks in the near future. More recently, also mixed approximation methods have been used for the discretization as integral transformations and boundary elements for time dependent problems and mixed Fourier and spline trial functions for rotational symmetric problems. Within the research program, particularly engineering groups have accomplished many results for modelling and algorithms of time depending problems.
- An important duty within the program was the exchange, nursing and availability of BEM software generated by the program. W. Hackbusch and his staff are now developing a boundary element package which should be available pretty soon. Also benchmark problems have been defined serving for the comparison of the different boundary element programs.

It should be mentioned that the program owes many great ideas and contributions to the numerous guests of the groups involved.

At the end of the program our referees drew rather positive conclusions. In particular, the cooperation between engineers and mathematicians was remarkably good which was very helpful for the transfer of knowledge between these two groups. At the end of six years work we can say that there is now a rigorous and innovative basis for the BEM. Nevertheless, there is a great need for the transfer of this fundamental knowledge to corresponding software tools usable by the industry. However, there is no institution to carry through this duty which might be a particular problem in Germany. In addition, we all are very worried about the current tendency to close down whole areas of industrial research.

During the six years period we organized annually one workshop for the whole group and, in addition, several particular colloquia and meetings on specific topics.

Herewith all the participants want to express their sincere gratitude to the German Research Foundation and, in particular, to all the people and the referees who helped to make this marvelous six years program happen.

In this volume we publish most of the final reports of this program which will serve as information basis on corresponding literature and on the research groups which participated. I am very grateful to the Springer Verlag for its patience and the opportunity to publish this book. Sincere thanks are due to H. Schulz and all the others who helped preparing the manuscript.

Stuttgart, November 1996 Wolfgang L. Wendland

Efficient Calculation of Acoustic Fields by Boundary Element Method

G. Tröndle, M. Jäger, and H. Antes

Dedicated to Prof. Dr.-Ing. Dr.-Ing. E.h. Dr. h.c.mult. E. Stein on the occasion of his 65th birthday

1 Introduction

A wide class of problems related to wave propagation phenomena in acoustics can be treated by applying the linear theory. The governing equations can be derived by combining the hydrodynamic equations of motion and the local mass balance, the continuity equation. In order to obtain a linearized form of the resulting equations, all the physical quantities and their spatial derivatives are considered to be small variations of their undisturbed position. Furthermore, the fluid should be in rest, i.e., there shouldn't be any steady flow. Assuming the medium to be an ideal gas and the constitutional change of the fluid at the wave front to be adiabatic leads to a relationship between the pressure and the density of the fluid. Introducing the latter to the linearized forms of the principles mentioned above finally yields the so called *wave-equation*

$$\operatorname{div} \mathbf{grad}\, p - \frac{1}{c^2}\frac{\partial^2}{\partial t^2}\, p = \operatorname{div} \boldsymbol{b} - \varrho \frac{\partial}{\partial t}\, a\,, \tag{1.1}$$

where p denotes the hydrodynamic pressure and c the sonic speed. The non-homogeneity on the right hand side depends on the body forces $\boldsymbol{b}$, on the source density a, and on the density ϱ of the fluid in the original state. Basically, there are two approaches for solving such kind of equation: a) solving the wave equation directly in the time domain by a time stepping scheme or b) after a transformation into frequency domain by solving the HELMHOLTZ equation. Especially, when dealing with time harmonic excitations, the formulation in frequency domain is well suited. This formulation can be found by separating the unknowns, where the time dependency is described by the function $p(\boldsymbol{x},t) := \widehat{p}(\boldsymbol{x})\mathrm{e}^{-\mathrm{i}\omega t}$. Herein, $\widehat{p}(\boldsymbol{x})$ denotes the complex amplitude, ω the circular frequency and i the imaginary unity $\mathrm{i} := \sqrt{-1}$. Introducing this to equation (1.1) yields the HELMHOLTZ equation mentioned above:

$$\operatorname{div} \mathbf{grad}\, \widehat{p} + \kappa^2\, \widehat{p} = \widehat{f}\,, \tag{1.2}$$

where κ denotes the wave number, $\kappa := \omega / c$. The paper deals with both, the direct approach in time domain as well as the solution process in the frequency domain.

2 Acoustic problems in frequency domain

For a wide class of problems arising in acoustics, domains extending to infinity are involved. For such kind of problems equation (1.2) is known to yield a unique solution provided the boundary conditions:

$$p(\boldsymbol{x})\big|_{\Gamma_p} = u(\boldsymbol{x})\,,\quad q(\boldsymbol{x}) := \frac{\partial p}{\partial n}(\boldsymbol{x})\Big|_{\Gamma_q} = g(\boldsymbol{x})\,,\quad \Gamma = \Gamma_p \cup \Gamma_q\,,\quad \Gamma_p \cap \Gamma_q = \emptyset$$

and special decay conditions, the SOMMERFELD radiation conditions:

$$\frac{\partial p}{\partial r} - \mathrm{i}\kappa \mathrm{p} = \mathrm{o}\Big(\frac{1}{\mathrm{r}}\Big)\,,\qquad \mathrm{p} = \mathcal{O}\Big(\frac{1}{\mathrm{r}}\Big)\,,\qquad \mathrm{r} \to \infty$$

are satisfied. The latter prevents outgoing waves from traveling back due to reflections at infinity.

Boundary element methods have turned out to be a very sophisticated tool for solving problems of this class. Especially one feature makes them superior to other numerical methods: The satisfaction of the SOMMERFELD radiation condition, i.e. energy radiation is contained in a natural way. Boundary integral formulation can be obtained in various ways, for example by applying GREEN's representation theorem. This leads to the so called *direct method* which reads as follows:

$$\int_\Gamma \Big(p(\boldsymbol{y})\frac{\partial \phi(\boldsymbol{x},\boldsymbol{y})}{\partial n_y} - \phi(\boldsymbol{x},\boldsymbol{y})q(\boldsymbol{x})\Big)\,\mathrm{d}\Gamma_y - \int_\Omega \phi(\boldsymbol{x},\boldsymbol{y})\,\mathrm{f}(\boldsymbol{y})\,\mathrm{d}\Omega_y$$

$$= \begin{cases} 0 & \text{for} \quad \boldsymbol{x} \in \mathbb{R}^3 \setminus \overline{\Omega} \\ \frac{1}{2}p(\boldsymbol{x}) & \text{for} \quad \boldsymbol{x} \in \Gamma \\ p(\boldsymbol{x}) & \text{for} \quad \boldsymbol{x} \in \Omega\,. \end{cases} \tag{2.1}$$

Herein, $\phi(\boldsymbol{x},\boldsymbol{y})$ denotes the integral kernel which represents the fundamental solution for the present case. The formulation (2.1) refers to the exterior problem. For points located on the boundary this relation is only valid for sufficiently smooth boundaries. With the exception of the integral over the known perturbation function $f(\boldsymbol{y})$, all integrals refer to the boundary only. This means that for the numerical approach nothing but the boundary has to be discretized.

Nevertheless, boundary element methods contain some remarkable drawbacks. Although only the boundary of the domain under consideration has to be discretized, the amount of required storage increases with $\mathcal{O}(h^4)$ when h denotes the discretization parameter. Compared to the finite element method where this quantity increases merely with $\mathcal{O}(h^3)$, the boundary element method admits only smaller problems to deal with. Especially in frequency domain where the mesh parameter h has to be adapted reciprocally proportional to the frequency, realistic problems can be handled only in a very restricted manner.

Another critical point in boundary element calculations is the extensive consumption of computing time. This is a result a) of the complicated integration involved and b) of the linear system of equations to be solved whose coefficient matrices are normally fully populated and non-symmetric. For a system discretized by the collocation method, figure 2.1 clearly demonstrates the dependence of computing time for the integration and for the solution of the algebraic system from the number of nodes. This typical example shows which actions have to be taken in order to increase the efficiency of the method: a fast solver should be employed to reduce the CPU-time from $\mathcal{O}(h^3)$ to a significant lower order.

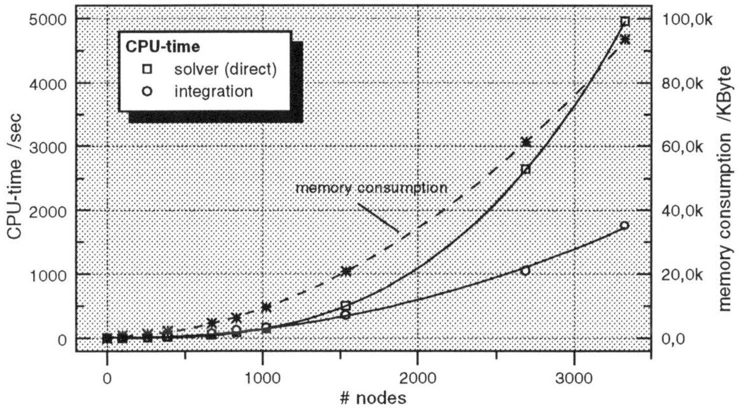

Fig. 2.1. Numerical effort: integration versus solution of the system

3 Multigrid solver

Substituting the direct solver by an iterative one appears to be convenient way to attain such a reduction. In most cases, iterative solvers base on matrix-vector multiplications with a numerical effort of $\mathcal{O}(h^2)$ per iteration sweep, provided that the matrices are fully populated. Clearly, the total numerical effort depends on the number of iterations required to achieve a desired accuracy.

Linear iterative methods for a system $A\,x = b$ can be written in various forms. Among these the first (I) and second (II) normal form:

$$\begin{aligned} x^{(\nu+1)} &= (I - SA)\,x^{(\nu)} + S\,b = M\,x^{(\nu)} + S\,b && \text{(I)} \\ x^{(\nu+1)} &= x^{(\nu)} + S\left(b - Ax^{(\nu)}\right) = x^{(\nu)} + S\,r && \text{(II)} \end{aligned} \qquad (3.1)$$

play an important role. Note that the type and the behavior of the iterative scheme is completely defined by the choice of S which turns out to be a smoothing operation later. In regard of the convergence analysis the iteration operator M has to be studied, i.e., its spectral radius $\varrho(M)$ has to be less than one, in order to have a convergent method.

Most iterative solvers show a specific behavior: they are well suited to reduce the short wave part of the error while the reduction of the long wave error is rather poor. One possibility to get rid of this drawback was proposed by BRANDT and HACKBUSCH in the early 70th. The idea is the combination of iterative and direct solvers, the iterative one to reduce the short-wave error, the direct one to correct the long-wave part. This procedure is known as the *multigrid method.*

The following sequence builds the most important multigrid scheme called coarse grid correction.

Coarse grid correction scheme

(a) smoothing of the *short-wave* error on the *fine* grid by means of an *iterative* method:

$$\tilde{\mathbf{x}}^{(\nu)} = \mathbf{x}^{(\nu)} + \mathbf{S}\left(\mathbf{b} - \mathbf{A}\,\mathbf{x}^{(\nu)}\right)$$

(b) calculation of the residuum on the *fine* grid:

$$\mathbf{r}^{(\nu)} = \mathbf{b} - \mathbf{A}\,\tilde{\mathbf{x}}^{(\nu)}$$

(c) *restriction*, that means the transformation of the residuum $\mathbf{r}^f$ onto the coarse grid with the restriction operator $\mathbf{R}_f$:

$$\mathbf{r}_c^{(\nu)} = \mathbf{R}_f\,\mathbf{r}_f^{(\nu)}$$

(d) solving the system to obtain $\mathbf{e}_c^{(\nu)}$ on the coarse grid by means of a direct method:

$$\mathbf{A}_c\,\mathbf{e}_c^{(\nu)} = \mathbf{r}_c^{(\nu)}$$

(e) *prolongation*: Interpolation of the error $\mathbf{e}_c^{(\nu)}$ from the coarse grid back to the fine grid with the prolongation operator: $\mathbf{P}_f$:

$$\mathbf{e}_f^{(\nu)} = \mathbf{P}_f\,\mathbf{e}_c^{(\nu)}$$

(f) Correction of the approximation vector on the fine grid:

$$\mathbf{x}^{(\nu+1)} = \tilde{\mathbf{x}}^{(\nu)} + \mathbf{e}_f^{(\nu)}.$$

This process just describes the *two-grid method.* Replacing the direct solution on the coarse grid in point (d) by a further application of this scheme on the next coarser grid leads to the real multigrid method. Here, different types of cycles can be defined, especially the V-cycle by applying the coarse grid correction once on each grid, or the W-cycle where the coarse grid correction is performed twice on each grid.

One of the most subtle points in multigrid analysis is the smoothing operation. It should be pointed out that the iterative process used as relaxation does not have to converge, but sufficiently reduce the short wave error. This is absolutely necessary because the residuum must be as smooth as it is required for the approximation on the coarse grid.

For most radiation and scattering problems in acoustics equation (2.1) leads to an integral equation of the second kind. Its discretized form can be described by the following matrix equation:

$$(\lambda\mathbf{I} - \mathbf{K})\,\mathbf{p} \;=\; \mathbf{L}\,\mathbf{q} + \mathbf{s} \quad \text{or, more general:} \quad \mathbf{A}\,\mathbf{x} \;=\; \mathbf{b}\,. \tag{3.2}$$

Now, when such an integral equation is considered, with $\mathbf{A} = \lambda\mathbf{I} - \mathbf{K}$, $\lambda = \frac{1}{2}$ and the PICARD-iteration ($\mathbf{S} = \mathbf{I}$) acting as smoother, the residuum on the fine grid reads:

$$\mathbf{r}^{(\nu)} \;=\; 2\,\mathbf{K}\big\{\mathbf{b} - \big(\tfrac{1}{2}\mathbf{I} - \mathbf{K}\big)\mathbf{x}^{(\nu)}\,\big\}\,. \tag{3.3}$$

This equation makes it evident that the smoothness of the residuum $\mathbf{r}^{(\nu)}$ depends directly on the smoothing property of the integral operator $\mathbf{K}$. Now, a comparison of equation (3.2) and (2.1) yields that $\mathbf{K}$ is made up by the double-layer potential. Therefore, $\mathbf{K}$ has such property only if the boundary is sufficiently smooth. Edges and corners destroy this property.

However, the smoothing effect can be regained if a suitable splitting for $\mathbf{K}$ can be established. Such kind of splitting was proposed for example by AMINI [1].

3.1 Operator splitting

A suitable splitting for K means that K is devided into one part that contains the smoothing property and a second one, whose norm is less than one:

$$K \;:=\; \widetilde{K} + \widehat{K} \qquad \text{with} \qquad \begin{cases} \|\widehat{K}\| \le p < 1 & \text{contracting} \\ \widetilde{K} & \text{smoothing}\,. \end{cases} \tag{3.4}$$

This splitting leads to a modification of the smoothing process according to the following form:

$$\begin{aligned} (\lambda I - K)\,x \;&=\; b & &\rightarrow & \underbrace{(\lambda I - \widehat{K})}_{S^{-1}}\,x \;&=\; \widetilde{K}\,x + b \\ x^{(\nu+1)} \;&=\; \lambda^{-1}K\,x^{(\nu)} + \lambda^{-1}\,b & &\rightarrow & x^{(\nu+1)} \;&=\; S\,\widetilde{K}\,x^{(\nu)} + S\,b\,. \end{aligned} \tag{3.5}$$

Here, λ^{-1} is replaced by S and K by $\widetilde{K}$, the smoothing part of K. Now $\widetilde{K}$ is smoothing by definition, and due to the condition that $\widehat{K}$ is contracting, S, the inverse of $(\lambda I - \widehat{K})$ exists such that the smoothing character of this modified iterative process is guaranteed.

For this operator splitting, the convergence considerations given by HACKBUSCH [2] have to be modified. To do this, the whole multigrid scheme is considered as an iterative process written in terms of the first normal form. Assuming the smoothing iteration to be carried out only once the iteration matrix for a coarse grid correction step over two grids reads as follows:

$$\mathbf{M}_l^Z = \Big[\mathbf{I} - \mathbf{P}_l\,(\lambda\mathbf{I} - \mathbf{K}_{l-1})^{-1}\,\mathbf{R}_l\,(\lambda\mathbf{I} - \mathbf{K}_l)\Big]\,\mathbf{S}_l\,\widetilde{\mathbf{K}}_l\,. \tag{3.6}$$

Since the multigrid method is represented by a recursive application of the two grid method, the formulation for the multigrid iteration matrix is also a recursive one. For the V- and W-cycle, the method can be described by the the relations:

$$\begin{aligned} \mathbf{M}_l^M &= \mathbf{M}_l^Z + \mathbf{P}_l\,\mathbf{M}_{l-1}^M(\lambda\mathbf{I} - \mathbf{K}_{l-1})^{-1}\mathbf{R}_l\,(\lambda\mathbf{I} - \mathbf{K}_l)\,\mathbf{S}_l\,\widetilde{\mathbf{K}}_l \qquad (\mathrm{V}) \\ \mathbf{M}_l^M &= \mathbf{M}_l^Z + \mathbf{P}_l\,(\mathbf{M}_{l-1}^M)^2(\lambda\mathbf{I} - \mathbf{K}_{l-1})^{-1}\mathbf{R}_l\,(\lambda\mathbf{I} - \mathbf{K}_l)\,\mathbf{S}_l\,\widetilde{\mathbf{K}}_l \quad (\mathrm{W}). \end{aligned} \tag{3.7}$$

As the spectral radius is bounded by some suitable norms, the convergence behavior of the two- and multigrid cycle is determined by the norm of the iteration matrix. For the two-grid method it can be derived from equation (3.6). After some algebraic manipulations (see TRÖNDLE [3]) one obtains:

$$\begin{aligned} \big\|\mathbf{M}_l^Z\big\| \;\le\; & \Big[\big\|\mathbf{I} - \mathbf{P}_l\,\mathbf{R}_l\big\| + \\ & + \big\|\mathbf{P}_l\big\|\,\big\|(\lambda\mathbf{I} - \mathbf{K}_{l-1})^{-1}\big\|\,\big\|(\mathbf{R}_l\,\mathbf{K}_l - \mathbf{K}_{l-1}\,\mathbf{R}_l)\big\|\Big]\,\big\|\mathbf{S}_l\big\|\,\big\|\widetilde{\mathbf{K}}_l\big\|. \end{aligned} \tag{3.8}$$

This relation shows clearly which conditions have to be satisfied in order to obtain a convergent two-grid method. In detail, these are:

- convergence of $\mathbf{P}_l\,\mathbf{R}_l$ towards the identity: $\big\|\mathbf{I} - \mathbf{P}_l\,\mathbf{R}_l\big\| < C_{pr}\,h_l^\beta$
- $\mathbf{P}_l$ has to be bounded, i.e., $\big\|\mathbf{P}_l\big\| < C_P$,
- the stability of the coarse grid matrix: $\big\|(\lambda\mathbf{I} - \mathbf{K}_{l-1})^{-1}\big\| < C_s$,
- the relative consistency: $\big\|(\mathbf{R}_l\,\mathbf{K}_l - \mathbf{K}_{l-1}\,\mathbf{R}_l)\big\| < C_k\,h_l^\beta$,
- the regularity of the smooth part of the integral operator: $\big\|\widetilde{\mathbf{K}}_l\big\| < C_r$,
- the existence of a bounded inverse for $(\lambda\mathbf{I} - \widehat{\mathbf{K}}_l)$: $\big\|\mathbf{S}_l\big\| < C_S$.

With the constants introduced above, the norm of the two-grid iteration matrix can be bounded by:

$$\big\|\mathbf{M}_l^Z\big\| \le (C_{pr} + C_P\,C_s\,C_k)\,C_S\,C_r\,h_l^\beta \;=\; C_Z\,h_l^\beta\,. \tag{3.9}$$

This means that the convergence rate is improved the more the grid parameter decreases.

Similar considerations lead to the norm of the iteration matrix for the multigrid W-cycle which, of course, is defined recursively:

$$\begin{aligned}\|\mathbf{M}_l^M\| &\le \|\mathbf{M}_l^Z\| + \|\mathbf{P}_l\|\,\|(\mathbf{M}_{l-1}^M)^2\| \times \\ &\times \Big[\|\mathbf{R}_l\|\,\|\mathbf{I}\| - \|(\lambda\mathbf{I}-\mathbf{K}_{l-1})^{-1}\|\,\|(\mathbf{R}_l\,\mathbf{K}_l - \mathbf{K}_{l-1}\,\mathbf{R}_l)\|\Big]\,\|\mathbf{S}_l\|\,\|\widetilde{\mathbf{K}}_l\| \qquad (3.10)\\ &\le C_Z\,h_l^\beta + C_P\,\|(\mathbf{M}_{l-1}^M)^2\|\,\big(C_S\,C_k\,h_l^\beta + C_R\,C_i\big)\,C_s\,C_r\,.\end{aligned}$$

Here, C_i denotes the estimation for $\|\mathbf{I}\|_{X_l \leftarrow Y_l}$ (the norm of the inclusion when $\mathbf{K}$ is a mapping $\mathbf{K}: X_l \to Y_l$): $\|\mathbf{I}\|_{X_l \leftarrow Y_l} < C_i$. The assumption of the above considerations was that $\mathbf{S}$ the inverse of $(\lambda\mathbf{I} - \widehat{\mathbf{K}}_l)$ is computed exactly. In practice, this would mean that at least an exact decomposition has to be performed, and for each iteration sweep the full algebraic system

$$\left(\lambda\mathbf{I} - \widehat{\mathbf{K}}_l\right)\widetilde{\mathbf{x}}_l^{(\nu)} = \widetilde{\mathbf{K}}_l\,\mathbf{x}_l^{(\nu)} - \mathbf{b}_l \qquad (3.11)$$

has to be solved by a backward and forward substitution, respectively. The numerical effort for these operation would result in a drastic reduction of the multigrid efficiency. Hence, the task is to find a suitable matrix $\widehat{\mathbf{K}}_l$ that can be decomposed with less numerical effort. One possible way to achieve this is to take into account only those nodes lying close to each collocation point under consideration. Thereby, the so-called near-field is defined which, for its part, defines a special kind of sparsity pattern for $\widehat{\mathbf{K}}_l$, i.e., besides the main diagonal each row contains only a few entries. Other types of sparsity patterns are introduced in [4]. No matter what kind of sparsity pattern is used, the size of the decomposition of $(\lambda\mathbf{I} - \widehat{\mathbf{K}}_l)$ can be nearly as large as the fully populated system matrix in the worst case. Hence, this would also increase the storage requirement of the method to a high degree.

In the following, we consider two ways as a practical solution of these problems:

- a restriction of $\widehat{\mathbf{K}}_l$ to the diagonal of $\mathbf{K}_l$
- the calculation of the inverse, resp. the decomposition is performed in an approximative sense only.

If the near-field is restricted to the diagonal of $\mathbf{K}_l$, the inverse can be obtained very easily. In this case, when $\mathbf{D}_l$ denotes the diagonal of the system matrix $\mathbf{A}_l$, $(\lambda\mathbf{I} - \widehat{\mathbf{K}}_l) = \mathbf{D}_l$ holds and the iteration scheme reads as follows:

$$\widetilde{\mathbf{x}}_l^{(\nu)} = \mathbf{D}_l^{-1}\widetilde{\mathbf{K}}_l\,\mathbf{x}_l^{(\nu)} + \mathbf{D}_l^{-1}\mathbf{b}\,. \qquad (3.12)$$

Converting this scheme to the second normal form in (3.1) leads to

$$\widetilde{\mathbf{x}}_l^{(\nu)} = \mathbf{x}_l^{(\nu)} + \mathbf{D}_l^{-1}\left(\mathbf{b}_l - \mathbf{A}_l\,\mathbf{x}_l^{\nu}\right) \qquad (3.13)$$

which shows that this equation means nothing but the JACOBI-iteration scheme.

The second method under consideration, in the following called "near-field ILU smoother", differs from the above described way by the effect, that the decomposition of the near-field matrix $(\lambda\mathbf{I}-\widehat{\mathbf{K}}_l)$ is performed only incompletely (indicated by the tilde): $(\lambda\mathbf{I}-\widehat{\mathbf{K}}_l) \approx (\widetilde{\mathbf{LU}})_l$. With this the smoothing iterations reads:

$$\widetilde{\mathbf{x}}_l^{(\nu)} = (\widetilde{\mathbf{LU}})_l^{-1}\left(\widetilde{\mathbf{K}}_l\,\mathbf{x}_l^{(\nu)} - \mathbf{b}_l\right). \tag{3.14}$$

The representation in the second normal form

$$\mathbf{x}_l^{(\nu+1)} = \mathbf{x}_l^{(\nu)} + (\widetilde{\mathbf{LU}})_l^{-1}\left(\mathbf{b}_l - \mathbf{A}_l\,\mathbf{x}_l^{(\nu)}\right). \tag{3.15}$$

turns out to be more convenient with regards to the implementation in a computer-code.

In order to demonstrate the efficiency of this type of smoother, we consider an example with a model which contains many edges and corners (Fig. 3.1). This study is based on a two-grid method, and the criterion for the comparison is given by the residuum $\mathbf{r}$ relative to the right hand side of the system of equations: $\mathbf{r}^* = \frac{\mathbf{r}}{||\mathbf{b}||}$. It is computed from time to time after 5 iteration sweeps and, therefore, represents some kind of truncation error. This truncation error is plotted versus the excitation frequency, given in figure 3.1 by the HELMHOLTZ-number κa. The diagram demonstrates clearly that the convergence behavior of the two-grid method is much improved when the ILU-near-field smoother or the JACOBI-smoother is used. Compared to these the two-grid convergence using PICARD-iteration which doesn't contain any operator splitting is rather poor. Moreover, for higher excitation frequencies the smoothing properties of the ILU-smoother are superior to the others, which leads to an improved convergence of the method.

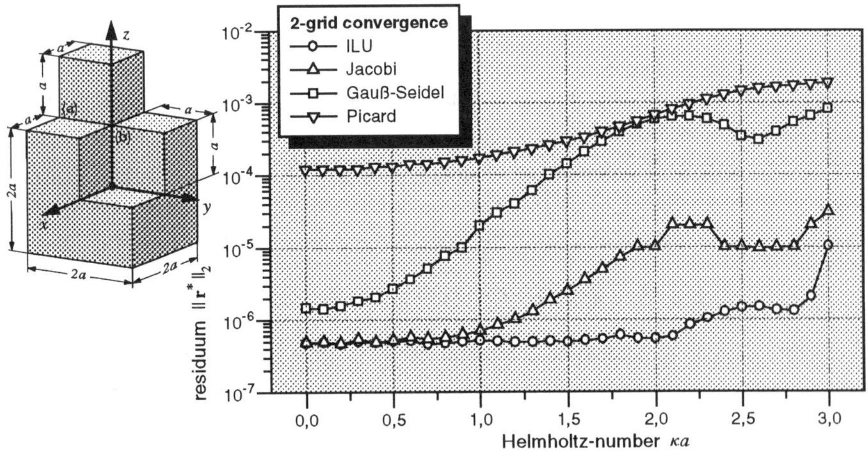

Fig. 3.1. Convergence of multigrid schemes with different smoothers

4 Adaptive mesh refinement

In the multigrid scheme considered above all the grids on finer levels are constructed by halving the grid parameter. This means that a uniform refinement has been performed. It's easy to imagine that such a procedure is not very economic because certainly there will be areas that are discretized too fine. A procedure where the error is distributed continuously over the whole grid would be much better. This can be achieved by refining those areas that contribute a large amount to the error. Of course, there are some instruments required to detect those regions. This is done by so-called error indicators.

Such an adaptive process is well suited to be combined with the multigrid method. In order to do this, a so-called nested iteration is modified such that each single grid is generated adaptively. Figure 4.1 may illustrate this process, which is referred to as an adaptive multigrid method. The process is started by computing the solution on the coarse grid, which is required to calculate the error indicator afterwards. Subsequently to this, the construction of the new grid controlled by the error indicator can take place as well as the the generation of the new system matrix. With a following prolongation step the coarse grid solution is transformed onto the adaptively refined grid, so to speak as a first approximation. This approximation is improved later by performing one or more V-cycles. In most cases only one single cycle is needed because the solution on the intermediate grids is just necessary to compute the error indicator and, hence, may not require the full accuracy. Now the process described above is extended to further grids until the mesh is fine enough. At this moment, a coarse grid correction can be applied repeatedly until the desired accuracy for the fine grid solution is reached.
The most crucial point within this refining procedure is the calculation of the error indicator.

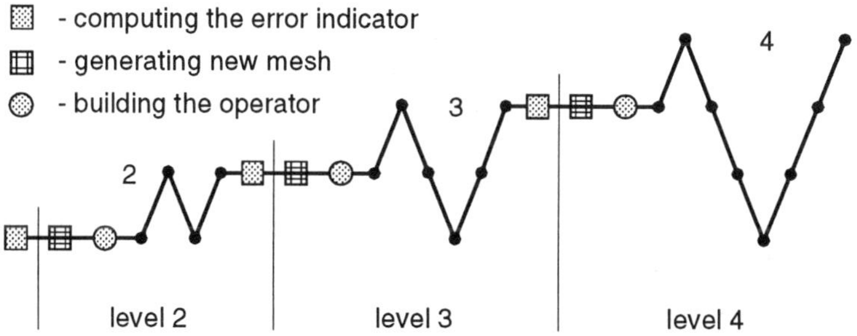

Fig. 4.1. Adaptive multigrid scheme

4.1 Error indicator

An error estimator can basically be written in the form

$$c_1\,\eta \;\le\; \|e\| \;\le\; c_2\,\eta\,, \tag{4.1}$$

where c_1 and c_2 are real constants, i.e., they are independent from the solution itself. Now, the task is to find a quantity satisfying these conditions. In [5], it is shown that the *residual error* provides the required properties.

In the following it is assumed, that there exists a bounded mapping $L: X \to Y$ with a bounded inverse. For an integral equation of the second kind ($L := \lambda I - K$) this holds at least for regular values of λ and compact K. Assuming p_n to be the approximation for the solution p of the system $Lp = g$, the residuum $r_n \in Y$ can be defined as:

$$r_n \;:=\; g - L\,p_n \;=\; L\,(p - p_n) \;=\; L\,e_n\,. \tag{4.2}$$

With the assumptions mentioned above, the norm of the residuum can be casted into the form of an error estimator:

$$c_1\,\|r_n\|_Y \;\le\; \|e_n\|_X \;\le\; c_2\,\|r_n\|_Y\,. \tag{4.3}$$

For the inhomogeneous HELMHOLTZ-equation the residuum reads as follows:

$$\begin{aligned} r(\boldsymbol{x}') \;=\; -\frac{1}{2}p(\boldsymbol{x}') \;&+\; \int\limits_{\Gamma} \frac{\mathrm{e}^{\mathrm{i}\kappa \mathrm{a}}}{4\pi \mathrm{a}} \Big(\frac{\mathrm{i}\kappa \mathrm{a} - 1}{\mathrm{a}}\,\frac{\partial a}{\partial n_y}\,p(\boldsymbol{y}) - \frac{\partial p(\boldsymbol{y})}{\partial n} \Big)\,\mathrm{d}\Gamma_y \\ &-\; \int\limits_{\Omega} \frac{\mathrm{e}^{\mathrm{i}\kappa \mathrm{a}}}{4\pi \mathrm{a}}\, f(\boldsymbol{y})\,\mathrm{d}\Omega_y \\ &\qquad\qquad \text{for}\quad \boldsymbol{x}' \in \Gamma \quad \text{with} \quad a = |\boldsymbol{x}' - \boldsymbol{y}|\,. \end{aligned} \tag{4.4}$$

As stated in [5] for the GALERKIN-approach, the H^0-Norm of the residuum can be used as an estimate of accuracy, although the real error is underestimated. On the other hand, this quantity is well suited to act as an error *indicator* in an adaptive process. This is done by taking into account only that portion of the "error" contributed by a single element under consideration. One obtains an "element related" error indicator η_k according to

$$\eta_k \;=\; \sqrt{\int_{\Gamma_k} r\,\overline{r}\;\mathrm{d}\Gamma_k}\,. \tag{4.5}$$

Obviously, two integrations have to be performed, one to get the residuum and the other to compute its norm. Due to the complicated kernel functions the first one has to be evaluated numerically. Therefore also the second integration requires numerical computation which, in practice, results in a large numerical effort.

In order to reduce the numerical effort, a slightly different procedure is proposed here. Its application provides the residuum to be sufficiently smooth at least in the range of one element. Thus a projection can be applied which maps the residuum onto a finite dimensional space X_n^* containing the solution space X_n as subspace. The additional degrees of freedom are posed on the element edges right between two vertex nodes. This leads to a special kind of interpolation which also takes into account that the residuum vanishes at the nodes related to X_n (the vertex nodes for linear elements).

This interpolation allows now the integration of the norm to be carried out analytically. Moreover, equation (4.4) has to be evaluated only at the additionally introduced nodes. Thus, the numerical effort is significantly reduced such that the method becomes applicable in practice.

4.2 Refinement strategy

In various cases it is desirable for the collocation method to have only conforming elements, i.e., no "hanging" nodes shall occur. This can easily be achieved when the elements were subdivided by halving the edges where the neighboring element containing the same edge is subdivided too. On the other hand, this method contains a certain drawback: if elements are refined such that all their edges are subdivided the refinement can easily extend to those regions that do not need any refinement. Hence, whenever possible only those edges should be subdivided whose residual value contribute considerably to the value of the error indicator.

Therefore, the element related error indicator (4.5) has to be converted to an edge related one. This can be done, for example, by the instruction

$$\eta_{Edge}(\boldsymbol{x}_i') = \sum_{j\in\Theta} \left(\eta_{Elem}\right)_j \frac{r(\boldsymbol{x}_i')}{\sum_{l=1}^{N_{loc}} r(\boldsymbol{x}_l')} \tag{4.6}$$

where Θ represents a set of elements containing point $\boldsymbol{x}_i'$ and N_{loc} denotes the number of residual points in the element under consideration.

In order to generate the new mesh, first, a refining rate is defined as the ratio between the number of subdivided edges and their total number. Then, by bisection a threshold value for the error indicator is determined such that the desired refining rate is reached. Edges whose error indicator exceed this value are marked for to be refined. Now, the new mesh can be generated by subdividing all the marked edges. In order to prevent the vertex angles of the elements from degeneration, all elements have to pass certain tests. Those elements not passing this tests get further edges marked and the mesh generation process is repeated until all elements have passed.

The adaptive multigrid method proceeding is demonstrated by an example. Here a sound-hard box is considered which is exposed to sound waves produced by a single point source right in front of the face ($x = 0.2a$, $y = 0.05a$, $z = 0.95a$ in Fig. 4.2). The geometry is shown in figure 4.2 as well as the initial discretization which is made of only 10 linear quadrilateral elements. Note that this special discretization is completely unsuitable for such kind of configuration. Now, the adaptive process is performed over several refinement steps in order to adapt the mesh to the given situation. The resulting mesh is also plotted in figure 4.2. Here, a strong refinement of the areas in the neighborhood of the source can be recognized. Moreover, an intensive reduction of the mesh size at the edges becomes visible, whereby the so called edge- and corner-singularities are taken into account. So, it seems that the physical situation has been approximated quite well. To rate the efficiency of the adaptively refined mesh, a comparison with a uniformly refined one is done. Figure 4.3 shows the reduction of the defect or residuum (related to the right hand side) versus an increasing number of nodes.

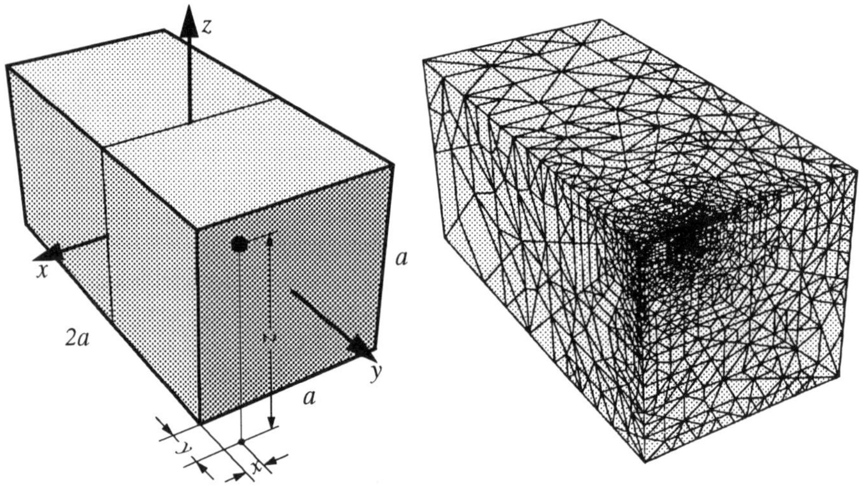

Fig. 4.2. Original discretization and mesh after 8 refining steps

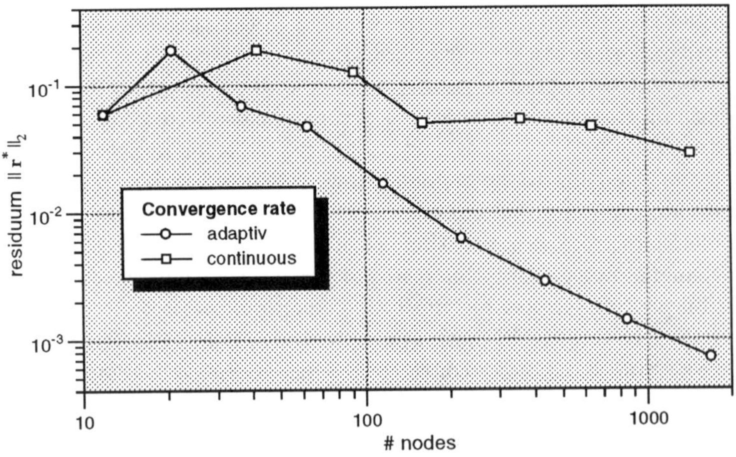

Fig. 4.3. Convergence for adaptive and uniform refinement

This figure clearly demonstrates the much better convergence rate of the adaptive method. Instead of approximately 2000 nodes for the uniform refinement the adaptive one needs only about 80 to 90 nodes to achieve the same accuracy. Similar results are obtained when the reduction of the defect is plotted versus the computing time (as it is done in Fig. 4.4). Although the evaluation of the error indicator is a very time consuming process, the convergence rate of the adaptive scheme related to the computing time is much better than in the uniform case. The same accuracy is reached with the uniform refinement in about 900 CPU-seconds while the adaptive method needs only 8 CPU-seconds.

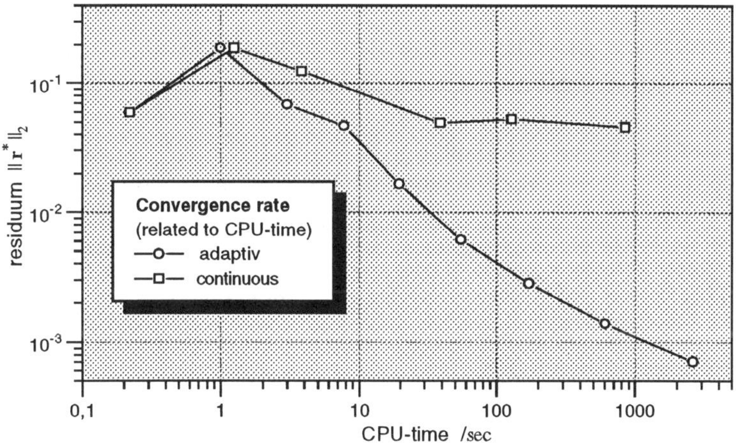

Fig. 4.4. Adaptive versus uniform refinement related to computing effort

5 Acoustic problems in time domain

It is well known that time-dependent acoustic wave propagation phenomena in homogeneous media may be described by KIRCHHOFF's integral equation. This integral formulation is especially advantageous when exterior radiation problems shall be considered. But for three-dimensional problems, its numerical evaluation via boundary element procedures needs high computational effort, and, hence, all procedures should be as efficient and also as stable as possible.

For this reason, in this paper, both the method of point-collocation and the Galerkin method, proposed by many mathematicians for having better convergence, are considered [6]. Another important point of the investigation is the instability arising when choosing the time-step too small. Here, following an advice of MORINO et al. [7] a new formulation of one of the boundary integral terms is derived and studied regarding to its efficiency.

Moreover, the use of adaptive meshes is studied with special regard to the problem of moving noise sources. A simple, but nevertheless effective error estimator is presented, and algorithms for local mesh refinements but also for its coarsening are developed. This results in the effect of a "moving" mesh which changes with time: a fine mesh in the neighborhood of a moving noise source and a coarser grid in a larger distance of it.

The performance of these algorithms is demonstrated. Finally, a rather complex example, the noise distribution due to a fast moving noise source in a housing area protected against noise by walls, is solved. Results may be found in [6].

6 Analysis of different boundary element methodologies

The KIRCHHOFF integral equation is a mathematical model of small-amplitude acoustic waves propagating with the speed c through homogeneous media:

$$d(\boldsymbol{\xi})p(\boldsymbol{\xi},t) = \int_0^t \Big\{ \oint_\Gamma \left[q(\boldsymbol{x},\tau)p^*(\boldsymbol{x},\boldsymbol{\xi};t-\tau) - p(\boldsymbol{x},\tau)q^*(\boldsymbol{x},\boldsymbol{\xi};t-\tau)\right] \mathrm{d}\Gamma_x + \rho \int_\Omega \frac{\partial}{\partial \tau}\gamma(\boldsymbol{x},\tau)p^*(\boldsymbol{x},\boldsymbol{\xi};t-\tau)\,\mathrm{d}\Omega_x \Big\}\,\mathrm{d}\tau \tag{6.1}$$

It is essentially a reciprocal relation between the acoustic pressure p and its flux q across the boundary Γ of the domain Ω and the corresponding states p^* and q^* of a unit point source in the full space [8]. The jump coefficient $d(\boldsymbol{\xi})$ is $\Delta\Omega/4\pi$ with the inner solid angle $\Delta\Omega$, and $\gamma(\boldsymbol{x},t)$ denotes the density of sound sources in the interior of the domain Ω. The mentioned full space solution is ($r = \mid \boldsymbol{x} - \boldsymbol{\xi} \mid$):

$$p^*(\boldsymbol{x},\boldsymbol{\xi};t-\tau) = \frac{\vartheta(t-\tau-\frac{r}{c})}{4\pi r} \tag{6.2}$$

$$q^*(\boldsymbol{x},\boldsymbol{\xi};t-\tau) = \frac{1}{4\pi r}\left[-\frac{1}{r}\vartheta\left(t-\tau-\frac{r}{c}\right) + \frac{\partial}{\partial r}\vartheta\left(t-\tau-\frac{r}{c}\right)\right]\frac{\partial r}{\partial n_x} \tag{6.3}$$

where ϑ is the Dirac function. Then, taking advantage from the relation

$$\frac{\partial}{\partial r}\vartheta\left(t-\tau-\frac{r}{c}\right) = c\frac{\partial}{\partial \tau}\vartheta\left(t-\tau-\frac{r}{c}\right) \tag{6.4}$$

the equation (6.1) is usually transformed to

$$
\begin{aligned}
d(\boldsymbol{\xi})p(\boldsymbol{\xi},t) &- \int_{\Gamma} \frac{\partial r}{\partial n_{\boldsymbol{x}}} \frac{1}{4\pi r} \left[\frac{\partial}{\partial t} p(\boldsymbol{x},t_r) + \frac{1}{r} p(\boldsymbol{x},t_r) \right] \mathrm{d}\Gamma_x \\
= &\int_{\Gamma} \frac{1}{4\pi r} q(\boldsymbol{x},t_r)\, \mathrm{d}\Gamma_{\mathrm{x}} + \rho \sum_{\mathrm{l}=1}^{\mathrm{L}} \frac{\partial}{\partial \mathrm{t}} \mathrm{g_l}\left(\mathrm{t} - \frac{\mid \boldsymbol{x}^{\mathrm{l}} - \boldsymbol{\xi} \mid}{\mathrm{c}}\right) \frac{1}{4\pi \mid \boldsymbol{x}^{\mathrm{l}} - \boldsymbol{\xi} \mid}
\end{aligned}
\tag{6.5}
$$

where $t_r = t - \frac{r}{c}$ denotes the so-called "retarded" time with the property $p(\boldsymbol{x}, t_r) = q(\boldsymbol{x}, t_r) = 0$ for $t_r \leq 0$. Here, as a special case, the sources are considered to be concentrated at an interior point $\boldsymbol{x}^l$ and to have a time dependent strength $g_l(t)$. This expression is also correct when the position $\boldsymbol{x}^l$ is moving with constant speed.

6.1 Comparison of Point Collocation and Galerkin Method

The discrete form of equation (6.5) is developed by the usual boundary element techniques where interpolation functions as simple as appropriate are implemented on the elements Γ_e of a triangular surface grid and in equal time intervals Δt, i.e., $\varphi_j(\boldsymbol{x})$ bilinear in the space variable $\boldsymbol{x}$ and $\theta(t)$ linear in time t. Then, one obtains from (6.5) without the source term by collocation at all time steps $t_k, k = 1(1)m$

$$
\begin{aligned}
4\pi d(\boldsymbol{\xi}) \sum_{j=1}^{n} p_j^k \varphi(\boldsymbol{\xi}) &- \sum_{l=0}^{m} \sum_{j=1}^{n} q_j^l \int_{\Gamma} \frac{1}{r}\, \theta_l(t_{kr}) \varphi_j(\boldsymbol{x})\, \mathrm{d}\Gamma_x \\
= \sum_{l=0}^{m} \sum_{j=1}^{n} p_j^l &\int_{\Gamma} \frac{1}{r} \frac{\partial r}{\partial n_x} \left\{ \left(\frac{1}{r} + \frac{1}{c\Delta t} \right) \theta_l(t_{kr}) - \frac{\theta_l(t_{k-1r})}{c\Delta t} \right\} \varphi_j(\boldsymbol{x})\, \mathrm{d}\Gamma_x
\end{aligned}
\tag{6.6}
$$

In order to obtain as many equations as unknowns, two methodologies are the most common ones: the point collocation method and the Galerkin method. In the first method, also collocation in space, i.e., at all nodes $\boldsymbol{x}_i$ of the shape functions has to be performed. This results in the following system ($r_i = \mid \boldsymbol{x}_i - \boldsymbol{x} \mid$) where $\varphi_j(\boldsymbol{x}_i) = \vartheta_{ij}$:

$$
\begin{aligned}
4\pi d(\boldsymbol{x}_i) \sum_{j=1}^{n} p_j^k \varphi_j(\boldsymbol{x}_i) &- \sum_{l=0}^{m} \sum_{j=1}^{n} q_j^l \int_{\Gamma} \frac{1}{r_i} \theta_l(t_{kr_i}) \varphi_j(\boldsymbol{x})\, \mathrm{d}\Gamma_x \\
= \sum_{l=0}^{m} \sum_{j=1}^{n} p_j^l &\int_{\Gamma} \frac{1}{r_i} \frac{\partial r_i}{\partial n_x} \left\{ \left(\frac{1}{r_i} + \frac{1}{c\Delta t} \right) \theta_l(t_{kr_i}) - \frac{\theta_l(t_{k-1r_i})}{c\Delta t} \right\} \varphi_j(\boldsymbol{x})\, \mathrm{d}\Gamma_x
\end{aligned}
\tag{6.7}
$$

In the Galerkin approach, the final system is obtained by determining the weighted residual of the approximated system (6.6) where the above introduced shape functions $\varphi_i(\boldsymbol{x})$ on the boundary elements are used as weighting functions, i.e., for $i = 1, .., n$:

$$
\begin{aligned}
0 \;=\;& 4\pi \sum_{j=1}^{n} p_j^k \int_{\Gamma} d(\boldsymbol{x})\varphi_j(\boldsymbol{x})\varphi_i(\boldsymbol{x})\,\mathrm{d}\Gamma_x \\
& - \sum_{l=0}^{m}\sum_{j=1}^{n} \Bigg\{ q_j^l \int_{\Gamma}\int_{\Gamma} \frac{1}{r}\,\theta_l(t_{kr})\varphi_j(\boldsymbol{y})\varphi_i(\boldsymbol{x})\,\mathrm{d}\Gamma_y\,\mathrm{d}\Gamma_x \\
& + p_j^l \int_{\Gamma}\int_{\Gamma} \frac{1}{r}\frac{\partial r}{\partial n_y} \left\{ \left(\frac{1}{r} + \frac{1}{c\Delta t}\right)\theta_l(t_{kr}) - \frac{\theta_l(t_{k-1r})}{c\Delta t} \right\} \\
& \times \varphi_j(\boldsymbol{y})\varphi_i(\boldsymbol{x})\,\mathrm{d}\Gamma_y\,\mathrm{d}\Gamma_x \Bigg\}
\end{aligned}
\tag{6.8}
$$

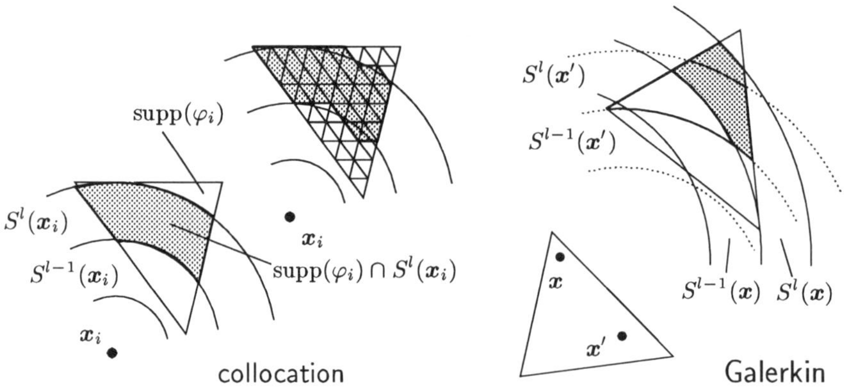

Fig. 6.1. Integration area and its subdividing for the point collocation method and integration area for the Galerkin method

The main difference between both the methodologies is the number of integrations needed. Besides, for the considered three-dimensional acoustic wave propagation problems, the integration area is dependent on the distance which a wave, started at $\boldsymbol{x}_i$, has already traversed: for each integration point $\boldsymbol{x}$, that area $S^l(\boldsymbol{x}_i)$ has to be found where $\boldsymbol{x}$ is between the sphere with radius $(l-1)c\Delta t$ and that with radius $lc\Delta t$. Then, for the time step $\mu\Delta t$, the integration area is the intersection of the spherical shell $S^\mu(\boldsymbol{x}_i)$ with "thickness" Δt and of the shape function's support supp$(\varphi_i) = \{\boldsymbol{x} \mid \varphi_i(\boldsymbol{x}) \neq 0\}$. Hence, for numerical integrations as exact as possible, e.g., by the Gaussian quadrature, enough Gaussian points must be placed in the interior of that "integration area" and, moreover, this area has to be approximated adequate to the applied quadrature formula. Here, an appropriate subdividing in sub-triangles is performed (for details, see [6]). This task is not easy in the point collocation method, but much more difficult in the Galerkin method. There, the integration area for the variable $\boldsymbol{x}$ is the support of φ_i and that for $\boldsymbol{y}$ the above defined area $spt(\varphi_j) \cap S^l(\boldsymbol{x})$ where here $\boldsymbol{x}$ is not a fixed collocation point but one of the Gaussian points in the support of φ_i (see Fig. 6.1).

It is known from theoretical analysis that boundary values are approximated by both the methods with a convergence like $O(h^2)$, but for results at interior points the convergence behavior of the point collocation method should be $O(h^2)$ while that of the Galerkin method is expected to be $\mathcal{O}(h^4)$. Numerical studies confirm this perfectly, but from an economical point of view, the numerical efforts or costs are much more important for the decision which method is "better". In this aspect, for errors not smaller than about one per cent, the point collocation method is much better (Fig. 6.2), e.g., for determining an interior value with an accuracy of 5 %, the Galerkin method needs twice the CPU-time of the point collocation method.

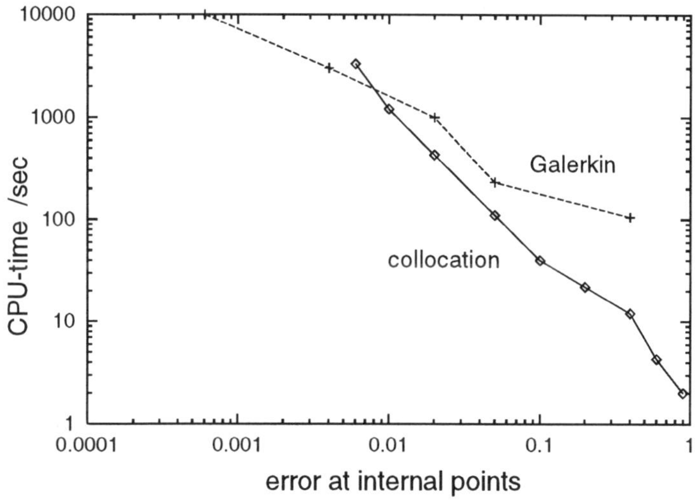

Fig. 6.2. Numerical effort versus error for interior point results: Comparison of point collocation method and Galerkin method

6.2 New BE realizations of Morino's modified Kirchhoff's equation

It is known from numerical studies that for too small time steps Δt the point collocation procedure (6.7) of equation (6.5) becomes unstable. In a mathematical analysis, FIGUIEREDO/MORINO [7] found some theoretical reasons and proposed to use the following transformation

$$\begin{aligned}\int_\Gamma \frac{\partial r}{\partial n_x}\frac{1}{cr}\frac{\partial}{\partial t}p\Big(\boldsymbol{x},t-\frac{r}{c}\Big)\,\mathrm{d}\Gamma_x &= -\int_\Gamma \frac{1}{r}\frac{\partial_2}{\partial n_x}p(\boldsymbol{x},t_r)\,\mathrm{d}\Gamma_x \\ &= -\int_\Gamma \frac{1}{r}\frac{\partial r}{\partial n_x}\frac{\partial}{\partial r}p\Big(\boldsymbol{x},t-\frac{r}{c}\Big)\,\mathrm{d}\Gamma_x\end{aligned} \tag{6.9}$$

where $\frac{\partial_2}{\partial n_x}p(y,t-\frac{r}{c})$ means the normal derivative of p only in its second variable $t-\frac{r}{c}$, and, moreover, to approximate it by

$$\frac{\partial_2}{\partial n_x}p\Big(\boldsymbol{x},t-\frac{r}{c}\Big) \approx \frac{1}{\Delta n}\left(p\Big(\boldsymbol{x},t-\frac{r}{c}\Big) - p\Big(\boldsymbol{x},t-\frac{|\,r+\Delta \boldsymbol{n}\cdot\boldsymbol{n}_x|}{c}\Big)\right) \tag{6.10}$$

Here, $\Delta \boldsymbol{n}$ has to chosen appropriately dependent on the boundary element discretization. Unfortunately, this modification has the defect that for cases where the collocation point $\boldsymbol{y}_k$ is in the same plane as the triangular integration domain, and $r_k = |\,\boldsymbol{x}-\boldsymbol{y}_k|$ is small, the approximation does not give the correct value zero, resulting from $\frac{\partial r}{\partial n_x}=0$. To overcome this defect, backward or central difference quotients for $\frac{\partial p}{\partial r}$ can be used [6] where higher order quotients are to prefer since they have less numerical damping, e.g., the second order backward difference quotient:

$$\begin{aligned}&\frac{\partial}{\partial r}p\Big(\boldsymbol{x},t-\frac{r}{c}\Big) \approx \\ &\frac{1}{2\Delta r}\left[3p\Big(\boldsymbol{x},t-\frac{r}{c}\Big) - 4p\left(\boldsymbol{x},t-\frac{r+\Delta r}{c}\right) + p\left(\boldsymbol{x},t-\frac{r+2\Delta r}{c}\right)\right]\end{aligned} \tag{6.11}$$

Moreover, with these modifications of KIRCHHOFF's integral equation, also small time steps Δt are numerically stable, but one has to take care that the resulting small integration areas (see the above explanation) are integrated exact enough. For studies of the numerically efficiency see [6].

7 Adaptive meshes for moving noise sources

The efficiency of a numerical procedure based on discretizations of either the domain or the boundary is very much dependent on the mesh refinement strategy. This is especially important if the sources of waves are moving, since then the mesh has to be fine in the neighborhood of the source and may be coarse in a larger distance. But, in any case, a simple but nevertheless effective error indicator is essential.

7.1 An Effective but Simple Local Error Indicator

From the literature, several types of error indicators are known also for boundary element procedures, mostly based on the residuum. Here, the error indicator is based on the often realistic assumption that the sought solution is smooth and its curvature does not change very much from element to element. Hence, in the case of linear shape functions over triangular element, one determines the change of the slope when crossing the boundary between two adjacent elements. If this angle is essentially smaller than π, both the adjacent triangular element are subdivided. Obviously (see Fig. 7.1), the numerical effort is much less for this simple error indicator than that when using a residuum based indicator.

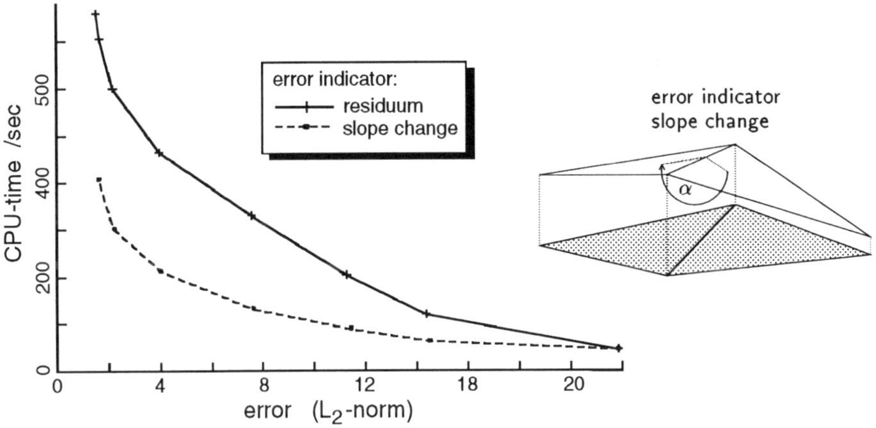

Fig. 7.1. Error indicator "slope change" and comparison of numerical effort

7.2 Strategy For Refining and Re-Coarsening Grids Creating a "Moving" Mesh

The needed subdivisions have to be performed appropriately in order to allow not only refinements but also the re-coarsening of the grid when the noise source producing the acoustic waves has moved away far enough. In the refinement procedure, all "neighbor" elements are marked which have been subdivided simultaneously, and only these "related" elements are permitted for a re-coarsening (Fig. 7.2). Obviously (Fig. 7.3), the numerical effort for solving a problem with the same accuracy is much less when adaptive grids are used.

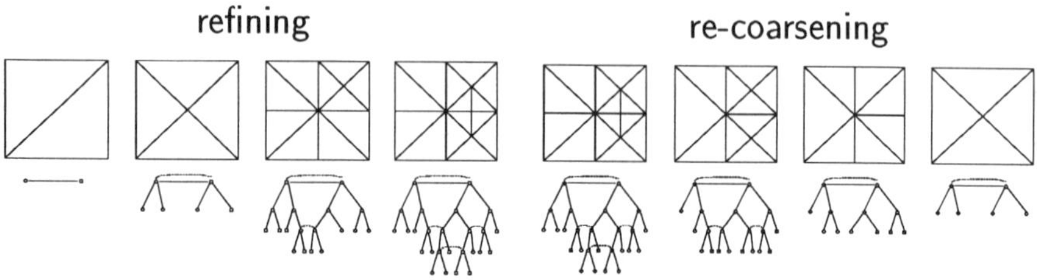

Fig. 7.2. Example for a refining and re-coarsening strategy for triangular grids

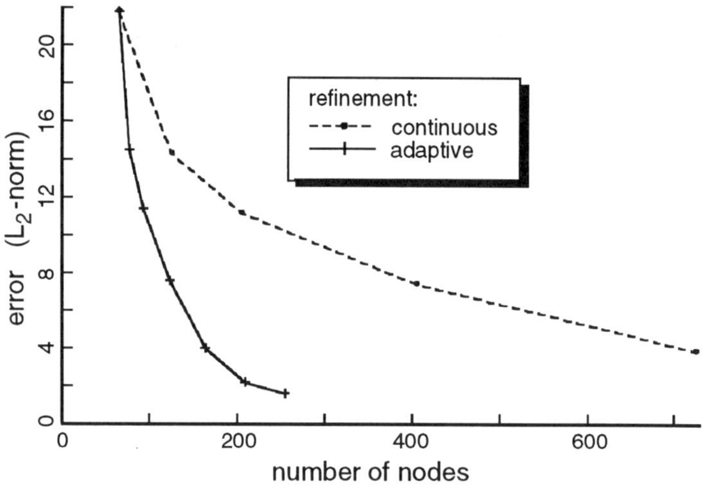

Fig. 7.3. Comparison of convergence: adaptively versus continuously refined grids

8 Conclusion

The paper demonstrates some methods to improve the efficiency of boundary element methods in frequency domain as well as in time domain. Particularly, for both methods, the adaptive mesh generation turned out to be a very efficient tool to speed up boundary element computation and reduce the amount of storage, too. For large systems which, in practice, can hardly be avoided, the numerical effort can be reduced considerably by combining the adaptive process with a multigrid solver. Moreover, by means of the ILU near-filed smoother proposed here the convergence behavior of the multigrid method is improved especially for higher frequencies.

On the other hand, due to the fully populated system matrix the required storage may be still too large for some realistic problems. Particularly in time domain, the distributed computation on a parallel computer [9] is well suited to overcome those difficulties to some extend. For the frequency domain, methods like *panel-clustering* [10] or the use of *wavelet* approximations [11] are very promising.

References

1. Amini, S., Chen Ke (1991): Multigrid solution of boundary integral equations with non-compact operators – application to the exterior acoustic problem. In: Hackbusch, W., Trottenberg, U. (eds.) Multigrid Methods: Special Topics and Applications II (GMD-Studien Nr. 189). GMD, Gesellschaft für Mathematik und Datenverarbeitung, Bonn, pp. 1–11
2. Hackbusch, W. (1985): Multigrid methods and applications. Springer, Series in Computational Methods 4
3. Tröndle, G. (1995): Efficient computation of sound fields by an adaptive multigrid method for the 3-D boundary element method (in german). Braunschweig Series on Mechanics 21-1995. Technical University of Braunschweig
4. Antes, H., Tröndle, G. (1992): Experience with multigrid in three-dimensional acoustic boundary element method. In: Proc. 14th Aeroacoustics Conference. Springer, Aachen
5. Hsiao, G.C., Kleinman, R.E. (1992): Feasible error estimates in boundary element methods. In: Brebbia, C.A., Ingber, M.S. (eds.) Boundary Element Technology VII. Computational Mechanics Publications, Southampton Boston & Elsevier Science Publishers, London New York, pp. 875–886
6. Jäger, M. (1994): Development of an efficient boundary element method for moving noise sources (in german). Braunschweig Series on Mechanics 17-1994. Technical University of Braunschweig
7. Figueiredo, A.M., Sipcic, Morino, L. (1992): Kirchhoff integral equation: Instability issues introduced by the spacial discretization. In: Proc. 14th Aeroacoustics Conference. Springer, Aachen
8. Antes, H., Panagiotopoulos, P.D. (1992): The Boundary Integral Approach to Static and Dynamic Contact Problems — Equality and Inequality Methods. Int. Series of Numerical Mathematics 108. Birkhäuser, Basel
9. Volk, K., Antes, H., Tröndle, G. (1996): A time domain boudary element method for hyperbolic equations on a workstation cluster. To appear in: International Journal for Numerical Methods in Engineering
10. Hackbusch, W., Nowak, Z.P. (1989): On the fast matrix multiplication in the boundary element method by panel clustering. Numer. Math. 54, 463–491
11. Dahmen, W., Prössdorf, S., Schneider, R. (1993): Wavelet approximation methods for pseudodifferential equations matrix compression and fast solution. Advances in Computational Mathematics 1, 259–335

A Boundary Element Formulation for Generalized Viscoelastic Solids in Time Domain

M. Schanz, L. Gaul, W. Wenzel, and B. Zastrau

1 Introduction

The boundary element method (BEM) provides a powerful tool for the calculation of elastodynamic response in frequency and time domain. Field equations of motion and boundary conditions are cast into boundary integral equations (BIE), which are discretized only on the boundary. The boundary data often are of primary interest because they govern the transfer dynamics of members and the energy radiation into a surrounding medium. Formulations of BEM currently include conventional viscoelastic constitutive equations in the frequency domain. Herein viscoelastic behaviour of materials in time domain is implemented in a boundary element formulation. The constitutive equations are generalized by taking time derivatives of fractional order. Instead of generating a viscoelastic fundamental solution in time domain used by the first two authors before, the present approach uses an analytical integration of the boundary integral equation in a time step. Viscoelastic constitutive properties are introduced after taking Laplace transform of an elastic-viscoelastic correspondence principle. The transient response is obtained through the inverse transformation in each time step and the elastic as well as a viscoelastic wave propagation in 3-d continuum are studied numerically.

The transient behaviour of 3-d continua in time domain described by boundary element formulations currently is limited to elastic solids. Behaviours of viscoelastic solids in frequency domain are treated effectively by BEM. Calculation of transient response via the frequency domain requires the inversion of Fourier transform. Since the frequency response is known only in a limited frequency range, truncation effects exist. On the other hand, direct calculations of the response of viscoelastic solids in time domain require the knowledge of fundamental solutions of viscoelastic materials. Such a solution can be obtained by means of an elastic-viscoelastic correspondence principle. In [8], the elastic fundamental solution was transformed into Laplace domain and then the elastic-viscoelastic correspondence principle was adopted. For a simple rheological material, an analytical inverse Laplace transformation is given which leads to the viscoelastic fundamental solution in time domain. For the implementation of this viscoelastic fundamental solution in a 3-d time domain BEM program, it is advantageous to perform the time integration in a time step analytically. This was carried out in [15], but leads to a very complicated series solution. An alternative approach for obtaining a viscoelastic boundary integral formulation in time domain is given in [7]. The generalization of this approach for establishing constitutive equations with better curve fitting properties of measured data is the objective of the present study.

It is known from linear acoustics [11] that time and space discretizations of BIE lead to instability, if the time step is smaller than a critical value. The present paper also displays this instability associated with numerical damping in elastodynamic response.

An extension of the stability range and reduction of numerical damping are achieved by modified representation of terms in the fundamental solution for traction according to a suggestion by Figueiredo *et al.* in acoustics [4].

2 Constitutive equation

Decomposition of the stress tensor σ_{ij} into the the hydrostatic part $\delta_{ij}\sigma_{kk}/3$ and the deviatoric part s_{ij} yields

$$\sigma_{ij} = \frac{1}{3}\sigma_{kk}\delta_{ij} + s_{ij} \qquad \text{where} \quad s_{ii} = 0. \tag{2.1}$$

The corresponding decomposition holds for the strain tensor ε_{ij}

$$\varepsilon_{ij} = \frac{1}{3}\varepsilon_{kk}\delta_{ij} + e_{ij} \qquad \text{where} \quad e_{ii} = 0 \tag{2.2}$$

which results in the following two independent sets of constitutive equations for viscoelastic materials:

$$\sum_{k=0}^{N} p'_k \frac{d^k}{dt^k} s_{ij} = \sum_{k=0}^{M} q'_k \frac{d^k}{dt^k} e_{ij} \qquad \sum_{k=0}^{N} p''_k \frac{d^k}{dt^k} \sigma_{ii} = \sum_{k=0}^{M} q''_k \frac{d^k}{dt^k} \varepsilon_{ii}. \tag{2.3}$$

More flexibility in fitting measured data in a large frequency range is obtained by replacing the time derivatives of integer order by fractional order [6]. The derivative of fractional order α is defined by

$$\frac{d^\alpha x(t)}{dt^\alpha} = \frac{1}{\Gamma(1-\alpha)} \frac{d}{dt} \int_0^t \frac{x(t-\tau)}{\tau^\alpha} d\tau \qquad 0 \le \alpha < 1 \tag{2.4}$$

in which $\Gamma(1-\alpha) = \int_0^\infty e^{-x} x^{-\alpha} dx$ is the Gamma function. The inverse operation of fractional integration is attributed to Riemann and Liouville [13]. A different definition based on generalized finite differences is given by Grünwald [10] as

$$\frac{d^{\alpha}x(t)}{dt^{\alpha}} = \lim_{N\to\infty}\left\{\left(\frac{t}{N}\right)^{-\alpha}\sum_{j=0}^{N-1}\frac{\Gamma(j-\alpha)}{\Gamma(-\alpha)\,\Gamma(j+1)} \quad x\left[t\left(1-\frac{j}{N}\right)\right]\right\}. \quad (2.5)$$

This discrete definition is more convenient in representing constitutive equations by time stepping algorithms, and can be shown to be equivalent to the definition given in equation (2.4). The fractional derivatives in equations (2.4) and (2.5) appear to be complicated in time domain. However, their Laplace transform reveals the useful result

$$\mathcal{L}\left\{\frac{d^{\alpha}x(t)}{dt^{\alpha}}\right\} = s^{\alpha}\mathcal{L}\{x(t)\}, \quad (2.6)$$

where s is the Laplace variable, and vanishing initial conditions are assumed.

With the definitions (2.4) and (2.5) the generalized viscoelastic constitutive equations are given by

$$\sum_{k=0}^{N} p_k' \frac{d^{\alpha_k}}{dt^{\alpha_k}} s_{ij} = \sum_{k=0}^{M} q_k' \frac{d^{\alpha_k}}{dt^{\alpha_k}} e_{ij} \qquad \sum_{k=0}^{N} p_k'' \frac{d^{\alpha_k}}{dt^{\alpha_k}} \sigma_{ii} = \sum_{k=0}^{M} q_k'' \frac{d^{\alpha_k}}{dt^{\alpha_k}} \varepsilon_{ii}. \quad (2.7)$$

For $N = M = 1$ and integer time derivatives, a corresponding three parameter rheological model with two springs and a viscous dashpot is shown in Fig. 2.1.

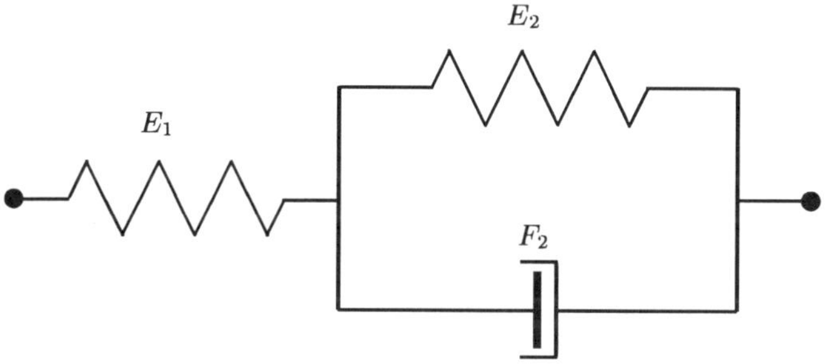

Fig. 2.1. Three Parameter Model

This is one of two possible representations of a uniaxial stress strain equation

$$p\frac{d}{dt}\sigma + \sigma = E\left(\varepsilon + q\frac{d}{dt}\varepsilon\right) \quad (2.8)$$

with

$$p = \frac{F_2}{E_1 + E_2} \qquad E = \frac{E_1 E_2}{E_1 + E_2} \qquad q = \frac{F_2}{E_2}. \quad (2.9)$$

The use of fractional derivatives into the model leads to the second possible representation of the generalized model with five parameters

$$p\frac{d^\alpha}{dt^\alpha}\sigma + \sigma = E\left(\varepsilon + q\frac{d^\beta}{dt^\beta}\varepsilon\right) . \tag{2.10}$$

According to [2], a credible model of the viscoelastic phenomenon should be able to predict non-negative internal work, a non-negative rate of energy dissipation, and the viscoelastic wave speeds

$$c_{1v} = \sqrt{\frac{K(0) + \frac{4}{3}G(0)}{\varrho}} \qquad c_{2v} = \sqrt{\frac{G(0)}{\varrho}} . \tag{2.11}$$

The initial modulus of the relaxation functions $K(0)$ and $G(0)$ have to be finite. These are fulfilled, if the orders of fractional derivatives are the same for all terms [14], i.e.

$$\alpha = \beta , \tag{2.12}$$

which is in accordance with experimental data. Setting α equal to β and satisfying the restrictions mentioned above, we arrive at the constraints [14]

$$\begin{aligned} &E \geq 0 q > p \\ &q \geq 0 0 < \alpha < 2 \\ &p > 0 \end{aligned} \tag{2.13}$$

and the viscoelastic wave speeds become

$$c_{1v} = \sqrt{\frac{K\frac{q_K}{p_K} + \frac{4}{3}G\frac{q_G}{p_G}}{\varrho}} \qquad c_{2v} = \sqrt{\frac{G\frac{q_G}{p_G}}{\varrho}} . \tag{2.14}$$

If the same damping mechanisms are assumed in hydrostatic and deviatoric stress strain state, the corresponding 3-d constitutive equations are obtained by replacing the uniaxial stress and strain by the hydrostatic and the deviatoric states. A powerful tool for calculating viscoelastic behaviour from a known elastic response is the elastic-viscoelastic correspondence principle. According to this principle [5], the viscoelastic solution is obtained from the analytical elastic solution when the elastic moduli in the Fourier or Laplace transformed domain is replaced by the transformed impact response functions of the viscoelastic material model. The viscoelastic solution is then obtained by the inverse transformation.

The elastic-viscoelastic correspondence for the above mentioned generalized three parameter model is given by

$$3K \to 3K\frac{1 + q_K s^{\alpha_K}}{1 + p_K s^{\alpha_K}} \qquad 2G \to 2G\frac{1 + q_G s^{\alpha_G}}{1 + p_G s^{\alpha_G}} \tag{2.15}$$

in which K is the elastic bulk modulus, G the elastic shear modulus, and the transformation property (2.6) is used.

3 Elastic BE-formulation

The boundary integral equation of elastodynamics in time domain is recalled for completeness. The Lamé field equations of a homogeneous isotropic elastic domain Ω are given by

$$\left(c_1^2 - c_2^2\right) u_{i,ij} + c_2^2\, u_{j,ii} + b_j = \ddot{u}_j \tag{3.1}$$

with displacement $u_j(\mathbf{x}, t)$ and wave speeds

$$c_1^2 = \frac{K + \frac{4}{3}G}{\varrho}, \qquad c_2^2 = \frac{G}{\varrho}. \tag{3.2}$$

The corresponding traction and displacement boundary conditions over the boundary Γ are

$$\begin{aligned} t_i\left(\mathbf{x}, t\right) = \sigma_{ik} n_k = p_i\left(\mathbf{x}, t\right) \qquad & \mathbf{x} \in \Gamma_t, \\ u_i\left(\mathbf{x}, t\right) = q_i\left(\mathbf{x}, t\right) \qquad & \mathbf{x} \in \Gamma_u \quad \text{with} \quad \Gamma = \Gamma_t \cup \Gamma_u \end{aligned} \tag{3.3}$$

and the initial conditions are

$$\begin{aligned} u_i\left(\mathbf{x}, 0\right) &= u_{0i}\left(\mathbf{x}\right), \\ \dot{u}_i\left(\mathbf{x}, 0\right) &= v_{0i}\left(\mathbf{x}\right) \qquad \mathbf{x} \in \Omega. \end{aligned} \tag{3.4}$$

The 3-dimensional Stokes fundamental displacement tensor of the Lamé equation (3.1) under the body force $b_j\left(\mathbf{x}, \boldsymbol{\xi}, t, \tau\right) = \delta\left(t - \tau\right) \delta\left(\mathbf{x} - \boldsymbol{\xi}\right) e_j$ in an unbounded space is given by (e.g. [6])

$$\begin{aligned} &\tilde{u}_{ij}\left(\mathbf{x}, \boldsymbol{\xi}, t, \tau\right) = \\ &\frac{1}{4\pi\varrho}\left\{\frac{t-\tau}{r^2}\left(\frac{3 r_i r_j}{r^3} - \frac{\delta_{ij}}{r}\right)\left[H\left(t - \tau - \frac{r}{c_1}\right) - H\left(t - \tau - \frac{r}{c_2}\right)\right]\right. \\ &\left. + \frac{r_i r_j}{r^3}\left[\frac{1}{c_1^2}\,\delta\left(t - \tau - \frac{r}{c_1}\right) - \frac{1}{c_2^2}\,\delta\left(t - \tau - \frac{r}{c_2}\right)\right] + \frac{\delta_{ij}}{r c_2^2}\,\delta\left(t - \tau - \frac{r}{c_2}\right)\right\} \end{aligned} \tag{3.5}$$

where $r = \sqrt{r_i r_i}$, $r_i = x_i - \xi_i$ is the Euclidian distance between the field point $\mathbf{x}$ and the load point $\boldsymbol{\xi}$. The capital H denotes the unit step function, and δ the Dirac distribution. The corresponding fundamental stress vector components are obtained by substituting (3.5) into the constitutive equation and adopting Cauchy's stress formula with the outward normal n_k

$$\begin{aligned} \tilde{t}_{ij}\left(\mathbf{x}, \boldsymbol{\xi}, t, \tau\right) = \frac{1}{4\pi}\Bigg\{ & g_0\left(r\right)\; \frac{\left(t-\tau\right) c_2^2}{r^2}\left[H\left(t - \tau - \frac{r}{c_1}\right) - H\left(t - \tau - \frac{r}{c_2}\right)\right] \\ & + g_1\left(r\right)\left[\delta\left(t - \tau - \frac{r}{c_1}\right) + \frac{r}{c_1}\delta'\left(t - \tau - \frac{r}{c_1}\right)\right] \end{aligned}$$

$$
\begin{aligned}
&+ g_2(r)\left[\delta\left(t-\tau-\frac{r}{c_2}\right)+\frac{r}{c_2}\delta'\left(t-\tau-\frac{r}{c_2}\right)\right] \\
&+ g_3(r)\left[\delta\left(t-\tau-\frac{r}{c_2}\right)-\left(\frac{c_2}{c_1}\right)^2\delta\left(t-\tau-\frac{r}{c_1}\right)\right] \\
&+ g_4(r)\left[\left(\frac{c_2}{c_1}\right)^3\delta'\left(t-\tau-\frac{r}{c_1}\right)-\delta'\left(t-\tau-\frac{r}{c_2}\right)\right]\Bigg\}
\end{aligned}
\tag{3.6}
$$

where $\delta'\left(t-\tau-\frac{r}{c_1}\right)$ denotes the derivative of the Dirac distribution with respect to the argument $\left(t-\tau-\frac{r}{c_1}\right)$. The functions depending on spatial coordinates are abbreviated by

$$g_0(r) = \frac{6}{r^2}\left(n_i r_{,j} + n_j r_{,i} + \delta_{ij} r_{,k} n_k - 5 r_{,i} r_{,j} r_{,k} n_k\right) \tag{3.7}$$

$$g_1(r) = -\left(1-2\left(\frac{c_2}{c_1}\right)^2\right)\frac{n_i r_{,j}}{r^2} \tag{3.8}$$

$$g_2(r) = -\frac{1}{r^2}\left(n_j r_{,i} + \delta_{ij} r_{,k} n_k\right) \tag{3.9}$$

$$g_3(r) = -\frac{2}{r^2}\left(n_i r_{,j} + n_j r_{,i} + \delta_{ij} n_k r_{,k} - 6 r_{,i} r_{,j} r_{,k} n_k\right) \tag{3.10}$$

$$g_4(r) = -\frac{2}{r c_2} r_{,i} r_{,j} r_{,k} n_k . \tag{3.11}$$

The dynamic extension of Betti's reciprocal work theorem [9] combines two states of displacements and tractions, $(\tilde{u}_{ij}, \tilde{t}_{ij})$ and (u_i, t_i) respectively, and leads to the integral equation

$$c_{ij} u_j = \int_\Gamma \left[\tilde{u}_{ij} * t_j - \tilde{t}_{ij} * u_j\right] d\Gamma + \int_\Omega \varrho\left[\tilde{u}_{ij} * b_j + \tilde{u}_{ij} v_{0j} + \dot{\tilde{u}}_{ij} u_{0j}\right] d\Omega \tag{3.12}$$

where $c_{ij} = \delta_{ij}/2$ for a smooth boundary. The $*$ denotes the convolution with respect to time and is defined by

$$\Phi * \Psi = \begin{cases} 0 & t < 0 \\ \int_0^t \Phi(t-\tau)\,\Psi(\tau)\,d\tau & t \geq 0 . \end{cases} \tag{3.13}$$

The integral equation (3.12) reduces to a boundary integral equation if the body forces b_j and the initial conditions vanish. Discretization of the boundary integral equation in the spatial and time coordinate leads to the boundary element formulation. Only the time discretization by n equal Δt steps is discussed here. The simplest nontrivial choice ensuring that no terms are dropped out in the boundary integral equation is using linear shape functions for the displacements u_i and constant shape functions for the tractions t_i in time domain

$$u_i(\mathbf{x},\tau) = \left(U_{il}^{em-1} \frac{t_m - \tau}{\Delta t} + U_{il}^{em} \frac{\tau - t_{m-1}}{\Delta t} \right) \eta_l^e(\mathbf{x}) \tag{3.14}$$

$$t_i(\mathbf{x},\tau) = 1 \cdot T_{il}^{em} \cdot \mu_l^e(\mathbf{x}) . \tag{3.15}$$

The spatial shape functions for the E elements are denoted by $\eta_l^e(\mathbf{x})$ and $\mu_l^e(\mathbf{x})$. The nodal values are U_{il}^{em}, T_{il}^{em} for the corresponding boundary element Γ_e at time $t_m = m\Delta t$. Substitution of equations (3.14) and (3.15) reduces the boundary integral (3.12) to

$$\int_0^t \int_\Gamma \left[t_i(\mathbf{x},\tau) \cdot \tilde{u}_{ij}(\mathbf{x},\boldsymbol{\xi},t-\tau) - \tilde{t}_{ij}(\mathbf{x},\boldsymbol{\xi},t-\tau) \cdot u_i(\mathbf{x},\tau) \right] d\Gamma d\tau$$

$$= \sum_{e=1}^{E} \sum_{l=1}^{L} \sum_{m=1}^{n} \int_{\Gamma_e} \int_{t_{m-1}}^{t_m} \Big[\tilde{u}_{ij}(\mathbf{x},\boldsymbol{\xi},t-\tau) \cdot 1 \cdot \mu_l^e(\mathbf{x}) \cdot T_{il}^{em}$$

$$\tilde{t}_{ij}(\mathbf{x},\boldsymbol{\xi},t-\tau) \cdot \eta_l^e(\mathbf{x}) \cdot \left(U_{il}^{em} \frac{\tau - t_{m-1}}{\Delta t} + U_{il}^{em-1} \frac{t_m - \tau}{\Delta t} \right) \Big] d\tau d\Gamma . \tag{3.16}$$

Time integration of equation (3.16) can be carried out analytically because of the time dependency of the Dirac and Heaviside functions. This integration leads to piecewise defined functions [16]. For the sake of brevity, the integral of the first term on the right hand side of equation (3.16) only is given here,

$$\int_{t_{m-1}}^{t_m} \tilde{u}_{ij}(\mathbf{x},\boldsymbol{\xi},t-\tau) d\tau =$$

$$\frac{1}{4\pi\varrho} \begin{cases} 0 & t < t_{m-1} + \frac{r}{c_1} \\ f_0(r) \frac{1}{2} \left[(t - t_{m-1})^2 - \left(\frac{r}{c_1} \right)^2 \right] + f_1(r) \frac{1}{c_1^2} & t_{m-1} + \frac{r}{c_1} < t < t_m + \frac{r}{c_1} \\ f_0(r) \left[t t_m - t t_{m-1} - \frac{t_m^2}{2} + \frac{t_{m-1}^2}{2} \right] & t_m + \frac{r}{c_1} < t < t_{m-1} + \frac{r}{c_2} \\ f_0(r) \frac{1}{2} \left[\left(\frac{r}{c_2} \right)^2 - (t - t_m)^2 \right] + f_2(r) \frac{1}{c_2^2} & t_{m-1} + \frac{r}{c_2} < t < t_m + \frac{r}{c_2} \\ 0 & t_m + \frac{r}{c_2} < t \end{cases} \tag{3.17}$$

with the following abbreviations depending only on spatial coordinates:

$$f_0(r) = \frac{3 r_{,i} r_{,j} - \delta_{ij}}{r^3}$$

$$f_1(r) = \frac{r_{,i} r_{,j}}{r}$$

$$f_2(r) = \frac{\delta_{ij} - r_{,i} r_{,j}}{r} .$$

The time intervals are valid only for the time step $\Delta t < \frac{r}{c_2} - \frac{r}{c_1}$. After substituting $t_m = m\Delta t$, $t_{m-1} = (m-1)\,\Delta t$ and $t = n\Delta t$ in equation (3.17), it can be seen that the functions depend on $(n-m)$. This implies that these functions depend only on the difference between the observation and the excitation time.

After time and space discretizations, a system of algebraic equations [1] is obtained

$$\left[\frac{1}{2}\mathbf{I} + \overset{(1)}{\mathbf{T}}\right]\overset{(n)}{\mathbf{u}} + \sum_{m=1}^{n-1} \overset{(n-m+1)}{\mathbf{T}}\overset{(m)}{\mathbf{u}} = \sum_{m=1}^{n} \overset{(n-m+1)}{\mathbf{U}}\overset{(m)}{\mathbf{t}}\,, \qquad (3.18)$$

where $\mathbf{I}$ is the identity matrix, $\overset{(m)}{\mathbf{T}}$ and $\overset{(m)}{\mathbf{U}}$ are the influence matrices of stresses and displacements at the time step m. The vectors $\overset{(m)}{\mathbf{u}}$ and $\overset{(m)}{\mathbf{t}}$ contain all nodal displacements and tractions of the time step m.

4 Viscoelastic BE-formulation

In order to obtain a viscoelastic boundary integral formulation from the elastic formulation (3.18), the elastic-viscoelastic correspondence principle (2.15) is applied. This requires the Laplace transformation of (3.18) first. The kernels of the matrices consist of the function (3.17) and the fundamental solutions of the tractions after time integration. For the sake of brevity the procedure of deducing a viscoelastic formulation is presented for the integral (3.17) only. More details can be found in [14].

The one sided Laplace transformation

$$\mathcal{L}(f) = F(s) = \int_0^\infty f(t)\, e^{-st} dt$$

$$f(t) = \frac{1}{2\pi i} \int_{x-i\infty}^{x+i\infty} F(s)\, e^{st} ds \quad t > 0 \qquad (4.1)$$

leads to the transformation of equation (3.17)

$$\int_0^\infty \sum_{m=1}^{n} \int_{t_{m-1}}^{t_m} \tilde{u}_{ij}(\mathbf{x}, \boldsymbol{\xi}, t-\tau)\, d\tau e^{-st} dt$$

$$= \frac{1}{4\pi\varrho} \sum_{m=1}^{n} \left\{ \int_{t_{m-1}+\frac{r}{c_1}}^{t_m+\frac{r}{c_1}} \left[f_0(r)\,\frac{1}{2}\left((t-t_{m-1})^2 - \left(\frac{r}{c_1}\right)^2 \right) + f_1(r) \right] e^{-st} dt \right.$$

$$+\int\limits_{t_m+\frac{r}{c_1}}^{t_{m-1}+\frac{r}{c_2}} f_0(r)\left(t\,t_m - t\,t_{m-1} - \frac{t_m^2}{2} + \frac{t_{m-1}^2}{2}\right)e^{-st}dt$$

$$+\int\limits_{t_{m-1}+\frac{r}{c_2}}^{t_m+\frac{r}{c_2}} \left[f_0(r)\left(t\,t_m - \frac{t_m^2}{2} - \frac{t^2}{2} + \frac{1}{2}\left(\frac{r}{c_2}\right)^2\right) + f_2(r)\right]e^{-st}dt\Bigg\}. \quad (4.2)$$

After integrating the time interval of (4.2), the elastic constants are replaced by the viscoelastic impact response functions. The elastic bulk and shear moduli appear only in the compression wave speed c_1 and the shear wave speed c_2. Thus the corresponding viscoelastic expressions

$$\begin{aligned}(c_{1v}^*)^2 &= \frac{1}{\varrho}\left(K\frac{1+q_K s^{\alpha_K}}{1+p_K s^{\alpha_K}} + \frac{4}{3}G\frac{1+q_G s^{\alpha_G}}{1+p_G s^{\alpha_G}}\right)\\ (c_{2v}^*)^2 &= \frac{1}{\varrho}G\frac{1+q_G s^{\alpha_G}}{1+p_G s^{\alpha_G}}\end{aligned} \quad (4.3)$$

are inserted into equation (4.2). After taking Laplace transformation of equation (4.2), this leads to

$$\mathcal{L}\left(\sum_{m=1}^{n}\int\limits_{t_{m-1}}^{t_m}\tilde{u}_{ij}(\mathbf{x},\boldsymbol{\xi},t-\tau)\,d\tau\right) =$$

$$\frac{1}{4\pi\varrho}\sum_{m=1}^{n}\left(e^{-st_{m-1}} - e^{-st_m}\right)\left\{f_0(r)\frac{1}{s^2}\left[\left(\frac{r}{c_{1v}^*}\right)e^{-\frac{r}{c_{1v}^*}s} - \left(\frac{r}{c_{2v}^*}\right)e^{-\frac{r}{c_{2v}^*}s}\right]\right.$$

$$\left. + f_1(r)\frac{1}{s}e^{-\frac{r}{c_{1v}^*}s} + f_2(r)\frac{1}{s}e^{-\frac{r}{c_{2v}^*}s}\right\}. \quad (4.4)$$

Causality of the solution implies that no response is present prior to the arrival of the viscoelastic waves (2.14). This physical requirement is assumed for the numerical inversion of equation (4.4). The correspondence relation

$$F(s)e^{-st_{m-1}} - F(s)e^{-st_m} \bullet\!\!-\!\!\circ\, f(t-t_{m-1})\,H(t-t_{m-1}) - f(t-t_m)\,H(t-t_m) \quad (4.5)$$

is taken into account as well. The inverse transformation therefore leads to a piecewise defined function in the time ranges of equation (3.17), where c_1 and c_2 are replaced by c_{1v} and c_{2v}. The inversion is carried out numerically by the method of Talbot [17].

5 Numerical example

The propagation of waves in a 3-d continuum has been studied by the present boundary element formulation in time domain for two examples. They involve a 3-d pressure bar and an elastic foundation on a viscoelastic soil.

5.1 3-d pressure bar

The problem geometry and the material data (corning glass at 550^oC) are shown in Fig. 5.1.

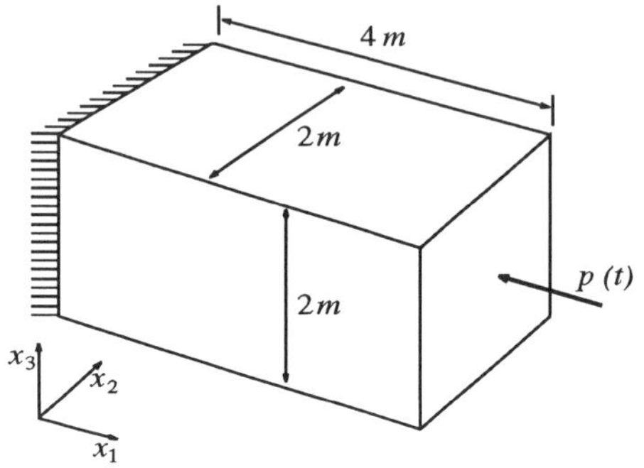

$$
\begin{aligned}
E &= 2.075 \cdot 10^9 \,\frac{\text{N}}{\text{m}^2}\\
\varrho &= 1000 \,\frac{\text{kg}}{\text{m}^3}\\
p &= 3.5 \text{ s}^{-\alpha}\\
q &= 3.5/5/10/26 \text{ s}^{-\alpha}\\
\nu &= 0/0.25\\
\alpha &= 0.635/1\\
p(t) &= H(t) \,\frac{\text{N}}{\text{m}^2}
\end{aligned}
$$

Fig. 5.1. Step function excitation of a free-fixed bar

The free end of the bar is excited by a unit step loading function $H(t)$ from a pressure pump. The opposite end is fixed at the nodes. The use of different spatial shape functions and different meshes are studied first. The three discretizations are shown in Fig. 5.2. The time step size Δt was chosen close to the time the viscoelastic compression wave c_{1v} takes to travel across the largest element. In Fig. 5.3 the displacements and tractions at the middle point of the free end (observation point) in longitudinal direction are plotted versus time.

It can be clearly seen in Fig. 5.3 that a finer mesh or spatial shape functions of higher order improve the result. Also, it is shown that the meshes for the tractions are too coarse.

As a second test, the results of the current BEM formulation are compared with a Finite Element (FE) formulation. In this test, the parameter α of the fractional derivative must be set to 1, because the authors have no available FE program for calculating viscoelastic response using fractional derivatives in time domain.

The FE discretization and the comparison of results are shown in Fig. 5.4. The displacements at the observation point in longitudinal direction are compared. Mesh B is used for the BE formulation. The comparison shows a good agreement of both formulations.

In Fig. 5.5, the behavior of the travelling waves in the bar is studied for several values of the damping parameter q in the constitutive equation (2.8).

The increase of damping parameter q stiffens the solid and increases the wave speed, resulting in shorter travel times and smaller deflections as shown in Fig. 5.5.

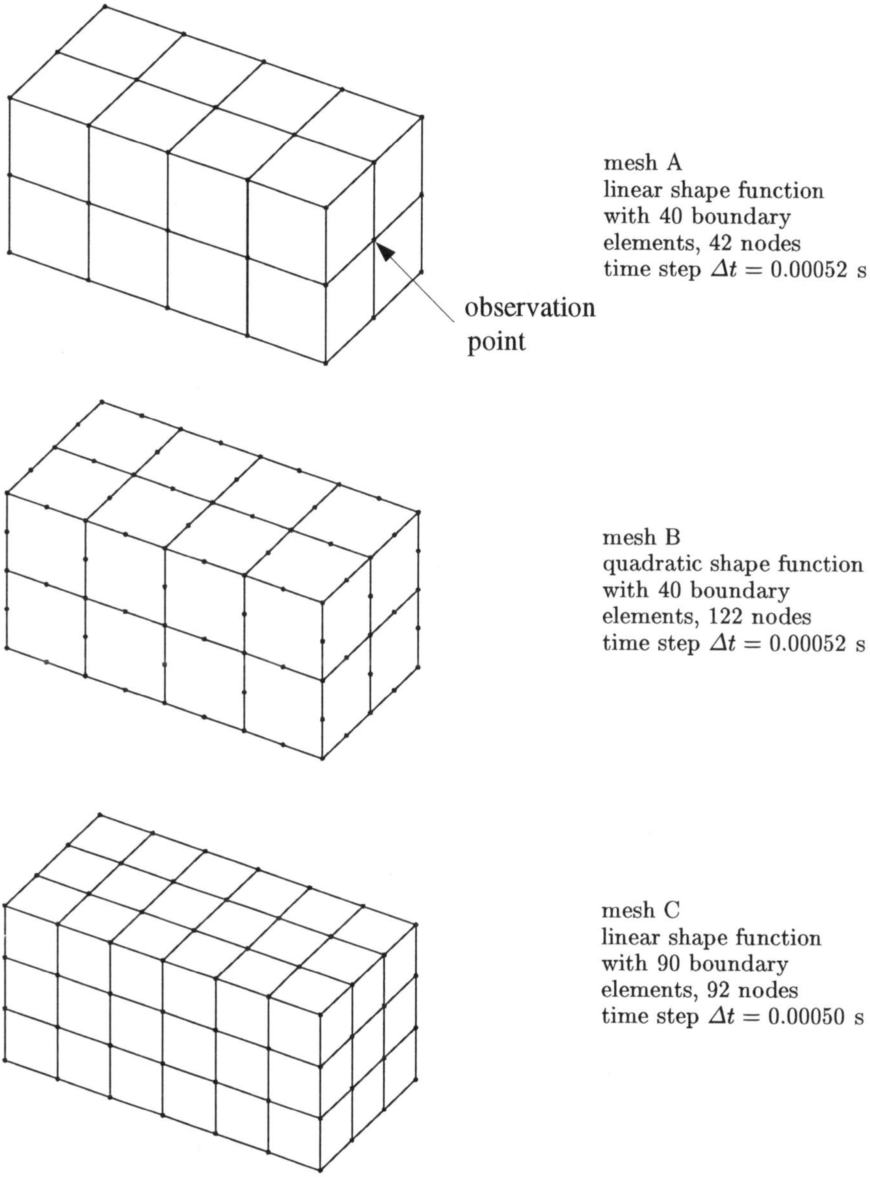

Fig. 5.2. Different meshes of a free fixed bar

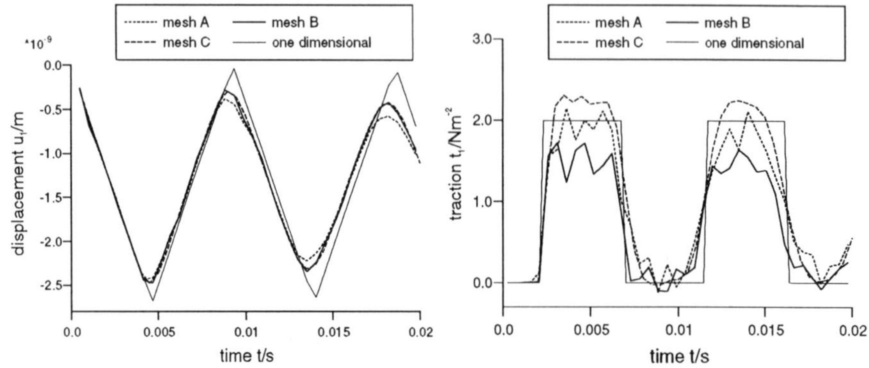

Fig. 5.3. Displacement u_1 and tractions t_1 at the observation point

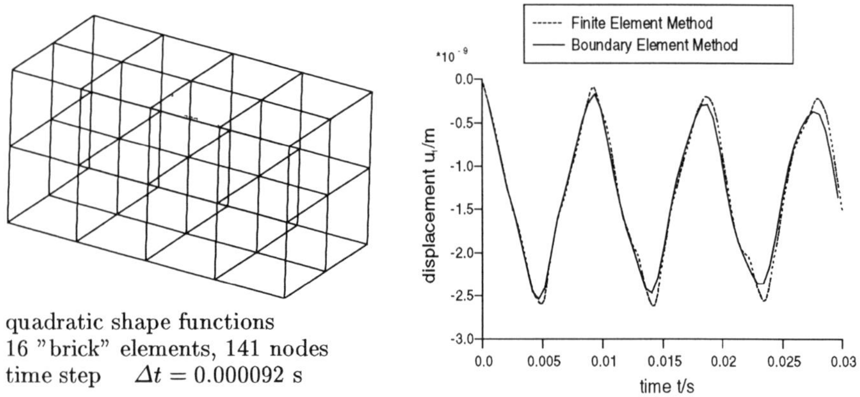

Fig. 5.4. Comparison of FE and BE calculation

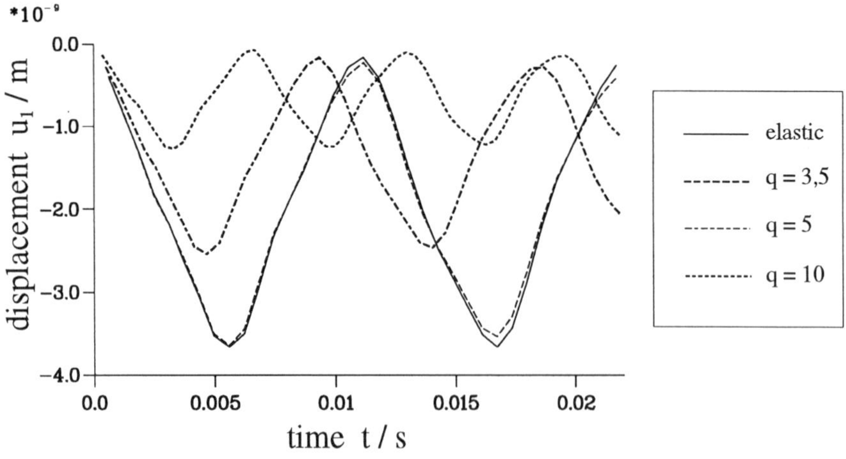

Fig. 5.5. Step function response of a free fixed bar

5.2 Foundation slab on soil

As a second example an elastic concrete foundation slab on a viscoelastic sand soil is considered. In the bonded contact area of the two domains continuity in stresses and displacements is assumed.
The problem geometry and the material parameters are shown in Fig. 5.6.

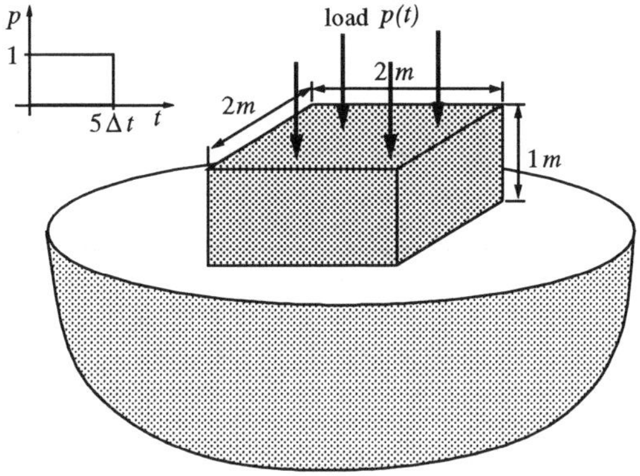

sand

$$E = 1.38 \cdot 10^8 \frac{N}{m^2}$$
$$\varrho = 1966 \frac{kg}{m^3}$$
$$\nu = 0.35$$
$$\alpha = 1.3$$
$$p = 1\ s^{-1.3}$$
$$q = 1.5/5/7.5/10 s^{-1.3}$$

concrete

$$E = 3 \cdot 10^8 \frac{N}{m^2}$$
$$\varrho = 2000 \frac{kg}{m^3}$$
$$\nu = 0.20$$

Fig. 5.6. Elastic foundation slab on viscoelastic soil

For the discretization (Fig. 5.7) of the surface of interest, linear shape functions are adopted. The time step is chosen such that the viscoelastic compression wave with velocity c_{1v} is nearly propagated over one element during a time step. The element diagonal is selected herein as characteristic length.

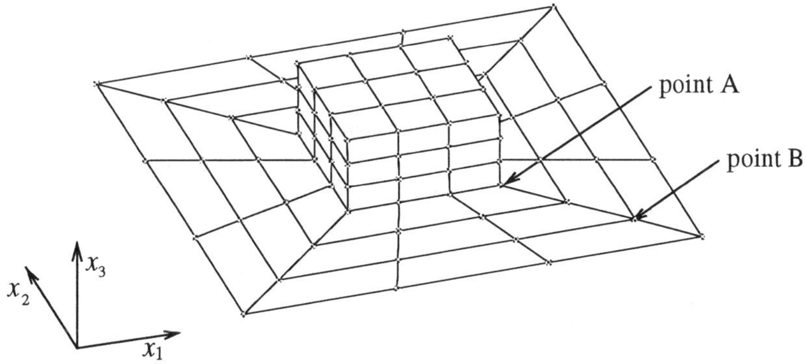

Fig. 5.7. Discretization of foundation slab and soil surface

The displacements u_3 perpendicular to the halfspace surface are plotted versus time for different values of the viscoelastic parameter q in Fig. 5.8.

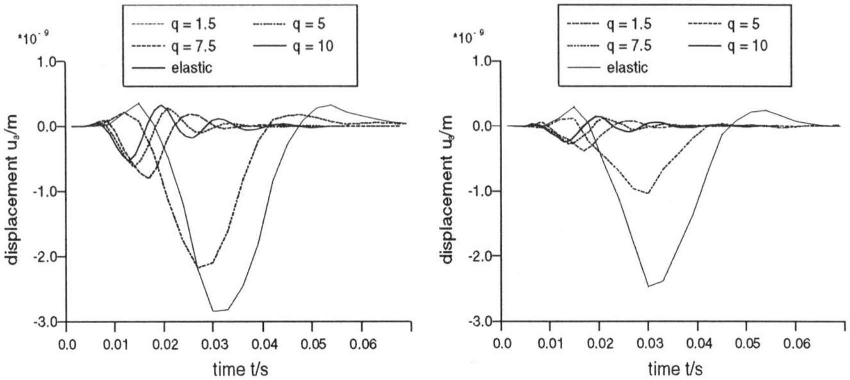

Fig. 5.8. Displacement u_3 at the points A and B

The figure on the left hand side presents the displacements at point A and the figure on the right hand side the displacement at point B. Similar to the response of the pressure bar discussed in section 5.1, the material is stiffened with increasing value of q. This is observed by smaller deflections and a shorter period of vibration. The propagation of waves on the surface is depicted in Fig. 5.9, where the magnitude of the displacement field at the nodes is shown.

6 Stability and numerical damping of the elastic solution

The time stepping algorithm (3.18) in conjunction with the viscoelastic formulation leads to instability of the solution, if the time step Δt is chosen too small. This problem can be observed by calculating the wave propagation in an elastic free-fixed rod made of steel as shown in (Fig. 5.1). The load function, geometry and spatial discretization by 40 linear elements are shown in Figs. 5.1 and 5.2. The longitudinal displacement of the centroid of the free end is plotted versus the time step number n in Fig. 6.1.

If the the time step size Δt is chosen below a critical value, an instable displacement response occurs as shown in Fig. 6.1. This behaviour differs from the time marching schemes applied to FE equations.

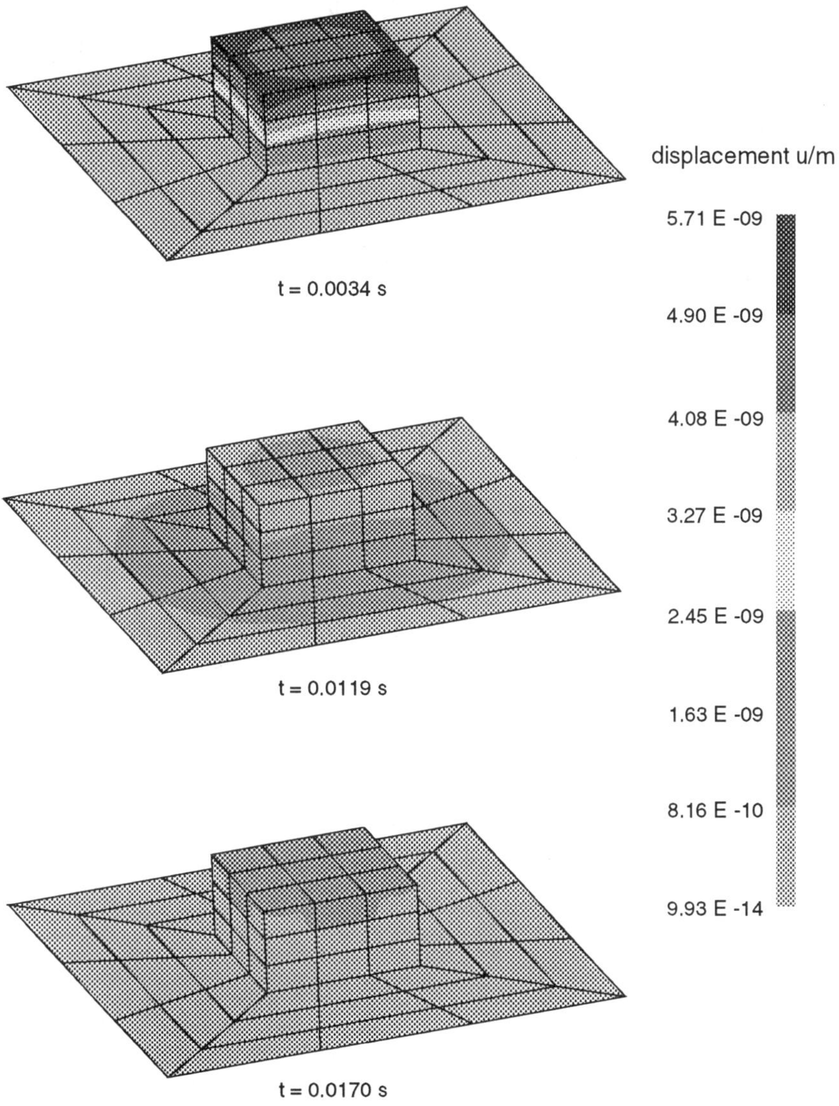

Fig. 5.9. Displacement field of propagating waves on the surfaces of foundation slab and halfspace

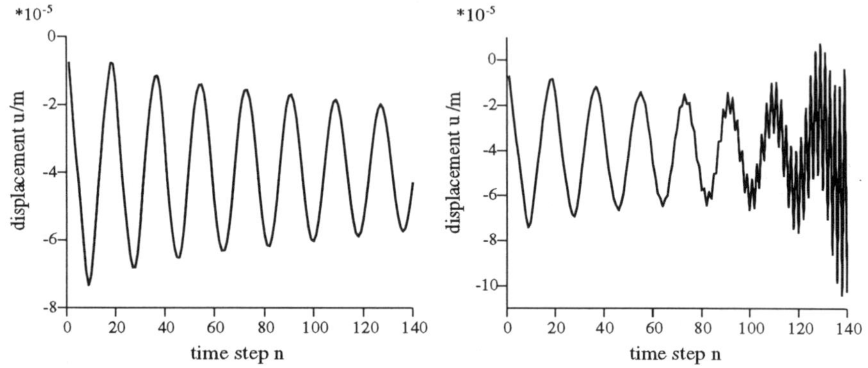

Fig. 6.1. Stable ($\Delta t = 0.002699$ s) and instable ($\Delta t = 0.002698$ s) rod displacements

The decay of the displacement amplitude as shown in Fig. 6.2 is due to numerical damping, because the material is elastic. For comparison purposes, the analytical 1-d elastic solution

$$u\left(l,t\right)_{\text{analytic}} = -\frac{1}{\varrho c}\left\{t + \sum_{n=1}^{\infty}\left(-1\right)^{n} 2\left(t - \frac{2nl}{c}\right) H\left(t - \frac{2nl}{c}\right)\right\} \qquad c = \sqrt{\frac{E}{\varrho}} \tag{6.1}$$

associated to the 3-d problem is plotted in Fig. 6.2. The numerical damping can be distinguished for different values of Δt. The reason for the numerical problems of instability and numerical damping is due to the term in equation (3.6) which contains the derivative of the Dirac distribution. It has been shown for the Kirchhoff Integral Equation, representing the wave equation of linear acoustics [4], that instabilities can be partly removed by approximating the derivative of the Dirac distribution by finite differences. In the present paper, this idea is adopted and modified for the application in elastodynamics by using finite differences of higher order. Equation (3.16) contains integrals of the type

$$\frac{1}{\Delta t}\int_{t_{m-1}}^{t_m} \frac{\partial}{\partial\left(t - \frac{r}{c_\beta} - \tau\right)}\delta\left(t - \frac{r}{c_\beta} - \tau\right)\tau d\tau \qquad \beta = 1, 2. \tag{6.2}$$

These integrals unfortunately cause the above mentioned problems in equation (3.16). A model problem with the structure of equation (3.16) treated by [4] indicates the possible loss of stability. The problems can be partly circumvented by replacing the derivative with respect to the argument by the spatial derivative

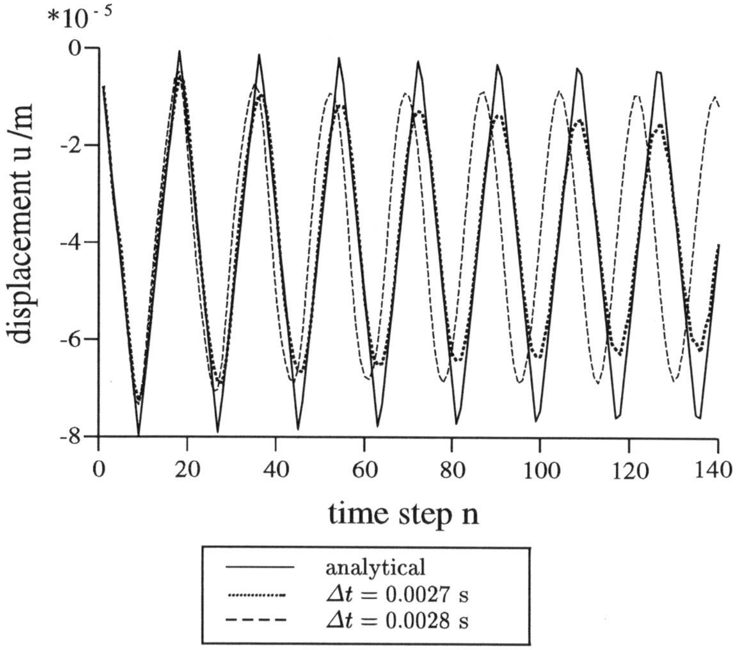

Fig. 6.2. Rod displacement of 3-d problem with numerical damping compared with 1-d analytical solution

$$\frac{\partial}{\partial r}\delta\left(t-\tau-\frac{r}{c_\beta}\right)=\frac{\partial\,\delta\left(t-\tau-\frac{r}{c_\beta}\right)}{\partial\left(t-\tau-\frac{r}{c_\beta}\right)}\cdot\underbrace{\frac{\partial\left(t-\tau-\frac{r}{c_\beta}\right)}{\partial r}}_{-\frac{1}{c_\beta}} \tag{6.3}$$

leading to

$$\frac{\partial}{\partial\left(t-\frac{r}{c_\beta}-\tau\right)}\delta\left(t-\tau-\frac{r}{c_\beta}\right)=-c_\beta\frac{\partial}{\partial r}\delta\left(t-\tau-\frac{r}{c_\beta}\right) \tag{6.4}$$

which is then approximated by a backward finite difference quotient

$$-c_\beta\frac{\partial}{\partial r}\delta\left(t-\tau-\frac{r}{c_\beta}\right)\approx -c_\beta\frac{\delta\left(t-\tau-\frac{r}{c_\beta}\right)-\delta\left(t-\tau-\frac{r-\Delta r}{c_\beta}\right)}{\Delta r}. \tag{6.5}$$

While a finite difference of first order is used in equation (6.5), further improvement can be obtained by using higher order finite differences. Results of the following finite difference representation are shown in Fig. 6.3

– first order backward

$$-c_\beta \frac{\partial}{\partial r}\delta\left(t-\tau-\frac{r}{c_\beta}\right) \approx -c_\beta \frac{\delta\left(t-\tau-\frac{r}{c_\beta}\right)-\delta\left(t-\tau-\frac{r-\Delta r}{c_\beta}\right)}{\Delta r}$$

– central

$$-c_\beta \frac{\partial}{\partial r}\delta\left(t-\tau-\frac{r}{c_\beta}\right) \approx -c_\beta \frac{\delta\left(t-\tau-\frac{r+\Delta r}{c_\beta}\right)-\delta\left(t-\tau-\frac{r-\Delta r}{c_\beta}\right)}{2\Delta r}$$

– second order backward

$$-c_\beta \frac{\partial}{\partial r}\delta\left(t-\tau-\frac{r}{c_\beta}\right) \approx$$

$$-c_\beta \frac{3\delta\left(t-\tau-\frac{r}{c_\beta}\right)-4\delta\left(t-\tau-\frac{r-\Delta r}{c_\beta}\right)+\delta\left(t-\tau-\frac{r-2\Delta r}{c_\beta}\right)}{2\Delta r}$$

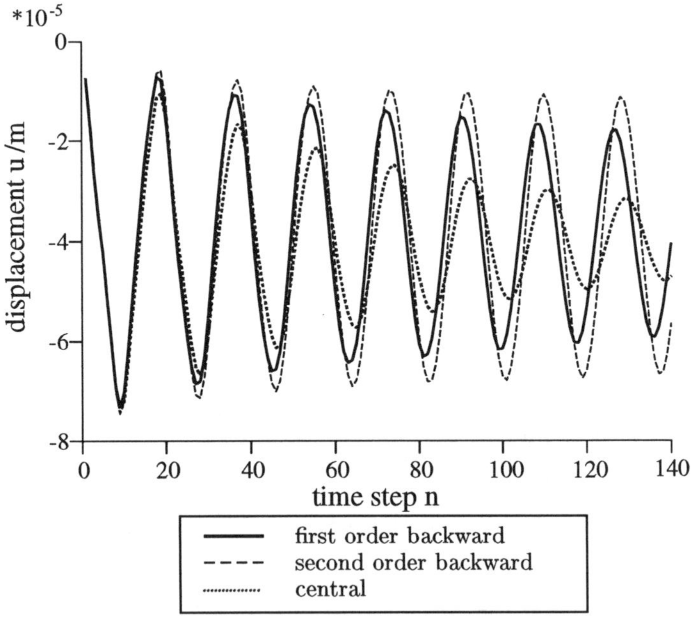

Fig. 6.3. Rod displacement with reduced numerical damping

The central difference leads to poor results, however, the backward difference scheme improves the computational behaviour. The new parameter Δr of the finite difference has to be properly linked to the space discretization, and does not depend on the time step size. Too large values of Δr lead to very stable but damped results, and too small values tend to give instable but undamped results before instability occurs. Therefore, stable and undamped results can be obtained by properly choosing Δt and Δr.

7 Conclusions

Formulations in elastodynamics by BEM have generally been used to treat viscoelastic solids in Laplace or Fourier domain in which elastic moduli are replaced by the corresponding viscoelastic moduli. The calculation of transient response in time domain involves inverse transformations [12].The accuracy of the solution depends strongly on the limited number of available frequencies. To avoid this problem, the present paper uses the BEM formulation in time domain and converts the elastic formulation to that for viscoelastic materials. This method turns out to be advantageous to convert the formulation after integration over a time step Δt rather than converting the fundamental solution itself. The conversion is done by applying the elastic-viscoelastic correspondence principle in Laplace domain. The inverse transformation in each time step is performed numerically and thus allows one to account for general constitutive equations involving fractional time derivatives. The wave propagation in two viscoelastic continua are presented by BEM in time domain as illustrative examples. The material behaviour is described by a three parameter rheological model, generalized by two fractional time derivatives.

Different from what is known in time marching schemes used in FE equations, too small time steps may lead to instability in the discretized BIE of elastodynamics of viscoelastic materials. The effect is shown to be more pronounced for elastic constitutive equations.

After having replaced the derivatives of Dirac distributions in the traction fundamental solution by finite differences, the stability range was extended to smaller time steps. A generalized application of using higher order backward finite differences improved the stability range and significantly reduced the effect of numerical damping as well.

References

1. Antes, H. (1988): Anwendungen der Methode der Randelemente in der Elastodynamik und der Fluiddynamik. Mathematische Methoden in der Technik 9. Teubner, Stuttgart
2. Bagley, R.L., Torvik, P.J. (1986): On the Fractional Calculus Model of Viscoelastic Behaviour. Journal of Rheology 30(1), 133–155
3. Eringen, A. C., Suhubi, E. S. (1975): Elastodynamics, Vol. II. Academic Press, New York San Francisco London
4. Figueiredo, A. M., Sipcic, S. R., Morino, L. (1992): Kirchhoff Integral Equation: Instability Issues Introduced by the Spacial Discretization. Proc. 14th Aeroacoustics Conference, Aachen

5. Flügge, W. (1975): Viscoelasticity. Springer, New York Heidelberg Berlin
6. Gaul, L., Klein, P., Kempfle, S. (1991): Damping Description Involving Fractional Operators. Mechanical Systems and Signal Processing, 5(2), 81–88
7. Gaul, L., Schanz, M. (1992): BEM Formulation in Time Domain for Viscoelastic Media Based on Analytical Time Integration. Proc. 14th Boundary Element International Conference, Vol. II. Sevilla, Spain, pp. 223–234
8. Gaul, L., Schanz, M., Fiedler, C. (1992): Viscoelastic Formulations of BEM in Time and Frequency Domain. Engineering Analysis with Boundary Elements 10, 137–141
9. Graffi, D. (1954): Über den Reziprozitätssatz in der Dynamik elastischer Körper. Ingenieur Archiv 22, 45–46
10. Grünwald, A. K. (1867): Über "begrenzte" Derivationen und deren Anwendung. Zeitschrift f. Mathematik u. Physik 12, 441–480
11. Jäger, M. (1994): Entwicklung eines effizienten Randelementverfahrens für bewegte Schallquellen. Braunschweiger Schriften zur Mechanik 17-1994, Technische Universität Braunschweig
12. Klein, P. (1990): Zur Beschreibung der dynamischen Wechselwirkung von Fundamentstrukturen mit dem viskoelastischen Baugrund durch dreidimensionale Randelementformulierungen. Bericht aus dem Institut für Mechanik, Universität der Bundeswehr Hamburg
13. Oldham, K. B., Spanier, J. (1974): The Fractional Calculus. Academic Press, New York London
14. Schanz, M. (1994): Eine Randelementformulierung im Zeitbereich mit verallgemeinerten viskoelastischen Stoffgesetzen. Bericht aus dem Institut A für Mechanik Heft 1, Universität Stuttgart
15. Schanz, M., Gaul, L. (1993): Implementation of viscoelastic behaviour in a time domain boundary element formulation. Applied Mechanics Reviews 46(11, Part 2), 41–46
16. Steinfeld, B. (1993): Numerische Berechnung dreidimensionaler Kontaktprobleme Bauwerk-Boden mittels zeitabhängiger Randintegralgleichungen der Elastodynamik. Technisch–wissenschaftliche Mitteilungen Mitteilung Nr. 93–1, Ruhr–Universität Bochum
17. Talbot, A. (1979): The Accurate Numerical Inversion of Laplace Transforms. Journal of the Institute of Mathematics and its Applications 23, 97–120

On the Efficient Realization of Sparse Matrix Techniques for Integral Equations with Focus on Panel Clustering, Cubature and Software Design Aspects

W. Hackbusch, C. Lage, and S.A. Sauter

1 Introduction

The boundary element method (BEM) is an elegant tool for solving elliptic boundary value problems numerically. First, the method of integral equations is applied transforming the PDE on a domain Ω into an integral equation defined on the boundary of Ω. This integral equation can be discretized by Petrov-Galerkin methods defined on the surface of Ω. Instead of the discretization of the whole domain, only the lower dimensional boundary of Ω has to be partitioned into a FE-grid. This is one of the major advantages of the BEM. Especially for 3-d problems, grid generation of the whole domain Ω in many cases is still an extremely time consuming step. As a consequence, the dimension of the stiffness matrix is much smaller as for the corresponding FE-discretization. Furthermore, the matrix condition number is smaller compared to the FEM-system governing the convergence speed of iterative solvers applied to the linear system. On the other hand, the major drawback of the BEM is that the arising system matrix is full and, in addition, the computation of the matrix entries requires the evaluation of complicated surface integrals. Recently, many attempts have been made to overcome these two difficulties. The full matrix can avoided by representing the integral operator on the discrete level in an alternative form, which is based on the approximation of the kernel function of the integral operator. The panel clustering, originally developed by Hackbusch and Nowak (see [9], [7]) for the collocation discretization of the Laplace equation, turned out to be also applicable to any Petrov-Galerkin method, where we emphasize the important case of the Galerkin method where test and trial space coincide. A related method is the multipole expansion of Rokhlin (see [15]) which, however, can be applied only to Nyström discretizations and not, e.g., to the important case of the hypersingular formulation of integral equations. A further approach in this direction was presented by Brandt and Lubrecht in [4], [5] where multigrid techniques together with high order inter-grid interpolation is used to condensate the system matrix.

Another method to condensate the system matrix is the application of wavelets to boundary integral equations. It turns out that, by using a suitable wavelet basis, the off-diagonal entries of the system matrix have a strongly decreasing behaviour with increasing distance from the diagonal (see [6], [20]).

All condensation strategies have in common that the ***nearfield matrix entries***, i.e., those elements lying close to the diagonal have to be computed in the standard matrix-oriented way. As mentioned above, this is a non-trivial and time-consuming task in BEM computations. The product of two basis functions with a possibly very complicated ***kernel function*** having a singular or nearly singular behaviour has to be integrated over surface pieces. Due to the singular behaviour of the kernel function, standard cubature techniques lose their accuracy. We will show that, for kernel functions arising by discretizing boundary value problems via the BEM, the application of suitable coordinate transforms enables us to integrate the variable containing the singularity analytically resulting in an integrand which has no or significantly reduced singular behaviour. For hypersingular integrals, we write the arising regularization in a form which can be evaluated in many cases analytically resulting in surprisingly easy formulae.

In this paper, we discuss thoroughly the software design aspects which came up by the realization of the developed numerical algorithms. Especially we explain how the complexity of implementing the BEM can be managed by object-oriented design methods. Also emphasize is taken on parallelization strategies.

The paper is organized as follows. In the next chapter, we go briefly through the setting of Petrov-Galerkin discretization of boundary integral equations and introduce the notations necessary for the sequel. Then, in Chapter 3, we investigate the computation of the surface integrals arising when computing the matrix entries of the linear system. First, we have to provide some analytical properties of the arising kernel functions which then will be used in the design of efficient cubature strategies. (Note that in more than one dimension the terminus *quadrature* is replaced by *cubature*). In Chapter 4, we explain the panel clustering algorithm. After having developed the basic principle, the algorithm is formulated and an error analysis is presented. Estimates of the asymptotic complexity of the algorithm follows and remarks on their relevance for practical problem sizes are given. Chapter 6 is devoted to the design of modern BEM software in C++. It will be explained which data structures are well suited in order to make the different algorithms efficient and flexible with respect to different kinds of integral equations, discretization schemes as, e.g., collocation or Galerkin BEM, geometries and cubature techniques.

2 Setting and Preliminaries

Let us consider an elliptic boundary value problem on a domain $\Omega \subset \mathbb{R}^d$ of order $2m$ of the form

$$\begin{aligned} Du &= 0, \qquad &&\text{in } \Omega \\ Bu &= r, \qquad &&\text{on } \Gamma := \partial\Omega \end{aligned} \tag{2.1}$$

with a system of m boundary differential operators B_j having mutually different order. If Ω is unbounded, suitable radiation conditions have to be imposed. We assume that the operator D has constant coefficients and the fundamental solution defined by

$$\begin{aligned} &DS = \delta_0, \qquad \text{in } \mathbb{R}^3, \\ &S \text{ satisfies the radiation conditions,} \end{aligned} \tag{2.2}$$

is known explicitly. Here and in the following, δ_x denote the Dirac-functional located at the point x. In order to exhibit the principal ideas, it is sufficient to make the following assumptions, while we state that all presented methods can be applied to the general situation without any significant modification.

- the space dimension d equals 3,
- the order $2m$ of the operator is 2,
- $B = (\partial/\partial n)^j$ with j is either 0 or 1, whereas $\partial/\partial n$ denotes the normal derivative,
- the surface Γ is Lipschitz continuous.

Using the (direct) method of integral equations, problem (2.1) is transformed into an integral equation which, in the variational formulation, is given by seeking $u \in X$ such that

$$\langle \lambda u, v \rangle + \langle K_1 u, v \rangle = \langle K_2 r, v \rangle , \qquad \forall v \in Y \tag{2.3}$$

is satisfied. Here, X, Y are suitable function spaces and $\langle \cdot, \cdot \rangle$ is a dual pairing. The operators K_1 and K_2 denote integral operators which will be specified later. The function λ is piecewise constant on Γ. The reason for presenting the formulation in this abstract setting is that most of the theory and algorithms discussed in the sequel can be presented in a more compact form as if discussed for each choice of the spaces and pairing above separately. The Petrov-Galerkin method is characterized by defining finite dimensional subspaces X_n and Y_n of X and Y, i.e, finding $u_n \in X_n$ such that (2.3) is satisfied for all $v_n \in Y_n$. The definition of the discrete spaces is typically based on a geometrical partitioning of the surface Γ. For this, let $T_N := \{\Delta_j : 1 \le j \le N\}$ be a partitioning of Γ into small surface pieces. Here, we assume that T_N is a triangulation in the sense that there exists a family of diffeomorphisms $\{\chi_i\}_{1 \le i \le N}$ mapping the *master element*, i.e., the triangle with vertices $(0,0)^T, (1,0)^T$, and $(1,1)^T$, onto the elements $\{\Delta_i\}_{1 \le i \le N}$. We assume that T_N satisfies the following conditions:

1. $\Gamma = \bigcup_{\Delta \in T_N} \Delta$,
2. $\Delta_i \cap \Delta_j$ is either empty, a vertex, a (curved) edge or a (curved) triangle,

3. $h := \max_{\Delta \in T_N} \operatorname{diam}(\Delta) \leq C \operatorname{diam}(\Delta), \qquad \forall \Delta \in T_N,$
4. $\max_{B \text{ is a ball contained in } \Delta} \operatorname{diam}(B) \geq Ch, \qquad \forall \Delta \in T_N.$

The finite element space X_n is given by lifting polynomials of degree p defined on the master element onto the surface, i.e.,

$$X_n := \left\{ u \in C^k(\Gamma) : u \circ \chi_i \mid_{\hat{\Delta}} \text{ is a polynomial of degree } p \text{ for all } 1 \leq i \leq N \right\}.$$

Let $\Theta_n = \{x_i : 1 \leq i \leq n\} \subset \Gamma$ denote the set of uni-solvent nodal points having the property that the interpolation problem of finding $u \in X_n$ such that $u(x_i) = f_i$ for all $1 \leq i \leq n$ has a unique solution for any given vector $f \in \mathbb{C}^n$. The example below illustrates typical choices of discrete spaces X_n, Y_n and dual pairings.

Example 2.1. 1. The Collocation Method is characterized by the following choices. X_n is defined as explained above and $Y_n := \operatorname{span}\{\delta_x : x \in \Theta_n\}$. The dual pairing is given by $\langle w, \delta_x \rangle = w(x)$. Explicitly, (2.3) takes the form: find $u \in X_n$ such that

$$\lambda(x) u(x) + K_1[u](x) = K_2[r](x), \qquad \forall x \in \Theta_n. \tag{2.4}$$

2. The Galerkin Method is characterized by choosing X_n as above, $Y_n = X_n$ and $\langle w, u \rangle = (w, u)_0$, where $(\cdot, \cdot)_0$ denote the L^2-scalar product on the surface Γ. This results in seeking $u \in X_n$ such that

$$(\lambda u, v)_0 + (K_1 u, v)_0 = (K_2 r, v)_0, \qquad \forall v \in X_n. \tag{2.5}$$

In the following we collect some properties of the integral operators $K_{1,2}$ which will be used in the next chapters. All integral operators appearing in the context of solving elliptic boundary value problem via the method of integral equations can be written in the form

$$K[u](x) = \text{p.f.} \int_\Gamma k(x, y, y - x) u(y) \, dy. \tag{2.6}$$

Some explanations are necessary. The kernel function $k(x, y, y - x)$ is a suitable Gâteau derivative (of order less or equal than $2m$) of the fundamental solution defined by (2.2). The fundamental solution behaves singularly only for $x = 0$ and is smooth elsewhere. In the three-dimensional case, we know (cf. [12]) that it can be estimated as

$$|S(x - y)| \leq C \|x - y\|^{2m-3}, \qquad \forall x, y \in \Gamma \text{ with } x \neq y.$$

Hence, it is natural to assume that

$$|k(x, y, y - x)| \leq C \|x - y\|^{-s}, \qquad \forall x, y \in \Gamma \text{ with } x \neq y, \tag{2.7}$$

where the *order of the singularity* s is an integer smaller or equal than 3. With respect to the first both variables, k is smooth in smooth parts of the surface Γ and may jump across edges and corners. An important case is the case of flat patches of the surface. In such regions, it is natural to assume that k is constant in the first two variables, i.e.,

$$k\left(x,y,y-x\right)=k\left(x-y\right). \tag{2.8}$$

For $s=2,3$ the kernel function is not Lebesgue-integrable. We have to apply the concept of regularized integrals in the sense of Hadamard. Let $B_\epsilon\left(x\right)$ denote a ball with radius ϵ about x and γ be a surface piece of Γ. Then, the functional J is well-defined by

$$J_\gamma\left[u\right]\left(\epsilon,x\right):=\int_{\gamma\backslash B_\epsilon(x)}k\left(x,y,y-x\right)u\left(y\right)dy.$$

It was shown in [22] that, for all kernels which arise by transforming elliptic boundary value problems into integral equations, the function J admits an expansion w.r.t ϵ of the form

$$J_\gamma\left[u\right]\left(\epsilon,x\right)=A_{-1}\left(x\right)\epsilon^{-1}+A_{\log}\left(x\right)\log\epsilon+A_0\left(x\right)+o\left(1\right).$$

The *finite part* of such an expansion is given by

$$\underset{\epsilon}{\text{p.f.}}\,J_\gamma\left[u\right]\left(\epsilon,x\right):=A_0\left(x\right).$$

In order to exhibit the dependency of A_0 on γ and u we alternatively write $A_\gamma\left[u\right]$. We define the regularized integral (2.6) by

$$\text{p.f.}\int_\gamma k\left(x,y,y-x\right)u\left(y\right)dy:=\underset{\epsilon}{\text{p.f.}}\,J_\gamma\left[u\right]\left(\epsilon,x\right)=A_0\left(x\right). \tag{2.9}$$

Now, all general assumptions are collected which will be needed for discussing cubature methods for the arising integrals and approximation of the kernel functions by means of the panel clustering method.

3 Cubature Techniques for the Approximation of Singular and Nearly Singular Surface Integrals Arising in BEM

In this chapter, we present cubature techniques for collocation and Galerkin discretizations of boundary integral equations. We start with the abstract Petrov-Galerkin formulation. Let $\{\phi_i\}_{1\leq i\leq n}$ and $\{\varphi_i\}_{1\leq i\leq n}$ denote a basis of X_n and Y_n respectively. Any function $u\in X_n$ is uniquely linked to a coefficient vector $\boldsymbol{u}\in\mathbb{C}^n$ by the basis representation

$$u\left(x\right)=\sum_{i=1}^n\boldsymbol{u}_i\phi_i\left(x\right). \tag{3.1}$$

The Petrov-Galerkin discretization is then equivalent to solving the linear system,

$$(\boldsymbol{M}+\boldsymbol{K})\,\boldsymbol{u}=\boldsymbol{f}$$

with $n \times n$ matrices $\boldsymbol{M}$ and $\boldsymbol{K}$ and the n-vector $\boldsymbol{f}$ defined by

$$\begin{aligned}\boldsymbol{M}_{i,j} &:= \langle \lambda\phi_j, \varphi_i\rangle\,,\\ \boldsymbol{K}_{i,j} &:= \langle K_1\phi_j, \varphi_i\rangle\,,\\ \boldsymbol{f}_i &:= \langle K_2 r, \varphi_i\rangle\,.\end{aligned}$$

The computation of the *mass matrix* $\boldsymbol{M}$ is trivial, since no integral operator is involved and $\boldsymbol{M}$ usually is sparse. The computation of $\boldsymbol{f}$ is similar as the computation of $\boldsymbol{K}$, thus, we restrict our attention on the approximation of the entries of $\boldsymbol{K}$. In the following we write K instead of K_1. In the rest of this paper we assume that

1. the integral operator K arises by transforming an elliptic boundary value problem into an integral equation,
2. the trial space X_n is a subspace of the (continuous) function space X,
3. $X_n \subset C^{0,1}(\Gamma)$ for hypersingular equations, i.e., $s=3$.

An important property of the application of K to basis functions $\phi_i \in X_n$ is stated in the following

Theorem 3.1. *Let γ be a simple connected surface piece which consists of some triangles of T_N. Recall the definition of the part-fini integral (see (2.9)) and the order s of the singularity of the kernel function (see (2.7)). Let*

$$A_\gamma[\phi_i](x) := \text{p.f.}\int_\gamma k(x,y,y-x)\,u(y)\,dy.$$

Then,

$$|A_\gamma[\phi_i](x)| \le \begin{cases} C & \text{if } s \le 1,\\ C\log \operatorname{dist}(x,\partial\gamma) & \text{if } s=2,\\ C\operatorname{dist}^{-1}(x,\partial\gamma) & \text{if } s=3,\end{cases} \tag{3.2}$$

with C independent of x.

For $\gamma=\Gamma$ and any $s\le 3$,

$$|A_\Gamma[\phi_i](x)| \le C, \qquad \forall x\in\Gamma$$

is satisfied.

Proof. The proof of this theorem can be found in [13, 22, or 14].

In the following subsection, we consider cubature techniques for the collocation method.

3.1 Cubature Techniques for the Collocation Method

From Theorem 3.1, it follows that it is problematic to use the collocation methods for strongly singular integral equations, i.e., $s \geq 2$, since, typically, one would have to evaluate $A_\gamma [\phi_i]$ at corner and edge points. However, estimate (3.2) does not necessarily imply that, for $x \in \partial\gamma$, the function $A_\gamma [u] (x)$ is infinite. Typically for Cauchy-singular integral equations, i.e., $s = 2$, the function $A_\gamma [u] (x)$ tends to infinity as x approaches $\partial\gamma$ but remains finite at points $x \in \gamma$. This behaviour is thoroughly analyzed and correction functionals can be computed (see [22] and [23]). In the cited papers, it was shown that it is always possible to split the integral into some point functionals and an integral having a weakly singular integrand. Then, either Duffy's triangle coordinates or polar coordinates smoothes the integrand such that it can be treated efficiently with properly scaled Gauß-Legendre formulae. Here, we do not go into the details but proceed with considering the nearly singular situation. The arising problem is to compute

$$I_\Delta (x_i) := \int_\Delta k (x_i, y, y - x_i) \phi_i (x) \, dx, \tag{3.3}$$

where x_i denote a collocation point having a positive distance from the surface panel Δ,

$$\operatorname{dist} (x_i, \Delta) = \delta > 0.$$

In this case, the part-fini integral reduces to the Rieman integral, since the integrand is regular. On the other hand, the nth derivative of the integrand behaves like $O(n! h^{-n})$ resulting in a very poor convergence speed of Gauß-Legendre formulae. Different attempts have been made to overcome this problem. Subdividing the triangle in the direction of the singularity was presented in [1] and [2]. This method was generalized by introducing variable cubature order in [21]. We present here a semi-analytical technique which was worked out in detail in [11]. It was shown that the integral (3.3) can be computed to any desired accuracy by computing integrals of the type

$$\int_\Delta \frac{(u_1 + \delta_\triangle)^{\nu_1} u_2^{\nu_2} \delta_\perp^{\nu_3}}{\sqrt{(u_1 + \delta_\triangle)^2 + u_2^2 + \delta_\perp^2}^{\,s+t}} \, d\Delta_u$$

where $\delta_\triangle$ denote the horizontal distance of x_i from Δ and $\delta_\perp$ the vertical distance. By introducing shifted polar coordinates of the form

$$u = (r - \delta_\triangle \cos\alpha) \begin{pmatrix} \cos\alpha \\ \sin\alpha \end{pmatrix}$$

the integral can be written as

$$\delta_{\perp}^{\nu_3}\sum_{i=0}^{\nu_1+\nu_2+1}\int_{\alpha_{\min}}^{\alpha_{\max}}\delta_{\triangle}^{\nu_1+\nu_2+1-i}c_{\nu_1,\nu_2,i}\left(\sin\alpha,\cos\alpha\right)$$
$$\int_{\delta_{\triangle}\cos\alpha}^{R(\alpha)+\delta_{\triangle}\cos\alpha}\frac{r^i}{\sqrt{r^2+\delta_{\triangle}^2\sin^2\alpha+\delta_{\perp}^2}^{\,s+t}}drd\alpha,$$

where $c_{\nu_1,\nu_2,i}\left(\cdot,\cdot\right)$ are polynomials, $R\left(\alpha\right)$ is a smooth function and $\alpha_{\min}$, $\alpha_{\max}\in\left[-\pi,\pi\right[$. The inner integration can be performed analytically by using

$$\int_a^b\frac{r^n}{\sqrt{r^2+\delta_\alpha^2}^{\,s}}dr=\frac{1}{\sigma_{n,s}}\frac{p\left(\delta_\alpha,r\right)}{\sqrt{r^2+\delta_\alpha^2}^{\,s-2}}+\kappa_{n,s}\log\left(\pm r+\sqrt{r^2+\delta_\alpha^2}\right)\Big|_a^b$$

where p is a polynomial of degree less or equal than $\frac{s+n-3}{2}$ and $\sigma_{n,s}$, $\kappa_{n,s}$ are some coefficients. It can be shown that the integrand of the outer integration has a reduced singular behaviour and can be computed by properly scaled one-dimensional Gauß-Legendre quadrature. A careful error analysis (see [11]) shows that the order of the angular integration has to be increased logarithmically with decreasing distance δ and triangle size diamΔ.

3.2 Cubature Techniques for the Galerkin Method

For the Galerkin discretization of BIE, the elements of the stiffness matrix are given by

$$K_{i,j}:=\int_{\mathrm{supp}\phi_i}\phi_i\left(x\right)\text{p.f.}\int_{\mathrm{supp}\phi_j}k\left(x,y,y-x\right)\phi_j\left(y\right)dydx.$$

The support of the basis functions ϕ_i consists of a few triangles. It would be desirable to split the computation of $K_{i,j}$ into a sum of integrals over pairs of panels $\Delta_x\times\Delta_y$. This is not trivial in view of the regularization process involved with the part fini integral. From Theorem 3.1, we know that

$$A_\Gamma\left[\phi_j\right]\left(x\right):=\text{p.f.}\int_\Gamma k\left(x,y,y-x\right)\phi_j\left(y\right)dydx$$

is bounded and hence,

$$\int_\Delta\phi_i\left(x\right)A_\Gamma\left[\phi_j\right]\left(x\right)dx=\int_\Delta\phi_i\left(x\right)\sum_{\Delta_y\in\mathrm{supp}\phi_j}A_{\Delta_y}\left[\phi_j\right]\left(x\right)dx.$$

Unfortunately, $A_{\Delta_y}\left[\phi_j\right]$ may contain strong singularities at corners and edges of Δ such that the sum may not be interchanged with the outer integration in general. The following concept of introducing an outer regularization process was developed in [14]. For this, let the δ-strip Δ^δ be defined by

$$\Delta^\delta:=\left\{x\in\Delta_x\mid\mathrm{dist}\left(x,\partial\Omega\right)\leq\delta\right\}.$$

Then, it was shown in the cited paper that

$$\int_{\Delta} \phi_i(x) A_\Gamma[\phi_j](x)\,dx$$

$$= \sum_{\Delta_y \in \mathrm{supp}\phi_j} \underset{\epsilon}{\mathrm{p.f.}} \underbrace{\int_{\Delta}\int_{\Delta_y \setminus B_\epsilon(x)} \phi_i(x)\, k(x,y,y-x)\, \phi_j(y)\, dy dx}_{=:K_{\Delta\times\Delta_y}(\epsilon)} \quad (3.4)$$

$$-\underset{\delta}{\mathrm{p.f.}}\,\underset{\epsilon}{\mathrm{p.f.}} \underbrace{\int_{\Delta^\delta}\int_{\Delta_y \setminus B_\epsilon(x)} \phi_i(x)\, k(x,y,y-x)\, \phi_j(y)\, dy dx}_{=:K_{\Delta\times\Delta_y}(\epsilon,\delta)}.$$

First, we consider the approximation of the term $\underset{\delta}{\mathrm{p.f.}}\,\underset{\epsilon}{\mathrm{p.f.}}\, K_{\Delta\times\Delta_y}(\epsilon,\delta)$. In [14], it was shown that this term vanishes for Cauchy-singular and weakly singular integrands and in any case if $\bar{\Delta}\cap\bar{\Delta}_y$ is either empty or a vertex, thus, we may restrict to the case $s = 3$ and $\bar{\Delta}\cap\bar{\Delta}_y$ is either an edge or a panel. Since $\Delta^\delta \to \emptyset$, we may introduce several simplification of the integrand and integration domain which leaves the limit unchanged. It can be shown that Δ^δ may be decomposed into a sum containing integrals over rectangles $e_m \times (0,\delta)$, where e_m denotes an edge of Δ. The triangle Δ_y may be replaced by a suitable square Q_m. Furthermore the basis functions may be replaced by constant continuation of the traces $\phi_i|_{e_m}$ vertical to e_m and the kernel function by its principal part. Finally, the (curved) triangles may be replaced by suitable flat triangles. Thus, it is sufficient to develop techniques to compute

$$I_m^{i,j} := \underset{\delta}{\mathrm{p.f.}}\,\underset{\epsilon}{\mathrm{p.f.}} \int_{e_m\times(0,\delta)}\int_{Q_m} \tilde{\phi}_{i,m}(x)\, k_{\mathrm{principal}}(y-x)\, \tilde{\phi}_{j,m}(y)\, dy dx.$$

We introduce relative coordinates by $z = y - x$ and define the polynomial $b_{i,j}(x,z) := \tilde{\phi}_{i,m}(x)\,\tilde{\phi}_{j,m}(z+x)$, yielding

$$I_m^{i,j} = \int_{D(\delta,\epsilon)} b_{i,j}(x,z)\, k_{\mathrm{principal}}(z)\, dz dx$$

with a suitable four-dimensional domain $D(\delta,\epsilon)$. It turns out that one integration can be carried out analytically due to the polynomial character of $b_{i,j}$. In many cases the remaining integrand can be computed analytically, too. This has been worked out in detail in [14] for the case that the principal part of the kernel function is given by

$$\frac{\partial^2}{\partial n_y \partial n_x}\frac{C}{\|x-y\|},$$

i.e., the underlying differential equation is the Laplace or the Helmholtz equation. The result is that for linear elements and triangles Δ, Δ_y with a common edge

$$I_m^{i,j} := c_{i,j} \left| e_m \right| \frac{2}{9} \left(\frac{\omega}{\tan \omega} + \frac{4}{9} + \frac{1}{3} \log \frac{\left| e_m \right|}{2} \right),$$

where ω denote the (smaller) angle between the planes through Δ and Δ_y. The coefficient $c_{i,j}$ is either $0, 1$, or 2 dependent on the indices i, j. This formulae can further be simplified if the case $\Delta = \Delta_y$ is taken into consideration. Here, we skip the details.

Now, we investigate the approximation of the quantity $K_{\Delta \times \Delta_y}(\epsilon)$. Since this integral is regular, we may use the usual rule of transformation of variables. In [10], it was shown how $K_{\Delta \times \Delta_y}(\epsilon)$ can be approximated by a sum of integrals of the same form over flat triangles. Hence, we restrict to this situation and in view of (2.8), we assume that $k(x, y, y - x)$ is constant in the first both variables on the integration domain $D := \Delta \times \Delta_y$. This means that $\tilde{k}(z) := k(x, y, z)$, where we skip the tilde in the sequel. We have to distinguish the following situation.

1. The singular case: $\bar{\Delta} \cap \bar{\Delta}_y \neq \emptyset$.
2. The nearly singular case: $\operatorname{dist}(\Delta, \Delta_y) = \delta > 0$ but "small".
3. The regular case: $\operatorname{dist}(\Delta, \Delta_y) \geq C > 0$ with $C = O(1)$.

These cases will be considered in the following subsections.

3.2.1 The Singular Case: $\bar{\Delta} \cap \bar{\Delta}_y \neq \emptyset$

Due to the space limitations of this paper, we consider here only the case of $\Delta = \Delta_y$, where the general situation is considered in [19], [10] and [14]. The restriction to the case that Δ lies in the (1,2)-plane is not essential but will simplify the notations. We consider the integral

$$Q := \int_{\substack{\Delta \times \Delta \\ \|y - x\| \geq \epsilon}} \phi_i(x)\, k(x - y)\, \phi_j(y)\, dy dx.$$

The key point of the transformation procedure is the use of relative coordinates:

$$y = z + x.$$

Let the polynomial b be defined by $b_{i,j}(x, z) := \phi_i(x)\, \phi_j(z + x)$. The integral above then can be written in the form

$$\int_{\substack{\tilde{D} \\ \|z\| \geq \epsilon}} b_{i,j}(x, z)\, k(z)\, dz dx$$

where the domain of integration is a four-dimensional polyhedron given by

$$\tilde{D} := \left\{ (x, z) \in \mathbb{R}^{2 \times 2} \mid \exists (x, y) \in \Delta \times \Delta : z = y - x \right\}.$$

This polyhedron can be parametrized also in the opposite way such that x depends on z. This means that the integration with respect to z can be interchanged with the x-integration. We obtain

$$Q := \int_{\substack{\tilde{D} \\ \|z\| \geq \epsilon}} k(z)\, b_{i,j}(x,z)\, dx dz.$$

Since b is a polynomial and $\tilde{D}$ is a polyhedron, the x-integration can be carried out analytically yielding a *piecewise* polynomial function $B_{i,j}(z) := \int b_{i,j}(x,z)\, dx$. Splitting the remaining integration domain into subdomains where $B_{i,j}(z)$ is a polynomial and further affine transformations results in

$$Q = \sum_{m=1}^{3} \text{p.f.} \int_{\substack{\Delta \\ h\|z - A_m\| \geq \epsilon}} k(A_m - z)\, H_m^{i,j}(z)\, dz,$$

where the functions $H_m^{i,j}$ are polynomials in z. The remaining integral can be computed using polar coordinates about A_m :

$$Q = \sum_{m=1}^{3} \text{p.f.} \int_{\alpha_{\min,m}}^{\alpha_{\max,m}} \int_{\epsilon/h}^{R(\alpha)} k\begin{pmatrix} r\cos\alpha \\ r\sin\alpha \end{pmatrix} H_m^{i,j}\left(A_m + \begin{pmatrix} r\cos\alpha \\ r\sin\alpha \end{pmatrix}\right) r dr d\alpha.$$

It turns out that the r-integration can be valuated analytically very easily and the limit process, too. The integrand of the remaining angular integration is smooth and can be approximated efficiently using properly scaled Gaussian quadrature rules.

3.2.2 The Nearly Singular Case

In the case that $\operatorname{dist}(\Delta, \Delta_y) =: \delta > 0$ is small the convergence of standard cubature formulae becomes very poor. To overcome this problem, it is possible to write the integral over $\Delta \times \Delta_y$ as a sum over domains containing the singularity and proceeding as explained for the singular case (see [19]). Here, we will explain briefly a further alternative which is very similar as described above for the collocation method (see [14]). Again the starting point is the introduce relative coordinates $y - x = z$ and shifted polar coordinates. The r-integration can be done explicitly resulting in simple formulae for the primitive integrals having a reduced singular behaviour. The remaining three-dimensional integral can be approximated by properly scaled Gauß-Legendre cubature, i.e., the order of integration again has to be increased logarithmically with respect to the diameter of the triangles and the distance δ.

3.2.3 The Regular Case

In this case the kernel function has no singular behaviour. Standard cubature techniques as tensor Gauß rules may be applied having the well known fast convergence properties. However, the question arises how accurate the integrals have to be computed such that the overall asymptotic discretization error is not reduced. In [17] it was shown what degree of exactness the inner integration and the outer integration must have, dependent on the order of the integral operator, on the order of approximation, on distance from the singularity, and the norm in which the error is measured, such that the asymptotic convergence order of the overall discretization is not influenced. In order to illustrate the results, let us assume that linear finite elements are employed, the distance δ from the singularity is $O(1)$ and the integral operator K maps the Sobolev space H^s onto H^{-s}. The error is measured in the H^ρ-norm. The following table shows pairs of numbers, where the first one denotes the degree of exactness of the outer integration and the second one for the inner integration.

Table 3.1. Required degree of exactness (γ_x, γ_y) in order to get an optimal discretization order with piecewise linear functions in the H^ρ-norm.

s	$\rho=+\frac{1}{2}$	$\rho=0$	$\rho=-\frac{1}{2}$	$\rho=-1$	$\rho=-\frac{3}{2}$	$\rho=-2$	$\rho=-\frac{5}{2}$	$\rho=-3$
$+\frac{1}{2}$	$(2,1)$	$(1,1)$	$(2,2)$	$(1,2)$	–	–	–	–
0	–	$(2,1)$	$(3,2)$	$(2,2)$	$(3,3)$	$(2,3)$	–	–
$-\frac{1}{2}$	–	–	$(2,2)$	$(3,2)$	$(4,3)$	$(3,3)$	$(3,4)$	$(2,4)$

4 The Panel Clustering Method

The boundary element method leads to a system of linear equation which contains a *full* coefficient matrix of dimension n. The reason lies in the fact that the kernel function $k(x,y,x-y)$ contains a term of the form $\|x-y\|$ linking every point $x \in \Gamma$ with a point $y \in \Gamma$. The panel clustering method was developed by Hackbusch and Nowak in [9], [8] for collocation discretizations and generalized to the Galerkin BEM in [19], [10]. The idea of the method is the approximation of k by a series of the form $k(x,y,y-x) \approx \sum \kappa_{\nu,\mu}\omega_\nu(x)\,\omega_\mu(y)$ with a suitable function system $\{\omega_\nu\}$. Hence, the x-integration is separated from the y-integration reducing the amount of work substantially. The efficiency of the method depends on how fast the kernel function can be approximated by the function system $\{\omega_\nu\}$, i.e., the number of term in the series and on the question how efficiently the arising quantities can be computed and organized. This will be discussed in the following. We recall here the setting of Petrov-Galerkin methods for boundary integral equations as defined in Chapter 2 and the notation of the trial space by $X_n = \operatorname{span}\{\phi_i\}_{1\le i\le n}$ and test space $Y_n = \operatorname{span}\{\varphi_i\}_{1\le i\le n}$.

4.1 Kernel Expansions

The kernel functions appearing in the definition of integral equations are related to the singularity function $s : \mathbb{R}^3 \to \mathbb{R}$ of the underlying boundary value problem. This function can be written in the form

$$s(y-x) = \sum_{|\nu|\ge t} c_\nu \frac{(y-x)^\nu}{\|y-x\|^{s+t}}$$

where $\nu \in \mathbb{N}_0^3$ denote multi-indices and $|\nu| := \nu_1 + \nu_2 + \nu_3$. Let $z_0 := y_0 - x_0$ be "large". Consequently, $s(z) = s(y-x)$ is smooth in a neighborhood $z \in \mathcal{U}(z_0)$. We expand s as a series of order m about z_0:

$$s(z) = \sum_{|\nu|=0}^{m-1} \sum_{\mu=0}^{\nu} \kappa_{\nu,\mu-\nu}(z_0)\, \omega_\nu(x)\, \omega_\mu(y) + R_m(z_0, z) \tag{4.1}$$

with some coefficients κ. The kernel function is a suitable Gâteau derivative of s, where we restrict here to normal derivatives. Hence, we can use the expansion

$$k(x, y, y-x) = \sum_{|\nu|=0}^{m-1} \sum_{\mu=0}^{\nu} \kappa_{\nu,\mu-\nu}(z_0) \left(\frac{\partial}{\partial n_x}\right)^{\alpha_1} \omega_\nu(x) \left(\frac{\partial}{\partial n_y}\right)^{\alpha_2} \omega_\mu(y) + \\ + \frac{\partial^{\alpha_1+\alpha_2}}{\partial n_x^{\alpha_1} \partial n_y^{\alpha_2}} R_m(z_0, y-x).$$

Example 4.1. Let $s(z) = \|z\|^{-1}$ which is (up to a constant factor) the singularity function of the 3-d Laplace operator. Taylor expansion about $z_0 = y_0 - x_0$ results in

$$s(z) = \sum_{|\nu|=0}^{m-1} \underbrace{\frac{1}{\nu!} \frac{\partial^{|\nu|}}{\partial z^\nu} s(z)\bigg|_{z=z_0}}_{=: \kappa_\nu(z_0)} (z - z_0)^\nu + R_m(z_0, z).$$

Putting $z = y - x$, expanding $(y - x - z_0)^\nu$ and reordering the terms results in

$$\sum_{|\nu|=0}^{m-1} \sum_{\mu=0}^{\nu} \kappa_{\nu,\mu-\nu}(z_0)\, y^\mu x^{\nu-\mu}$$

which is of the form (4.1).

Remark 4.1. For highly oscillatory kernel as, e.g., for the Helmholtz problem where $s(z) = e^{ik\|z\|}/\|z\|$, Taylor expansion is converging slowly. The situation can be improved by first performing a 1-d expansion in the radial direction, i.e.,

$$\frac{e^{ikr}}{r} \approx \frac{1}{r} \sum_{l=0}^{m-1} \frac{(ik)^l}{l!} e^{ikr_0} (r - r_0)^l$$

and then applying standard Taylor expansion to the right hand side above. The error term behaves like $\frac{((r-r_0)k)^m}{m!}$ and hence, a minimal requirement is that the order of the expansion has to satisfy

$$m \geq k(r - r_0). \tag{4.2}$$

On the other hand the function system consists still of polynomials and the algorithmic apparatus developed for the Laplace operator can be used. An alternative to this expansion is the use of spherical Hankel functions, i.e., writing

$$\frac{e^{ikr}}{r} = \sum_{l=0}^{m-1} \alpha_l h_l \left(k \left(r - r_0\right)\right)$$

which has a better convergence behaviour. However, (4.2) is the minimal requirement for convergence, too. For details see [16]. The use of this function system for Petrov-Galerkin discretizations makes the development of new algorithms necessary in order to compute the required quantities efficiently.

In order to define the panel clustering algorithm, we have to define appropriate regions on the surface Γ where the expansion above gives good approximation.

Definition 4.1. *Let T_N denote the panelization of the Γ as, e.g., the triangulations explained in the previous chapters. A "cluster" is the union of one or more panels: $\tau = \bigcup_{j=1}^{q} \Delta_{i_j}$. The size of a cluster is given by the "cluster radius" $\rho(\tau)$ which is defined by the radius of the minimal ball containing τ. The centre z_τ of this balls is called "cluster center" .*

Definition 4.2. *The relative distance of a cluster from the support of a basis function of the test space $\varphi_i \in Y_n$ is given by*

$$d\left(\varphi_i, \tau\right) := \frac{\rho\left(\tau\right)}{\operatorname{dist}\left(\operatorname{supp}\varphi_i, z_\tau\right)},$$

where for the Dirac functional δ_x, i.e., the collocation method, $\operatorname{supp}\delta_x = x$.

Our aim is to determine the size of clusters such that the replacement of the kernel function by the expansion on those clusters has an accuracy of a given value $\epsilon > 0$. In this light, for $\epsilon > 0$ and arbitrary expansion order m, a cluster is said to be "admissible" with respect to a basis function $\varphi_i \in Y_n$, if

$$\left|k\left(x, y, y - x\right) - k_m\left(x, y\right)\right| \leq \epsilon \frac{1}{\|x - y\|}, \qquad \forall y \in \tau, x \in \operatorname{supp}\varphi_i. \tag{4.3}$$

The size of the clusters has to be linked to the approximation property of the kernel expansion in the following way.

Assumption 4.3. *For given ϵ and expansion order m, there exists $0 < \eta < 1$ such that, for all $1 \leq i \leq N$, the condition*

$$d\left(\varphi_i, \tau\right) \leq \eta < 1$$

implies that τ is admissible with respect to φ_i.

Definition 4.4. *Let ϵ and an expansion order $m \in \mathbb{N}$ be given. Let the relative size η of admissible clusters be determined as explained above. A set of clusters $\{\tau_i\}_{1 \leq i \leq k}$ with disjoint interiors is called a covering of Γ, if $\bigcup_{i=1}^{k} \tau_i = \Gamma$.*

A covering is called admissible with respect to a basis function $\varphi_i \in Y_n$, if either

$$d(\varphi_i, \tau) < \eta \qquad \text{(admissible cluster)}$$

or

$$\tau \text{ is a panel.}$$

The admissible covering which contains a minimal number of clusters is called minimal admissible covering C_i. The nearfield C_i^{near} and the farfield C_i^{far} are defined by

$$C_i^{near} := \{\tau \in C_i \mid \tau \text{ is non-admissible with respect to } \varphi_i\},$$
$$C_i^{far} := \{\tau \in C_i \mid \tau \notin C_i^{near}\}.$$

4.2 The Algorithm

Let $\boldsymbol{K}$ denote the matrix corresponding to Petrov-Galerkin discretization of the integral operator. Then, a matrix vector multiplication can be written in the form

$$\sum_{j=1}^{n} \boldsymbol{K}_{i,j} \boldsymbol{u}_j = \langle \varphi_i, K[u] \rangle = \left\langle \varphi_i, \text{p.f.} \int_\Gamma k(x,y,y-x)\, u(y)\, dy \right\rangle \tag{4.4}$$

with the dual pairing $\langle \cdot, \cdot \rangle$ of (2.3) and the correspondence (3.1) of a vector $\boldsymbol{u}$ and a FE-function u. Splitting the surface into the nearfield and the farfield part and using (4.1) shows that (4.4) equals

$$\sum_{j=1}^{n} \boldsymbol{u}_j \underbrace{\left\langle \varphi_i, \text{p.f.} \int_{C_i^{near}} k(x,y,y-x)\, \phi_j(y)\, dy \right\rangle}_{=: \boldsymbol{K}_{i,j}^{near}}$$
$$+ \sum_{\tau \in C_i^{far}} \left\langle \varphi_i, \int_\tau k(x,y,y-x)\, u(y)\, dy \right\rangle$$
$$\approx \sum_{j=1}^{n} \boldsymbol{K}_{i,j}^{near} \boldsymbol{u}_j + \sum_{\tau \in C_i^{far}} \sum_{|\nu|=0}^{m-1} \sum_{\mu=0}^{\nu} \underbrace{\kappa_{\nu,\mu-\nu}(z_{i,\tau}) \left\langle \varphi_i, \left(\frac{\partial}{\partial n_x}\right)^{\alpha_1} \omega_\nu(x) \right\rangle}_{=: \kappa_{\nu,\mu-\nu}(\varphi_i,\tau)}$$
$$+ \underbrace{\int_\tau \left(\frac{\partial}{\partial n_y}\right)^{\alpha_2} \omega_\mu(y)\, u(y)\, dy}_{J_\tau^\mu(u)}$$

$$= \sum_{j=1}^{n} \boldsymbol{K}_{i,j}^{\boldsymbol{near}} \boldsymbol{u}_j + \sum_{\tau \in \mathcal{C}_i^{far}} \sum_{|\nu|=0}^{m-1} \sum_{\mu=0}^{\nu} \kappa_{\nu,\mu-\nu} \left(\varphi_i, \tau\right) J_\tau^\nu \left(u\right). \tag{4.5}$$

This is the panel clustering representation of a matrix vector multiplication. The coefficient κ are called the *expansion coefficient* and $J_\tau^\nu(u)$ the *farfield coefficient.*

The definition of the so-called nearfield matrix $\boldsymbol{K}_{i,j}^{\boldsymbol{near}}$ looks identical as the original matrix $\boldsymbol{K}$. However, since the integral over $\mathcal{C}_i^{near}$ vanishes if $\text{supp}\phi_j \cap \mathcal{C}_i^{near} = \emptyset$, the matrix $\boldsymbol{K}_{i,j}^{\boldsymbol{near}}$ is sparse dependent on the size of $\mathcal{C}_i^{near}$.

The panel clustering can be structured in the following way.

<u>Phase I:</u>

- Choose the accuracy ϵ for the kernel approximation.
- Choose the expansion order m and relative size of the admissible clusters $\eta = \eta(\epsilon, m)$ such that the storage/computational consumptions (which are known a priori) are minimal.
- Organize the cluster in a binary tree.
- For each basis function φ_i of Y_n, compute the near- and farfield of the minimal covering.

<u>Phase II:</u>

- Compute and store the nearfield matrix $\boldsymbol{K}_{i,j}^{\boldsymbol{near}}$, the expansion coefficients $\kappa_{\nu,\mu-\nu}(\varphi_i, \tau)$, and the *basis* nearfield coefficients $J_\Delta^\nu(\phi_i)$.

<u>Phase III:</u>

- Compute the farfield coefficients $J_\tau^\nu(u)$ from the coefficients $J_\Delta^\nu(\phi_i)$ by using the tree structure of the clusters.
- Approximate a matrix vector multiplication by (4.5).

5 Error Analysis of the Panel Clustering Method

The panel clustering method introduces a further error in the discretization process which can be treated similarly as, e.g., errors due to numerical cubature. In the present situation local analysis has to be done to determine the dependence of the accuracy of the kernel expansion on the size of the clusters and the order of the expansion. As a typical example, we consider Taylor approximation of the singularity function of the Laplace operator in 3-d. For given $\epsilon \geq 0$ and expansion order m, the bound for the relative size of admissible cluster η has to be chosen as $\eta = \sqrt[m]{\epsilon}$ in order to obtain

$$\left|\frac{1}{\|z\|} - T_m(z_0)\right| \leq \frac{\epsilon}{\|z\|}, \qquad \forall z : \frac{\|z - z_0\|}{\|z\|} \leq \eta$$

where T_m denotes the Taylor expansion of order m of $\|z\|^{-1}$. This estimate and also error estimates of Taylor-based expansions of general kernel functions are worked out in [18], [19], [9] and [14].

In the second step, the influence of these local errors (valid on the admissible clusters) to the consistency error, namely a matrix vector multiplication with and without panel clustering, has to be investigated. In [9], [19] and [10] it is shown that, under moderate assumption on the surface and the triangulation, the estimate

$$|\langle \varphi_i, (K - K_{PC})\, u\rangle| \leq C\epsilon\gamma_h \langle \varphi_i, 1\rangle \|u\|_\infty, \qquad \forall u \in X_n,$$

is satisfied with γ_h equals 1 for weakly singular kernels, is $O(\log h)$ for Cauchy-singular and h^{-1} for hypersingular kernels. K_{PC} denote the panel clustering operator where the kernel is replaced by expansions satisfying Assumption 4.3 and (4.3).

5.1 Complexity of the Panel Clustering Algorithm

Let N denote the number of panels of the panelization T_N. An asymptotic complexity analysis of the panel clustering method (cf. [9]) shows that the choice of $m = \lfloor \frac{\kappa}{4} \log N \rfloor$ and $\eta = const$ implies that, again under moderate assumption on the surface and triangulation, the number of nearfield panels is bounded independent of N, while the number of farfield clusters grows like $O(\log N)$. Furthermore, the asymptotic amount of work for the initialization Phase II of the panel clustering algorithms is proportional to $O(N \log^r N)$ with $r = 5$ for the collocation and 4 for the Galerkin method. The evaluation Phase III requires $O(N \log^t N)$ with $t = 4$ for the collocation and $t = 7$ for the Galerkin method. The storage consumption for the collocation method are of order $O(N \log^4 N)$ and for the Galerkin BEM of order $O(N \log^3 N)$. The powers of the logarithmic terms are not ultimative. The algorithm can be structured such that Phase II becomes cheap and the storage consumptions are low (relatively few quantities are computed) but Phase III becomes expansive since more preparation steps have to be performed and vice versa.

The asymptotic gain of the panel clustering method is obvious compared to the standard matrix techniques where both the CPU-time and the storage consumption behaves like $O(N^2)$. Nevertheless it is important to consider the constants of the $O(\cdot)$ estimates in detail to study the amount of work used by the panel clustering algorithm for "small" problem sizes. As a test problem we examined the Galerkin discretization of the double-layer potential for the Laplace equation on the surface of the unit cube scaled by $(3:1:0.3)$. For the panel clustering and the conventional matrix oriented version the amount of time and space required to assemble the corresponding discrete operator $\tilde{\boldsymbol{K}}$ satisfying a given relative error ϵ, i.e., $\|\boldsymbol{K}\boldsymbol{u} - \tilde{\boldsymbol{K}}\boldsymbol{u}\|_0 \leq \epsilon\|\boldsymbol{K}\boldsymbol{u}\|_0$ with $\boldsymbol{K}$ denoting the exact matrix, was measured. The accuracy in the case of $\tilde{\boldsymbol{K}} = \tilde{\boldsymbol{K}}^{\mathrm{pc}}$ of the panel clustering was controlled by the parameters η and m.

In order to adjust the accuracy of $\tilde{\boldsymbol{K}} = \tilde{\boldsymbol{K}}^{\mathbf{m}}$ of the matrix oriented version the size of the nearfield, where high order cubature formulae are used, and the farfield, where only one-point formulae are applied, was changed. The results for a problem size of $N = 1154$ vertices (2304 panels) are shown in Figure 5.1 illustrating the amount of time needed to assemble the operators and in Figure 5.2 depicting the number of entries to represent these operators.

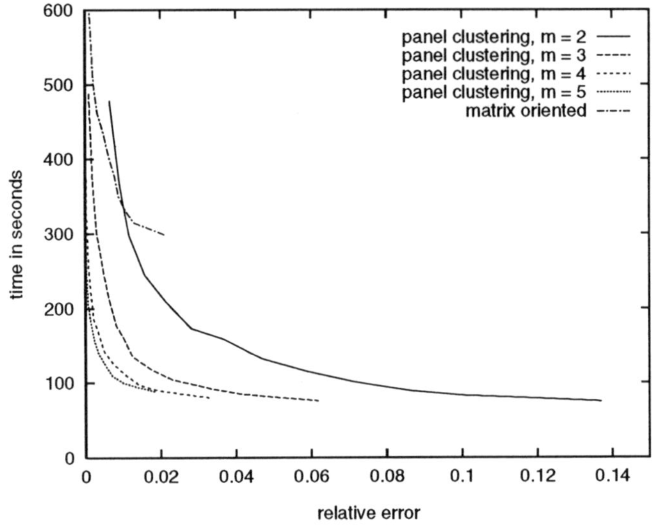

Fig. 5.1. Time consumptions assembling the discrete operators.

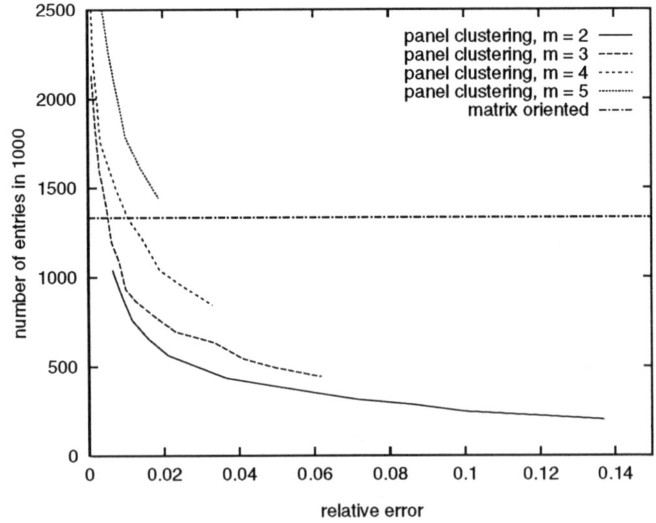

Fig. 5.2. Storage consumptions of the discrete operators.

The calculations were carried out on a SUN SPARC-10/40. It turns out that even for small problem sizes the panel clustering method could reduce the CPU-time and storage consumptions considerably. Further results are reported in [19] and [14].

6 Software Design Aspects

Regarding the software activities for finite element methods (FEM) there are already complex and powerful packages to solve various problems. It would be quite convenient to reuse this software or at least the design aspects for the boundary element method as well. Though even the conception of FEM and BEM are nearly the same, the aspects concerning design decisions differ considerably. The system matrices of the FEM, for instance, are sparse. Adequate data structures for this type of matrices are lists or other containers taking the local character into consideration. However, the BEM leads to full matrices making it quite plain to use arrays. There are a lot of other examples like this showing that it is unavoidable to develop BEM software from the beginning.

The difficulty one has to encounter is the great variety of tasks: There are, among others, different discretization schemes like collocation or Galerkin methods, different kinds of operators (single layer, double layer, hypersingular) and quite a lot of cubature techniques to consider. A design strategy which isolates the common parts and 'focus upon the essential characteristics' [3] to build the basic abstract data types would be very useful. These properties are given by object-oriented methods known to be a suitable tool, to manage complex systems. In the following we specify an object-oriented approach to BEM which has been elaborated as a class library implemented in C++.

In our situation the object-oriented decomposition could be extracted from the formulation of Petrov-Galerkin schemes introduced in Chapter 2 The foundation of this scheme are the subspaces X_n and Y_n defining the test and trial functions. A basic information necessary to construct these spaces is a description of the geometry, i.e. the surface of the physical object. For this purpose it is advantageous to have a (continuous) description which offers the possibility to resolve the geometry with any necessary accuracy – an important fact, especially in the case of multilevel methods where hierarchies of spaces must be generated. So we can state a first class of the library called `Surface` representing a formal description of the surface:

```
class Surface {                     class SurfaceApx {
        ⋮                                   ⋮
  public:                             public:
    Surface(char* fname);               SurfaceApx(const Surface& sf, int level);
        ⋮                                   ⋮
};                                  };
```

In the next step, we require an approximation of the surface, i.e. a set of panels (triangulation) that meets several conditions listed in Chapter 2 To maintain this set, a second class, called **SurfaceApx**, has to be introduced into the context of geometry abstraction. In the class definition above the parameter level is used to distinguish between particular levels of approximation. Instances of the class SurfaceApx for different levels form a hierarchy of panels.

To classify a basis of the subspaces two well-known alternatives are available. The first one, the node-oriented version, characterizes a single basis function by its support in conjunction with a related node. The basis is described by the set of all basis functions. This definition is obvious and close to the discrete mathematical formulation of the Petrov-Galerkin scheme. Nevertheless the second, panel-oriented alternative is more efficient in respect of the time-consuming integrations. It uses the corresponding shape function for each panel, i.e., the set of all non-vanishing basis functions restricted to the panel. The abstraction of the entire basis is formed by the recombination of all shape functions. An entity, representing a single basis function, no longer exists.

The properties of common shape functions are reflected in the definition of an abstract base class:

```
class ShapeFunction {
        ⋮
  public:
    virtual const Panel&  support() const = 0;
    virtual int  dimension() const = 0;
    virtual const int*  index() const = 0;
        ⋮
};
```

It gives information about the related panel, the number of restricted basis functions and their associated indices. By the mechanism of inheritance we can generate subclasses of ShapeFunction to module different basis functions, as constant, linear or dirac functions, on different types of panels (Figure 6.1).

For the instantiation and maintenance of shape functions a further class, Space, is established. This provider is responsible for the correct combination of shape functions to form a basis and thus the desired subspace of the Petrov-Galerkin method; especially it has to support the management of conforming indices. To realize various subspaces, several specializations of the base class `Space` are used. Analogous to the subclasses of `Shapefunction` we could define `ConstSpace`, `LinSpace` or `DiracSpace`.

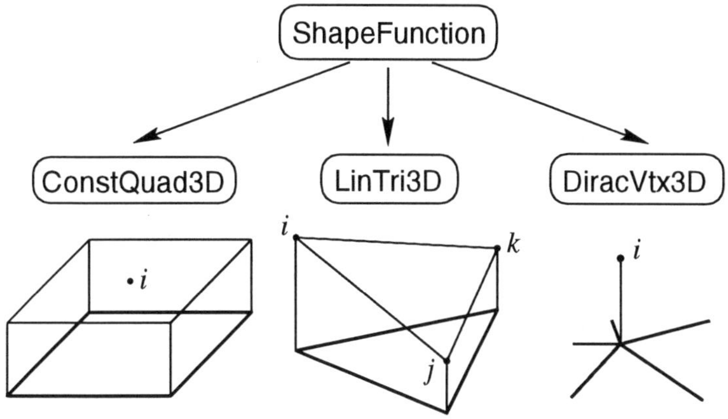

Fig. 6.1. Specializations of `ShapeFunction`.

The next class in our library addresses the dual forms of (2.3). Corresponding to the construction of spaces the abstraction of a dual form is a function that assigns to each pair of shape functions (Φ, Ψ) an element matrix of the form

$$(m)_{\phi\psi} := \langle K\psi, \phi \rangle , \quad \phi \in \Phi, \psi \in \Psi.$$

Embeded in a class definition we get

```
class DualForm {
  public:
    virtual const real* operator()(const ShapeFunction& trialfnc,
                                   const ShapeFunction& testfnc) const;
};
```

Before the assignment can be executed an appropriate integration or cubature rule dependent on the particular type of shape functions has to be chosen, e.g., to distinguish between collocation and Galerkin schemes. For this purpose run-time type information (RTTI) is an essential tool and used in our implementation to solve this kind of multi-method problem (see [24]). The class `DualForm` acts as an interface class which 'adjusts the appearance' [24] for further class definitions derived from it. With these subclasses a variety of integral operators and associated integrations strategies could be arranged.

To assemble local element matrices in a global one the class `MatrixOp` is introduced into the design. It is a specialization of the base class `Operator` (Figure 6.2) which first of all is responsible for the abstraction of integral operators.This is done by defining a virtual member function of the class `Operator` common to all subclasses which declares the mapping of the considered operator:

```
virtual void operator()(const Function& trialfnc, const Function& testfnc);
```

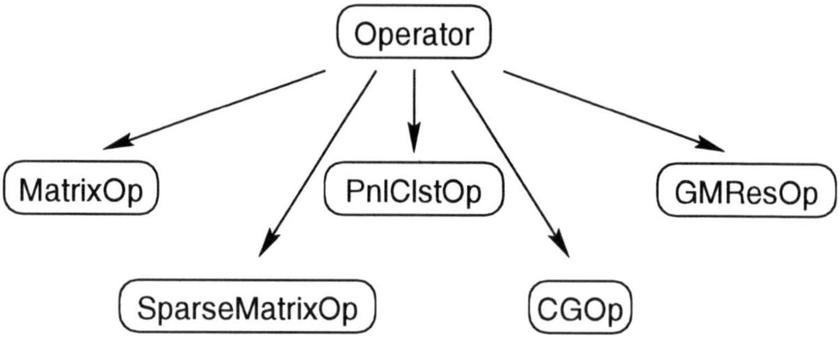

Fig. 6.2. Specializations of `Operator`.

For the class `MatrixOp` for instance this function is to be overloaded by a simple matrix-vector product. The class `Function` in the definition above just represents a valuation of the related basis. It can be implemented by a vector of floats or doubles.

The subclass `MatrixOp` is used to specify the integral operator on the discrete level by a full coefficient matrix. Another alternative offers the panel clustering method described in Chapter 4 Its complex algorithmic issues can be totaly hidden in the implementation of an additional subclass of `Operator` denoted by `PnlClstOp`. The external behaviour or usage resp. of `PnlClstOp` is nearly the same as for the simple `MatrixOp`. Only the information needed during the instantiation process must be extended: a specialized version of the class `DualForm` to describe the far field dual forms $\langle \partial^{\alpha_1}/\partial n_x\, w_\mu(x),\, \varphi_i \rangle$ and $\langle \phi_i,\, \partial^{\alpha_2}/\partial n_y\, w_\nu(y) \rangle$ as well as an instruction to calculate the expansion coefficients $\kappa_\nu(z)$.

Even the solver of the linear system interpreted as inverse operator meets this context. For example overloading the `operator()` by the gmres algorithm leads to the class `GMResOp`. An instance of `GMResOp` is initialized with the operator that should be inverted.

To summarize the rough introduction of our class library we present a short application:

```
int main() {
  // initialize the geometry given in file "cube"
  Surface      sf("cube");
  SurfaceApx   sfapx(sf, 2);
```

```
    // construct test and trial spaces for collocation
    DiracSpace  test(sfapx);
    LinSpace    trial(sfapx);

    // get the necessary informations about dual forms and
    // expansion coefficients
    DlpDF            df(3);
    FarFieldDF       ftstdf(2);
    FarFieldDrvdDF   ftrldf(2);
    LaplaceExp       exp(2);

    // assemble the stiffness matrix and the panel clustering operator
    MatrixOp   op(trial, test, df);
    PnlClstOp  pcop(trial, test, df, ftrldf, ftstdf, exp, 1-e2);

    // initialize two functions; one matching the trial space,
    // another the test space
    Function f(trial); f = "sin(x*y)";
    Function g(test);

    // calculate ((Kf, φ_i))_j with both operators
    op(f, g);
    pcop(f, g);

    // solve using the panel cluster operator
    GMResOp solve(pcop, 1e-8, 100);
    solve(g, f);
}
```

Finally the approach to involve parallel architectures, in particular SPMD schemes (same program multiple data), should be mentioned. The most time-consuming task of the BEM is to assemble the matrices. A distribution of this task among several processors can be achieved by decomposing the subspaces X_n and Y_n leading to a two dimensional torus as proper topology of processors (Figure 6.3). To be independent of the given physical architecture an interface class `ProcessNode` is created which establishs basic operations in the terms of the underlying torus (e.g. to broadcast along rows or columns, the upper, lower, left or right neighbour):

```
class ProcessNode {
        ⋮
    public:
        virtual int broadcast(...) = 0;
        virtual int reduce(...) = 0;
        virtual int shift(...) = 0;
        ⋮
};
```

Specializations of this interface class realize the connection between the physical layer (hypercube, workstation cluster, etc.) and the application layer (2d torus).

To make the information and operations of `ProcessNode` available to the concerned classes of the library a reference to an instance of `ProcessNode` could be used. The virtual function call ensures the expected behaviour. Parameterized classes (templates) offer another, more efficient way of embedding the derived classes of `ProcessNode` in the library. In this alternative the member functions `broadcast()`, `reduce()`, etc. are bounded statically resulting in an efficient function call.

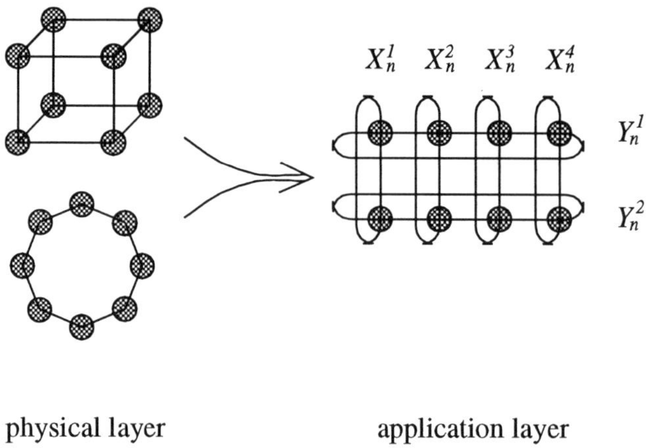

Fig. 6.3. Abstraction of parallel architectures.

References

1. Atkinson, K. (1985): Solving Integral Equations on Surfaces in Space. In: Hämmerlin, G., Hoffmann, K. (eds.) Constructive Methods for the Practical Treatment of Integral Equations, pages 20–43. Birkhäuser: ISNM
2. Atkinson, K., Chien, D. (1995): Piecewise Polynomial Collocation for Boundary Integral Equations. SIAM J. Sci. Comp. 16, 651-681
3. Booch, G. (1994): Object-oriented Analysis and Design, Second Edition. Benjamin/Cummings Publishing Company, Redwood City, California
4. Brandt, A. (1990): Multilevel computations of integral transforms and particle interactions with oscillatory kernels. In: IMACS 1st Int. Conf. On Comp. Phys.. Boulder, Colorado
5. Brandt, A., Lubrecht, A. (1990): Multilevel Matrix Multiplication and Fast Solutions of Integral Equations. J. Comp. Physics 90, 348–370
6. Dahmen, D., Prößdorf, S., Schneider, R. (1993): Wavelet Approximation Methods for Pseudodifferential Equations II: Matrix Compression and Fast Solutions. Advances in Comp. Math. 1, 259–335
7. Hackbusch, W. (1995): Integral Equations. Birkhäuser, Basel: ISNM 120

8. Hackbusch, W. (1991): The Solution of Large Systems of BEM Equations by the Multi-Grid and Panel Clustering Technique. Rend. Sem. Mat. Uni. Pol. Torino, Fasc. Spec.: Numerical Methods, pp. 163–187
9. Hackbusch, W., Nowak, Z. (1989): On the Fast Matrix Multiplication in the Boundary Element Method by Panel-Clustering. Numerische Mathematik 54, 463–491
10. Hackbusch, W., Sauter, S. (1993): On the Efficient Use of the Galerkin Method to Solve Fredholm Integral Equations. Application of Mathematics (formerly: aplikace matematiky) 38(4-5), 301–322
11. Hackbusch, W., Sauter, S.A. (1994): On Numerical Cubatures of Nearly Singular Surface Integrals arising in BEM Collocation. Computing 52, 139–159
12. John, F. (1955): Plane Waves and Spherical Means. Springer, New York
13. Kieser, R. (1990): Über einseitige Sprungrelationen und hypersinguläre Operatoren in der Methode der Randelemente. PhD thesis, Mathematisches Institut A, Universität Stuttgart
14. Lage, C. (1995): Analyse, Entwurf und Implementation von Randelementmethoden. PhD thesis, Inst. f. Prakt. Math., Universität Kiel, to appear
15. Rokhlin, V. (1985): Rapid solutions of integral equations of classical potential theory. Journal of Computational Physics 60(2), 187–207
16. Rokhlin, V. (1993): Diagonal Forms of Translation Operators for the Helmholtz Equation in Three Dimensions. Appl. and Comp. Harm. Anal. 1(1), 82-93
17. Sauter, S., Krapp, A. (1995): On the Effect of Numerical Integration in the Galerkin Boundary Element Method. Technical Report 95-4, Lehrstuhl Praktische Mathematik, Universität Kiel, to appear in Numer. Math.
18. Sauter, S.A. (1991): Der Aufwand der Panel-Clustering-Methode für Integralgleichungen. Technical Report 9115, Institut für Praktische Mathematik, University of Kiel, 24118 Kiel
19. Sauter, S.A. (1992): Über die effiziente Verwendung des Galerkinverfahrens zur Lösung Fredholmscher Integralgleichungen. PhD thesis, Inst. f. Prakt. Math., Universität Kiel
20. Schneider, R., von Petersdorff, T., Schwab, C.: Multiwavelets for Second Kind Integral Equations. To appear in SIAM, J. Numer. Anal.
21. Schwab, C. (1995): Variable Order Composite Quadrature of Singular and Nearly Singular Integrals. Computing 53, 173-194
22. Schwab, C., Wendland, W.L. (1992): Kernel Properties and Representations of Boundary Integral Operators. Math. Nachr. 156, 187–218
23. Schwab, C., Wendland, W.L. (1992): On Numerical Cubatures of Singular Surface Integrals in Boundary Element Methods. Numerische Mathematik, pp. 343–369
24. Stroustrup, B. (1991): The C++ Programming Language, Second Edition. Addison-Wesley Publishing Company, Reading Massachusetts
25. Stroustrup, B. (1994): The Design and Evolution of C++. Addison-Wesley Publishing Company, Reading Massachusetts

Bimetal Problems

L. Jentsch and D. Mirschinka

1 Introduction

The topic of our investigations are special boundary-transmission problems for the equations of mathematical physics, the so-called bimetal problems. This is a special class of problems for piece-wise homogeneous bodies where the interface between the homogeneous parts intersects the outer boundary. The following two pictures illustrate the domains which will be considered.

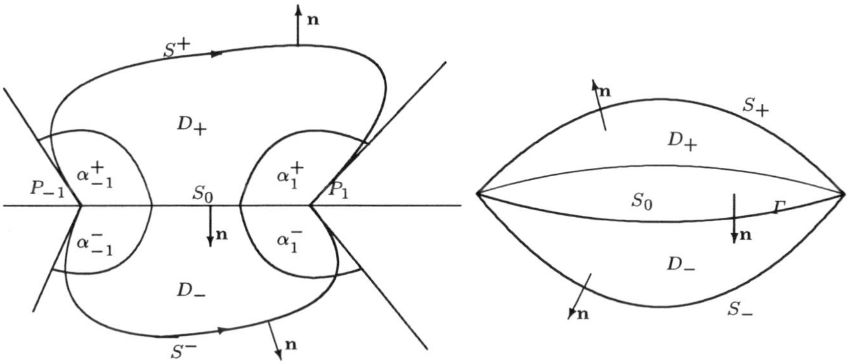

A useful tool for the solution of such kind of problems is the so-called Green's contact tensor or contact function method. These Green's functions were constructed for various problems of elasticity and stationary heat conduction [7-10]. It follows from their properties that the associated boundary transmission problem can be reduced to a boundary integral equation over the outer boundary, i.e. the transmission conditions on the interface are a priori fulfilled. Our main investigations are concerned with the Fredholmness, invertibility, asymptotic behaviour of the solutions and the numerical analysis of the resulting boundary integral equations. In the first section the plane problems will be shortly discussed and we give some references for further informations. The main tool for the investigation of the boundary integral operators is the Mellin transform. The remaining parts of this paper are devoted to the three-dimensional case. We will discuss the analytical and numerical treatise of the boundary integral equations for the problem of stationary heat conduction. In contrast to boundary value problems in non-smooth domains the theory of three-dimensional transmission problems (or interface problems) is rather incomplete. Only recently, Nicaise and Sändig studied such problems in a more general context (s. [21], [22]). But their results does not cover the case of three-dimensional interface edges. Moreover, the topics of these two papers are solvability in certain function spaces and the asymptotics of the solution, while we are interested in the boundary integral equation method for the solution of the transmission problem. In [25] solvability and asymptotics of general elliptic interface problems in an infinite wedge are investigated. While Mellin techniques are successfully used in the plane theory the three-dimensional problem seems to be more difficult with respect to the analytical and numerical treatment. The double layer potential operator for our problem is an integral operator with fixed singularities along the edge of the boundary surface. Similar operators were considered by Duduchava and Schneider [3], [26]. In contrast to the smooth case the double layer potential operator is no longer compact in the considered function spaces. With a local principle the treatise of this operator can be reduced to the study of a family of integral operators on the half-plane. Such operators were studied by the authors [20], [13] via a harmonic analysis approach. We will prove the invertibility of the boundary integral operators in a scale of weighted Sobolev spaces extending our earlier results [19]. Besides the analytical treatment of the boundary integral equations we will study spline approximation methods for the numerical solution of the boundary transmission problem. To obtain optimal order convergence results we use an appropriate mesh refinement near the interface edge. Similar results for the harmonic double layer potential operator on polyhedral boundaries were obtained by Elschner [5],[6]. The problems of the linear elastostatics are much more difficult. Nevertheless, it was shown in [13] that the associated operators can be studied with the same approach as in the case of the Laplacian. But the question of Fredholmness of the boundary integral operators would lead to an invertibility problem for a family of 6×6-matrices with nonhomogeneous Mellin pseudodifferential operators as entries.

For the special geometry of a bimetal ball Poisson formulae for boundary transmission problems of stationary heat conduction were obtained by one of the authors in [15].

Together with D. Natroshvili one of the authors considered non-classical interface and mixed interface problems for anisotropic elastic bodies in [16],[17]. The main tool for the solution of such problems via boundary integral equations is the theory of pseudodifferential operators on manifolds without and with boundary. The case of stationary elastic oscillations is investigated in [18].

2 Two-dimensional Problems

In the case of elastostatics the following boundary transmission problem will be investigated.

a) Differential equations for the displacement vector $\mathbf{u}(x)$:

$$\begin{aligned} \mu_1 \Delta \mathbf{u} + (\lambda_1 + \mu_1)\text{grad div } \mathbf{u} &= \mathbf{0} \quad \text{in } D^+, \\ \mu_0 \Delta \mathbf{u} + (\lambda_0 + \mu_0)\text{grad div } \mathbf{u} &= \mathbf{0} \quad \text{in } D^-. \end{aligned} \tag{2.1}$$

b) Transmission conditions at S_0 :

$$\{\mathbf{u}\}^+ - \{\mathbf{u}\}^- = \mathbf{0} \qquad \text{at } S_0, \tag{2.2}$$

$$\Big\{\underset{(1)}{T}(\partial_x, \mathbf{n})\mathbf{u}\Big\}^+ = \Big\{\underset{(0)}{T}(\partial_x, \mathbf{n})\mathbf{u}\Big\}^- \quad \text{at } S_0, \tag{2.3}$$

where $\underset{(i)}{T}$ is the traction operator with the Lamé constants λ_i, μ_i

$$\underset{(i)}{T}(\partial_x, \mathbf{n})\mathbf{u} = 2\mu_i \frac{\partial \mathbf{u}}{\partial \mathbf{n}} + \lambda_i \mathbf{n} \text{ div } \mathbf{u} + \mu_i [\mathbf{n} \times \text{ rot } \mathbf{u}] \,,$$

and $\{.\}^+, \{.\}^-$ denote the traces at S_0 from D_+ and D_-, respectively.

c) Boundary conditions of the Dirichlet or Neumann type at $S = S^+ \cup S^-$.

In the sequel we restrict ourselves to the Neumann problem, i.e. the traction at the boundary is given. We will apply the indirect boundary integral method for the solution of this problem. Then the program for this method is the following:

α) First, we make a single layer ansatz over S with the contact tensor $G(x, y)$ in the kernel of the potential

$$\mathbf{u}(x) = \int_S G(x, y)\varphi(y)dS_y, \tag{2.4}$$

which fulfils (2.1)-(2.3) a priori.

β) Applying the traction operator to (2.4) and after that the trace operator to S we receive a boundary integral equation $A\varphi = \mathbf{f}$, with a singular matrix integral operator A which has additional fixed singularities at the points P_1 and P_{-1}.

γ) Treatise of the operator A : Fredholmness and index of A, solvability conditions for the equation $A\varphi = \mathbf{f}$.

δ) Study of the asymptotics of φ at the corner points P_1 and P_{-1}.

ϵ) Asymptotics of the solution $\mathbf{u}(x)$ at P_1 and P_{-1} by computing the potential (2.4).

A localization procedure reduces this program to the study of two Mellin-convolution operators on the interval $[0,1]$ associated with the two corner points P_1 and P_{-1}. The properties of the symbol matrices are connected with the essential properties of the operator A. If the symbols are regular on the line Re $z = \frac{1}{2}$ the operator A is Fredholm in $L_2(S)$ and the index can be determined by the function-theoretic index of the determinants of the matrices. The singular exponents of the asymptotics of φ at the corner points can be computed via the zeros of the determinants of the symbol matrices in the half-plane Re $z < \frac{1}{2}$ (s. [12]). Then the asymptotics of the solution $\mathbf{u}(x)$ of the boundary transmission problem can be computed from (2.4) (s. [14]). In the case of perfect and non-perfect stationary heat conduction this program was studied in [10].

The investigation of numerical methods for the solution of the boundary integral equations will be restricted to the case of stationary heat conduction in a polygonally bounded bimetal domain, i.e. we look for a solution of the following problem

$$\Delta u = 0 \qquad \text{in } D_+ \cup D_- \tag{2.5}$$

$$\gamma_+ u = \gamma_- u \qquad \text{at } S_0 \tag{2.6}$$

$$\lambda_+ \gamma_+ \frac{\partial u}{\partial \mathbf{n}} = \lambda_- \gamma_- \frac{\partial u}{\partial \mathbf{n}} \qquad \text{at } S_0 \tag{2.7}$$

$$\gamma u = g \qquad \text{at } S\,, \tag{2.8}$$

where $\gamma, \gamma_+, \gamma_-$ denote the traces at S and at S_0 from D_+ and D_-, respectively. For the solution of this problem we consider a double layer potential ansatz

$$u(x) = \frac{1}{\pi} \int_S \frac{\partial G(x,y)}{\partial \mathbf{n}_y} v(y)\, dS_y, \tag{2.9}$$

such that the differential equation and the transmission conditions are fulfilled. Here $G(x,y)$ is the contact function

$$G(x,y) = \begin{cases} -\ln|x-y| - \dfrac{\lambda_+ - \lambda_-}{\lambda_+ + \lambda_-} \ln|x-\bar{y}| & x,y \in H_+ := \{x_2 > 0\} \\[2ex] -\ln|x-y| - \dfrac{\lambda_- - \lambda_+}{\lambda_+ + \lambda_-} \ln|x-\bar{y}| & x,y \in H_- := \{x_2 < 0\} \\[2ex] -\dfrac{2\lambda_-}{\lambda_+ + \lambda_-} \ln|x-y| & x \in H_+\,, y \in H_- \\[2ex] -\dfrac{2\lambda_+}{\lambda_+ + \lambda_-} \ln|x-y| & x \in H_-\,, y \in H_+ \end{cases} \tag{2.10}$$

with $\bar{y} = (y_1, -y_2)$.

The boundary condition yield the integral equation

$$(-I+K)(v)(x) := -v(x)+\frac{1}{\pi}\int\limits_S \frac{\partial G(x,y)}{\partial \mathbf{n}_y}\, v(y)\, dS_y = g(x), \qquad x \in S \quad (2.11)$$

for the unknown density v. The boundary integral operator K is at the corner points of Mellin type. To obtain optimal convergence results we use an appropriate mesh refinement near the corner points. The graduation depends on the corner angles and the material parameters or more precisely on the asymptotics of the solution of the problem at the corner points. On a mesh Δ_n we consider spline spaces $S^d(\Delta_n)$ consisting of discontinuous piece-wise polynomials of degree less than d which are constant in the subintervals containing the corner points. To obtain a fully discrete scheme we define an approximate operator K_n by replacing the integral in K by a quadrature formula. If we put Q_n for the interpolation operator with respect to the nodal points of the quadrature formula we obtain the following approximation scheme:
Find $v \in S^d(\Delta_n)$, such that

$$-v_n + Q_n K_n v_n = Q_n g\,. \qquad (2.12)$$

This is a linear system for the values of v_n at the nodal points of the quadrature formula. The Nyström solution v_n^* results from

$$v_n^* = -g + K_n v_n\,. \qquad (2.13)$$

If R is the order of the quadrature formula, the grading parameter q is appropriately chosen and the right-hand side g belongs to a weighted Sobolev space (s. [4]) the Nyström method (2.13) converges with the order of $O((\log n/n)^R)$ in the L_∞-Norm.
In our experiments we used piece-wise linear splines and the trapezoidal rule, i.e. $R = 1$. The considered domain is a quadrangle with the corner points $(1,0), (1.5,-1), (2,0), (1.5,1)$, the interface between the first and the third point and the boundary data is given by the function $G(x,y)$ with $y = (2,2)$. The following tabular contains the L_∞-error on $D_+ \cup D_-$ between the exact solution of the boundary transmission problem and the approximate potential $K_n v_n$ in the interior and $2N$ is the number of collocation points at each side of the quadrangle.

$\lambda_- = 1$	$\lambda_+ = 2$		$\lambda_- = 1$	$\lambda_+ = 10$		$\lambda_- = 1$	$\lambda_+ = 100$	
q	N	Error	q	N	Error	q	N	Error
1	5	0.446022	1	5	0.608212	1	5	0.662410
	10	0.017703		10	0.024141		10	0.026292
	19	0.006906		19	0.009417		19	0.010256
	40	0.002293		40	0.003127		40	0.003406
	80	0.000947		80	0.001292		80	0.001407
	161	0.000416		161	0.000567		161	0.000617
2	6	0.056906	2	6	0.077600	2	6	0.084515
	11	0.253609		11	0.345830		11	0.376646
	21	0.007644		21	0.010423		21	0.011352
	39	0.002200		39	0.003001		39	0.003268
	81	0.000495		81	0.000675		81	0.000735
	160	0.000190		160	0.000259		160	0.000282
3	4	0.258974	3	4	0.353147	3	4	0.384615
	9	0.253569		9	0.345776		9	0.376587
	24	0.046240		24	0.063054		24	0.068673
	40	0.009005		40	0.012280		40	0.013374
	79	0.000861		79	0.001175		79	0.001279
	156	0.000362		156	0.000494		156	0.000538

3 Three-dimensional Problems

Let D be a three-dimensional domain with piece-wise smooth Lipschitz boundary consisting of two parts D_+, D_- and a common plane interface S_0 (s. the picture above).

The surfaces S_+, S_- and the curve Γ are assumed to be smooth but for the whole boundary S an edge along Γ with nonconstant edge angle will be allowed. The normal vector $\mathbf{n}$ at S and S_0 is directed outside from D and D_+, respectively.

We consider the following boundary-transmission problem of bimetal type:

$$\Delta u = 0 \qquad \text{in } D \setminus S_0 \tag{3.1}$$

$$\eta\{\gamma_+ u - \gamma_- u\} = \lambda_+ \gamma_+ \frac{\partial u}{\partial x_3} \qquad \text{at } S_0 \tag{3.2}$$

$$\lambda_+ \gamma_+ \frac{\partial u}{\partial x_3} = \lambda_- \gamma_- \frac{\partial u}{\partial x_3} \qquad \text{at } S_0 \tag{3.3}$$

$$\gamma_S u = g \qquad \text{at } S \tag{3.4}$$

where λ_+, λ_-, η are positive reals and $\gamma_\pm$ and γ_S denote the trace operators from D_+ and D_- on S_0 and S, respectively. The function g is a given function from a suitable function space. The problem (3.1)-(3.4) will be solved using certain Green's functions for contacting half-spaces such that the transmission conditions are a priori fulfilled. We include the case $\eta = \infty$ of perfect heat contact where (3.2) becomes

$$\gamma_+ u = \gamma_- u \qquad \text{at } S_0.$$

In the latter case the Green's contact function is given by

$$G_\infty(x,y) = \begin{cases} \dfrac{1}{\mid x-y \mid} + \dfrac{\lambda_+ - \lambda_-}{(\lambda_+ + \lambda_-) \mid x - \bar{y} \mid} & \text{for } x, y \in \mathbb{R}^3_+ \\ \dfrac{1}{\mid x-y \mid} + \dfrac{\lambda_- - \lambda_+}{(\lambda_+ + \lambda_-) \mid x - \bar{y} \mid} & \text{for } x, y \in \mathbb{R}^3_- \\ \dfrac{2\lambda_-}{(\lambda_+ + \lambda_-) \mid x - y \mid} & \text{for } x \in \mathbb{R}^3_+ , y \in \mathbb{R}^3_- \\ \dfrac{2\lambda_+}{(\lambda_+ + \lambda_-) \mid x - y \mid} & \text{for } x \in \mathbb{R}^3_- , y \in \mathbb{R}^3_+ \end{cases} \tag{3.5}$$

where $\bar{y}$ denotes the point $(y_1, y_2, -y_3)$, and $\mathbb{R}^3_+ := \{x \in \mathbb{R}^3 \ : \ x_3 > 0\}$, $\mathbb{R}^3_- := \mathbb{R}^3 \setminus \overline{\mathbb{R}^3_+}$ denote the upper and lower half-space, respectively. For nonperfect heat contact, i.e. $\eta < \infty$, the Green's function was calculated in [10] as

$$G_\eta(x,y) = G_\infty(x,y) + K(x,y) \quad ,$$

with

$$K(x,y) = \begin{cases} \dfrac{1}{2\pi} \dfrac{2\lambda_-}{\lambda_+ + \lambda_-} I(x,y) & \text{if } x, y \in \mathbb{R}^3_+ \\ \dfrac{1}{2\pi} \dfrac{2\lambda_+}{\lambda_+ + \lambda_-} I(x,y) & \text{if } x, y \in \mathbb{R}^3_- \\ -\dfrac{1}{2\pi} \dfrac{2\lambda_-}{\lambda_+ + \lambda_-} I(x,y) & \text{if } x \in \mathbb{R}^3_+ , y \in \mathbb{R}^3_- \\ -\dfrac{1}{2\pi} \dfrac{2\lambda_+}{\lambda_+ + \lambda_-} I(x,y) & \text{if } x \in \mathbb{R}^3_- , y \in \mathbb{R}^3_+ \end{cases} \tag{3.6}$$

and

$$I(x,y) = 2\pi \int_0^\infty \frac{t}{t+a} \exp[-(|x_3| + |y_3|)t] J_0 \left(t\sqrt{(x_1 - y_1)^2 + (x_2 - y_2)^2} \right) dt \tag{3.7}$$

where $a = \dfrac{\eta(\lambda_+ + \lambda_-)}{\lambda_+ \lambda_-}$. We note that $I(x,y)$ is the Hankel transform of order zero of the function $\dfrac{2\pi \exp(-(|x_3| + |y_3|)t)}{t+a}$. Since this function decreases exponentially at infinity a fast Hankel transform technique can be applied to compute (3.7) approximately [23]. In [10] it was shown that

$$I(x,y) = \frac{2\pi}{\sqrt{(x_1-y_1)^2+(x_2-y_2)^2+(|x_3|+|y_3|)^2}} + R(x,y) \quad , \tag{3.8}$$

where $R(x,y)$ has only logarithmic singularities at the plane $\{x_3 = 0\}$. This implies that the Green's function admits a representation $G(x,y) = G_0(x,y) + R_0(x,y)$ with

$$G_0(x,y) = \begin{cases} \dfrac{1}{|x-y|} + \dfrac{1}{|x-\bar{y}|} & x,y \in \mathbb{R}^3_+ \\ \dfrac{1}{|x-y|} + \dfrac{1}{|x-\bar{y}|} & x,y \in \mathbb{R}^3_- \\ 0 & x \in \mathbb{R}^3_+ \,, y \in \mathbb{R}^3_- \\ 0 & x \in \mathbb{R}^3_- \,, y \in \mathbb{R}^3_+ \end{cases} \tag{3.9}$$

and $R_0(x,y)$ has only logarithmic singularities. We note that the main part of the Green's function in the case of nonperfect heat contact does not depend on the material parameters. Moreover, the interaction between the two half spaces is only weak.

For the solution of our boundary-transmission problem (3.1)-(3.4) we will make a double layer potential ansatz over the boundary with the Green's contact function, i.e. we look for a solution in the form

$$u(x) = \frac{1}{2\pi} \int\limits_S \frac{\partial G_\eta(x,y)}{\partial \mathbf{n}_y} v(y)\, dS_y \ , \ x \in D_+ \cup D_- . \tag{3.10}$$

We note that this ansatz fulfils the transmission conditions at S_0 a priori. The jump relation of the double layer potential yields the boundary integral equation

$$(Av)(x) := -v(x) + \frac{1}{2\pi} \int\limits_\Gamma \frac{\partial G_\eta(x,y)}{\partial \mathbf{n}_y} v(y)\, d\Gamma_y = g(x) \ , \ x \in S. \tag{3.11}$$

The integral operator in (3.11) is in generally not a compact operator in $L_2(S)$ due to the non-smoothness of S. Therefore, the theory of second kind integral equations cannot be applied to the analytical treatment, and the theory of collectively compact operators for the numerical analysis does not work. The main topic of this paper will be the analytical and numerical treatment of equation (3.11).

We will finish this section with some results concerning existence and uniqueness of the solution of the problem (3.1)-(3.4). The following lemma is a consequence of the Green's formula.

Lemma 3.1. *Let u fulfil (3.1)-(3.3) and let* $u|_{D_\pm} \in H^1_\Delta(D_\pm) := \{u \in H^1(D_\pm)\colon\ \Delta u|_{D_\pm} \in L_2(D_\pm)\}$. *Then u admits the representation*

$$u(x) = \frac{1}{4\pi} \int\limits_D G_\eta(x,y) \Delta u(y)\, dx$$

$$+ \frac{1}{4\pi} \int\limits_S \left(G_\eta(x,y) \frac{\partial u(y)}{\partial n_y} - \frac{\partial G_\eta(x,y)}{\partial n_y} u(y) \right) dS_y,$$

Theorem 3.1. *Let the solution u of the homogeneous Dirichlet problem fulfil the assumptions of the previous lemma. Then u is zero.*

Proof. Applying Green's formula in D_+ and D_-, respectively, we obtain

$$\lambda_+ \int\limits_{D_+} |\text{grad } u|^2 dx = -\lambda_+ \int\limits_{S_0} u \frac{\partial u}{\partial x_3} dS$$

$$\lambda_- \int\limits_{D_-} |\text{grad } u|^2 dx = \lambda_- \int\limits_{S_0} u \frac{\partial u}{\partial x_3} dS\ .$$

The transmission conditions on S_0 imply

$$\lambda_+ \int\limits_{D_+} |\text{grad } u|^2 dx + \lambda_- \int\limits_{D_-} |\text{grad } u|^2 dx = -\eta^{-1} \lambda_+ \int\limits_{S_0} \left(\frac{\partial u}{\partial x_3} \right)^2 dS \leq 0$$

which finishes the proof. □

Theorem 3.2. *Let $g \in H^{1/2}(S)$. Then for $\eta = \infty$ and $\eta < \infty$ the Dirichlet problem is uniquely solvable in $H^1(D)$ and $H^1(D_+) \times H^1(D_-)$, respectively.*

Proof. Since the uniqueness follows from the previous theorem we only have to show the existence of a solution. For the case of perfect heat contact let us consider the form

$$\Phi_i(v,w) := \lambda_+ \int\limits_{D_+} \text{grad } v\, \overline{\text{grad } w} dx + \lambda_- \int\limits_{D_-} \text{grad } v\, \overline{\text{grad } w} dx$$

with $v, w \in H^1(D)$. Then the estimate

$$\Phi_i(v,v) \geq c \|v\|_1^2$$

holds for $v \in H^1_0(D)$, i.e. Φ_i is elliptic on $H^1_0(D)$. Hence, the variational problem

$$\Phi_i(u,v) = (f,v) \quad \text{for all } v \in H^1_0(D)$$

is uniquely solvable in $H_0^1(D)$. Let $\tilde{g} \in H^1(D)$ be an extension of g into D and let w be the solution of

$$\Phi_i(w, v) = -\Phi_i(\tilde{g}, v) \quad \text{for all } v \in H_0^1(D).$$

Then the function $u := w + \tilde{g}$ is a solution of the problem (3.1)-(3.4).
In the case of nonperfect heat contact the variational formulation is the following: Find $u \in V$ such that

$$\Phi_i(u, v) + \int_{S_0} \frac{\lambda_+^2}{\eta} \left(\gamma_+ \frac{\partial u}{\partial n}\right) \left(\gamma_+ \frac{\partial v}{\partial n}\right) dS = 0 \quad \text{for all } v \in V_0, \tag{3.12}$$

where $V \subset H_\Delta^1(D_+) \times H_\Delta^1(D_-)$ consists of functions which fulfils the boundary condition and the transmission condition (3.2) on S_0 and V_0 is the subspace of V with zero boundary data. The coercivity of the form (3.12) is obvious. To show the boundedness we have to consider the integral over S_0. With the transmission condition we obtain

$$\int_{S_0} \frac{\lambda_+^2}{\eta} \left(\gamma_+ \frac{\partial u}{\partial n}\right) \left(\gamma_+ \frac{\partial v}{\partial n}\right) dS = -\lambda_+ \left(\gamma_+ u - \gamma_- u, \gamma_+ \frac{\partial v}{\partial n}\right)_{L_2(S_0)},$$

and since the trace of the function and of its normal derivative is a continuous map from the H_Δ^1-spaces to $H^{1/2}(S_0)$ and $H^{-1/2}(S_0)$, respectively, the last quantity is bounded up to a constant by the product of the $H^1(D_+) \times H^1(D_-)$-norms of u and v. The construction of the solution is now the same as in the case of perfect heat contact which finishes the proof. □

4 Boundary integral operators

This section is devoted to the study of the boundary integral operators associated with the problem (3.1)-(3.4). First, we will introduce the single and double layer potential with the Green's contact function

$$(Vw)(x) = \frac{1}{2\pi} \int_S G_\eta(x, y)\, w(y)\, dS_y\,, \qquad x \notin S \tag{4.1}$$

$$(Ww)(x) = \frac{1}{2\pi} \int_S \frac{\partial G_\eta(x, y)}{\partial n_y}\, w(y)\, dS_y\,, \qquad x \notin S. \tag{4.2}$$

The trace of V on S will be denoted by V_0, i.e. $(V_0 w)(x) := (\gamma_S V)(x)$, $x \in S$. For the double layer and the normal derivative of the single layer potential the well-known jump relations hold

$$\left(\gamma_S^{\pm} \frac{\partial}{\partial n_S} V w\right)(x) = \pm w(x) + (K_N w)(x) , \qquad x \in S \tag{4.3}$$

$$\left(\gamma_S^{\pm} W w\right)(x) = \mp w(x) + (K_D w)(x) , \qquad x \in S, \tag{4.4}$$

with the direct values K_D and K_N of the double layer potential and the normal derivative of the single layer potential, respectively.

For $\lambda := (\lambda_+, \lambda_-)$ with $\lambda_+ > 0, \lambda_- > 0$ we define an equivalent inner product on $L_2(S)$ by

$$(u,v)_\lambda := \lambda_+ \int_{S_+} u\bar{v} dS + \lambda_- \int_{S_-} u\bar{v} dS .$$

Then the following theorem holds.

Theorem 4.1. *The operator V_0 is an isomorphism between $H^{-1/2}(S)$ and $H^{1/2}(S)$. Moreover, it is selfadjoint with respect to $(.,.)_\lambda$, i.e.*

$$(V_0 u, v)_\lambda = (u, V_0 v)_\lambda \qquad \forall u, v \in H^{-1/2}(S).$$

Proof. The selfadjointness with respect to the modified duality product is obvious. To prove the invertibility we will show that the operator V_0 is positive definite. We restrict ourselves to the case $\eta < \infty$ (The proof for the perfect contact is similar [19].). Consider the forms

$$\Phi_i(u,w) := \lambda_+ \int_{D_+} \operatorname{grad} u \, \overline{\operatorname{grad} w} dx + \lambda_- \int_{D_-} \operatorname{grad} u \, \overline{\operatorname{grad} w} dx$$

$$\Phi_a(u,w) := \lambda_+ \int_{\mathbb{R}^3_+ \setminus \overline{D_+}} \operatorname{grad} u \, \overline{\operatorname{grad} w} dx + \lambda_- \int_{\mathbb{R}^3_- \setminus \overline{D_-}} \operatorname{grad} u \, \overline{\operatorname{grad} w} dx$$

for u, w belonging to a subspace of $H^1(\mathbb{R}^3_+) \times H^1(\mathbb{R}^3_-)$ such that the interface condition (3.2) is fulfilled on the plane $\{x_3 = 0\}$. Then we have obviously (s. section 3.)

$$\Phi_i(w,w) + \Phi_a(w,w) + \int_{\{x_3=0\}} \frac{\lambda_+^2}{\eta} \left(\gamma_+ \frac{\partial w}{\partial n}\right) \left(\gamma_+ \frac{\partial w}{\partial n}\right) dS \geq c||w||^2_{H^1(\mathbb{R}^3_+) \times H^1(\mathbb{R}^3_-)} \cdot \tag{4.5}$$

We put $\Psi(w,w)$ for the left-hand side of (4.5). Let be $u := Vv$. The properties of the Green's function imply that u belongs to the space $H^1_{loc}(\mathbb{R}^3_+) \times H^1_{loc}(\mathbb{R}^3_-)$ and fulfils the interface conditions. Furthermore, the gradient of u is in $L_2(\mathbb{R}^3)$. Green's formula together with the jump relation for the normal derivative of the single layer potential yields now

$$\Psi(u,u) = (V_0 v, v)_\lambda .$$

Since the trace map $\gamma\frac{\partial}{\partial n}$ is continuous from $H^1_\Delta(\mathbb{R}^3_+) \times H^1_\Delta(\mathbb{R}^3_-)$ to $H^{-1/2}(S)$ we have

$$||v||_{H^{-1/2}(S)} \le c\left(||u||_{H^1(\mathbb{R}^3_+)\times H^1(\mathbb{R}^3_-)}\right) ,$$

which together with (4.5) implies the desired estimate. □

It follows from the previous theorem that V_0 generates an equivalent norm $|v|^2_{V_0} := (V_0 v, v)_\lambda$ on $H^{-1/2}(S)$.

Theorem 4.2. *The operator $I + K_N$ is a Fredholm operator with index zero in $H^{-1/2}(S)$ and the a priori estimate*

$$(V_0 v, (I + K_N)v)_\lambda \ge c_0||v||_{-1/2} - c_1\left|\int\limits_S V_0 v dS\right|^2$$

holds.

Proof. The previous theorem and the continuity of the trace map imply

$$||v||^2_{H^{-1/2}(S)} \le c_1||V_0 v||^2_{H^{1/2}(S)} \le c_2||Vv||^2_{H^1(D)} .$$

For the last quantity we obtain with Poincaré's inequality

$$||Vv||^2_{H^1(D)} \le c\left\{\int\limits_D |\text{grad}(Vv)|^2 dx + \left|\int\limits_S V_0 v\, dS\right|^2\right\} . \tag{4.6}$$

We consider the form $\Phi_i(Vv, Vv)$. Green's formula applied in D_+ and D_- yields

$$\Phi_i(Vv, Vv) = (V_0 v, (I + K_N)v)_\lambda . \tag{4.7}$$

Since $\Phi_i(Vv, Vv)$ and $\int\limits_D |\text{grad}(Vv)|^2 dx$ are equivalent we get the estimate of the theorem. It follows now that $I + K_N$ is a semi-Fredholm operator with finite dimensional kernel in $H^{-1/2}(S)$. Using the associated version of (4.7) for the outer domain we obtain that $-I + K_N$ is semi-Fredholm. Obviously, for any $a \in \mathbb{C}$, $|a| \ge 1$ the estimate

$$|(V_0 v, (aI + K_N)v)_\lambda| \ge c_0||v||_{-1/2} - c_1\left|\int\limits_S V_0 v dS\right|^2$$

holds. But $aI + K_N$ is invertible for sufficiently large a which implies the assertion of the theorem. □

Theorem 4.3. *The operator $-I + K_D$ is invertible in $H^{1/2}(S)$.*

Proof. The operator K_D is the adjoint operator of K_N with respect to the inner product $(.,.)_\lambda$. This together with the previous theorem implies that the operator $-I + K_D$ is Fredholm with index zero in the space $H^{1/2}(S)$. Since the kernel of $-I + K_D$ is trivial due to the uniqueness of the Dirichlet problem the proof is finished. □

It follows from the previous theorem that the operator $-I+K_D$ is strongly elliptic in $H^{1/2}(S)$ with respect to the dual norm of $|.|_{V_0}$. This implies the stability of any Galerkin method with an appropriate space X_n of ansatz functions, but with test functions of the form $V_0(X_n)$. Since this is not very applicable we will study the double layer potential in L_p-spaces on the boundary. While the previous considerations were motivated by the boundary value problem, basically, the following approach is based on the theory of singular integral operators. Thus, the techniques used now are local principles and convolution type operators.

We will consider the indirect boundary integral method using the double layer ansatz (3.10) for the Dirichlet problem (3.1)-(3.4). The jump relation for the double layer potential yields the integral equation over the outer boundary in $L_2(S)$

$$-v(x) + \frac{1}{2\pi} \int\limits_S \frac{\partial G_\eta(x,y)}{\partial \mathbf{n}_y} v(y)\, dS_y = g(x) \quad . \tag{4.8}$$

The assumptions on the boundary S imply the following theorem.

Theorem 4.4. *Let be $x \in S \setminus \Gamma$. Then the operator K_D is locally compact (s. [23]) in x.*

Proof. The theorem is a consequence of the local compactness of the usual harmonic potential operators at smoothness points of the surface. □

Let be $x_0 \in \Gamma$ fixed. We put $\alpha = \alpha(x_0)$ and $\beta = \beta(x_0)$ for the angles between the tangential half-planes at x_0 and the plane S_0. Let $U \subset S$ be a neighbourhood of x_0 consisting of two parts $U_+ = S_+ \cap U$ and $U_- = S_- \cap U$ such that U_+ and U_- are diffeomorph to a neighbourhood of zero in the half-plane $\mathbb{R}^2_+$ and the image of x_0 under booth mappings is the point zero. We denote by τ_+ and τ_- two such diffeomorphisms and by χ_+ and χ_- the characteristic function of U_+ and U_-, respectively. Then the restriction of the operator $-I + K_D$ on U has the following matrix representation:

$$\begin{pmatrix} -I + \chi_+ K_D \chi_+ & \chi_+ K_D \chi_- \\ \chi_- K_D \chi_+ & -I + \chi_- K_D \chi_- \end{pmatrix}$$

where I is the identity at U_+ and U_-, respectively. Composing this matrix operator from the left with

$$\begin{pmatrix} \tau_+ & 0 \\ 0 & \tau_- \end{pmatrix}$$

and from the right with

$$\begin{pmatrix} \tau_+^{-1} & 0 \\ 0 & \tau_-^{-1} \end{pmatrix},$$

we obtain an operator $\tilde{A}_D$ up to now defined only on a zero neighbourhood of the half-plane $\mathbb{R}^2_+$. In the terminology of the local principle of Simonenko [23] the operators $-I + K_D$ and $\tilde{A}_D$ are $x_0 - 0$-quasi-equivalent. Extending the operator A_D on the whole half-plane by zero this assertion remains true. Using the arguments of [1], pp. 209-211, one can show that the operator $\tilde{A}_D$ is 0-equivalent to the following operator A_D

$$A_D = \begin{pmatrix} -I - G_{2\alpha} & 0 \\ 0 & -I + G_{2\beta} \end{pmatrix} \tag{4.9}$$

in the case $\eta < \infty$ and

$$A_D = \begin{pmatrix} -I - \dfrac{\lambda_+ - \lambda_-}{\lambda_+ + \lambda_-} G_{2\alpha} & -\dfrac{2\lambda_-}{\lambda_+ + \lambda_-} G_{\alpha+\beta} \\ -\dfrac{2\lambda_+}{\lambda_+ + \lambda_-} G_{\alpha+\beta} & -I + \dfrac{\lambda_+ - \lambda_-}{\lambda_+ + \lambda_-} G_{2\beta} \end{pmatrix} \tag{4.10}$$

for $\eta = \infty$, where

$$(G_\phi u)(x_1, x_2) := \frac{1}{2\pi} \int\limits_{\mathbb{R}^2_+} \frac{x_2 \sin\phi \, u(y_1, y_2) \, dy_1 dy_2}{[(x_1 - y_1)^2 + x_2^2 - 2x_2 y_2 \cos\phi + y_2^2]^{3/2}} .$$

In [20] the operator G_ϕ was studied in weighted L_p-spaces on the half-plane $\mathbb{R}^2_+$. In particular, it was shown that

$$||G_\phi||_{L_2(\mathbb{R}^2_+)} = \left|\cos\frac{\phi}{2}\right| .$$

Moreover, the function $\phi \longmapsto G_\phi$ is continuous in the operator norm topology. This yields the following two theorems.

Theorem 4.5. *Let be $\eta < \infty$. Then the boundary integral operator for the Dirichlet problem, $-I + K_D$, is a Fredholm operator with index zero in $L_2(S)$. The essential norm of the operator K_D is smaller than one.*

Proof. For all $x \in \Gamma$ the operator $-I + K_D$ is quasi-equivalent to invertible 2×2-matrix operators on the half-plane. This remains true for the operator $-I + \nu K_D$ with $\nu \in [0, 1]$. Since the index is homotopy-invariant we obtain that $-I + K_D$ has zero index. □

Theorem 4.6. *Let be $\eta = \infty$. The boundary integral operator $-I + K_D$ is a Fredholm operator with index zero in $L_2(S)$ if one of the following conditions is fulfilled:*

(i)

$$\max_{x\in\Gamma}\left\{\frac{1-k}{1+k}|\cos(\alpha(x))| + \frac{2\sqrt{k}}{1+k}\left|\cos\left(\frac{\alpha(x)+\beta(x)}{2}\right)\right|,\right.$$

$$\left.\frac{1-k}{1+k}|\cos(\beta(x))| + \frac{2\sqrt{k}}{1+k}\left|\cos\left(\frac{\alpha(x)+\beta(x)}{2}\right)\right|\right\} < 1, \quad (4.11)$$

where $\lambda_+ = k\lambda_-$, $k \in (0,1]$,

(ii)

$$\alpha(x) = \beta(x)\ , \quad \textit{for all } x \in \Gamma\ , \quad (4.12)$$

(iii)

$$\alpha(x) + \beta(x) = \pi \quad \textit{for all } x \in \Gamma\ . \quad (4.13)$$

The proof is the same as for the previous theorem using the results on the invertibility of the local operators on the infinite wedge (s. [20]).

The assertion of the theorem remains valid if we replace the $L_2(S)$-space by any $L_{p,\rho}(S)$-space with $p \geq 2$ and $0 \leq \rho \leq 1/p$, where

$$L_{p,\rho}(S) := \{u \in \ \text{dist}\ (x,\Gamma)^\rho L_p(S)\}$$

and $\text{dist}(x,\Gamma)$ denotes the distance of $x \in S$ to the edge Γ.

We note that the existence theorem of the previous section is not applicable since an L_2-density in the double layer potential does not produce a H^1-solution of the boundary value problem. Nevertheless, the following uniqueness result holds.

Theorem 4.7. *The operator* $-I + K_D$ *is invertible in* $L_2(S)$.

Proof. Since the operator $-I + K_D$ is Fredholm with zero index we have only to show that its adjoint has a trivial kernel. Let $f \in L_2(S)$ be a nontrivial function with $(-I + K_D^*)f = 0$. Then the function

$$v(x) := \frac{1}{2\pi}\int_S G(y,x)f(y)dS_y$$

is a solution of the following outer Neumann problem

$$\Delta v = 0 \qquad \text{in } \left(\mathbb{R}^3_+ \setminus \bar{D}_+\right) \cup \left(\mathbb{R}^3_- \setminus \bar{D}_-\right) \quad (4.14)$$

$$\eta\{\lambda_1^{-1}\gamma_+ v - \lambda_0^{-1}\gamma_- v\} = -\gamma_+\frac{\partial v}{\partial x_3} \qquad \text{at } S_1 \quad (4.15)$$

$$\gamma_+\frac{\partial v}{\partial x_3} = \gamma_-\frac{\partial v}{\partial x_3} \qquad \text{at } S_1 \quad (4.16)$$

$$\gamma\frac{\partial v}{\partial n} = 0 \qquad \text{at } S, \quad (4.17)$$

with $S_1 = \{x_3 = 0\} \setminus \bar{S}_0$. It follows from the mapping properties of the single layer potential that the function v belongs to the space $H^{3/2-\epsilon}_{loc}$ in the subdomains $\mathbb{R}^3_+ \setminus \bar{D}_+$ and $\mathbb{R}^3_- \setminus \bar{D}_-$. We consider a ball B_R with sufficiently large radius R and apply Green's formula to the domains

$$B_R^+ := (\mathbb{R}^3_+ \setminus \bar{D}_+) \cap B_R$$

and

$$B_R^- := (\mathbb{R}^3_- \setminus \bar{D}_-) \cap B_R \quad .$$

Then for the gradient of v we get

$$\lambda_1^{-1} \int\limits_{B_R^+} |\text{grad } v|^2 dx + \lambda_0^{-1} \int\limits_{B_R^-} |\text{grad } v|^2 dx$$

$$= \lambda_1^{-1} \int\limits_{\partial B_R \cap \mathbb{R}^3_+} \frac{\partial v}{\partial n} v dS + \lambda_0^{-1} \int\limits_{\partial B_R \cap \mathbb{R}^3_-} \frac{\partial v}{\partial n} v dS +$$

$$+ \lambda_1^{-1} \int\limits_{S_+} \frac{\partial v}{\partial n} v dS + \lambda_0^{-1} \int\limits_{S_-} \frac{\partial v}{\partial n} v dS +$$

$$- \lambda_1^{-1} \int\limits_{S_1 \cap B_R} \gamma_+ \frac{\partial v}{\partial x_3} \gamma_+ v dS + \lambda_0^{-1} \int\limits_{S_1 \cap B_R} \gamma_- \frac{\partial v}{\partial x_3} \gamma_- v dS \, .$$

The first two integrals are asymptotically $O(R^{-1})$. The third and fourth integral vanish due to the homogeneous boundary condition. For the sum of the remaining two integrals we obtain

$$-\lambda_1^{-1} \int\limits_{S_1 \cap B_R} \gamma_+ \frac{\partial v}{\partial x_3} \gamma_+ v dS + \lambda_0^{-1} \int\limits_{S_1 \cap B_R} \gamma_- \frac{\partial v}{\partial x_3} \gamma_- v dS$$

$$= -\eta^{-1} \int\limits_{S_1 \cap B_R} \left| \gamma_+ \frac{\partial v}{\partial x_3} \right|^2 dS \leq 0,$$

whence $v =$ const and since $v(x) \to 0$ for $x \to \infty$, v vanishes identically in the outer domain. The continuity of the single layer potential through S implies that v vanishes in the interior domain, too, i.e. it is zero everywhere which yields that f vanishes at S. This contradicts our assumption and the proof is finished. □

We note that the conditions (4.11),(4.13) imply that the essential norm of the operator K_D is smaller than one if the space $L_2(S)$ is endowed with the modified norm generated by the inner product $(.,.)_\lambda$. Under the condition (4.12) this in generally not true.

For the numerical treatment of equation (4.8) we need some additional weighted Sobolev spaces. For $k \in \mathbb{N}$ and $\rho \in [0, 1/2)$ we define the spaces

$$X^k_\rho(\mathbb{R}^2_+) := \left\{ u \in x_2^\rho L_2(\mathbb{R}^2_+) : \ x_2^{|\alpha|-\rho} D^\alpha_x u \in L_2(\mathbb{R}^2_+), \ 1 \leq |\alpha| \leq k \right\}$$

$$Y^k_\rho(\mathbb{R}^2_+) := \left\{ u \in x_2^\rho L_2(\mathbb{R}^2_+) : \ x_2^{i-\rho} D^i_{x_2} D^j_{x_1} u \in L_2(\mathbb{R}^2_+), \ 1 \leq i + j \leq k \right\}$$

on the half-plane $\mathbb{R}^2_+$. The space $X^0_\rho(\mathbb{R}^2_+)$ $(= Y^0_\rho(\mathbb{R}^2_+)$ will be denoted by $L_{2,\rho}(\mathbb{R}^2_+)$. The associated Sobolev spaces on the surface S are defined either by choosing a partition of unity and a family of diffeomorphisms or directly in the following way. Let dist (x, Γ) be the geodetic distance between a point at S and the edge Γ. Then, the definition of the spaces $X^k_\rho(S)$ and $Y^k_\rho(S)$ is the same as above if we replace the weight function x_2 by dist (x, Γ) and the differential operators by an appropriate system of tangential differential operators.

For the infinite wedge problem the following regularity result holds.

Theorem 4.8. *Let the operator A_D be invertible in $L^2_{2,\rho}(\mathbb{R}^2_+) := L_{2,\rho}(\mathbb{R}^2_+) \times L_{2,\rho}(\mathbb{R}^2_+)$. Then it is invertible in $(X^k_\rho)^2(\mathbb{R}^2_+)$ and $(Y^k_\rho)^2(\mathbb{R}^2_+)$ for any $k \in \mathbb{N}$.*

Proof. We will show that the operator G_ϕ maps the space $L_{2,\rho}(\mathbb{R}^2_+)$ $(= X^0_\rho(\mathbb{R}^2_+))$ into $X^k_\rho(\mathbb{R}^2_+)$ for any $k \in \mathbb{N}$. Let be $k = 1$. We have to show that the functions $x_2 D_{x_1} G_\phi v$ and $x_2 D_{x_2} G_\phi v$ belong to $L_{2,\rho}(\mathbb{R}^2_+)$. The kernel functions of these two integral operators are linear combinations of functions of the form

$$k(x_1 - y_1, x_2, y_2) = \frac{(x_1 - y_1)^k x_2^l y_2^m}{[(x_1 - y_1)^2 + x_2^2 - 2x_2 y_2 \cos\phi + y_2^2]^{(m+k+l+2)/2}}$$

with natural numbers m, k, l. Theorem 2 in [13] gives the boundedness of such operators in $L_{2,\rho}(\mathbb{R}^2_+)$. The cases $k = 2, 3, ...$ can be done in the same way. The mapping property together with the invertibility in $L^2_{2,\rho}(\mathbb{R}^2_+)$ yields now the invertibility in the X^k_ρ-spaces. To prove the second part of the theorem we first note that the operator G_ϕ commutes with the operator of differentiation with respect to the x_1-variable. This is due to the convolution structure of the operator G_ϕ with respect to the first variable. We denote by $\hat{u}$ the partial Fourier transform of u with respect to x_1 (the co-variable of x_1 is denoted by ξ). Then the invertibility of A_D is equivalent to the invertibility of the operator function $A_D(\xi) := \mathcal{F}_{x_1 \to \xi} A_D$. Let us consider the equations

$$A_D(\xi)\xi^l \hat{v} = \xi^l \hat{f} \quad , \ l = 0, ..., k, \tag{4.18}$$

where $\hat{f} \in \mathcal{F}\left((Y_\rho^k)^2(\mathbb{R}_+^2)\right)$. For $l = 0$ (4.18) has a unique solution $\hat{v} \in \mathcal{F}\left((L_{2,\rho})^2(\mathbb{R}_+^2)\right)$.
Since $\xi^l \hat{f} \in \mathcal{F}\left((L_{2,\rho})^2(\mathbb{R}_+^2)\right)$, we obtain $\xi^l \hat{v} \in \mathcal{F}\left((L_{2,\rho})^2(\mathbb{R}_+^2)\right)$ for $l \geq 1$. The proof of the first part of the theorem shows now that $A_D(\xi)\hat{v}$ belongs to $\mathcal{F}\left((Y_\rho^k)^2(\mathbb{R}_+^2)\right)$ which finishes the proof. □

5 Meshes and Spline spaces

It is well-known that the solution of the boundary integral equation has a singular behavior at the edge even if the right-hand side is arbitrary smooth. This implies that the convergence of standard spline approximation methods is only of low order. To obtain similar results as for the approximate solution at smooth boundaries we have to adapt the method to the asymptotics of the solution. This can be done either by adding some special singular functions to the space of trial functions or by a special mesh grading. While the first method does not seem to be very applicable in the three-dimensional situation, since the exact asymptotic behavior of the solution has to be computed, the mesh grading is often used in the literature (s. [6], [24]).
For $q \geq 1$ we define a sequence Δ_n, $n \in \mathbb{N}$, of graded meshes on the boundary S. We first cut off two strips F_+, F_- of width one on both sides of the edge Γ. Taking closed curves with distance $(i/n)^q$, $i = 1, ..., n-1$, to Γ on both strips, we obtain a graded partition of the boundary near the edge. Furthermore, we choose a uniform partition of Γ with mesh-size of the order $1/n$. Taking curves from these mesh points in the normal direction to Γ, we get a quadrangulation of the strips F_+, F_-. Dividing each quadrangle into two triangles by non-intersecting diagonals and choosing a regular family of partitions into triangles of diameter of the order of $1/n$ for the remaining two parts of the boundary, we finally obtain our graded mesh Δ_n. We assume that for any triangle $\tau_i \in \Delta_n$ there is a smooth bijection π_i on a reference triangle $\tilde{\tau}_i$ in the plane.
For a sequence of triangulations Δ_n we define the spline spaces $S_d(\Delta_n)$ consisting of the pre-images of all piece-wise polynomials of total degree not greater than d subordinate to Δ_n. Let P_n denotes the orthogonal projection of $L_2(S)$ onto $S_d(\Delta_n)$. We note that the sequence $\{P_n\}$ converges strongly to the identity.
The Galerkin method for the approximate solution of equation (4.8) is the following: Find $u_n \in S_d(\Delta_n)$ such that

$$P_n A u_n = P_n g \ . \tag{5.1}$$

The stability of the method (5.1) is closely connected to the stability of the Galerkin method for the approximate solution of the boundary integral equation for the infinite wedge problem. Therefore, we define a associated triangulation of the half-plane $\mathbb{R}_+^2$. On the strip $F = \{x \ : \ 0 < x_2 < 1\}$ we take parallels to the x_1-axes at the distance $(i/n)^q$, $i = 1...n$. The remaining part of the half-plane is divided by parallels with the uniform distance $1/n$. Taking half-lines in the normal direction to the x_1-axes with distance $1/n$ we finally get a rectangular grid with an anisotropic mesh grading with respect to the boundary of the half-plane. Lifting that partition to the boundary of the infinite wedge and dividing each rectangle by nonintersecting diagonals we finally obtain the grid $\tilde{\Delta}_n$. The spline spaces will be defined as for the bounded surface. The associated Galerkin projection will be denoted by $\tilde{P}_n$. We note that the Galerkin method

$$\tilde{P}_n \tilde{A} u_n = \tilde{P}_n g \tag{5.2}$$

in the case of the infinite wedge has an infinitedimensional system matrix. Therefore, for practical computations a finite section method should be applied at first.

We finish this section with a well-known result about the approximation of functions in Sobolev spaces by polynomials (s. [27]). Let τ be a triangle in the plane. We put

$$h_k = \sup\{|x_k - y_k| \; : \; x, y \in \tau\}, \quad k = 1, 2,$$

and for any multi-index $\alpha = (\alpha_1, \alpha_2)$ we set $h^\alpha := h_1^{\alpha_1} h_2^{\alpha_2}$.

Proposition 5.1. *Let $d \in \mathbb{N}$ and $f \in H^d(\tau)$. Then there exists a polynomial p of total degree less than d such that*

$$||f - p||_{L_2(\tau)} \leq c \sum_{|\alpha|=d} h^\alpha ||D^\alpha f||_{L_2(\tau)} \tag{5.3}$$

6 Stability and error estimates for the Galerkin method

This section is devoted to the L_2-stability of the Galerkin method (5.1) and the associated method on the infinite wedge. For $\eta < \infty$ the essential norm of the double layer potential operator is less than one which implies the L_2-stability of any Galerkin method. This remains true for $\eta = \infty$ if the conditions (4.11) or (4.13) are fulfilled. Under the condition (4.12) the operator is not strongly elliptic, in generally. We consider the Galerkin method for the infinite wedge with $\alpha = \beta$.

Theorem 6.1. *Let be $\alpha = \beta$, then the Galerkin method (5.2) is stable.*

Proof. We have to show that the sequence $\{\tilde{P}_n A_D \tilde{P}_n\}$ is stable, where A_D is given by (4.10) with $\alpha = \beta$. Since all entries of the matrix operators in this sequence commute with each other, the determinant is well defined. So the stability is equivalent to the stability of

$$\{ \det (\tilde{P}_n A_D \tilde{P}_n)\} = \{\tilde{P}_n I \tilde{P}_n - \tilde{P}_n G_{2\alpha} \tilde{P}_n G_{2\alpha} \tilde{P}_n\}$$

which is obvious since the operators of this sequence are even invertible on the image of P_n. □

From the previous theorem follows now the stability of the Galerkin method (5.1).

Theorem 6.2. *Let be $\eta < \infty$ or let be $\eta = \infty$ and the invertibility conditions (4.11), (4.12) or (4.13) are fulfilled. Then the Galerkin method (5.1) is stable in $L_2(S)$.*

It follows in a standard way from the stability theorem that the Galerkin error is equivalent to the error of the best polynomial approximation, i.e.

$$||u - u_n||_{L_2(S)} \leq c||(I - P_n)u||_{L_2(S)} \text{ for all } n \geq n_0. \tag{6.1}$$

This reduces the question of the convergence rate to a question of approximation.

Theorem 6.3. *Let the assumptions of Theorem 6.2 be fulfilled. If $g \in Y_\rho^{d+1}(S)$, $0 < \rho < 1/2$, and the grading parameter fulfils $q \geq (d+1)/\rho$, then the Galerkin method (5.1) converges with the error estimate*

$$||u - u_n||_{L_2(S)} = O(n^{-d-1}),\ n \to \infty\,. \tag{6.2}$$

Proof. Under the assumptions of the theorem the solution u belongs to the space $Y_\rho^{d+1}(S)$. We have to show that the best L_2-approximation of an element $u \in Y_\rho^{d+1}(S)$ by splines from $S^d(\Delta_n)$ is of order n^{-d-1} if the grading parameter satisfies $q \geq (d+1)/\rho$. It suffices to prove that for any triangle τ of the mesh Δ_n there holds

$$||(I - P_n)u||_{L_2(\tau)} \leq cn^{-d-1} \sum_{i+j \leq d+1} ||\text{dist}^{j-\rho}(x, \Gamma) D_1^i D_2^j u||_{L_2(\tau)} \tag{6.3}$$

where D_1 and D_2 denote locally acting tangential differential operators in the edge direction and normal to the edge, respectively. First, let τ be a triangle from the strip of unit width around the edge such that $\text{dist}(\tau, \Gamma) = 0$. It follows from the uniform boundedness of the projections P_n that

$$\begin{aligned} ||(I - P_n)u||^2_{L_2(\tau)} &\leq c||u||^2_{L_2(\tau)} \\ &\leq \int_\tau \text{dist}^{2\rho}(x, \Gamma)\ |\text{dist}^\rho(x, \Gamma)u|^2 dS \\ &\leq c\left(\frac{1}{n}\right)^{2q\rho} ||u||^2_{L_{2,\rho}(\tau)} \end{aligned}$$

which gives the desired order since $q\rho \geq d+1$.

Let now τ be a triangle from this strip with positive distance to the edge. It follows from proposition 5.1 that

$$\|(I - P_n)u\|_{L_2(\tau)} \le c\|u\| \sum_{i+j=d+1} h_1^i h_2^j \|D_1^i D_2^j u\|_{L_2(\tau)}$$

$$\le c \sum_{i+j=d+1} h_1^i h_2^j \mathrm{dist}^{\rho-j}(\tau, \Gamma)\|\mathrm{dist}^{j-\rho}(x, \Gamma) D_1^i D_2^j u\|_{L_2(\tau)} .$$

By the definition of the grid , we have $h_1^i \sim 1/n^i$ and there is an index k such that

$$h_2^j \mathrm{dist}^{\rho-j}(\tau, \Gamma) \le c\frac{1}{n^j}\left(\frac{k}{n}\right)^{(q-1)j}\left(\frac{k}{n}\right)^{q(\rho-j)} \le c\frac{1}{n^j}$$

which implies (6.3).

If τ belongs to the remaining parts of the boundary we apply the proposition 5.1 again and obtain

$$\|(I - P_n)u\| \le cn^{-d-1} \sum_{|\alpha|=d+1} \|D^\alpha u\|_{L_2(\tau)},$$

We note that the partition is uniform and the norms of the spaces $Y_\rho^k(\tau)$ and $H^k(\tau)$ are equivalent to each other. Combining all estimates we obtain the convergence result of the theorem. □

References

1. Duduchava, R. (1984): On multidimensional singular integral operators. I:The half-space case. J. Operator Theory 11, 41–76
2. Duduchava, R. (1987): On algebras generated by continuous and discontinuous functions. Integral Equations and Operator Theory 10(4), 505–530
3. Duduchava, R., Schneider, R. (1987): The algebra of nonclassical singular integral operators on halfspace. Integral Equations and Operator Theory 10(4), 531–553
4. Elschner, J. (1990): On spline approximation for a class of non-compact integral equations. Math. Nachr. 146, 271–321
5. Elschner, J. (1992): The double layer potential operator over polyhedral domains. I. Solvability in weighted Sobolev spaces. Appl. Anal. 45, 117–134
6. Elschner, J. (1992): The double layer potential operator over polyhedral domains. II. Spline Galerkin Methods. Math. Meth. Appl. Sci. 15, 23–37
7. Jentsch, L. (1972): Über Wärmespannungen in Körpern mit stückweise konstanten Lamé'schen Elastizitätsmoduln. Schriftenreihe d. Zentralinst. f. Math. und Mech. der AdW
8. Jentsch, L. (1978): Die Greensche Matrix für zwei aneinander reibungsfrei gleitende elastische Halbräume mit verschiedenen Lamé'schen Moduln. ZAMM 58, 209–224
9. Jentsch, L. (1980): Zur Theorie der Wärmespannungen in Bimetallkörpern. Wiss. Zeitschr. Univ. Leipzig, Math.-Naturw. Reihe, 29. Jg., H.1, 49–58
10. Jentsch, L. (1991): Green's contact function for heat conduction in two half spaces and half planes with applications to several boundary contact problems. Math. Meth. Appl. Sci. 14, 633–661

11. Jentsch, L. (1992): On a bimetal problem for heat conduction. In: Symposium "Analysis on Manifolds with Singularities", Breitenbrunn 1990, volume 131, pages 69–78. Teubner-Texte zur Mathematik
12. Jentsch, L. (1994): On the boundery transmission problem of thermoelastostatics in a plane domain with interface corners. Math. Meth. Appl. Sci. 17, 1089–1114
13. Jentsch, L. (1995): On the boundedness of some integral operators with fixed singularities arising in elasticity. To be published in Demonstratio Mathematica
14. Jentsch, L. (1995): On the elastic potentials at corners. Submitted to Math. Meth. Appl. Sci.
15. Jentsch, L. (1995): Poissonsche Formeln für die Bimetallkugel. ZAMM 75, 105–114
16. Jentsch, L., Natroshvili, D. (1995): Non-classical interface problems for piecewise homogeneous anisotropic elastic bodies. Math. Meth. Appl. Sci. 18, 27–49
17. Jentsch, L., Natroshvili, D. (1995): Non-classical mixed interface problems for anisotropic bodies. To be published in Math. Nachr.
18. Jentsch, L., Natroshvili, D., Wendland, W.L. (1995): General Transmission Problems in the Theory of Elastic Oscillations of Anisotropic Bodies. DFG-Schwerpunkt Randelementmethoden, Bericht 95-7, to be published in Journ. Math. Anal. Appl.
19. Mirschinka, D. (1993): Randintegralmethoden für stationäre Wärmeleitprobleme in stückweise homogenen Körpern. PhD thesis, Technische Universität Chemnitz-Zwickau
20. Mirschinka, D. (1995): The double layer potential method for a boundary transmission problem for the Laplace operator in an infinite wedge. Math. Meth. Appl. Sci. 18, 515–528
21. Nicaise, N., Sändig, A.-M. (1994): General Interface Problems-I . Math. Meth. Appl. Sci. 17, 395–429
22. Nicaise, N., Sändig, A.-M. (1994): General Interface Problems-II . Math. Meth. Appl. Sci. 17, 431–450
23. Petry, M. (1993): Über die Streuung zeitharmonischer Wellen im geschichteten Raum. PhD thesis, Universität Göttingen
24. Rathsfeld, A. (1989): The invertibility of the double layer potential operator in the space of continuous functions defined on a polyhedron. The panel method. Complex Variables Theory Appl. 12(6), 835–854
25. Schmutzler, B. (1995): Edge Asymptotics for general Elliptic Interface Problems in a Double Wedge. TU-BA Freiberg, Preprint 95-09
26. Schneider, R. (1989): Contributions to the Theory of Elliptic Pseudodifferential Operators on Manifolds with Nonsmooth Boundaries. PhD thesis, Technische Hochschule Darmstadt
27. Schumaker, L.L. (1981): Spline Functions: Basic Theory. Wiley
28. Simonenko, I.B. (1964): A new general method for the investigation of linear operator integral equations. Izv. AN SSSR 29, 567–586 and 757–782. [Russian]

Analysis of 3D Elastoplastic Notch and Crack Problems using Boundary Element Method

G. Kuhn and P. Partheymüller

1 Introduction

It is a permanent challenge for designers and engineers in mechanical engineering to design technical structures appropriate to their future use and to correctly assess how safe they are against failure. In order to solve this problem, basically three points must be considered:

- How can the behaviour of a given material be described ?
- Which mechanisms of failure does this material show, and which are the corresponding quantitative criteria of failure ?
- How can stresses and strains at every position of the structure be determined effectively, taking into consideration the known material behaviour and the external load, and how can the characteristics for the underlying failure be obtained ?

Due to the complex geometries and the loads imposed on them, evaluation of stresses and strains is only possible using numerical methods, even if the simplest material behaviour - linear elasticity - is considered. In practice, mainly the Finite Element Method (FEM), the Boundary Element Method (BEM) and the Finite Differences Method (FDM) are applied.

For linear elastic material behaviour, BEM has meanwhile developed into a real alternative to FEM. Owing to transformation of the problem to the boundary, time required for pre- and post-processing of the computational models, especially for three-dimensional problems, is reduced considerably. Furthermore, experience has shown that high gradients of field quantities, occurring for example in case of stress concentration, can be approximated very exactly.

If BEM is applied in the framework of time-independent small strain elastoplasticity, the advantage of discretization in BEM in comparison to the volume-orientated methods FEM and FDM is largely preserved for problems of local plasticity, which are of great importance in mechanical engineering (for example notch problems).

Apart from the standard proof of strength, it may be necessary for safety-relevant structures in mechanical engineering to do an assessment in terms of fracture mechanics, especially when cracks occur due to manufacturing or during operation. Since a crack is the most stringent form of stress concentration, BEM has proved to be a powerful method especially in linear elastic fracture mechanics (LEFM), but also in nonlinear fracture mechanics.

At the beginning of the project the Boundary Element Method provided promising results in two-dimensional linear as well as nonlinear continuum and fracture mechanics [39, 31, 38] and for linear elastic applications also in 3D [26]. Thus the main aim of this project was to apply BEM to three-dimensional elastoplastic notch and crack problems and to realize it in terms of a computer code, because efficient numerical methods are of great practical interest especially for those problems. During the last period of granting this project, the aim was extended to investigate the potential of Boundary Element Method for numerical simulation of fatigue crack growth in 3D. For definite decoupling of crack surfaces, the formerly used *subregion technique* was replaced by a more efficient method, the *DUAL-BEM* [36].

2 Basics of Continuum Mechanics

A three-dimensional, homogeneous, isotropic body (domain Ω, boundary Γ) with linear elastic or inelastic material behaviour is considered. It is loaded as shown in Fig. 2.1 in Ω with body forces b_j, on Γ_t with tractions $\bar{t}_j$ and on Γ_u with displacements $\bar{u}_j$.

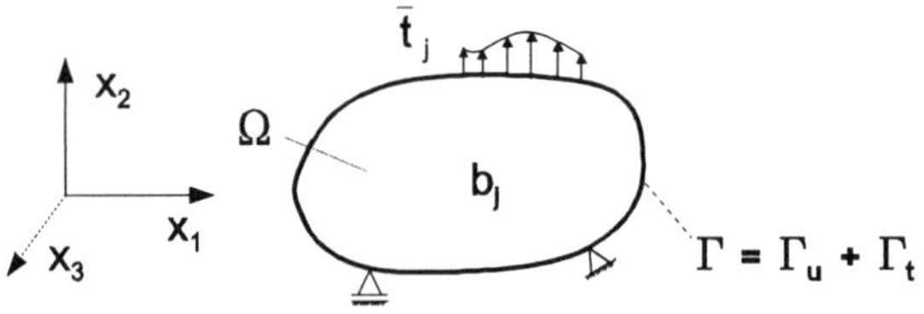

Fig. 2.1. Sketch of problem

2.1 Elasticity Theory

The theory of linear elasticity is based on the following equations:

$$\text{equilibrium conditions:} \quad \sigma_{ij,i} + b_j = 0 \ , \quad \sigma_{ij} = \sigma_{ji} \ , \quad t_j = \sigma_{ji}\, n_i$$

$$\text{kinematic relation:} \quad \varepsilon_{ij} = \tfrac{1}{2}(u_{i,j} + u_{j,i})$$

$$\text{constitutive relation:} \quad \sigma_{ij} = 2G\left[\varepsilon_{ij} + \tfrac{\nu}{1-2\nu}\,\varepsilon_{kk}\,\delta_{ij}\right] = C_{ijkl}\,\varepsilon_{kl} \ . \tag{2.1}$$

C_{ijkl} is the elasticity tensor with the material constants *shear modulus* G, and *Poisson's ratio* ν, n_i is the outward directed normal vector of the boundary Γ. The usual elimination of stresses σ_{ij} and strains ε_{ij} leads to the well-known Lamé-Navier's equation

$$u_{j,ii} + \frac{1}{1-2\nu} u_{l,lj} = -\frac{b_j}{G} \quad , \tag{2.2}$$

which, in combination with the boundary conditions $\bar{u}_j$ on Γ_u and $\bar{t}_j$ on Γ_t, completely describe the elasticity problem.

2.2 Plasticity Theory

Time-independent, elastoplastic material behaviour in the framework of small strain plasticity is assumed. Here, the total strain tensor

$$\varepsilon_{ij} = \varepsilon^e_{ij} + \varepsilon^p_{ij} \tag{2.3}$$

can be split up additively into an elastic and a plastic component. Analogously to elasticity (see equations (2.1)), the equilibrium conditions and the kinematic relation as well as Hooke's law for the elastic component of the strain tensor are valid:

$$\sigma_{ij} = 2G \left[(\varepsilon_{ij} - \varepsilon^p_{ij}) + \frac{\nu}{1-2\nu} (\varepsilon_{kk} - \varepsilon^p_{kk}) \, \delta_{ij} \right] = C_{ijkl} \, (\varepsilon_{kl} - \varepsilon^p_{kl}) \quad . \tag{2.4}$$

If again stresses and strains are eliminated, one obtains a *generalized* form of Lamé-Navier's equation

$$u_{j,ii} + \frac{1}{1-2\nu} u_{l,lj} = -\frac{\hat{b}_j}{G} \quad \text{with} \quad \begin{array}{lcll} u_j & = & \bar{u}_j & \text{on} \;\; \Gamma_u \\ t_j & = & \bar{t}_j & \text{on} \;\; \Gamma_t \end{array} \tag{2.5}$$

and the *pseudo body force vector*

$$\hat{b}_j = b_j - 2G \left[\varepsilon^p_{ij,i} + \frac{\nu}{1-2\nu} \varepsilon^p_{ll,j} \right] \quad , \tag{2.6}$$

which contains the influence of the nonlinear material behaviour.

For complete description of the problem, additionally the constitutive equations for time-independent elastoplastic material behaviour must be provided. For this purpose, the internal variables $\Phi = (\Phi_1, \Phi_2, ... \Phi_n)$ are introduced. Together with the variables stress and temperature, they describe the current state of the material. For the internal variables, suitable evolution equations are necessary and must be available.

For time-independent, incremental plasticity, the flow law usually consists of

- the yield criterion: $F(\sigma, T, \Phi) = 0$,

- the flow rule: $d\varepsilon_{ij}^{p} = d\lambda \frac{\partial Q}{\partial \sigma_{ij}}$ and (2.7)

- the consistency condition: $d\lambda\, F = 0$, $d\lambda \geq 0$ mit $F \leq 0$,

written in Kuhn-Tucker form. F denotes the yield function, $d\lambda$ denotes the plastic proportionality factor, and Q denotes the plastic potential. In case of associated plasticity, the yield function is considered as the plastic potential.

The classic yield functions (Tresca, von Mises, Drucker-Prager, Mohr--Coulomb) were applied both for kinematic and isotropic hardening by Dallner in his PhD-thesis [9]. In this thesis, the phenomenological and physical basics as well as the respective literature are given.

3 Basics of Fracture Mechanics

3.1 Linear Elastic Fracture Mechanics

In fracture mechanics, there are three basic modes of crack opening - *Mode I, Mode II* and *Mode III.* Every mode creates an independent asymptotic *crack near-field* in an elastic body with a crack. Superposition of all three modes leads to the general three-dimensional crack near-field. In the crack front coordinate system (see Fig. 3.1), the components of the asymptotic displacement field

$$u_i = \frac{K_I}{2G}\sqrt{\frac{r}{2\pi}}\, f_i^{I}(\varphi,\nu) + \frac{K_{II}}{2G}\sqrt{\frac{r}{2\pi}}\, f_i^{II}(\varphi,\nu) + \frac{2K_{III}}{G}\sqrt{\frac{r}{2\pi}}\, f_i^{III}(\varphi,\nu) \tag{3.1}$$

and of the asymptotic stress field

$$\sigma_{ij} = \frac{K_I}{\sqrt{2\pi r}}\, g_{ij}^{I}(\varphi) + \frac{K_{II}}{\sqrt{2\pi r}}\, g_{ij}^{II}(\varphi) + \frac{K_{III}}{\sqrt{2\pi r}}\, g_{ij}^{III}(\varphi) \tag{3.2}$$

can be described in dependence on a local polar coordinate system (r, φ) [31]. For $r \to 0$ the asymptotic stress field shows a $1/\sqrt{r}$- singularity. Its intensity is described by the ***stress intensity factors (K-values)*** K_I, K_{II} and K_{III} for the three modes of crack opening. An exception to this are those points of the crack front located on the free surface of the cracked body. Here, an edge singularity is superposed.

If under single-mode or mixed-mode load the K-value or a representative K-value $K_v = K_v(K_I, K_{II}, K_{III})$ exceeds a critical material-specific value (*fracture toughness*), crack propagation becomes instable. Fracture criteria are then

$$K_i \geq K_{ic} \quad \text{for } i = I, II, III \quad \text{and} \quad K_v \geq K_{vc}(K_{Ic}, K_{IIc}, K_{IIIc}). \tag{3.3}$$

In the following, the path independent J-integral (Rice) is investigated. In LEFM it corresponds to the energy release rate (Griffith, Irwin), but can be also extended to nonlinear material behaviour in the framework of deformation theory. Within the LEFM for 3D problems, the J-integral on a point s of the crack front can be written as [33]:

$$J^n_{x_1}(s) = \int_{\Gamma} (Wn_1 - \sigma_{ij} n_j u_{i,1})\, d\Gamma - \int_{A(\Gamma)} (\sigma_{i3} u_{i,1})_{,3}\, dA \quad . \tag{3.4}$$

Here, n_j denotes the outward directed normal vector to the contour Γ coplanar with $A(\Gamma)$ (see Fig. 3.1). W denotes the elastic strain energy density. While in 2D, the J-integral can be described by the contour integral alone, in 3D additionally a surface integral must be considered. This surface integral includes both the singularity on the crack front and the boundary values on the crack surfaces.

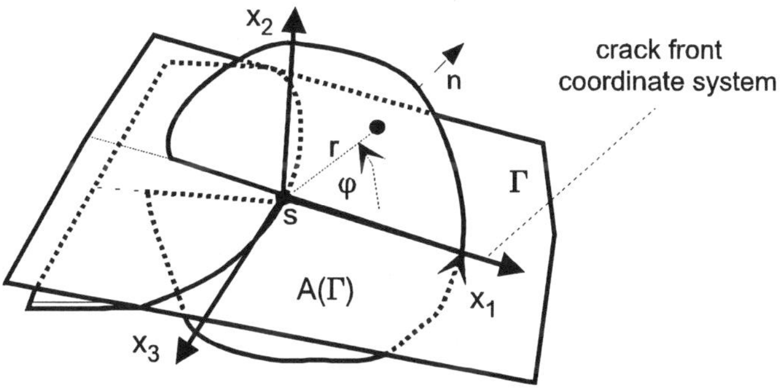

Fig. 3.1. 3D crack problem: contours and area configuration for J-integral

Within the crack near-field, the J-integral can be decomposed into the three modes of crack opening ($J^n_{x_1} = J_I + J_{II} + J_{III}$) [28], [34]. Then the K-values

$$K_I = \sqrt{\frac{8G}{\kappa+1} J_I}\,, \quad K_{II} = \sqrt{\frac{8G}{\kappa+1} J_{II}}\,, \quad K_{III} = \sqrt{2G\, J_{III}} \tag{3.5}$$

can be determined. G denotes the shear modulus, and $\kappa = 3 - 4\nu$ stands for the plain strain state.

Within the crack near-field, not only the J-integral itself but also the three modes J_I, J_{II}, J_{III} are path independent. In the far-field, path independence can be proved only separately for the symmetric and the antisymmetric part of the J-integral [20].

3.2 Nonlinear Fracture Mechanics

In the framework of the deformation theory, due to Carpenter et al. [7] one obtains the J-integral

$$\tilde{J}(s) = \int_{\Gamma} (W n_1 - \sigma_{ij} n_j u_{i,1}) d\Gamma - \int_{A(\Gamma)} (\sigma_{i3} u_{i,1})_{,3} dA + \int_{A(\Gamma)} (\sigma_{ij} \varepsilon^p_{ij,1} - W^p_{,1}) dA \tag{3.6}$$

for 3D crack problems, which is also path independent. The strain energy density ($W = W^e + W^p$) consists of an elastic and a plastic component.

Following Hutchinson [21] and Rice and Rosengren [37], in analogy to the linear elastic case singular crack near-field equations (HRR field) can be developed by application of the deformation theory. For material hardening in the form of a *Power Law Strain Hardening* stress-strain relation due to Ramberg-Osgood

$$\frac{\varepsilon_v}{\varepsilon_0} = \frac{\sigma_v}{\sigma_0} + \alpha \left(\frac{\sigma_v}{\sigma_0} \right)^N \tag{3.7}$$

and on condition that near the crack front the elastic strains compared to the plastic strains can be neglected, the order of singular behaviour and the respective angle distribution depend on the hardening exponent N. In the crack front coordinate system (see Fig. 3.1) the HRR field can be written as

$$\sigma_{ij}(r, \varphi) = \sigma_0 \left(\frac{J}{\alpha \sigma_0 \varepsilon_0 I_N r} \right)^{\frac{1}{1+N}} \tilde{\sigma}_{ij}(\varphi, N) \tag{3.8}$$

$$\varepsilon_{ij}(r, \varphi) = \alpha \varepsilon_0 \left(\frac{J}{\alpha \sigma_0 \varepsilon_0 I_N r} \right)^{\frac{N}{1+N}} \tilde{\varepsilon}_{ij}(\varphi, N), \tag{3.9}$$

where I_N is dependent on N and is a known scaling factor. The J-integral describes the intensity of the singular fields.

4 Boundary Element Method in Elastoplasticity

For the sake of simplicity, body forces are neglected in the boundary integral equations for elastoplastic material behaviour. Furthermore, a *rate*-formulation is not necessary, because in correspondence to Section 2.2 time-independent small-strain plasticity is assumed.

4.1 Boundary Integral Formulations

In analogy to elasticity, for elastoplastic material behaviour the *principle of weighted residuals* can be applied. Thus, the *generalized* Somigliana identity

$$u_i(\xi) = \int_\Gamma U_{ij}(\xi,x)\,t_j(x)\,d\Gamma - \int_\Gamma T_{ij}(\xi,x)\,u_j(x)\,d\Gamma + \int_\Omega \Sigma_{ijk}(\xi,x)\,\varepsilon^p_{jk}(x)\,d\Omega \tag{4.1}$$

can be derived. U_{ij}, T_{ij} and Σ_{ijk} are the well-known Kelvin fundamental solutions. The domain integral, which arises in addition to the linear elastic case, describes the influence of nonlinear material behaviour and is weakly singular. For this reason, only the strongly singular boundary integral with the kernel T_{ij} must be considered for the limiting procedure. Thus, one obtains the elastoplastic displacement boundary integral equation

$$C_{ij}(\xi)u_j(\xi) = \int_\Gamma U_{ij}(\xi,x)t_j(x)d\Gamma - \fint_\Gamma T_{ij}(\xi,x)u_j(x)d\Gamma + \int_\Omega \Sigma_{ijk}(\xi,x)\varepsilon^p_{jk}(x)d\Omega \tag{4.2}$$

for source points on the boundary [1, 6, 9].

Analogously to elasticity an integral equation can be given also for displacement derivatives on source points inside the domain, if the generalized Somigliana identity (4.1) is differentiated with respect to the source point coordinate ξ. Due to the strong singularity of the additional domain integral, a thorough limiting procedure and a regularization has to be carried out. Here, one distinguishes *indirect*, *approximative direct* and *exact direct* methods of calculation, for which detailed literature studies can be found in the PhD-thesis of Dallner [9].

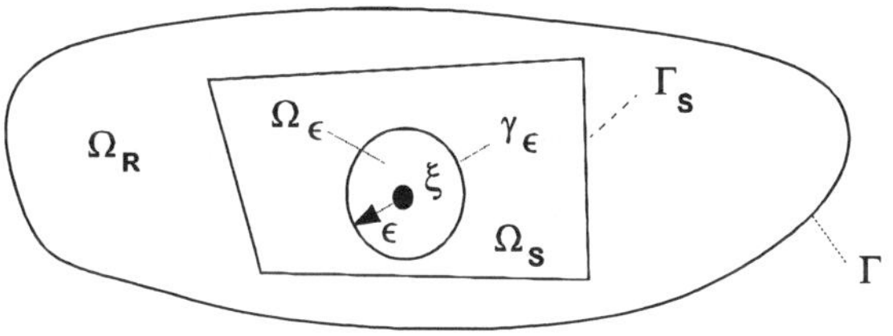

Fig. 4.1. Source point ξ located inside the domain

A method suggested by Dallner, Kuhn [10] provides the formulation

$$\begin{aligned}\frac{\partial u_i(\xi)}{\partial \xi_l} &= \int_\Gamma \frac{\partial U_{ij}}{\partial \xi_l}\,t_j(x)\,d\Gamma \;-\; \int_\Gamma \frac{\partial T_{ij}}{\partial \xi_l}\,u_j(x)\,d\Gamma \;+\; \int_{\Omega_R} \frac{\partial \Sigma_{ijk}}{\partial \xi_l}\,\varepsilon^p_{jk}(x)\,d\Omega_R \\ &+ \int_{\Omega_S} \frac{\partial \Sigma_{ijk}}{\partial \xi_l}\left[\varepsilon^p_{jk}(x) - \varepsilon^p_{jk}(\xi)\right] d\Omega_S - \varepsilon^p_{jk}(\xi) \int_{\Gamma_S} \Sigma_{ijk} n_l(x) d\Gamma_S \end{aligned} \tag{4.3}$$

for the displacement derivatives on source points inside the domain, if the limiting procedure $\epsilon \to 0$ (see Fig. 4.1), the Leibniz formula, a regularization via Taylor series expansions and the Gauss theorem are taken into consideration.

The advantage of this formulation is that the strongly singular domain integral is replaced by regular and weakly singular domain integrals as well as regular boundary integrals over Γ_s. These integrals can be evaluated both in 2D and in 3D using standard numerical integration schemes. If, in view of discretization of the domain, all cells which contain the source point ξ are summed up to Ω_s, the displacement derivatives on source points inside the domain can be determined efficiently for any cell types.

An integral formulation for the displacement derivatives can be derived also for source points on the boundary. Due to the hypersingular boundary integrals, this had been a serious problem for a long time. After the hypersingular integrals in the framework of linear elasticity had been solved [14, 19], Leitao et al. [29] for the first time presented a hypersingular formulation for elastoplastic material behaviour in 2D. The strongly singular domain integral is evaluated directly by using the method of Guiggiani [14]. Parallel to this, Huber, Dallner, Partheymüller, Kuhn [18] extended the above derived formulation (4.3) to source points on the boundary and applied it to 3D elastoplastic problems. The advantage of this approach is that direct evaluation of the strongly singular domain integral, which is especially costly for three-dimensional problems, can be avoided.

Starting from a formulation of the *generalized* Somigliana identity (4.1) where the source point is excluded from the domain (see Fig. 4.2), the limiting procedure $\epsilon \to 0$ and the derivation to the source point coordinate ξ can be done separately for the boundary and the domain integral:

$$\lim_{\epsilon\to 0}\left\{\frac{\partial}{\partial \xi_l}\int\limits_{\Gamma-\Gamma_\epsilon+\gamma_\epsilon}[U_{ij}t_j(x)-T_{ij}u_j(x)]\,d\Gamma\right\}+\lim_{\epsilon\to 0}\left\{\frac{\partial}{\partial \xi_l}\int\limits_{\Omega-\Omega_\epsilon}\Sigma_{ijk}\varepsilon^p_{jk}(x)d\Omega\right\}=0. \tag{4.4}$$

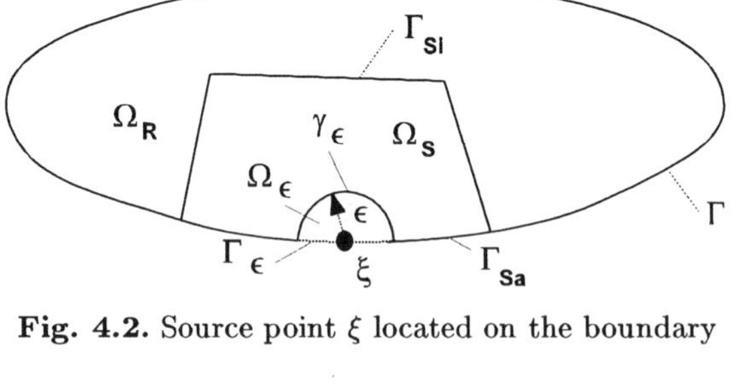

Fig. 4.2. Source point ξ located on the boundary

The first limit of equation (4.4) is the initial formulation for the derivation of the displacement derivatives at source points located on the boundary for linear elastic material behaviour. For 3D problems, this was first applied for smooth boundaries by Huber, Lang, Kuhn [19]. Evaluation of the hypersingular and strongly singular integrals is based on the Guiggiani [14] method. For this, $C^{1,\alpha}$-continuity is required for the displacements at the source point.

The limiting procedure for the domain integral in equation (4.4) can be done analogous to the source points inside the domain. When both parts are summed, the hypersingular formulation of the displacement derivatives for elastoplastic material behaviour can be rewritten as

$$\frac{1}{2}\frac{\partial u_i(\xi)}{\partial \xi_l} = \lim_{\epsilon \to 0}\left\{ \int\limits_{\Gamma-\Gamma_\epsilon} \left[\frac{\partial U_{ij}}{\partial \xi_l} t_j(x) - \frac{\partial T_{ij}}{\partial \xi_l} u_j(x)\right] d\Gamma - u_j(\xi)\frac{b_{ijl}}{\epsilon}\right\} \tag{4.5}$$
$$+ \int\limits_{\Omega_R} \frac{\partial \Sigma_{ijk}}{\partial \xi_l} \varepsilon^p_{jk}(x)\, d\Omega_R + \int\limits_{\Omega_S} \frac{\partial \Sigma_{ijk}}{\partial \xi_l} \left[\varepsilon^p_{jk}(x) - \varepsilon^p_{jk}(\xi)\right] d\Omega_S$$
$$- \varepsilon^p_{jk}(\xi)\left\{ \int\limits_{\Gamma_{Si}} \Sigma_{ijk} n_l(x)\, d\Gamma_{Si} + \lim_{\epsilon \to 0} \int\limits_{\Gamma_{Sa}-\Gamma_\epsilon} \Sigma_{ijk} n_l(x)\, d\Gamma_{Sa}\right\}$$

for a smooth boundary [18].

The techniques applied for numerical evaluation of all integrals occurring in the boundary integral equations for elastoplastic material behaviour are described in the report [25]. More details on this topic can be found in the articles [2, 10, 14, 18, 19, 35].

4.2 Elastoplastic Solution

For numerical solution of the elastoplastic problem, both the geometry (boundary Γ, domain Ω) and the relevant field quantities (displacements u_j, tractions t_j, plastic strains ε^p_{jk}) must be discretized and approximated. This is done via discrete nodal values and interpolating shape functions.

Although the domain must be meshed, the advantage of discretization for problems with local plasticity (notch and crack problems) in comparison to volume-oriented methods (FEM, FDM) is preserved to a large extent. Only those areas where plastification occurs must be considered.

Using a *collocation method*, the discretized form of the displacement boundary integral equation (4.2) is evaluated on the n boundary nodes. For 3D problems one obtains a set of $3\,n$ algebraic equations, which can be written in matrix form. Rearrangement according to the known boundary values $\boldsymbol{y}$ and the unknown boundary values $\boldsymbol{x}$

$$\boldsymbol{A}\,\boldsymbol{u} = \boldsymbol{B}\,\boldsymbol{t} + \boldsymbol{D}\,\boldsymbol{\varepsilon^p} \quad \rightarrow \quad \tilde{\boldsymbol{A}}\,\boldsymbol{x} = \tilde{\boldsymbol{B}}\,\boldsymbol{y} + \boldsymbol{D}\,\boldsymbol{\varepsilon^p} \tag{4.6}$$

provides a direct relation between the unknown plastic strains $\boldsymbol{\varepsilon^p}$ on the cell nodes and the unknown boundary values $\boldsymbol{x}$. In comparison to the linear elastic case there is additionally the inelastic part with the vector ε^p and the corresponding matrix $\boldsymbol{D}$. This matrix consists of the regular and the weakly singular integral parts over the cells.

Since within the flow law (2.7) the plastic strains are locally linked with the stresses, the stress tensor must be provided on every cell node.

For cell nodes inside the domain, the formulation for displacement derivatives (4.3) in its discretized form can be applied. Via the kinematic relation (2.1) and the stress-strain relation (2.4) the stress tensor can be derived.

There are three different approaches to determine the stress tensor for cell nodes on the boundary. The first is to use *the shape function derivatives.* This method is realized in the existing boundary element program for three-dimensional elastoplastic problems because it requires little time for computation and is well-known from elasticity.

The other approaches are based on integral formulations of the displacement derivatives. Other than the local shape function derivatives, they can easily be extended to geometrically nonlinear problems and general stress-strain relations (see Foerster, Kuhn [11]). One approach is to evaluate equation (4.3) for source points inside the domain located close to the boundary to avoid the evaluation of hypersingular integrals. This is more or less a heuristic method but it works if there are adaptive integration algorithms to determine the arising *nearly singular integrals.* The other approach is to evaluate the hypersingular identity for the displacement derivatives (4.6) in its discretized form. Since this equation is available for smooth boundaries only and additionally, $C^{1,\alpha}$-continuity of the displacements is required, at the moment domain discretization and approximation of plastic strains must be selected in such a way that the nodal points of the shape functions of ε^p_{jk} are not located on edges, corners or boundary element edges. This can be done efficiently by decoupling of boundary and domain meshes together with introduction of discontinuous cells to avoid edges and corners (see Herding, Kuhn [15]).

When evaluating the stress tensor at all cell nodes using the above methods and rearranging the matrix equation with respect to the boundary value problem

$$\boldsymbol{\sigma} = \tilde{\boldsymbol{H}}\boldsymbol{u} + \tilde{\boldsymbol{G}}\boldsymbol{t} + \tilde{\boldsymbol{D}}\boldsymbol{\varepsilon}^p \quad \rightarrow \quad \boldsymbol{\sigma} = \hat{\boldsymbol{A}}\boldsymbol{x} + \hat{\boldsymbol{B}}\boldsymbol{y} + \tilde{\boldsymbol{D}}\boldsymbol{\varepsilon}^p \ , \tag{4.7}$$

one receives a direct relation between the stress tensor, the boundary values and the plastic strains. Elimination of the unknown boundary values $\boldsymbol{x}$ using equation (4.6) supplies the stress tensor on all cell nodes

$$\boldsymbol{\sigma} = (\hat{\boldsymbol{A}}\tilde{\boldsymbol{A}}^{-1}\tilde{\boldsymbol{B}} + \hat{\boldsymbol{B}})\boldsymbol{y} + (\hat{\boldsymbol{A}}\tilde{\boldsymbol{A}}^{-1}\boldsymbol{D} + \tilde{\boldsymbol{D}})\boldsymbol{\varepsilon}^p = \boldsymbol{\sigma}^e + \boldsymbol{K}\boldsymbol{\varepsilon}^p \tag{4.8}$$

in dependence on the external load $\boldsymbol{y}$ on the boundary nodes and the plastic strains $\boldsymbol{\varepsilon}^p$ on the cell nodes. $\boldsymbol{\sigma}^e$ can be identified as that part of the stress tensor $\boldsymbol{\sigma}$ which would occur under the current load $\boldsymbol{y}$ at linear elastic material behaviour. $\boldsymbol{K}\,\boldsymbol{\varepsilon}^p$ is the plastic corrector term.

For consideration of loading history, Equation (4.8) must be rewritten as an incremental formulation, and external load has to be applied in steps. This in combination with the flow law and the corresponding evolution equations results in a nonlinear set of equations which must be solved iteratively for every single load step. Due to this, after each load step the accumulated plastic strains and the stress tensor on the cell nodes are known. From this, the unknown boundary values can be determined via relation (4.6). In a post-processing procedure, displacements and stresses on arbitrary source points inside the domain and on the boundary can be determined via the boundary integral equations provided in Section 4.1.

A detailed description of the elastoplastic algorithm and numerical examples can be found in the PhD-thesis of Dallner [9].

5 Crack Problems and Boundary Element Method

5.1 Crack Modelling in the Boundary Element Method

In order to derive stresses and strains in a cracked body by using Boundary Element Method, the crack surfaces must be definitely decoupled. Basically, there are three approaches: Application of Green's fundamental solutions [40, 31, 22, 38], subregion technique and the *DUAL BEM*.

If Green's fundamental solutions are applied, neither in linear elastic nor in elastoplastic cases discretization of the crack surfaces with boundary elements is necessary; application is, however, possible only for 2D problems. Recently, Telles et al. [41] introduced an extension to 3D problems in which Green's solutions for linear elastic material are determined numerically. The advantage that crack discretization is not necessary, however, cannot be maintained. For this reason, the subregion technique and the DUAL-BEM were realized in the framework of this project.

5.1.1 Subregion technique

The subregion technique, introduced in the Boundary Element Method by Lachat and Watson [27], was first used by Blandford et al. [5] for decoupling of crack surfaces. When the cracked body is split into two subregions at the ligament (see Fig. 5.1), the crack surfaces are definitely decoupled in separate domains. Using subregion technique BEM can be applied to both linear elastic and elastoplastic crack problems [9, 17].

Another advantage of the subregion technique is that a block diagonal structure of the system matrix is generated. This leads to a considerable reduction of computing time if suitable block solvers are used [3, 4]. Since different material behaviour can be assigned for each subdomain, interface cracks can be modelled, too. In nonlinear cases, a *multiple node concept* is necessary for the plastic strains [9].

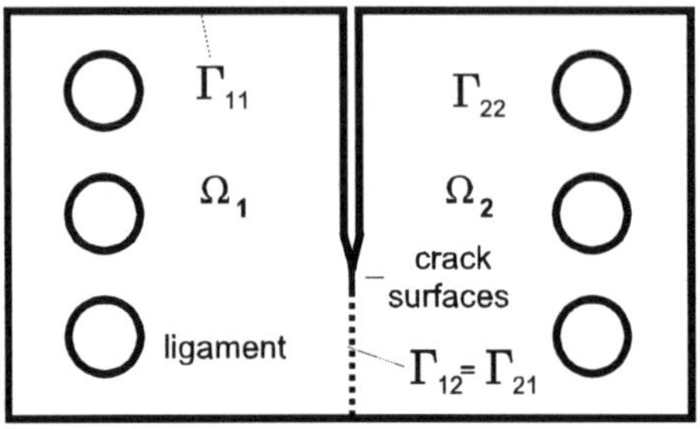

Fig. 5.1. Decoupling of crack surfaces by subregion technique

A drawback is that the subregion interface intersects the cracked body in that zone where the highest stress gradients occur. There are two approaches to approximate the singular crack near-field well. Either the discretization is refined towards the crack, which is very costly in terms of time, or, following FEM, special elements are used to describe the singular crack near-field.

In linear elastic cases, this means for boundary elements along the crack front that on the crack surfaces the shape functions of the displacements have to reproduce $\sqrt{r}$-behaviour, and that the shape functions of the tractions on the ligament have to reproduce $1/\sqrt{r}$-behaviour. Cruse and Wilson [8] were the first ones to use the *quarter-point elements* known from FEM with the respective extension for approximation of tractions in 3D Boundary Element Method. Later Luchi and Poggliani [30] introduced crack front elements with special shape functions and a suitable regularization for the approximation of the tractions in 3D BEM (*traction singular mid-point elements*).

For elastoplastic problems, singularity in the HRR field depends on the hardening exponent (see Section 3.2). For this reason, the *traction singular mid-point elements* were used and extended to general singularity orders. Additionally, singular volume elements were derived for approximation of plastic strains around the crack front. Detailed description of the shape functions of these singular boundary and volume elements as well as the respective numerical integration can be found in the reports [23, 24] and the PhD-thesis of Huber [17].

The most apparent drawback of crack modelling using subregion technique is that it requires more time and effort for discretization. This is especially costly when multiple cracks are modelled and when remeshing becomes necessary in case of crack propagation.

5.1.2 DUAL BEM

The idea to use different boundary integral equations for definite decoupling of the crack surfaces was introduced by Watson [42] for elasticity and by Gray and Giles [12] for potential problems. Hong and Chen [16] first applied the traction equation as the additional boundary integral equation for crack modelling. This technique was taken up by Portela, Aliabadi and Rooke [36], who created the name *Dual Boundary Element Method* (*DUAL BEM*) and used this method for 2D crack growth simulation. There are also applications [32], [13] for 3D elastic crack problems and in a recent paper an extension to elastoplastic material behaviour in 2D [29].

Boundary Integral Equations on Crack surfaces

Both the boundary integral equation for displacements and for tractions for linear elastic material behaviour can be derived on the crack surfaces analogous to that for source points located on the *non-cracked* boundary (see Fig. 5.2). If a formulation where the source point is excluded from the domain is assumed and if the boundary on the source point is smooth, the limiting procedure $\epsilon \to 0$ can be evaluated separately for each crack surface. With respect to a well-defined description, all parameters of one crack surface are marked by the upper index u, those of the opposite surface are marked by t.

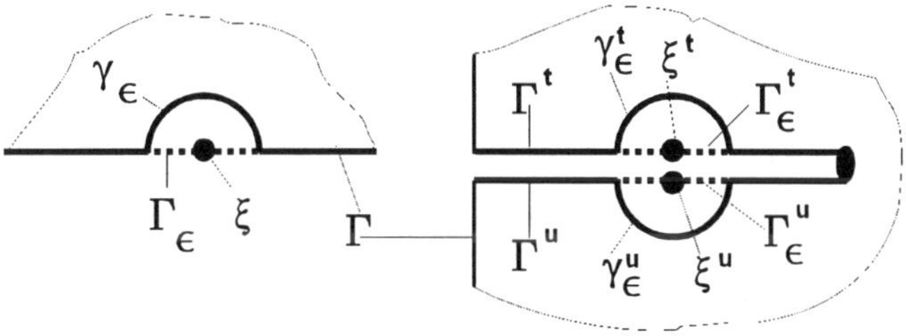

Fig. 5.2. Source point located on non-cracked boundary or crack surface

Thus the strongly singular displacement boundary integral equation for a source point located on a crack surface can be written as

$$\frac{1}{2}u_i(\xi^u)+\frac{1}{2}u_i(\xi^t)+\lim_{\epsilon\to 0}\int\limits_{\bar{\Gamma}-\Gamma^u_\epsilon-\Gamma^t_\epsilon} T_{ij}(\xi,x)u_j(x)d\Gamma(x) = \int\limits_{\bar{\Gamma}} U_{ij}(\xi,x)t_j(x)d\Gamma(x) \tag{5.1}$$

with $\bar{\Gamma} = \Gamma + \Gamma^u + \Gamma^t$. The two integral-free terms for the displacements on the one and on the opposite crack surface with the corresponding strongly singular integrals are characteristic. From equation (5.1) one recognizes immediately that evaluation on both crack surfaces leads to an identical result. For this reason, this equation for itself is not sufficient for crack modelling.

In this case also the traction boundary integral equation is necessary which can be obtained from the formulation for the displacement derivatives on the crack surfaces. The limiting procedure (see Fig. 5.2) and the regularization of the strongly singular and hypersingular integrals (see Section 4.1) can be done separately for each crack surface. Multiplication with the elasticity tensor C_{abil} and the normal vector $n_b(\xi^t)$ at the source point on the crack surface Γ^t provides the traction boundary integral equation for Γ^t

$$\frac{1}{2}\,t_a(\xi^t) \;-\; \frac{1}{2}\,t_a(\xi^u) \;= \tag{5.2}$$

$$\lim_{\epsilon\to 0}\left\{\int\limits_{\Gamma-\Gamma_\epsilon^u-\Gamma_\epsilon^t}\left[\bar{U}_{aj}t_j(x)-\bar{T}_{aj}u_j(x)\right]d\Gamma - u_j(\xi^u)\frac{\bar{b}_{aj}(\xi^u)}{\epsilon} - u_j(\xi^t)\frac{\bar{b}_{aj}(\xi^t)}{\epsilon}\right\}$$

with the kernels

$$\bar{U}_{aj} \;=\; \frac{\partial U_{ij}}{\partial \xi_l}(\xi,x)\,C_{abil}\,n_b(\xi^t) \;\text{ and }\; \bar{T}_{aj} \;=\; \frac{\partial T_{ij}}{\partial \xi_l}(\xi,x)\,C_{abil}\,n_b(\xi^t). \tag{5.3}$$

Again, the two integral-free terms are characteristic. Since the normals on the two crack surfaces are directed opposite to each other, one obtains the traction vector $t_a(\xi^u)$ with negative sign. The two divergent parts $\bar{b}_{aj}(\xi^u)$ and $\bar{b}_{aj}(\xi^t)$ must be treated together with their corresponding hypersingular integrals and cancel out [14, 19]. For validity of this formulation, C^1-continuity must be assumed for the displacements.

Numerical solution

In *DUAL BEM*, for linear elastic problems the complete boundary of the cracked body including the two crack surfaces is discretized within one subregion and the boundary values are approximated via shape functions. In the framework of a *collocation procedure*, the evaluation of the displacement boundary integral equation (5.1) at all nodes on Γ^u, evaluation of the traction boundary integral equation (5.2) on Γ^t and evaluation of the standard displacement boundary integral equation on the remaining non-cracked boundary Γ leads to a system of algebraic equations $\boldsymbol{Au} = \boldsymbol{Bt}$ (see Section 4). After rearranging according to known and unknown boundary values, one obtains the linear system of equations $\tilde{\boldsymbol{A}}\boldsymbol{x} = \boldsymbol{b}$ to determine the unknown boundary values.

Due to symmetry of the kernel T_{ij} on crack surfaces, it is not possible to compute the strongly singular integrals in equation (5.1) on the crack surfaces by using the *rigid body condition* [36]. For this reason, all strongly singular and hypersingular integrals on the crack surfaces from equations (5.1), (5.2) are evaluated directly by using the method of Guiggiani [14, 19].

In order to meet the continuity requirements for the hypersingular integrals and to solve problems also with kinked cracks, the crack surfaces are meshed using *discontinuous elements* (see Fig. 5.3 a). In case of edge cracks, *edge-discontinuous elements* for surfaces intersecting a crack surface together with *node-discontinuous elements* at the crack front are used (see Fig. 5.3 b). Using this modelling strategy for the cracks the non cracked-boundary can be discretized with *continuous elements* to keep the number of degrees of freedoms as low as possible [25]. In the existing boundary element program quadratic quadrilateral elements are implemented.

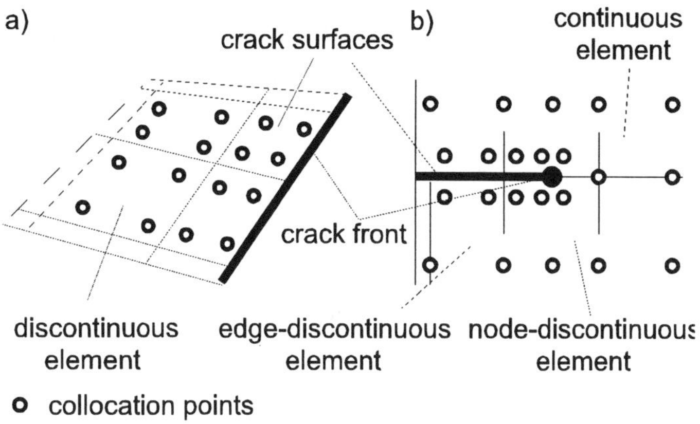

Fig. 5.3. Discretization: a) crack surface - b) edge crack

Although the number of degrees of freedom increases because of the discontinuous elements, it is, however, the most practicable method to deal with hypersingular boundary integral equations within a collocation solution procedure.

Extension to elastoplastic material behaviour

The main point in the extension of *DUAL BEM* to elastoplastic material behaviour is to provide the *dual* boundary integral equations on the crack surfaces. Analogous to elasticity (equation (5.1)), one obtains

$$\frac{1}{2}u_i(\xi^u) + \frac{1}{2}u_i(\xi^t) + \lim_{\epsilon\to 0} \int_{\bar{\Gamma}-\Gamma_\epsilon^u-\Gamma_\epsilon^t} T_{ij}(\xi,x)\,u_j(x)\,d\Gamma(x) =$$

$$\int_{\bar{\Gamma}} U_{ij}(\xi,x)\,t_j(x)\,d\Gamma(x) + \int_{\Omega} \Sigma_{ijk}(\xi,x)\,\varepsilon^p_{jk}(x)\;d\Omega(x) \tag{5.4}$$

for the displacements, because the additional domain integral is only weakly singular (see Section 4.1).

Derivation of the integral identity of the displacement derivatives on the crack surfaces is more complicated, because due to their singularity not only the boundary integrals but also the domain integral must be considered (see equation 4.4). For the boundary integrals on the crack surfaces, the limiting procedure and the regularization is the same as for elastic cases (equation (5.2)). Following equation (4.6), the strongly singular domain integral of each crack surface (see Fig. 5.4) can be transformed into a weakly singular domain integral and into regular and strongly singular cell surface integrals. This leads to the hypersingular formulation of the displacement derivatives

$$\frac{1}{2}\frac{\partial u_i}{\partial \xi_l}(\xi^u) + \frac{1}{2}\frac{\partial u_i}{\partial \xi_l}(\xi^t) = \tag{5.5}$$

$$\lim_{\epsilon \to 0} \left\{ \int\limits_{\bar{\Gamma}-\Gamma_\epsilon^u-\Gamma_\epsilon^t} [U_{ijl} t_j(x) - T_{ijl} u_j(x)]\, d\Gamma(x) - u_j(\xi^u) \frac{b_{ijl}(\xi^u)}{\epsilon} - u_j(\xi^t) \frac{b_{ijl}(\xi^t)}{\epsilon} \right\}$$

$$+ \int\limits_{\Omega_r} \Sigma_{ijkl}(\xi, x)\, \varepsilon^p_{jk}(x)\, d\Omega(x)$$

$$+ \int\limits_{\Omega_s^u} \Sigma_{ijkl} \left[\varepsilon^p_{jk}(x) - \varepsilon^p_{jk}(\xi^u) \right] d\Omega - \varepsilon^p_{jk}(\xi^u) \left[\int\limits_{\Gamma_r^u} \Sigma_{ijk} n_l(x) d\Gamma + \lim_{\epsilon \to 0} \int\limits_{\Gamma_s^u - \Gamma_\epsilon^u} \Sigma_{ijk} n_l(x) d\Gamma \right]$$

$$+ \int\limits_{\Omega_s^t} \Sigma_{ijkl} \left[\varepsilon^p_{jk}(x) - \varepsilon^p_{jk}(\xi^t) \right] d\Omega - \varepsilon^p_{jk}(\xi^t) \left[\int\limits_{\Gamma_r^t} \Sigma_{ijk} n_l(x) d\Gamma + \lim_{\epsilon \to 0} \int\limits_{\Gamma_s^t - \Gamma_\epsilon^t} \Sigma_{ijk} n_l(x) d\Gamma \right]$$

for source points on a crack surface in case of elastoplastic material. With the help of the nonlinear stress-strain-relation (2.4) and multiplication with the normal vector $n_b(\xi^t)$, Equation (5.5) can be transformed into the traction boundary integral equation with respect to Γ^t.

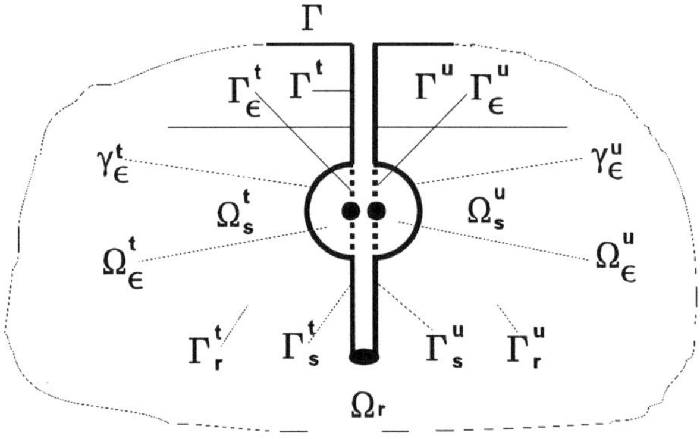

Fig. 5.4. Source point located on a crack surface for an elastoplastic problem

By decoupling of the crack surfaces with the *dual* boundary integral equations and use of discontinuous cells, 3D elastoplastic crack problems can be evaluated within one subregion (see Section 4.2). For 2D problems, this was introduced by Leitao et al. [29] for the first time.

5.2 Evaluation of Fracture Mechanics Parameters

5.2.1 Stress Intensity Factors

For direct evaluation of the stress intensity factors, the numerically determined crack near-field of displacements or stresses respectively is compared with the corresponding asymptotic crack near-field (see (3.1) and (3.2)). In principle, there are two different approaches.

In the *extrapolation method*, *Pseudo-K-values* are computed from the numerically determined stresses and displacements on points along a straight line towards the crack front. From these values, the required K-value on the crack front is extrapolated.

In the *correlation method*, special elements are required describing the $\sqrt{r}$-behaviour of the displacements and the $1/\sqrt{r}$-behaviour of the stresses. From the nodal values of the boundary data on the crack front, the K-values can be identified directly.

In the framework of this project, both methods were realized and tested on various numerical examples [17, 23, 24]. The subregion technique was used to decouple the crack surfaces. For evaluation of the crack near-field, the nodal values of the boundary elements on the crack surfaces and the ligament were used. This method requires very little time, but the drawback is that the element nodes on the crack surfaces and the ligament used for extrapolation must be perpendicular to the crack front. This implies increased discretization especially if crack geometry is complex.

In order to use the extrapolation method independently of the boundary element discretization and in view of *DUAL BEM*, where only the crack surfaces are meshed, the K-values were also evaluated via stresses determined on source points inside the domain.

5.2.2 J-integral

For 3D crack problems, the J-integral was evaluated both for elastic (3.4) and for elastoplastic material behaviour (3.6). Detailed documentation of the results with numerical examples can be found in the DFG reports [23, 24] and the PhD-thesis of Huber [17]. Only the most important aspects are summarized here.

For numerical evaluation of the J integral, the integration plane and the corresponding boundary contour are discretized with elements A^k and Γ^k (see Fig. 3.1), which are parameterized via shape functions. The field quantities to be integrated are also approximated using shape functions. If traction-free and plane cracks are considered, the contour integral on the crack surfaces does not contribute to the J-integral and therefore need not be discretized. For the surface elements around the crack front, singular shape functions are used, for all other elements standard quadratic shape functions.

The required nodal values for stresses and displacement derivatives can be determined by evaluation of source points inside the domain for both elastic and elastoplastic material behaviour. For the nodal values on the ligament, source points inside the domain located close to the boundary are used. For determination of the nodal values on the crack front for the singular shape functions, source points near the crack front are evaluated. Via the singular shape functions, they are re-calculated to provide the values on the crack front.

For elastic cases, the nodal values of the higher derivatives ($u_{i,lb}$, $\sigma_{ij,b}$) can also be determined via evaluation of source points inside the domain with the respective kernels. For plastic cases, this is not yet possible, because hypersingular domain integrals occur. For this reason, difference quotients are used here. The plastic component of the strain energy density can be integrated using the trapezoidal rule for each load step.

With these approximations, each element A^k and Γ^k can be integrated numerically via Gauss-quadrature. Summing up the results, one obtains the J-integral at the point s on the crack front. For the linear elastic case, additionally the J-integral is decomposed into its modes, and the K-values are calculated.

6 Conclusion

In conclusion, there are three main scientific results obtained in the project over the period from 1989 to 1995.

It could be proved [9, 10, 24] that the Boundary Element Method can be used successfully to solve 3D thermo-elastoplastic problems and that it is especially suitable for problems with local plasticity.

By means of subregion technique for decoupling of crack surfaces, the Boundary Element Method could be extended to 3D elastoplastic crack problems [23, 24, 17]. Especially in LEFM, it is an efficient tool for exact determination of fracture mechanics parameters such as stress intensity factors and J-integral. But also in the framework of deformation theory, the J-integral as a fracture parameter in the HRR field can be evaluated efficiently by using the Boundary Element Method.

In order to develop the potential of the Boundary Element Method for numerical simulation of fatigue crack growth, a new technique for crack modelling was applied - the *DUAL BEM*. Apart from detailed incorporation of theoretical basics, a general formulation of the *dual boundary integral equations* on crack surfaces could be developed for elastoplastic material behaviour in 3D. In the framework of LEFM, *DUAL BEM* was realized in terms of a computer code for 3D problems. Its efficiency was demonstrated by a benchmark [25].

References

1. Bausinger, R., Kuhn, G. (1983): Die Boundary-Element-Methode. Expert-Verlag, Ehningen
2. Białecki, R.A., Dallner, R., Kuhn, G. (1994): Minimum distance calculation between a source point and a boundary element. Eng. Anal. Boundary Elements 12, 211-218
3. Białecki, R.A., Kuhn, G., Mews, H. (1990): Replacing standard solver by block solver routines. Interner Bericht Nr. 019 0653 111 49 am Lehrstuhl für Technische Mechanik, Erlangen
4. Białecki, R.A., Merkel, M., Mews, H., Kuhn, G.: In- and Out-of-Core Solver with Parallel and Nonlinear Options. Submitted for publication
5. Blandford, G.E., Ingraffea, A.R., Liggett, J.A. (1981): Two-Dimensional Stress Intensity Factor Computations Using the Boundary Element Method. Int. J. Num. Meth. Eng. 17, 387-404
6. Brebbia, C.A., Telles, J.C.F., Wrobel, L.C. (1984): Boundary Element Techniques - Theory and Applications in Engineering. Springer, Berlin
7. Carpeneter, W.C., Read, D.T., Dodds Jr., R.H. (1986): Comparison of several path independent integrals including plasticity effects. Int. J. Fract. 31, 303-323
8. Cruse, T.A., Wilson, R.B. (1978): Advanced Applications of Boundary-Integral Equation Methods. Nuclear Eng. Design 46, 223-234
9. Dallner, R. (1993): Behandlung inelastischer 3D-Probleme mittels Randelementmethode. VDI Fortschrittberichte Reihe 18: Mechanik/Bruchmechanik Nr. 126. VDI-Verlag, Düsseldorf
10. Dallner, R., Kuhn, G. (1993): Efficient evaluation of volume integrals in the boundary element method. Comp. Meth. Appl. Mech. Engng. 109, 95-109
11. Foerster, A., Kuhn, G. (1994): A Field Boundary Element Formulation for Material Nonlinear Problems at Finite Strains. Int. J. Solids Structures 31 No. 12/13, 1777-1792
12. Gray, L.J., Giles, G.E. (1988): Application of the Thin Cavity Method to Shield Calculations in Electroplating. In: Proc. of 10th Int. Conf. on the Boundary Element Method (2), Southampton
13. Gray, L.J., Martha, L.F., Ingraffea, A.R. (1990): Hypersingular integrals in boundary element fracture analysis. Int. J. Num. Meth. Eng. 29, 1135-1185
14. Guiggiani, M., Krishnasamy, G., Rudolphi, T.J., Rizzo, F.J. (1992): A general algorithm for the numerical solution of hypersingular boundary integral equations. ASME J. Appl. Mech. 59, 604-614
15. Herding, U., Kuhn, G.: A Field-Boundary Element Formulation for Damage Mechanics. Submitted for publication.
16. Hong, H., Chen, J. (1988): Derivations of Integral Equations of Elasticity. Journ. Engng. Mech. 114 (6), 1028-1044
17. Huber, O. (1994): Behandlung elastoplastischer 3D-Rißprobleme mittels der Randelementmethode. VDI Fortschrittberichte Reihe 18: Mechanik/ Bruchmechanik Nr. 153. VDI-Verlag, Düsseldorf
18. Huber, O., Dallner, R., Partheymüller, P., Kuhn, G.: Evaluation of the stress tensor in 3D elastoplasticity by direct solving of hypersingular integrals. Accepted for publication in Int. J. Num. Meth. Eng.
19. Huber, O., Lang, A., Kuhn, G. (1993): Evaluation of the Stress Tensor in 3D Elastostatics by Direct Solving of Hypersingular Integrals. Comp. Mech. 12, 39-50
20. Huber, O., Nickel, J., Kuhn, G. (1993): On the Decomposition of the J-Integral for 3D Crack Problems. Int. J. Fract. 64, 339-348

21. Hutchinson, J.W. (1968): Singular behaviour at the end of a tensile crack in a hardening material. J. Mech. Phys. Solids. 16, 13-31
22. Kuhn, G. (1981):' Numerische Behandlung von Mehrfachrissen in ebenen Scheiben. ZAMM 61, T105-T106
23. Kuhn, G., Huber, O. (1991): Behandlung elastoplastischer Kerb- und Rißprobleme mittels Randelementmethode. 1. Zwischenbericht zum DFG-Projekt: Ku 360/8, Lehrstuhl für Technische Mechanik, Erlangen
24. Kuhn, G., Huber, O. (1993): Behandlung elastoplastischer Kerb- und Rißprobleme mittels Randelementmethode. 2. Zwischenbericht zum DFG-Projekt: Ku 360/8, Lehrstuhl für Technische Mechanik, Erlangen
25. Kuhn, G., Huber, O., Dallner, R., Merkel, M., Partheymüller, P. (1995): Behandlung elastoplastischer Kerb- und Rißproblem mittels Randelementmethode. Abschlußbericht zum DFG-Projekt: Ku 360/8, Lehrstuhl für Technische Mechanik, Erlangen
26. Kuhn, G., Neureiter, W., Drexler, W., Möhrmann, W. (1982): Theoretische Grundlagen BETSY und Benutzerhandbuch. Forschungsberichte der Forschungsvereinigung Verbrennungskraftmaschinen e.V. Teil I - III (Heft 310-1/2/3), Frankfurt
27. Lachat, J.C., Watson, J.O. (1976): Effective numerical treatment of boundary integral equations: a formulation for three-dimensional elastostatics. Int. J. Num. Meth. Eng. 10, 991-1005
28. Ishikawa, H., Kitagawa, H., Okamura, H. (1979): J-integral of mixed mode crack and its application. In: 3rd Int. Conf. on mechanical behaviour of materials (3), Cambridge
29. Leitao, V., Aliabadi, M.H., Rooke, D.P. (1995): The Dual Boundary Element Formulation for Elastoplastic Fracture Mechanics. Int. J. Numer. Meth. Engn. 38, 315-333
30. Luchi, M.L., Poggialini, A. (1983): Computation of 3-dimensional stress intensity factors using special boundary elements. In: 5th Int. Conf. on Boundary Element Methods, Berlin
31. Mews, H. (1988): Berechnung von Spannungsintensitätsfaktoren mittels der Boundary-Element-Methode unter Verwendung von Greenschen Funktionen. Diss. Universität Erlangen-Nürnberg
32. Mi, Y., Aliabadi, M.H. (1992): Dual Boundary Element Method for Three-dimensional Fracture Mechanics Analysis. Eng. Anal. Boundary Elements 10, 161-171
33. Miyamoto, H., Kikuchi, M. (1980): Three-dimensional J-Integral. Theor. Appl. Mech. 28, 195-204
34. Nikishkov, G.P., Atluri, S.N. (1987): Calculation of fracture mechanics parameters for an arbitrary three-dimensional crack, by the 'equivalent domain integral' method. Int. J. Num. Meth. Eng. 24, 1801-1821
35. Partheymüller, P., Białecki, R.A., Kuhn, G. (1994): Self-adapting algorithm for evaluation of weakly singular integrals arising in the boundary element method. Eng. Anal. Boundary Elements 14, 285-292
36. Portela, A., Aliabadi, M.H., Rooke, D.P. (1992): The Dual Boundary Element Method: Effective Implementation for Crack Problems. Int. J. Numer. Meth. Engn. 33, 1269-1287
37. Rice, J.R., Rosengren, G.F. (1968): Plain strain deformation near a crack tip in a power-law hardening material. J. Mech. Phys. Solids. 16, 1-12
38. Rußwurm, S. (1992): Berechnung elastoplastischer Rißprobleme mittels der Randelementmethode. VDI-Fortschrittsberichte Reihe 18: Mechanik/ Bruchmechanik Nr. 104. VDI-Verlag, Düsseldorf

39. Sippel, H. (1988): Elastoplastische Berechnungen mittels der Boundary-Element-Methode. Diss. Universität Erlangen-Nürnberg
40. Snyder, M.D., Cruse, T.A. (1975): Boundary Integral Equation Analysis of Cracked Anisotropic Plates. Int. J. Fract. 11, 315-328
41. Telles, J.C.F., Castor, G.S., Guimaraes, S. (1995): A Numerical Green's Function Approach for Boundary Elements Applied to Fracture Mechanics. Int. J. Num. Meth. Eng. 38, 3259-3274
42. Watson, J.O. (1984): Hermitian Cubic Boundary Elements for the Analysis of Cracks of Arbitrary Geometry. In: Advanced Boundary Element Methods. Springer

Adaptive Domain Decomposition Methods for Finite and Boundary Element Equations

G. Haase, B. Heise, M. Kuhn, and U. Langer

1 Introduction

The use of the FEM and BEM in different subdomains of a non–overlapping Domain Decomposition (DD) and their coupling over the coupling boundaries (interfaces) brings about several advantages in many practical applications. The paper presents parallel solvers for large-scaled coupled FE–BE–DD equations approximating linear and nonlinear plane magnetic field problems as well as plane linear elasticity problems. The parallel algorithms presented are of asymptotically optimal, or, at least, almost optimal complexity and of high parallel efficiency.

The Domain Decomposition (DD) approach offers many opportunities to marry the advantages of the Finite Element Method (FEM) to those of the Boundary Element Method (BEM) in many practical applications. For instance, in the magnetic field computation for electric motors, we can use the BEM in the air subdomains including the exterior of the motor more successfully than the FEM which is prefered in ferromagnetic materials where non–linearities can occur in the partial differential equation (PDE), or in subdomains where the right–hand side does not vanish [23, 34]. The same is true for many problems in solid mechanics [31] and in other areas of research. A very straightforward and promising technique for the coupling of FEM and BEM was proposed by M. Costabel [7] and others [6, 33] In the different subdomains of a non–overlapping domain decomposition, we use either the standard finite element (FE) Galerkin method or a mixed–type boundary element (BE) Galerkin method which are weakly coupled over the coupling boundaries (interfaces) Γ_C. The mixed BE Galerkin method makes use of the full Cauchy data representation on the BE subdomain boundaries via the Calderón projector.

The main aim of the project was the design, analysis and implementation of fast and well adapted parallel solvers for large-scale coupled FE/BE–equations approximating plane, linear and nonlinear magnetic field problems including technical magnetic field problems (e.g. electric motors). To be specific, we consider a characteristical cross-section, which lies in the (x, y)-plane of the $\mathbf{R}^3$, of the original electromagnetic device that is to be modelled. Let us assume that $\Omega_0 \subset \mathbf{R}^2$ is a bounded simply connected domain and that homogeneous Dirichlet boundary conditions are given on $\Gamma_D = \partial\Omega_0$. Formally the nonlinear magnetic field problem can be written as follows [25]

$$-\text{div}\ (\nu(x,|\nabla u(x)|)\nabla u(x)) = \text{S}(x) + \frac{\partial \text{H}_{0\text{y}}(x)}{\partial \text{x}} - \frac{\partial \text{H}_{0\text{x}}(x)}{\partial \text{y}}, \quad x \in \Omega \qquad (1.1)$$

$$u(x) = 0, \quad x \in \Gamma_D \qquad (1.2)$$

$$|u(x)| \to 0 \quad \text{for } |x| \to \infty, \qquad (1.3)$$

with $\Omega := R^2 \setminus \bar{\Omega}_0$. The solution u is the z–component A_z of the vector potential $\mathbf{A} = (A_x, A_y, A_z)^T$ introduced in the Maxwell equation. The component of the current density, which acts orthogonal to the cross-section being considered, is represented by $\text{S}(x)$, whereas $\text{H}_{0\text{x}}$ and $\text{H}_{0\text{y}}$ stand for sources associated with permanent magnets that may occur and $\nu(.)$ denotes a coefficient depending on the material and on the gradient $|\nabla u(x)|$ (induction). Now, we introduce the exterior domain Ω_+ by defining a, so called, coupling boundary $\Gamma_+ := \partial\Omega_+$. The definition of Γ_+ is restricted by the conditions

$$\nu(x) = \nu_p \ \forall x \in \Omega_+, \quad (\text{supp S} \cup \text{supp H}_0) \subset \bar{\Omega}_-, \quad \text{diam}(\bar{\Omega}_0 \cup \bar{\Omega}_-) < 1, \quad (1.4)$$

where $\Omega_- := R^2 \setminus (\bar{\Omega}_+ \cup \bar{\Omega}_0)$. Note that the condition $\text{diam}(\bar{\Omega}_0 \cup \bar{\Omega}_-) < 1$ is only technical and can be fulfilled by scaling the problem appropriately. Besides the decomposition $\bar{\Omega} = \bar{\Omega}_- \cup \bar{\Omega}_+$, we allow the inner domain Ω_- to be decomposed further following the natural decomposition of Ω_- according to the change of data:

$$\bar{\Omega}_- = \bigcup_{j=1}^{N_M} \bar{\hat{\Omega}}_j, \quad \text{with} \quad \hat{\Omega}_i \cap \hat{\Omega}_j = \emptyset \quad \forall i \neq j. \qquad (1.5)$$

In Section 2 (see also Appendix A), we present an automatic and adaptive domain decomposition procedure providing such a decomposition of Ω into p subdomains (p = number of processors to be used) and such controlling data for the distributed mesh generator [11] that we can expect a well load-balanced performance of our solver.

In Section 3, we consider linear plane magnetic field problems for which a domain decomposition according to Section 2 is available. Now, we can make use of the advantages of a mixed variational DD FE/BE discretization and propose an algorithm for solving the linear coupled FE/BE equations. First of all, the coupled FE/BE equations can be reformulated as a linear system with a symmetric, but indefinite system matrix. We provide a preconditioning and a parallelization of Bramble/Pasciak's Conjugate Gradient (CG) method [2] applied to the symmetric and indefinite system (3.12). The components of the preconditioner can be chosen such that the resulting algorithm is, at least, almost asymptotically optimal with respect to the operation count and quite robust with respect to complicated geometries, jumping coefficients and mesh grading near singularities (see numerical results given in Sect. 3 and in [30]). Using a special DD data distribution, we parallelize the preconditioning equation and the remaining algorithm in such a way that the same amount of communication is needed as in the earlier introduced and well studied parallel

PCG for solving symmetric and positive definite FE equations [17, 18] (see Appendix B).

Section 4 is devoted to the description of the Full-DD-Newton–Solver for nonlinear magnetic field problems. In every nested Newton step we use basically the linear DD–solver given in Section 3.

In Section 5, we apply our linear DD–solver to plane linear elasticity problems modelled by Lamè's system of PDEs. An appropriate adaptation of the components of the DD–preconditioner results in a parallel solver the efficiency of which is comparable to that of the solver for the linear magnetic field problems described in Section 3.

In Section 6, we briefly describe the software package FEM⊕BEM [14] and draw some conclusions. All numerical results presented in this paper were obtained by the use of the package FEM⊕BEM. The code runs on various parallel computers and programming platforms including PVM (see, e.g., [28]).

2 Adaptive Domain Decomposition Preprocessing

2.1 The DD-Data Partitioning

In this section, we focus our interest on how a decomposition of the domain Ω into a given number of subdomains can be obtained from the natural decomposition into domains according to the change of materials (1.5). We are interested in well load-balanced decompositions especially in the case of discretizations which are adapted to singularities.

We assume that a triangular-based description of the geometry of the problem under consideration is given. Besides the geometrical data each triangle is characterized by a parameter pointing to that of the N_M material-regions the triangle belongs to. Note that interfaces between different materials, i.e. the boundaries of the $\hat{\Omega}_j$'s (cf. (1.5)), are represented by edges of the triangulation. We are interested in a decomposition into $p \geq N_M$ subdomains

$$\bar{\Omega} = \bigcup_{i \in \mathcal{I}} \bar{\Omega}_i, \quad \text{where} \quad \Omega_p := \Omega_+ \quad \text{and} \quad \bar{\hat{\Omega}}_j = \bigcup_{i \in \mathcal{I}_j} \bar{\Omega}_i \quad \forall j = 1, \ldots, N_M \tag{2.1}$$

where the sets of indices are given by $\mathcal{I} := \{1, \ldots, p\}$ and

$$\mathcal{I}_j \subset \mathcal{I}^\star := \{1, \ldots, p-1\}, \quad \bigcup_{j=1}^{N_M} \mathcal{I}_j = \mathcal{I}^\star, \quad \mathcal{I}_j \cap \mathcal{I}_k = \emptyset \quad \forall j \neq k,$$

i.e., the subdomains $\hat{\Omega}_j$ determined by the materials may be decomposed further (see, e.g., [19]). We assume that there exist open balls $B_{\underline{r}_i}$ and $B_{\overline{r}_i}$ ($i \in \mathcal{I}^\star$) with positive radii $\underline{r}_i$ and $\overline{r}_i$, such that $B_{\underline{r}_i} \subset \Omega_i \subset B_{\overline{r}_i}$ and $0 < \underline{c} \leq \overline{r}_i / \underline{r}_i \leq \overline{c} \quad \forall i \in \mathcal{I}^\star$ with fixed (i-independent) constants $\underline{c}$ and $\overline{c}$. Note, in the case of Ω being bounded we had $\mathcal{I}^\star := \{1, \ldots, p\}$ and in the following all terms induced by p, which then stands for the exterior domain, would vanish.

Although the algorithm being used for decomposing Ω is based upon a given triangulation of the domain it is of advantage to use a special DD-data structure as input for DD-based algorithms running on massively parallel computers. Thus, starting off with a given triangular-based geometrical description ($*$.*tri*–file) we wish to end up with a well-balanced decomposition of our problem which is described by using some DD-data format ($*$.*dd*–file). Figure 2.1 shows the interactions between the preprocessing codes *Decomp, Tri2DD* and *AdapMesh* and the file-types $*$.*tri*, $*$.*dd*, and $*$.*fb* being involved.

In the simplest case the process starts, on the left of the diagram, with applying *Decomp* to a $*$.*tri*–file which results in a decomposition as defined in (2.1), i.e. each triangle is assigned to one of the Ω_i's, $i \in \mathcal{I}$. The output of this process is also a $*$.*tri*–file which then is converted into a $*$.*dd*–file by applying the program *Tri2DD.* In our case, such a $*$.*dd*–file is the input for the parallel code FEM⊘BEM [14]. Because of the simple structure of the DD-data format being used it is quite easy to implement additional refinement information concerning, e.g., singularities into the $*$.*dd*–file. In the latter case the mesh created from this file may differ significantly from the one described by the original $*$.*tri*–file. As a consequence we have to expect a bad load balance. This problem can be solved by applying the program *AdapMesh* which simulates the mesh generation as it occurs in the parallel program using, optionally, adaptivity information which may be obtained from a coarse-grid computation. Thus, *AdapMesh* creates a $*$.*tri*–file which is the input for restarting the cycle with *Decomp.* Note, in the optimal case with respect to the load balance the mesh used for computations (i.e. the one created from a $*$.*dd*–file within FEM⊘BEM) would coincide with the mesh being used for the decomposition.

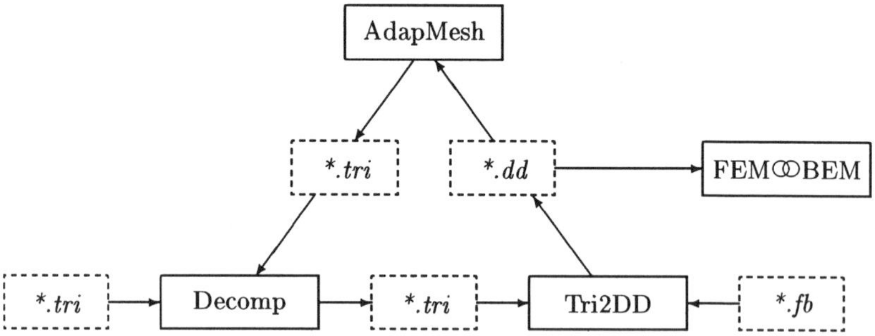

Fig. 2.1. The components of the preprocessing.

2.2 A Short Description of the Preprocessing Codes

At this place, we are going to explain the codes and the main ideas they are based on. More information and technical details can be found in the documentation [12]. First we give a short description of the codes.

Decomp decomposes single-material domains using the spectral bisection method (sbm) [42]. That is, as long as there are less than p subdomains the largest subdomain according to the number of triangles is divided into two new subdomains by the sbm. As a result, each triangle is assigned to one of the subdomains.

Tri2DD converts triangular-based data into the DD-data format. The algorithm is based on the definition of edges which then define the faces. Note, interfaces between different materials will be maintained as they were given in the original $*$.*tri*-file. On the other hand the artificially created boundaries within one material are smoothed.

AdapMesh creates a mesh based on DD-data using, optionally, adaptivity information. The resulting $*$.*tri*-file can be used as input for *Decomp.*

During the preprocessing we are concerned with two types of describing data. The first one ($*$.*tri*-files) is based on nodes (characterized by their coordinates) which define the edges (straight lines or arcs of circles are allowed) and, finally, triangles defined by their edges. Each of the triangles is characterized by two additional parameters, where only one of them pointing to the material the triangle belongs to has to be initialized from the very beginning. The second one describes the mapping of the triangles on the subdomains and it is defined as a result of decomposing Ω.

The second type of data ($*$.*dd*-files) follows the DD-data format described in [14]. It is based on defining cross-points numbered globally which then define edges. Main objects are faces described by edges. The faces, or a union of them, are mapped onto the array of processors. The $*$.*fb*-files are auxiliary files (optional) and contain controlling data or fixed cross-points as it occurs in our example (see below).

2.3 Example: Direct Current Motor

Now we apply our preprocessing tools to a model of a direct current motor (dc-motor) with permanent excitation (see also Fig. 2 and 3 in [24] for a detailed description of the machine). We start with a $*$.*tri*-file with 18 different material regions as shown on the left in Figure A.1 (Appendix A) . After applying *Decomp, Tri2DD* and one full cycle we have obtained the $*$.*dd*-file which represents the decomposition as shown on the right in Figure A.1. Note the additional cross-points especially on the outer circle which have been pre-defined in a $*$.*fb*–file. In our case no additional refinement information have been used so that the second decomposition (Figure A.1) did not differ significantly from the first one which were already obtained after applying

Decomp and *Tri2DD*. Figure A.2 (Appendix A) shows, on the left, the mesh being finally used as initial mesh for computations and, on the right, the equipotential lines of the solution.

3 Parallel Solution of Linear Coupled BE/FE–Equations Approximating Linear Magnetic Field Problems

3.1 A Mixed Variational Formulation

Let us first consider a linear ($\nu = \nu(x)$) magnetic field problem of the form (1.1) - (1.3) for which a domain decomposition according to Section 2 is available. In particular, we assume that the index set $\mathcal{I} = \mathcal{I}_F \cup \mathcal{I}_B$ can be decomposed into two disjoint sets of indices $\mathcal{I}_F$ and $\mathcal{I}_B$ such that

$$p \in \mathcal{I}_B, \tag{3.1}$$

$$(\mathrm{supp}\ \mathrm{S}(.) \cup \mathrm{supp}\ \mathrm{H}_0(.)) \cap \Omega_i = \emptyset \quad \forall i \in \mathcal{I}_B, \tag{3.2}$$

$$\nu(x) = \nu_i = const \quad \forall i \in \mathcal{I}_B. \tag{3.3}$$

For each Ω_i $(i \in \mathcal{I})$ the index i belongs to one of the two index sets $\mathcal{I}_B$ and $\mathcal{I}_F$ according to the discretization method applied to Ω_i, where $\mathcal{I}_B$ and $\mathcal{I}_F$ stand for BEM and FEM, respectively.

Following M.Costabel [7], G.C.Hsiao and W.L.Wendland [33] and others, we can rewrite the weak formulation of the boundary value problem (1.1) - (1.3) by means of partial integration in the boundary element subdomains Ω_i, $i \in \mathcal{I}_B$ and by the use of Calderón's representation of the full Cauchy data as a mixed DD coupled domain and boundary integral variational problem: Find $(\lambda, u) \in \mathbf{V} := \Lambda \times \mathbf{U}_0$ such that

$$a(\lambda, u; \eta, v) = \langle F, v \rangle \quad \forall (\eta, v) \in \mathbf{V} \tag{3.4}$$

where

$$\begin{aligned}
a(\lambda, u; \eta, v) &:= a_B(\lambda, u; \eta, v) + a_F(u, v) \\
a_B(\lambda, u; \eta, v) &:= \sum_{i \in \mathcal{I}_B \setminus \{p\}} \nu_i \Big\{ \langle \mathcal{D}_i u_i, v_i \rangle_{\Gamma_i} + \frac{1}{2} \langle \lambda_i, v_i \rangle_{\Gamma_i} + \langle \lambda_i, \mathcal{K}_i v_i \rangle_{\Gamma_i} + \\
&\qquad \langle \eta_i, \mathcal{V}_i \lambda_i \rangle_{\Gamma_i} - \langle \eta_i, \mathcal{K}_i u_i \rangle_{\Gamma_i} - \frac{1}{2} \langle \eta_i, u_i \rangle_{\Gamma_i} \Big\} \\
&\qquad + \nu_p \Big\{ \langle \mathcal{D}_p u_p, v_p \rangle_{\Gamma_+} - \frac{1}{2} \langle \lambda_p, v_p \rangle_{\Gamma_+} + \langle \lambda_p, \mathcal{K}_p v_p \rangle_{\Gamma_+} + \\
&\qquad \langle \eta_p, \mathcal{V}_p \lambda_p \rangle_{\Gamma_+} - \langle \eta_p, \mathcal{K}_p u_p \rangle_{\Gamma_+} + \frac{1}{2} \langle \eta_p, u_p \rangle_{\Gamma_+} \Big\}
\end{aligned}$$

$$a_F(u,v) := \sum_{i\in\mathcal{I}_F} \int_{\Omega_i} \nu(x)\nabla^T u(x)\nabla v(x)\,dx$$

$$\langle F,v\rangle := \sum_{i\in\mathcal{I}_F} \int_{\Omega_i} \left[\mathrm{S}(x)v(x) - \mathrm{H}_{0\mathrm{y}}(x)\frac{\partial v(x)}{\partial \mathrm{x}} + \mathrm{H}_{0\mathrm{x}}(x)\frac{\partial v(x)}{\partial \mathrm{y}}\right] dx$$

$$\langle\lambda_i, v_i\rangle_{\Gamma_i} := \int_{\Gamma_i} \lambda_i v_i\,ds \text{ and } v_i = v|_{\partial\Omega_i},\ u_i = u|_{\partial\Omega_i},\ \Gamma_i := \partial\Omega_i,$$

with the well-known boundary integral operators $\mathcal{V}_i, \mathcal{K}_i, \mathcal{D}_i$ defined by the relation

$$\begin{aligned}
\mathcal{V}_i\lambda_i(x) &:= \int_{\Gamma_i} \mathcal{E}(x,y)\lambda_i(y)\,ds_y\\
\mathcal{K}_i v_i(x) &:= \int_{\Gamma_i} \partial_y\mathcal{E}(x,y)v_i(y)\,ds_y\\
\mathcal{D}_i u_i(x) &:= -\partial_x \int_{\Gamma_i} \partial_y\mathcal{E}(x,y)u_i(y)\,ds_y
\end{aligned} \tag{3.5}$$

and with the fundamental solution

$$\mathcal{E}(x,y) = -\frac{1}{2\pi}\log|x-y| \tag{3.6}$$

of the Laplacian. The mapping properties of the boundary integral operators (3.5) on Sobolev spaces are now well known [8]. The spaces $\mathbf{U}_0$ and Λ are defined by the relations

$$\begin{aligned}
\mathbf{U}_0 &:= \{u \in \mathrm{H}^1(\Omega_-) : u|_{\Gamma_{BE}} \in \mathrm{H}^{1/2}(\Gamma_{BE}), u|_{\partial\Omega_0} = 0\}\\
\Lambda &:= \{\lambda = (\lambda_i)_{i\in\mathcal{I}_B} : \lambda_i \in \mathrm{H}^{-1/2}(\Gamma_i),\ i \in \mathcal{I}_B\} = \textstyle\prod_{i\in\mathcal{I}_B}\Lambda_i,
\end{aligned} \tag{3.7}$$

with $\Lambda_i = \mathrm{H}^{-1/2}(\Gamma_i)$, $i \in \mathcal{I}_B$. Further we use the notation $\Gamma_{BE} := \bigcup_{i\in\mathcal{I}_B}\partial\Omega_i \setminus \Gamma_D, \Gamma_{FE} := \bigcup_{i\in\mathcal{I}_F}\partial\Omega_i \setminus \Gamma_D, \Gamma_C := \Gamma_{BE}\cup\Gamma_{FE}$ and $\Omega_F := \bigcup_{i\in\mathcal{I}_F}\Omega_i$. Introducing in $\mathbf{V} := \Lambda \times \mathbf{U}_0$ the norm

$$\|(\lambda,u)\|_{\mathbf{V}} := \left(\|\lambda\|_\Lambda^2 + \|u|_{\Gamma_{BE}}\|^2_{\mathrm{H}^{1/2}(\Gamma_{BE})} + \|u\|^2_{\mathrm{H}^1(\Omega_F)}\right)^{1/2} \tag{3.8}$$

with

$$\|\lambda\|_\Lambda^2 = \sum_{i\in\mathcal{I}_B}\|\lambda_i\|^2_{\mathrm{H}^{-1/2}(\Gamma_i)} \quad\text{and}\quad \|u|_{\Gamma_{BE}}\|^2_{\mathrm{H}^{1/2}(\Gamma_{BE})} = \sum_{i\in\mathcal{I}_B}\|u_i\|^2_{\mathrm{H}^{1/2}(\Gamma_i)},$$

then one can prove that the bilinear form a(.,.) is $\mathbf{V}$–elliptic and $\mathbf{V}$–bounded provided that the domain decomposition satisfies the conditions imposed on above (see also [33]). Therefore, the existence and uniqueness of the solution are a direct consequence of the Lax–Milgram theorem.

3.2 The Coupled BE/FE Discretization

Now, we can define the nodal FE/BE basis of piecewise linear trial functions based upon a regular triangulation of the subdomains $\Omega_i, i \in \mathcal{I}_F$ and the according discretization of the boundary pieces $\Gamma_{ij} = \bar{\Omega}_i \cap \bar{\Omega}_j$, $i, j \in \mathcal{I}$:

$$\Phi = [\phi_1, \ldots, \phi_{N_\Lambda}, \phi_{N_\Lambda+1}, \ldots, \phi_{N_\Lambda+N_C}, \phi_{N_\Lambda+N_C+1}, \ldots, \phi_N],$$

where $N = N_\Lambda + N_C + N_I$ and $N_I = \sum_{i\in\mathcal{I}_F} N_{I,i}$, $N_\Lambda = \sum_{i\in\mathcal{I}_B} N_{\Lambda,i}$. Here, $\phi_1, \ldots, \phi_{N_\Lambda}$ are the basis functions for approximating λ on Γ_i, $i \in \mathcal{I}_B$, $\phi_{N_\Lambda+1}, \ldots, \phi_{N_\Lambda+N_C}$ represent u on Γ_C and $\phi_{N_\Lambda+N_C+1}, \ldots, \phi_{N_\Lambda+N_C+N_I}$ approximate u in Ω_i, $i \in \mathcal{I}_F$. The definition of the finite dimensional subspaces of $\Lambda, \mathbf{U}_0$ and $\mathbf{V}$

$$\begin{aligned}
\Lambda_h &:= \text{span } [\phi_1, \phi_2, \ldots, \phi_{N_\Lambda}], \\
\mathbf{U}_h &:= \text{span } [\phi_{N_\Lambda+1}, \ldots, \phi_{N_\Lambda+N_C}, \phi_{N_\Lambda+N_C+1}, \ldots, \phi_N], \\
\mathbf{V}_h &:= \Lambda_h \times \mathbf{U}_h
\end{aligned}$$

allows us to formulate the discrete problem as follows: Find $u_h \in \mathbf{V}_h$ such that

$$a(u_h, v_h) = \langle F, v_h \rangle \quad \forall v_h \in \mathbf{V}_h. \tag{3.9}$$

The isomorphism $\Phi : \mathbf{R}^N \to \mathbf{V}_h$ leads to the linear system:

$$\begin{pmatrix} K_\Lambda & -K_{\Lambda C} & 0 \\ K_{C\Lambda} & K_C & K_{CI} \\ 0 & K_{IC} & K_I \end{pmatrix} \begin{pmatrix} \mathbf{u}_\Lambda \\ \mathbf{u}_C \\ \mathbf{u}_I \end{pmatrix} = \begin{pmatrix} \mathbf{f}_\Lambda \\ \mathbf{f}_C \\ \mathbf{f}_I \end{pmatrix}, \tag{3.10}$$

where the block entries are defined by

$$\begin{aligned}
(K_\Lambda \mathbf{u}_\Lambda, \mathbf{v}_\Lambda) &= \sum_{i\in\mathcal{I}_B} \nu_i \langle \eta_i, \mathcal{V}_i \lambda_i \rangle_{\Gamma_i} \qquad \text{with } \lambda_i = \Phi_{\Lambda_i} \mathbf{u}_{\Lambda_i}, \eta_i = \Phi_{\Lambda_i} \mathbf{v}_{\Lambda_i}, \\
(K_{C\Lambda} \mathbf{u}_\Lambda, \mathbf{v}_C) &= \sum_{i\in\mathcal{I}_B\setminus\{p\}} \nu_i \{ \langle \lambda_i, \mathcal{K}_i v_i \rangle_{\Gamma_i} + \frac{1}{2} \langle \lambda_i, v_i \rangle_{\Gamma_i} \} \\
&\quad + \nu_p \{ \langle \lambda_p, \mathcal{K}_p v_p \rangle_{\Gamma_p} - \frac{1}{2} \langle \lambda_p, v_p \rangle_{\Gamma_p} \} \\
K_{\Lambda C} &= K_{C\Lambda}^T \\
K_C &= K_{CB} + K_{CF}, \text{ with} \\
(K_{CB} \mathbf{u}_C, \mathbf{v}_C) &= \sum_{i\in\mathcal{I}_B} \nu_i \langle \mathcal{D}_i u_i, v_i \rangle_{\Gamma_i}, \; u_i = \Phi_{C_i} \mathbf{u}_{C_i}, \; v_i = \Phi_{C_i} \mathbf{v}_{C_i} \text{ and} \\
\left(\begin{pmatrix} K_{CF} & K_{CI} \\ K_{IC} & K_I \end{pmatrix} \begin{pmatrix} \mathbf{u}_C \\ \mathbf{u}_I \end{pmatrix}, \begin{pmatrix} \mathbf{v}_C \\ \mathbf{v}_I \end{pmatrix} \right) &= \sum_{i\in\mathcal{I}_F} \int_{\Omega_i} \nu(x) \nabla^T u \, \nabla v \, dx,
\end{aligned}$$

where $u|_{\Omega_F} = \Phi_F \mathbf{u}_F$, $v|_{\Omega_F} = \Phi_F \mathbf{v}_F$. Here, Φ_{Λ_i} $(i \in \mathcal{I}_B)$ and Φ_{C_i} $(i \in \mathcal{I})$ contain the basis functions for approximating λ and u on $\partial\Omega_i$, respectively. The basis functions in Φ_F are used to approximate u in Ω_i $(i \in \mathcal{I}_F)$. The FE entries, especially K_I, are sparse matrices, whereas the BE blocks are fully populated.

3.3 The Parallel Solver

3.3.1 Bramble-Pasciak's Transformation and Spectral Equivalence Results

The nonsymmetric, positive definite system (3.10) can be solved approximately by Bramble/Pasciak's CG method [2]. The method requires a preconditioner C_Λ which can be inverted easily and which fulfills the spectral equivalence inequalities

$$\underline{\gamma}_\Lambda C_\Lambda \leq K_\Lambda \leq \overline{\gamma}_\Lambda C_\Lambda, \quad \text{with} \quad \underline{\gamma}_\Lambda > 1. \tag{3.11}$$

With the definitions

$$K_1 = K_\Lambda, \quad \mathbf{f}_1 = \mathbf{f}_\Lambda, \quad K_{12} = K_{21}^T = (-K_{\Lambda C} \quad 0)$$
$$K_2 = \begin{pmatrix} K_C & K_{CI} \\ K_{IC} & K_I \end{pmatrix}, \quad \mathbf{f}_2 = \begin{pmatrix} -\mathbf{f}_C \\ -\mathbf{f}_I \end{pmatrix}.$$

we can reformulate (3.10) as a symmetric but indefinite system:

$$\begin{pmatrix} K_1 & K_{12} \\ K_{21} & -K_2 \end{pmatrix} \begin{pmatrix} \mathbf{u}_1 \\ \mathbf{u}_2 \end{pmatrix} = \begin{pmatrix} \mathbf{f}_1 \\ \mathbf{f}_2 \end{pmatrix}. \tag{3.12}$$

Following Bramble and Pasciak [2] this system can be transformed into

$$G\,\mathbf{u} = \mathbf{p}, \text{ where} \tag{3.13}$$

$$G := \begin{pmatrix} C_\Lambda^{-1} K_1 & C_\Lambda^{-1} K_{12} \\ K_{21} C_\Lambda^{-1}(K_1 - C_\Lambda) & K_2 + K_{21} C_\Lambda^{-1} K_{12} \end{pmatrix}, \quad \mathbf{p} = \begin{pmatrix} C_\Lambda^{-1}\mathbf{f}_1 \\ K_{21} C_\Lambda^{-1}\mathbf{f}_1 - \mathbf{f}_2 \end{pmatrix}.$$

Then, the matrix G is self-adjoint and positive definite with respect to the scalar product $[.,.]$ which is defined by

$$[\mathbf{w}, \mathbf{v}] := ((K_1 - C_\Lambda)\,\mathbf{w}_1, \mathbf{v}_1) + (\mathbf{w}_2, \mathbf{v}_2). \tag{3.14}$$

Moreover, G is spectrally equivalent to the regularisator R, where

$$R := \begin{pmatrix} I & 0 \\ 0 & K_2 + K_{21} K_1^{-1} K_{12} \end{pmatrix}.$$

Bramble and Pasciak [2] proved the spectral equivalence inequalities

$$\underline{\lambda}[R\mathbf{v}, \mathbf{v}] \leq [G\mathbf{v}, \mathbf{v}] \leq \overline{\lambda}[R\mathbf{v}, \mathbf{v}] \quad \forall \mathbf{v} \in \mathbf{R}^N, \tag{3.15}$$

where

$$\underline{\lambda} = \left(1 + \frac{\alpha}{2} + \sqrt{\alpha + \frac{\alpha^2}{4}}\right)^{-1} \quad \text{and} \quad \overline{\lambda} = \frac{1 + \sqrt{\alpha}}{1 - \alpha} \tag{3.16}$$

with $\alpha = 1 - (1/\overline{\gamma}_\Lambda)$. Thus, we have to find a preconditioner C_2 for the matrix

$$K_2 + K_{21}K_1^{-1}K_{12} = \begin{pmatrix} K_C + K_{C\Lambda}K_\Lambda^{-1}K_{\Lambda C} & K_{CI} \\ K_{IC} & K_I \end{pmatrix}. \tag{3.17}$$

The DD preconditioner defined by

$$C_2 = \begin{pmatrix} I_C & K_{CI}B_I^{-T} \\ 0 & I_I \end{pmatrix} \begin{pmatrix} C_C & 0 \\ 0 & C_I \end{pmatrix} \begin{pmatrix} I_C & 0 \\ B_I^{-1}K_{IC} & I_I \end{pmatrix} \tag{3.18}$$

is spectrally equivalent to $K_2 + K_{21}K_1^{-1}K_{12}$ if we have preconditioners C_I and C_C fulfilling the inequalities

$$\underline{\gamma}_C C_C \leq \tilde{S}_C + K_{C\Lambda}K_\Lambda^{-1}K_{\Lambda C} \leq \overline{\gamma}_C C_C, \tag{3.19}$$

$$\underline{\gamma}_I C_I \leq K_I \leq \overline{\gamma}_I C_I, \tag{3.20}$$

where $\tilde{S}_C = K_C - K_{CI}K_I^{-1}K_{IC} + K_{CI}(K_I^{-1} - B_I^{-T})K_I(K_I^{-1} - B_I^{-1})K_{IC}$, and B_I is an appropriately chosen non-singular matrix [37].

Lemma 3.1. *If the symmetric and positive definite block preconditioners $C_I = diag(C_{I,i})_{i\in\mathcal{I}_F}$ and C_C satisfy the spectral equivalence inequalities (3.19) and (3.20) with positive constants $\underline{\gamma}_C$, $\overline{\gamma}_C$, $\underline{\gamma}_I$, $\overline{\gamma}_I$, then the spectral equivalence inequalities*

$$\underline{\gamma}_2 C_2 \leq K_2 + K_{21}K_1^{-1}K_{12} \leq \overline{\gamma}_2 C_2 \tag{3.21}$$

hold for the preconditioner C_2 defined in (3.18) with the constants

$$\underline{\gamma}_2 = \min\{\underline{\gamma}_C, \underline{\gamma}_I\}\left(1 - \sqrt{\tfrac{\mu}{1+\mu}}\right), \quad \overline{\gamma}_2 = \max\{\overline{\gamma}_C, \overline{\gamma}_I\}\left(1 + \sqrt{\tfrac{\mu}{1+\mu}}\right). \tag{3.22}$$

Here $\mu = \rho(S_C^{-1}T_C)$ denotes the spectral radius of $S_C^{-1}T_C$, with the FE Schur complement S_C and the operator T_C being defined by

$$S_C = \overset{o}{K}_C - \overset{o}{K}_{CI}K_I^{-1}\overset{o}{K}_{IC} \quad \text{and} \quad T_C = \overset{o}{K}_{CI}(K_I^{-1} - B_I^{-T})K_I(K_I^{-1} - B_I^{-1})\overset{o}{K}_{IC},$$

respectively. $\overset{o}{K}_C$, $\overset{o}{K}_{CI}$ and $\overset{o}{K}_{IC}$ denote the non-zero FE blocks of K_C, K_{CI} and K_{IC}, respectively.

The proof is given in [37], it applies the classical FE DD spectral equivalence result proved in [17, 18]. With (3.16), we conclude the following theorem.

Theorem 3.1. *If the conditions imposed on C_Λ, C_C, C_I, and B_I, especially (3.11), (3.19) and (3.20) are satisfied, then the FE/BE DD preconditioner*

$$C = diag(I_1, C_2) \tag{3.23}$$

is self-adjoint and positive definite with respect to the inner product $[.,.]$ and satisfies the spectral equivalence inequalities

$$\underline{\gamma}[C\mathbf{v}, \mathbf{v}] \leq [G\mathbf{v}, \mathbf{v}] \leq \overline{\gamma}[C\mathbf{v}, \mathbf{v}] \quad \forall \mathbf{v} \in \mathbf{R}^N, \tag{3.24}$$

with the constants

$$\underline{\gamma} = \underline{\lambda} \min\left\{1, \underline{\gamma}_2\right\} \quad \textit{and} \quad \overline{\gamma} = \overline{\lambda} \max\left\{1, \overline{\gamma}_2\right\},$$

where $\underline{\lambda}$, $\overline{\lambda}$, $\underline{\gamma}_2$, $\overline{\gamma}_2$ *are given in (3.16) and (3.22), respectively.*

3.3.2 The Parallel PCG Algorithm

For the vectors belonging to the inner coupling boundary Γ_C we define two types of distribution called overlapping (type 1) and adding (type 2):

type 1: $\mathbf{u}_C, \mathbf{w}_C, \mathbf{s}_C$ are stored in P_i as $\mathbf{u}_{C,i} = A_{C,i}\mathbf{u}_C$ (analogous $\mathbf{w}_{C,i}$, $\mathbf{s}_{C,i}$)

type 2: $\mathbf{r}_C, \mathbf{v}_C, \mathbf{f}_C$ are stored in P_i as $\mathbf{r}_{C,i}, \mathbf{v}_{C,i}, \mathbf{f}_{C,i}$ such that $\mathbf{r}_C = \sum_{i=1}^{p} A_{C,i}^T \mathbf{r}_{C,i}$ (analogous $\mathbf{v}_C, \mathbf{f}_C$),

where the matrices $A_{C,i}$ are the "C-block" of the Boolean subdomain connectivity matrix A_i which maps some overall vector of nodal parameters into the superelement vector of parameters associated with the subdomain $\bar{\Omega}_i$ only. P_i denotes the ith processor.

Using this notation and the operators introduced in the previous section we can formulate an improved version of the PCG-algorithm presented in [37] with a given accuracy ε as stopping criterion. This parallel PCG-algorithm is given in Appendix B.

Note the vectors $\mathbf{z}_i = (\mathbf{z}_{\Lambda,i}, \mathbf{z}_{C,i}, \mathbf{z}_{\Lambda C,i})^T$ and $\mathbf{h}_i = (\mathbf{h}_{\Lambda,i}, \mathbf{h}_{C,i}, \mathbf{h}_{\Lambda C,i})^T$ which have been inserted additionally in order to achieve a synchronization between the FEM and BEM processors especially in step 1 (matrix-times-vector operation). Without this synchronization one has to expect a computation time per iteration which is, depending on the problem, up to 30 per cent higher. The definition of the vector $\mathbf{p}_i$ avoids the computation of $C_{\Lambda,i}\mathbf{r}_{\Lambda,i}$ (occurred originally in step 4) which is not necessarily available ($C_{\Lambda,i}$ is defined such that the inverse operation $C_{\Lambda,i}^{-1}\mathbf{w}_{\Lambda,i}$ can be performed easily).

3.3.3 On the Components of the Preconditioner

The performance of our algorithm depends heavily on the right choice of the components C_Λ, C_C, C_I and B_I defining the preconditioner C (see Theorem 1). C_Λ, C_I and B_I are block-diagonal matrices with the blocks $C_{\Lambda,i}$, $C_{I,i}$ and $B_{I,i}$, respectively. In our experiments, the following components have turned out to be the most efficient ones:

$C_{I,i}$: **(Vmn)** Multigrid V-cycle with m pre- and n post-smoothing steps in the Multiplicative Schwarz Method [18, 15].

$C_{\Lambda,i}$: **(Circ)** Scaled single layer potential BE matrix for a uniformly discretized circle. This matrix is circulant and easily invertible [39]. **(Hyp)** $C_{\Lambda,i}^{-1} = T_i^T \tilde{M}_{h,i}^{-1} K_{C,i} \tilde{M}_{h,i}^{-1} T_i$ as proposed by Steinbach [43].

$B_{I,i}$: **(HExt)** Implicitly defined by hierarchical extension (formally $E_{IC,i} = -B_{I,i}^{-1} K_{IC,i}$) [20].

C_C: **(S-BPX)** Bramble/Pasciak/Xu type preconditioner [44].
(BPS-D) Bramble/Pasciak/Schatz type preconditioner [3, 9].
(mgD) $K_{C^*}(I_C - M_C)^{-1}$, $(K_{C^*}\mathbf{u}_C, \mathbf{v}_C) = \sum_{i \in \mathcal{I}} \nu_i \langle \mathcal{D}_i u_i, v_i \rangle$, as described in [5].

These preconditioners C_I, C_C, C_Λ, and the basis transformation B_I satisfy the conditions stated in Theorem 3.1. In particular, inequalities (3.20) for C_I are fulfilled with constants $\underline{\gamma}_I$, $\overline{\gamma}_I$ independent of the discretization parameter h [18, 15, 20]. The preconditioner C_Λ is scaled such that $\underline{\gamma}_\Lambda > 1$, and $\overline{\gamma}_\Lambda$ in (3.11) remains independent of h for both, (Circ) and (Hyp). In the case (Hyp), $C_{\Lambda,i}^{-1}$ involves a basis transformation T_i, a modified mass-matrix $\tilde{M}_{h,i}$ and the hypersingular operator $K_{C,i}$, and property (3.11) for C_Λ is due to properties of the corresponding continuous operators $\mathcal{V}_i^{-1}$ and $\mathcal{D}_i$ [43].

With respect to B_I, the constant μ in (3.22) can be estimated by

$$\mu \leq \eta^{2k}(1 + c_1\, l)^2 \leq \eta^{2k}(1 + c_2(\ln h^{-1}))^2,$$

cf. [20], with k being the number of local multigrid iterations, l being the number of grids, the h-independent multigrid rate $\eta < 1$, and the h-independent constants c_i. Thus, μ is independent of h if $k = \mathcal{O}(\ln \ln h^{-1})$.

In the (S-BPX) case, the inequalities in (3.19) hold with an h-independent constant $\underline{\gamma}_C$, and $\overline{\gamma}_C \leq c_3(1+\mu)$ [20, 44]. Therefore, we can prove for (S-BPX) that $\overline{\gamma}/\underline{\gamma} \leq c_4(1+\mu)(\sqrt{\mu} + \sqrt{1+\mu})^2 = \mathcal{O}(1)$ if $k = \mathcal{O}(\ln \ln h^{-1})$. However, in the range of practical applications, this means $k = 1$! For (BPS-D), the estimate $\overline{\gamma}_C/\underline{\gamma}_C \leq c_5(1 + (\ln h^{-1})^2)(1+\mu)$ has been proved [3]. In the case (mgD), C_C arises from the hypersingular operator and C_C^{-1} is realized via a standard multi-grid procedure applied to the global operator (assembled over the subdomains) K_{C^*} which is the discretization of a pseudo-differential operator of order one. K_{C^*} becomes positive definite after implementing the Dirichlet boundary conditions. We can get an estimate of the same type as for (S-BPX). Note that the FE/BE–Schur-complement energy is equivalent to the $\|.\|^2_{H^{1/2}(\Gamma_C)}$–norm (see, e.g., [3, 5, 9]).

Consequently, we can estimate the numerical effort Q to obtain a relative accuracy ε by $Q = \mathcal{O}(h^{-2} \ln h^{-1} \ln \ln h^{-1} \ln \varepsilon^{-1})$ for the (BPS-D) case, and by $Q = \mathcal{O}(h^{-2} \ln \ln h^{-1} \ln \varepsilon^{-1})$ in the (S-BPX) case, i.e. almost optimal. If a BPX-type extension [38] is applied instead of (HExt) in a nested iteration approach [21], we can prove that $Q = \mathcal{O}(h^{-2})$, i.e., we obtain an optimal method.

Preconditioners $C_{\Lambda,i}$ for $K_{\Lambda,i}$ and C_C for the FE/BE Schur complement derived on the basis of boundary element techniques can also be found in [35, 39]. The construction of efficient FE Schur complement preconditioners was one of the main topics in the research on FE-DD-methods (see Proceedings of the annual DD-conferences since 1987).

3.4 Numerical Results

The electromagnet as shown in Figure 3.1 serves now as test example for exterior magnetic field problems which lead to the variational form (3.4).

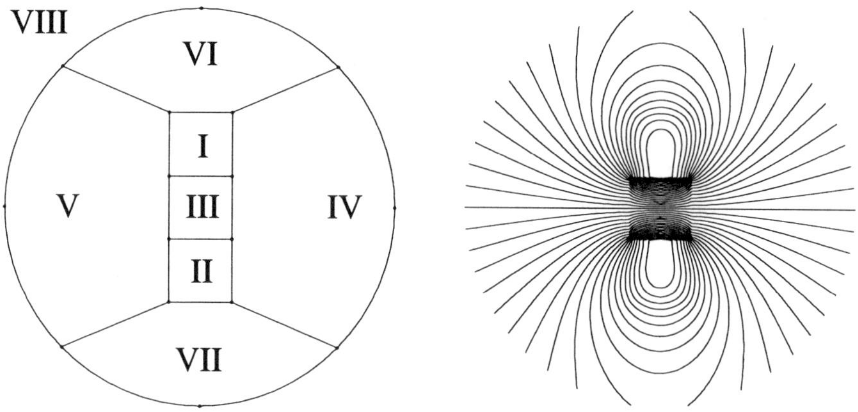

Fig. 3.1. The magnet and the subdomains being used (left) and the equipotential lines of the solution (right).

The copper domains (I, II), where we have a current density of the strength S and −S, respectively, and the iron domain (III), are squares with the edges being 16cm long. The material dependent coefficients (air: IV–VIII) are given by $\nu_{Cu} = 795779.0\,\mathrm{AmV^{-1}s^{-1}}$, $\nu_{air} = 795774.4\,\mathrm{AmV^{-1}s^{-1}}$, $\nu_{Fe} = 1000.0\,\mathrm{AmV^{-1}s^{-1}}$.

Table 3.1. Number of unknowns (N), iteration count ($I(\varepsilon)$, $\varepsilon = 10^{-6}$), CPU time in seconds. The experiments were carried out on a Power-XPlorer using 8 or 4 processors, respectively.

	FEM: BEM:	I-VII VIII	FEM: BEM:	I-III IV-VIII	FEM: BEM:	I-III exterior
l	I(ε)	CPU	I(ε)	CPU	I(ε)	CPU
1	15	0.6	17	0.5	14	0.3
2	19	0.9	18	1.0	19	0.6
3	20	1.6	19	1.9	20	1.7
4	21	5.2	21	5.2	21	5.9
5	23	20.5	22	18.8	22	22.5
N(5)	67329		18429		16129	

We will compare two different coupling procedures. On the one hand we use the natural boundary (of the metallic material) as coupling boundary and on the other hand we introduce a circle with the radius 50cm as coupling boundary. The advantage of the second method is that we obtain circulant matrices which can be generated very fast whereas the first method requires the generation of fully populated matrices. A disadvantage of the second method is that additional subdomains and, thus, additional unknowns have to be introduced. For this example, the uniqueness of the solution is guaranteed by the radiation condition, even if no Dirichlet boundary Γ_D is present. The radiation condition is implicitly contained in our BE discretization.

Numerical results are given in Table 3.1. The operators C_Λ, C_C, C_I, B_I have been chosen as follows:

$C_{\Lambda,i}$	:	Circ ($i=8$) and Hyp ($i \in \mathcal{I}_B \setminus \{8\}$)	C_C	:	S-BPX
$C_{I,i}$	:	V11 ($i \in \mathcal{I}_F$)	$B_{I,i}$	:	HExt ($i \in \mathcal{I}_F$).

We will compare two different coupling procedures. Comparing the three choices of the subdomains and their discretization, we observe that in the first choice (column 1) much time is spent for handling the FEM subdomains IV, V with many interior nodes. We may conclude that the BEM (column 2) is recommended for subdomains with a high ratio between the numbers of interior FEM nodes and the boundary nodes. The choice of the rectangular coupling boundary (column 3) increases the BEM system generation effort, but the total time remains nearly the same since the number of BEM unknowns (and subdomains) is reduced. Finally we observe that the number of iterations is independent of the combination of BE/FE discretizations being used.

4 Parallel Solution of Coupled BE/FE–Equations Approximating Nonlinear Magnetic Field Problems

4.1 The Nested-DD-Newton-Solver

Let us consider now a nonlinear magnetic field problem, i.e., we allow ferromagnetic materials with non-constant permeability ν to be in the FE subdomains. Then, the mixed DD coupled variational problem can be written as follows: Find $(\lambda, u) \in \mathbf{V} := \Lambda \times \mathbf{U}_0$ such that

$$a_N(\lambda, u; \eta, v) = \langle F, v \rangle \quad \forall (\eta, v) \in \mathbf{V}, \tag{4.1}$$

where

$$\begin{aligned} a(\lambda, u; \eta, v) &:= a_B(\lambda, u; \eta, v) + a_{FN}(u, v) \\ a_{FN}(u, v) &:= \sum_{i \in \mathcal{I}_F} \int_{\Omega_i} \nu(x, |\nabla u|) \nabla^T u(x) \nabla v(x) \, dx \end{aligned}$$

and a_B being defined as in (3.4). We refer to [25, 30] for the analysis of nonlinear magnetic field problems. Consequently, the discretization results in a nonlinear system [26]

$$K \begin{pmatrix} \mathbf{u}_\Lambda \\ \mathbf{u}_C \\ \mathbf{u}_I \end{pmatrix} = K_F \begin{pmatrix} \mathbf{u}_C \\ \mathbf{u}_I \end{pmatrix} + K_B \cdot \begin{pmatrix} \mathbf{u}_\Lambda \\ \mathbf{u}_C \end{pmatrix} = \begin{pmatrix} \mathbf{f}_\Lambda \\ \mathbf{f}_C \\ \mathbf{f}_I \end{pmatrix}, \tag{4.2}$$

where the nonlinear operator $K : \mathbf{R}^N \longrightarrow \mathbf{R}^N$ can be split up into the nonlinear operator $K_F : \mathbf{R}^{N_C+N_I} \longrightarrow \mathbf{R}^N$ originating from the nonlinear form a_{FN} and the linear operator $K_B : \mathbf{R}^{N_A+N_C} \longrightarrow \mathbf{R}^N$ originating from a_B [26].

The nonlinear system (4.2) is solved by Newton's method, see [26]. In this algorithm, the linear Newton defect system (4.5) with the Jacobi matrix $K'[\mathbf{u}]$ can be written as

$$K'\left[\begin{pmatrix}\mathbf{u}_A\\ \mathbf{u}_C\\ \mathbf{u}_I\end{pmatrix}\right]\cdot\begin{pmatrix}\mathbf{w}_A\\ \mathbf{w}_C\\ \mathbf{w}_I\end{pmatrix} = K_F'\left[\begin{pmatrix}\mathbf{u}_C\\ \mathbf{u}_I\end{pmatrix}\right]\cdot\begin{pmatrix}\mathbf{w}_C\\ \mathbf{w}_I\end{pmatrix} + K_B\cdot\begin{pmatrix}\mathbf{w}_A\\ \mathbf{w}_C\end{pmatrix} = \begin{pmatrix}\mathbf{d}_A\\ \mathbf{d}_C\\ \mathbf{d}_I\end{pmatrix}. \quad (4.3)$$

It can be rewritten in a block form similar to (3.10),

$$\begin{pmatrix}K_A & -K_{AC} & 0\\ K_{CA} & J_C & J_{CI}\\ 0 & J_{IC} & J_I\end{pmatrix}\begin{pmatrix}\mathbf{w}_A\\ \mathbf{w}_C\\ \mathbf{w}_I\end{pmatrix} = \begin{pmatrix}\mathbf{d}_A\\ \mathbf{d}_C\\ \mathbf{d}_I\end{pmatrix}, \quad (4.4)$$

and can be solved by a PCG method as described in Subsection 3.3. Further, we apply the "nested iteration" method [21, 24], i.e. we generate a multilevel sequence of coupled FE/BE discretizations denoted by the grid numbers $q = 1, \ldots, l$. We begin with solving the nonlinear system by Newton's method on the coarsest grid $q = 1$. Then we take the approximate solution on the grid $q-1$ interpolated onto the finer grid q as an initial approximation for Newton's method on the grid q, for $q = 2, \ldots, l$. This allows us to "catch" the nonlinearity on the coarsest grid, see [24, 26, 28, 30].

The result of the parallel algorithm presented in the following will be the approximate solution $\mathbf{u}_l^\star$ on the fine grid $q = l$ with relative accuracy ε (nested iteration ε).

Algorithm PNN (Parallel Nested Newton)

Step 0

Initialization of the grid number:

(0.)(P) $q := 1$.

Step 1

Set the initial solution for grid q:

(1.1)(P) IF $q = 1$ THEN $\mathbf{u}_q^0 = \mathbf{0}$;

(1.2)(P) IF $q > 1$ THEN $\mathbf{u}_q^0 = \tilde{I}_{q-1}^q \mathbf{u}_{q-1}^\star$;

the initial solution is the interpolation of the best solution on grid $q-1$.

(1.3)(P) Initialize the Newton iteration number $j := 0$.

Step 2

Compute the initial Jacobi matrix and the defect vector

(2.)(P) $J_q^0 = K_q'[\mathbf{u}_q^0]$; $\mathbf{d}_q^1 = \mathbf{f}_q - K_q\mathbf{u}_q^0$.

Step 3

(3.)(P) Choose a relaxation parameter τ_q^j with $0 < \tau_q^j \leq 1$ and a relative accuracy parameter ε_{lin} with $0 < \varepsilon_{\text{lin}} < 1$.

Step 4

(4.)(S) Solve the linear defect system

$$J_q^j \mathbf{w}_q^{j+1} = \mathbf{d}_q^{j+1} \tag{4.5}$$

approximately (with relative accuracy ε_{lin}) using a PCG solver as described in Subsection 3.3. The result is $\tilde{\mathbf{w}}_q^{j+1}$.

Step 5
Correct the solution:

(5.)(P) $\mathbf{u}_q^{j+1} = \mathbf{u}_q^j + \tau_q^j \tilde{\mathbf{w}}_q^{j+1}$.

Step 6
Control the convergence (parameter c_τ is chosen a priori with $c_\tau < 1$):

(6.1)(P) Compute the new defect vector and the new Jacobi matrix
$\mathbf{d}_q^{j+2} = \mathbf{f}_q - K_q \mathbf{u}_q^{j+1}; \quad J_q^{j+1} = K_q'[\mathbf{u}_q^{j+1}]$;
(6.2)(C) Compute defect norms
$d_q^{j+1} = \|\mathbf{d}_q^{j+1}\|; \quad d_q^{j+2} = \|\mathbf{d}_q^{j+2}\|$;

(6.3)(P) $\texttt{IF } d_q^{j+2} \geq d_q^{j+1} \texttt{ THEN } \left(\begin{array}{l} \tau_q^j := \min\left\{ c_\tau \tau_q^j, \dfrac{\tau_q^j d_q^{j+1}}{d_q^{j+1} + d_q^{j+2}} \right\}; \\ \texttt{GOTO } \textit{Step 5} \end{array} \right)$;

(6.4)(P) $\texttt{IF } d_q^{j+2} \leq \varepsilon d_q^1 \texttt{ THEN } \left(\begin{array}{l} \mathbf{u}_q^\star := \mathbf{u}_q^{j+1}; \\ \texttt{IF } q < l \texttt{ THEN } (q := q+1; \texttt{ GOTO } \textit{Step 1}); \\ \texttt{IF } q = l \texttt{ THEN EXIT}; \end{array} \right)$;

(6.5)(P) Perform a further Newton step:
$j := j + 1$;
GOTO *Step 3*.

In this description, (P) indicates that the step is performed completely in parallel, i.e., independently on the processors. The solver (S) includes parallel independent parts, communication between processors handling neighbouring subdomains, and global communication, cf. Subsection 3.3. Note, (C) indicates that global communication is necessary. Obviously, the only additional communication (compared with solving a linear problem) is the computation of global defect norms.

4.2 Numerical Results

A direct current motor designed for electronic devices (electronic motor) which is excited by permanent magnets serves as a first real-life test example. The interior of the machine is discretized by finite elements (cf. Figure

4.1). Calculations have been made for both the machine with homogeneous Dirichlet conditions on its boundary and the infinite domain with Sommerfeld's radiation condition where the infinite exterior domain is discretized by BEM. In order to obtain efficiency results, we consider additionally two model problems representing a quarter and a half of the whole machine, i.e. we discretize only sectors of 90 and 180 degrees, respectively, and impose Dirichlet conditions on the boundary.

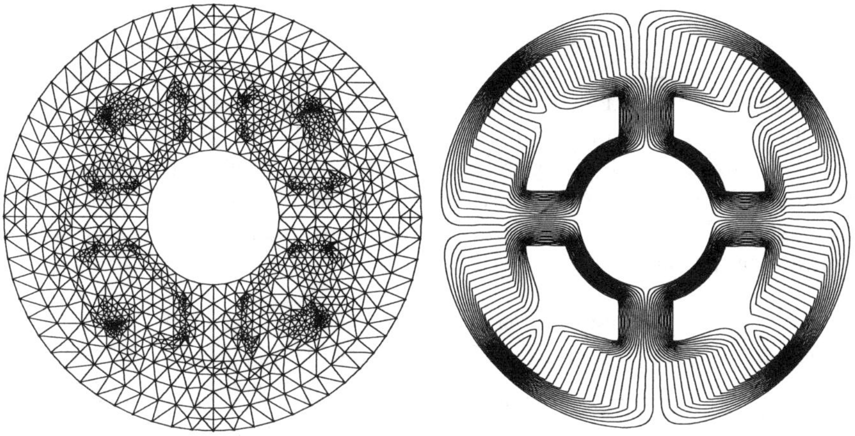

Fig. 4.1. FEM discretization of the electronic motor (coarse grid) and equipotential lines of the solution.

Table 4.1. Performance for a practical problem (electronic motor).

Example	Dirichlet b. c.	Dirichlet b. c.	Dirichlet b. c.	radiation condition
subdomains	16 FEM	32 FEM	64 FEM	63 FEM 1 BEM
No. of unknowns	374 129	734 199	1 514 008	1 489 416
Newton it.	4	5	4	4
CG iter. 1st grid	6,11/16,11	6,11/14,11,12	6,11/15,11	6,11/16,13
CG iter. 2nd grid	9,10	9,11	9,11	9,11
CG iter. 3rd grid	10,12	10,12	10,12	11,13
CG iter. 4th grid	11,13	11,13	11,13	11,14
CG iter. 5th grid	11,15	11,15	11,15	12,15
Newton it.	4	4	4	4
CG iter. 6th grid	12,16,9,17	12,17,10,16	13,16,10,16	13,16,10,16
generation	21.8	21.6	23.2	26.4
linear solver	63.4	64.9	71.7	85.6
Total time	85.2	86.5	94.9	112.0
Scale-up (norm.)	1.0 →	1.933 →	3.633	
Scaled eff. (rel.)	1.0 →	0.966 →	0.908	

Time in seconds, scale-up (normalized) and scaled efficiency (relative) on a GC-Power Plus using 16, 32, 64 processors, respectively; 2 Newton iterations on the grids 2–5, relative accuracy $\varepsilon = 10^{-6}$.

The components of the PNN algorithm are chosen in the standard way [23, 24, 28]. In particular, the parameter ε_{lin} can be adapted to the quadratic convergence speed of the Newton method [24, 30]. Here, a slash (/) marks the change from $\varepsilon_{\text{lin}} = 10^{-2}$ to $\varepsilon_{\text{lin}} = 10^{-4}$ in the accuracy of the CG solver. The components of the PCG solver have been chosen as follows:

$$C_{A,i} \;:\; \text{Circ}\ (i \in \mathcal{I}_B) \qquad C_C \;:\; \text{BPS-D}$$
$$C_{I,i} \;:\; \text{V11}\ (i \in \mathcal{I}_F) \qquad B_{I,i} \;:\; \text{HExt}\ (i \in \mathcal{I}_F).$$

The components involved and the numerical effort for nonlinear problems are discussed in [26]. Numerical results are given in Table 4.1. We present further results, in particular with respect to efficiency, in [30]. Computations with up to 128 processors are documented in [26, 27]. In [28], the application of a parallelized global multigrid method based on DD ideas in Step 4 of the algorithm PNN is discussed. Further, in [29], we demonstrate that the CG with a global BPX or a global multigrid preconditioner yields a robust solver for practical problems.

Table 4.2. Performance for the dc motor.

Example	Dirichlet b. c.	Dirichlet b. c.	radiation condition	radiation condition
Choice for C_C	BPS-D	S-BPX	BPS-D	S-BPX
subdomains	32 FEM	32 FEM	31 FEM 1 BEM	31 FEM 1 BEM
No. of unknowns	417 328	417 328	414 568	414 568
Newton it. 1st grid CG iter. 1st grid	7 16,12,13, 13,13,17,24	6 10,12,12 12,12,11	7 19,14,16, 16,12,19,28	7 10,14,14 14,14,13,17
Newton it. 2nd grid CG iter. 2nd grid	2 16,23	2 13,16	2 18,25	2 15,20
Newton it. 3rd grid CG iter. 3rd grid	2 17,27	2 15,17	2 18,31	2 19,24
Newton it. 4th grid CG iter. 4th grid	2 17,35	2 16,18	2 18,34	2 19,29
Newton it. 5th grid CG iter. 5th grid	4 18,38,22,30	4 16,21,18,20	4 18,42,22,33	4 20,32,27,33
Time (generation) Time (linear solver)	25.2 168.9	25.2 137.8	33.6 178.8	33.6 198.5
Total time	194.1	163.0	212.4	232.1

Time in seconds, GC-Power Plus, 32 processors; relative accuracy $\varepsilon = 10^{-6}$.

The second practical example, a technical direct current motor (dc motor, see Subsection 2.3), is to demonstrate the complete algorithm. Starting with a user mesh, we apply an automatic domain decomposition procedure (see Subsection 2.3 and Appendix A) and a parallel mesh generator, the basic ideas of which are presented in [11], to obtain the initial mesh ($q = 1$), which is to be refined four times to get the final mesh for our computations ($q = l = 5$). We present the numerical results in Table 4.2 and the level lines of the solution in Figure A.2. Again, we have done calculations for both, the machine with homogeneous Dirichlet conditions on its boundary and the infinite domain with the radiation condition. Best results with respect to the total computing time have been achieved with $\varepsilon_{\text{lin}} = 0.01$. All other components of the algorithm, except C_C, are chosen as for the electronic motor example.

5 Generalization to Linear Elasticity Problems

5.1 The Mixed Boundary and Domain Integral Variational Formulation

We now want to extend the ideas discussed above to problems of plane linear elasticity in which the displacement $u(x) = (u_1(x), u_2(x))^T$ satisfies formally the system of Lamé equations

$$\begin{array}{l} -\mu(x)\Delta u(x) - (\lambda(x) + \mu(x))\text{grad div}\, u(x) = f(x) \quad \text{in } \Omega \\ u(x) = 0 \text{ on } \Gamma_D, \quad \sum_{l=1}^{2} \sigma_{kl}(u(x))n_l = g_k(x) \text{ on } \Gamma_N, \ (k = 1,2) \end{array} \tag{5.1}$$

where Ω is a bounded Lipschitz domain, $\sigma_{kl}(u)$ are the components of the stress tensor $\sigma(u)$ and $n(x) = (n_1(x), n_2(x))^T$ is the outward normal vector to $\Gamma_D \cup \Gamma_N = \Gamma := \partial\Omega$ ($\Gamma_D \neq \emptyset$) and λ and μ, $\lambda, \mu > 0$, are the Lamé coefficients of the elastic material. In (5.1), $f = (f_1, f_2)^T$ is the vector of volumic forces, $g = (g_1, g_2)^T$ is the vector of boundary tractions.

As before, Ω is being decomposed into non-overlapping subdomains Ω_i, $i = 1, \ldots, p$, cf. (2.1). We then have μ_i and λ_i as Lamé coefficients for each of the Ω_i's. Similar to (3.2) we assume that the volumic forces vanish for $x \in \Omega_i$ with $i \in \mathcal{I}_B$. For simplicity we assume additionally that $\Gamma_N \cap \partial\Omega_i = \emptyset$ for $i \in \mathcal{I}_B$. Then we can write the variational formulation as follows: Find $(\sigma, u) \in \mathbf{V} := \Lambda \times \mathbf{U}_0$:

$$a(\sigma, u; \tau, v) = \langle F, v \rangle \quad \forall\, (\tau, v) \in \mathbf{V}, \tag{5.2}$$

where

$$a(\sigma, u; \tau, v) := a_B(\sigma, u; \tau, v) + a_F(u, v)$$

$$a_B(\sigma, u; \tau, v) := \sum_{i \in \mathcal{I}_B} \nu_i \Big\{ \langle \mathcal{D}_i u_i, v_i \rangle_{\Gamma_i} + \frac{1}{2} \langle \sigma_i, v_i \rangle_{\Gamma_i} + \langle \sigma_i, \mathcal{K}_i v_i \rangle_{\Gamma_i}$$

$$+ \langle \tau_i, \mathcal{V}_i \sigma_i \rangle_{\Gamma_i} - \langle \tau_i, \mathcal{K}_i u_i \rangle_{\Gamma_i} - \frac{1}{2} \langle \tau_i, u_i \rangle_{\Gamma_i} \Big\}$$

$$a_F(u, v) := \sum_{i \in \mathcal{I}_F} \int_{\Omega_i} \left(\lambda_i \, div\, u(x) \, div\, v(x) + 2\, \mu_i \sum_{k,l=1}^{2} \epsilon_{kl}(u)\, \epsilon_{kl}(v) \right) dx$$

$$\langle F, v \rangle := \sum_{i \in \mathcal{I}_F} \int_{\Omega_i} f(x)\, v(x)\, dx + \int_{\Gamma_N} g(x)\, v(x)\, ds,$$

with the duality pairing $\langle .,. \rangle$, the traces $u_i = u|_{\partial\Omega_i}$ and the boundary tractions $\sigma = [\sigma_i]_{i\in\mathcal{I}_B}$ belonging to $\partial\Omega_i$ and the strain $\epsilon_{kl}(u) := (\partial u_k/\partial x_l + \partial u_l/\partial x_k)/2$. Note, for simplicity we have assumed that Ω is bounded such that, in contrary to Section 3.1, special terms for the exterior domain do not occur in the definition of $a_B(.;.)$. Nevertheless, the ideas presented above concerning exterior problems can be applied analogously.

Let the spaces $\mathbf{U}_0$ and Λ be defined as follows:

$$\begin{aligned}\mathbf{U}_0 &:= \{u \in [\mathrm{H}^1(\Omega)]^2 : u|_{\Gamma_{BE}} \in \mathrm{H}^{1/2}(\Gamma_{BE}), u|_{\Gamma_D} = 0\} \\ \Lambda &:= \textstyle\prod_{i\in\mathcal{I}_B} \left[\mathrm{H}^{-1/2}(\partial\Omega_i)\right]^2,\end{aligned}$$

where $\Gamma_{BE} := \bigcup_{i\in\mathcal{I}_B} \partial\Omega_i \setminus \Gamma_D, \Gamma_{FE} := \bigcup_{i\in\mathcal{I}_F} \partial\Omega_i \setminus \Gamma_D, \Gamma_C := \Gamma_{BE} \cup \Gamma_{FE}$. Let $\Omega_F := \cup_{i\in\mathcal{I}_F} \Omega_i$, then we consider the following norm in $\mathbf{V}$:

$$\|(\sigma, u)\|_{\mathbf{V}} := (\|\sigma\|_\Lambda^2 + \|u\|^2_{\mathrm{H}^{1/2}(\Gamma_{BE})} + \|u\|^2_{\mathrm{H}^1(\Omega_F)})^{1/2}. \tag{5.3}$$

The boundary integral operators $\mathcal{V}_i$, $\mathcal{K}_i$ and $\mathcal{D}_i$ are defined as in (3.5), where $\partial(.)$ has to be replaced by the operator $\mathcal{T}(.)$ which is defined in its strong form as $\mathcal{T}(.) := 2\mu\partial(.) + \lambda n\, div\, u + \mu\, n \times curl(.)$ and in its weak form by the first Green formula. For $\mathcal{E}(x, y)$ we now have to insert the well known Kelvin fundamental solution (see, e.g., [6]). The variational formulation (5.2) has a unique solution provided that the single layer potential operators $\mathcal{V}_i$ are positive definite ($\mathrm{H}^{-1/2}(\partial\Omega_i)$–elliptic) [6, 32, 40].

Similar to Section 3.2 we discretize (5.2) to obtain a system of equations which is manipulated in the same way as discussed in Section 3.3.1. This leads again to a symmetric, positive definite system matrix. Thus, the parallel solution can be performed in a similar fashion provided suitable preconditioners C_Λ, C_C, C_I and B_I are known (see Section 5.2 and [40]).

5.2 Numerical Results

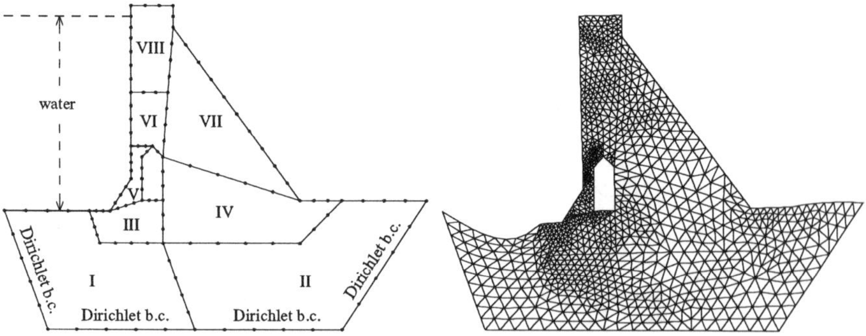

Fig. 5.1. The subdomains and the BE discretization of the 1st level (left) and the deformed (magnification factor 100) FE grid of the 2nd level (right).

Table 5.1. Levels (l), pure solution time (s), number of unknowns (N), iteration count ($I(\varepsilon)$, $\varepsilon = 10^{-6}$), CPU time in seconds for the dam-problem. The experiments were carried out on a Power-XPlorer using 8 processors.

	BEM: I-VIII		FEM: III-VIII BEM: I-II		FEM: I-VIII	
l	I(ε)	CPU	I(ε)	CPU	I(ε)	CPU
2	27	4.9	25	4.6	25	3.9
3	27	8.8	28	8.8	28	7.9
4	28	24.1	30	27.9	31	22.8
5	29	85.9	31	104.9	32	81.3
s(5)	49.5		68.5		63.8	
N(5)	6470		78130		119318	

As a test problem we consider a model of dam filled with water as sketched in Figure 5.1. As indicated there, boundary conditions are given on Γ_D (zero displacement) and on Γ_N (zero or according to the water pressure, respectively). The Lamé constants are given for rock (I-II) by $\mu_r = 7.265e4MPa$, $\lambda_r = 3.743e4MPa$ and for concrete (III-VIII) by $\mu_c = 9.2e6MPa$, $\lambda_c = 9.2e6MPa$. For the results presented in Table 5.1, the operators C_Λ, C_C, C_I, B_I have been chosen as follows:

$$C_{\Lambda,i} \; : \; \text{mgV } (i \in \mathcal{I}_B) \qquad C_C \; : \; \text{mgD}$$
$$C_{I,i} \; : \; \text{V11(HExt+S) } (i \in \mathcal{I}_F) \qquad B_{I,i} \; : \; \text{HExt+S } (i \in \mathcal{I}_F).$$

Here, (mgV) stands for a multigrid-based preconditioner for the single layer potential (see [1]). Furthermore, new algorithms for $C_{I,i}$ and $B_{I,i}$ have been used. In particular, the hierarchical extension (HExt+S) has been improved by a coarse grid solver and smoothing on the other levels [13].

The BE discretization of the 1st level and the FE discretization of the 2nd level (deformed mesh) are shown in Figure 5.1. We use piecewise linear trial functions for the displacements and piecewise constants for the boundary tractions. The entries of the BE matrices are computed fully analytically.

In Table 5.1 we present several combinations of FE/BE discretizations. Looking at the CPU-time we observe that the FE discretization (column 3) leads to the best results. However, if we are interested in the pure solving time $s(.)$ ($s(5)$ for the 5th level is given in Table 5.1) the BE discretization (first column) is of advantage.

6 Concluding Remarks and Generalizations

The DD-method has turned out to be a powerful tool for establishing the coupled FE/BE variational formulation and for solving the discrete systems efficiently on massively parallel computers. The results presented here have been obtained using the code FEM⊕BEM [14] which can solve linear and non-linear magnetic field problems as well as problems arising in linear elasticity. The high efficiency and the scalability of the algorithm has been demonstrated [5, 16, 30, 26, 28, 36].

Comparison of "local" DD methods described in this paper with "global" multigrid methods implemented on massively parallel machines as well as workstation clusters is given in [28]. The use and the parallelization of "global" methods is also discussed in [4]. Other coupling and solution techniques are studied in [10, 41, 43].

The techniques presented here can be generalized to the 3D case provided that fast matrix-by-vector multiplication routines for the BE matrices, e.g. based on Panel-Clustering-Techniques developed in [22], and asymptotically optimal, or almost optimal components C_I (e.g. multigrid preconditioners), B_I [38], C_C (e.g. BPX) and C_Λ [35, 39, 40] of the preconditioners are available.

A The Direct Current Motor

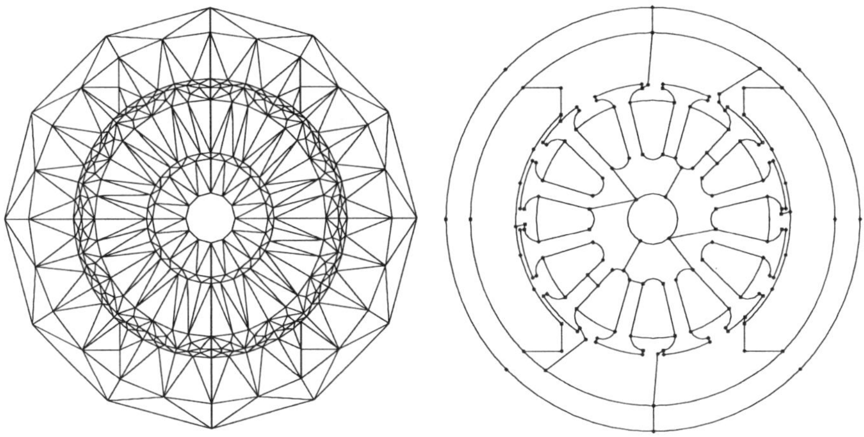

Fig. A.1. Initial triangular-based description and final decomposition of the dc-motor into 32 subdomains.

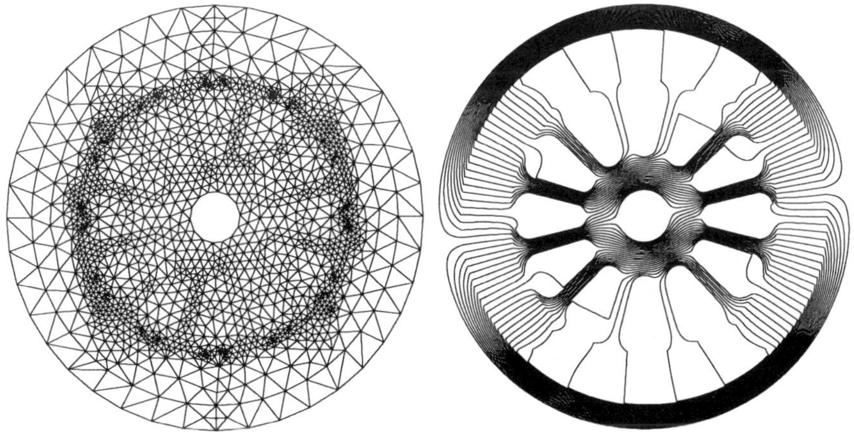

Fig. A.2. The mesh of the 1st level and the equipotential lines of the solution (dc-motor).

B The Parallel Algorithm

	FEM ($i \in \mathcal{I}_F$)	BEM ($i \in \mathcal{I}_B$)
	Starting Step	
0.	Choose an initial guess $\mathbf{u} = \mathbf{u}_0$	
	$\mathbf{u}_i = [\mathbf{u}_{C,i}\ \mathbf{u}_{I,i}]^T$	$\mathbf{u}_i = [\mathbf{u}_{\Lambda,i}\ \mathbf{u}_{C,i}]^T$
	$\mathbf{r}_{I,i} = \mathbf{f}_{I,i} - K_{IC,i}\mathbf{u}_{C,i} - K_{I,i}\mathbf{u}_{I,i}$ $\mathbf{r}_{C,i} = \mathbf{f}_{C,i} - K_{C,i}\mathbf{u}_{C,i} - K_{CI,i}\mathbf{u}_{I,i}$	$\mathbf{v}_{\Lambda,i} = \mathbf{f}_{\Lambda,i} - K_{\Lambda,i}\mathbf{u}_{\Lambda,i} + K_{\Lambda C,i}\mathbf{u}_{C,i}$ $\mathbf{r}_{\Lambda,i} = C_{\Lambda,i}^{-1}\mathbf{v}_{\Lambda,i}$ $\mathbf{r}_{C,i} = \mathbf{f}_{C,i} - K_{C,i}\mathbf{u}_{C,i} - K_{C\Lambda,i}\mathbf{u}_{\Lambda,i}$ $\mathbf{r}_{C,i} = \mathbf{r}_{C,i} - K_{C\Lambda,i}\mathbf{r}_{\Lambda,i}$
	$\mathbf{v}_{C,i} = \mathbf{r}_{C,i} - K_{CI,i}B_{I,i}^{-T}\mathbf{r}_{I,i}$	$\mathbf{w}_{\Lambda,i} = \mathbf{r}_{\Lambda,i};\ \mathbf{p}_i = \mathbf{v}_{\Lambda,i}$ $\mathbf{v}_{C,i} = \mathbf{r}_{C,i};\ \mathbf{z}_{\Lambda,i} = K_{\Lambda,i}\mathbf{w}_{\Lambda,i}$
	$\mathbf{w}_{C,i} = A_{C,i}\mathbf{w}_C \leftarrow \mathbf{w}_C = C_C^{-1}\sum_{i=1}^p A_{C,i}^T\mathbf{v}_{C,i} \rightarrow \mathbf{w}_{C,i} = A_{C,i}\mathbf{w}_C$	
	$\mathbf{w}_{I,i} = C_{I,i}^{-1}\mathbf{r}_{I,i} - B_{I,i}^{-1}K_{IC,i}\mathbf{w}_{C,i}$	$\mathbf{z}_{\Lambda C,i} = K_{\Lambda C,i}\mathbf{w}_{C,i};\ \mathbf{z}_{C,i} = K_{C,i}\mathbf{w}_{C,i}$
	$\mathbf{s} = \mathbf{w}$	$\mathbf{s} = \mathbf{w}$
	$\sigma_i = \mathbf{r}_{C,i}^T\mathbf{w}_{C,i} + \mathbf{r}_{I,i}^T\mathbf{w}_{I,i}$	$\sigma_i = \mathbf{r}_{C,i}^T\mathbf{w}_{C,i} + \mathbf{w}_{\Lambda,i}^T(\mathbf{z}_{\Lambda,i} - \mathbf{p}_i)$
	$\sigma = \sigma^0 = \sum_{i=1}^p \sigma_i$	
	Iteration	
1.	$\mathbf{v}_{I,i} = K_{IC,i}\mathbf{s}_{C,i} + K_{I,i}\mathbf{s}_{I,i}$ $\mathbf{v}_{C,i} = K_{C,i}\mathbf{s}_{C,i} + K_{CI,i}\mathbf{s}_{I,i}$ $\delta_i = \mathbf{v}_C^T\mathbf{s}_C + \mathbf{v}_{I,i}^T\mathbf{s}_{I,i}$	$\mathbf{w}_{\Lambda,i} = \mathbf{z}_{\Lambda,i} - \mathbf{z}_{\Lambda C,i}$ $\mathbf{v}_{\Lambda,i} = C_{\Lambda,i}^{-1}\mathbf{w}_{\Lambda,i}$ $\mathbf{v}_{C,i} = \mathbf{z}_{C,i} + K_{C\Lambda,i}(\mathbf{s}_{\Lambda,i} - \mathbf{v}_{\Lambda,i})$ $\delta_i = \mathbf{v}_{C,i}^T\mathbf{s}_{C,i} + \mathbf{v}_{\Lambda,i}^T\mathbf{z}_{\Lambda,i} - \mathbf{w}_{\Lambda,i}^T\mathbf{s}_{\Lambda,i}$
	$\alpha = \sigma/\delta \quad \leftarrow \quad \delta = \sum_{i=1}^p \delta_i \quad \rightarrow \quad \alpha = \sigma/\delta$	
2.	$\hat{\mathbf{u}}_i = \mathbf{u}_i + \alpha\mathbf{s}_i;\ \hat{\mathbf{r}}_i = \mathbf{r}_i - \alpha\mathbf{v}_i$	$\hat{\mathbf{u}}_i = \mathbf{u}_i + \alpha\mathbf{s}_i;\ \hat{\mathbf{r}}_i = \mathbf{r}_i - \alpha\mathbf{v}_i$
3.	$\hat{\mathbf{v}}_{C,i} = \hat{\mathbf{r}}_{C,i} - K_{CI,i}B_{I,i}^{-T}\hat{\mathbf{r}}_{I,i}$	$\hat{\mathbf{v}}_{C,i} = \hat{\mathbf{r}}_{C,i};\ \mathbf{h}_{\Lambda,i} = K_{\Lambda,i}\hat{\mathbf{r}}_{\Lambda,i}$
	$\mathbf{w}_{C,i} = A_{C,i}\mathbf{w}_C \leftarrow \mathbf{w}_C = C_C^{-1}\sum_{i=1}^p A_{C,i}^T\mathbf{v}_{C,i} \rightarrow \mathbf{w}_{C,i} = A_{C,i}\mathbf{w}_C$	
	$\hat{\mathbf{w}}_{I,i} = C_{I,i}^{-1}\hat{\mathbf{r}}_{I,i} - B_{I,i}^{-1}K_{IC,i}\hat{\mathbf{w}}_{C,i}$	$\hat{\mathbf{w}}_{\Lambda,i} = \hat{\mathbf{r}}_{\Lambda,i},\ \hat{\mathbf{p}}_i = \mathbf{p}_i - \alpha\mathbf{w}_{\Lambda,i}$ $\mathbf{h}_{\Lambda C,i} = K_{\Lambda C,i}\hat{\mathbf{w}}_{C,i};\ \mathbf{h}_{C,i} = K_{C,i}\hat{\mathbf{w}}_{C,i}$
4.	$\sigma_i = \hat{\mathbf{r}}_{C,i}^T\hat{\mathbf{w}}_{C,i} + \hat{\mathbf{r}}_{I,i}^T\hat{\mathbf{w}}_{I,i}$	$\sigma_i = \hat{\mathbf{r}}_{C,i}^T\hat{\mathbf{w}}_{C,i} + \mathbf{h}_{\Lambda,i}^T\hat{\mathbf{r}}_{\Lambda,i} - \hat{\mathbf{p}}_i^T\hat{\mathbf{r}}_{\Lambda,i}$
	$\beta = \hat{\sigma}/\sigma \quad \leftarrow \quad \hat{\sigma} = \sum_{i=1}^p \sigma_i \quad \rightarrow \quad \beta = \hat{\sigma}/\sigma$	
5.	$\hat{\mathbf{s}}_i = \hat{\mathbf{w}}_i + \beta\mathbf{s}_i$	$\hat{\mathbf{s}}_i = \hat{\mathbf{w}}_i + \beta\mathbf{s}_i;\ \hat{\mathbf{z}}_i = \mathbf{h}_i + \beta\mathbf{z}_i$
6.	If $\hat{\sigma} \leq \varepsilon^2 * \sigma^0$, then STOP else goto step 1.	

References

1. J. H. Bramble, Z. Leyk, and J. E. Pasciak. The analysis of multigrid algorithms for pseudodifferential operators of order minus one. *Math. Comp.*, 63:461–478, 1994.
2. Bramble, J.H., Pasciak, J.E. (1988): A preconditioning technique for indefinite systems resulting from mixed approximations of elliptic problems. Mathematics of Computation 50(181), 1–17
3. Bramble, J.H., Pasciak, J.E., Schatz, A.H. (1986,1987,1988,1989): The construction of preconditioners for elliptic problems by substructuring I – IV. Mathematics of Computation 47, 103–134, 49, 1–16, 51, 415–430, 53, 1–24
4. Brink, U., Kreienmeyer, M., Stein, E. (1995): Different methodologies for coupled BEM and FEM with implementation on parallel computers. In this proceedings
5. Carstensen, C., Kuhn, M., Langer, U. (1995): Fast parallel solvers for symmetric boundary element domain decomposition equations. Report 499, Johannes Kepler University Linz, Institute of Mathematics
6. Carstensen, C., Wriggers, P. (1996): On the symmetric boundary element method and the symmetric coupling of boundary elements and finite elements. To appear in IMA Journal of Numerical Analysis
7. Costabel, M. (1987): Symmetric methods for the coupling of finite elements and boundary elements. In: Brebbia, C.A., Wendland, W.L., Kuhn, G. (eds.) Boundary Elements IX. Springer, pp. 411–420
8. Costabel, M. (1988): Boundary integral operators on Lipschitz domains: Elementary results. SIAM J. Math. Anal. 19, 613–626
9. Dryja, M. (1982): A capacitance matrix method for Dirichlet problems on polygonal regions. Numerische Mathematik 39(1), 51–64
10. Funken, S.A., Stephan, E.P. (1995): Fast solvers with multigrid preconditioners for linear FEM–BEM coupling. Preprint, Universität Hannover, Institut für Angewandte Mathematik
11. Globisch, G. (1995): PARMESH - a parallel mesh generator. Parallel Computing 21(3), 509–524
12. Goppold, M., Haase, G., Heise, B., Kuhn, M. (1996): Preprocessing in BE/FE domain decomposition methods. Technical Report, Johannes Kepler University Linz. In Preparation.
13. Haase, G. (1995): Hierarchical extension operators plus smoothing in domain decomposition preconditioners. Report 497, Johannes Kepler University Linz, Institute of Mathematics
14. Haase, G., Heise, B., Jung, M., Kuhn, M. (1994): FEM⊕BEM - a parallel solver for linear and nonlinear coupled FE/BE-equations. DFG-Schwerpunkt "Randelementmethoden", Report 94-16, University Stuttgart
15. Haase, G., Langer, U. (1992): The non-overlapping domain decomposition multiplicative Schwarz method. International Journal of Computer Mathematics 44, 223–242
16. Haase, G., Langer, U. (1999): The efficient parallel solution of PDEs. To appear in Computers Math. Applic.
17. Haase, G., Langer, U., Meyer, A. (1990): A new approach to the Dirichlet domain decomposition method. In: Hengst, S. (ed.) Fifth Multigrid Seminar, Eberswalde 1990, Berlin. Karl–Weierstrass–Institut, pp. 1–59. Report R–MATH–09/90

18. Haase, G., Langer, U., Meyer, A. (1991): The approximate dirichlet domain decomposition method. Part I: An algebraic approach. Part II: Applications to 2nd-order elliptic boundary value problems. Computing 47, 137–151 (Part I), 153–167 (Part II)
19. Haase, G., Langer, U., Meyer, A. (1992): Domain decomposition preconditioners with inexact subdomain solvers. J. of Num. Lin. Alg. with Appl. 1, 27–42
20. Haase, G., Langer, U., Meyer, A., Nepomnyaschikh, S.V. (1994): Hierarchical extension operators and local multigrid methods in domain decomposition preconditioners. East-West J. Numer. Math. 2(3), 173–193
21. Hackbusch, W. (1985): Multi–Grid methods and applications, volume 4 of Springer Series in Computational Mathematics. Springer, Berlin
22. Hackbusch, W., Lage, L., Sauter, S.A. (1995): On the efficient realization of sparse matrix techniques for integral equations with focus on panel clustering, cubature and software design aspects. In this proceedings
23. Heise, B. (1991): Mehrgitter–Newton–Verfahren zur Berechnung nichtlinearer magnetischer Felder. Wissenschaftliche Schriftenreihe 4/1991, Technische Universität Chemnitz
24. Heise, B. (1993): Nonlinear field calculations with multigrid-Newton methods. IMPACT of Computing in Science and Engineering 5, 75–110
25. Heise, B. (1994): Analysis of a fully discrete finite element method for a nonlinear magnetic field problem. SIAM J. Numer. Anal. 31(3), 745–759
26. Heise, B. (1995): Nonlinear field simulation with FE domain decomposition methods on massively parallel computers. Submitted to Surveys on Mathematics for Industry
27. Heise, B. (1995): Nonlinear simulation of electromagnetic fields with domain decomposition methods on MIMD parallel computers. To appear in Journal of Computational and Applied Mathematics 62
28. Heise, B., Jung, M. (1995): Comparison of parallel solvers for nonlinear elliptic problems based on domain decomposition ideas. Report 494, Johannes Kepler University Linz, Institute of Mathematics
29. Heise, B., Jung, M. (1995): Robust parallel multilevel methods. Submitted for Publication
30. Heise, B., Kuhn, M. (1995): Parallel solvers for linear and nonlinear exterior magnetic field problems based upon coupled FE/BE formulations. To appear in Computing. Also Report No. 486, Inst. of Mathematics, Univ. Linz
31. Holzer, S. (1992): On the engineering analysis of 2D problems by the symmetric Galerkin boundary element method and coupled BEM/FEM. In: Kane, J.H., Maier, G., Tosaka, N., Atluri, S.N. (eds.) Advances in boundary element techniques. Springer, Berlin New York Heidelberg
32. Hsiao, G.C., Khoromskij, B.N., Wendland, W.L. (1994): Boundary integral operators and domain decomposition. Preprint 94–11, Universität Stuttgart, Mathematisches Institut A
33. Hsiao, G.C., Wendland, W.L. (1990): Domain decomposition in boundary element methods. In: Glowinski, R., Kuznetsov, Y.A., Meurant, G., Périaux, J., Widlund, O.B. (eds.) Proc. of IV Int. Symposium on Domain Decomposition Methods, Moscow, May 1990. Philadelphia, 1991. SIAM Publ., pp. 41–49
34. Khoromskij, B.N., Mazurkevich, G.E., Zhidkov, E.P. (1990): Domain decomposition method for magnetostatics nonlinear problems in combined formulation. Sov. J. Numer. Anal. Math. Modelling 5(2), 111–136
35. Khoromskij, B.N., Wendland, W.L. (1992): Spectrally equivalent preconditioners for boundary equations in substructuring techniques. East–West Journal of Numerical Mathematics 1(1), 1–26.

36. Kuhn, M., Langer, U. (1995): Parallel algorithms for symmetric boundary element equations. ZAMM, submitted for publication.
37. Langer, U. (1994): Parallel iterative solution of symmetric coupled FE/BE-equations via domain decomposition. Contemporary Mathematics 157, 335–344
38. Nepomnyaschikh, S.V. (1995): Optimal multilevel extension operators. Preprint SPC 95–3, Technische Universität Chemnitz–Zwickau, Fakultät für Mathematik
39. Rjasanow, S. (1990): Vorkonditionierte iterative Auflösung von Randelementgleichungen für die Dirichlet–Aufgabe. Wissenschaftliche Schriftenreihe 7/1990, Technische Universität Chemnitz
40. Rjasanow, S. (1995): Optimal preconditioner for boundary element formulation of the dirichlet problem in elasticity. Math. Methods in the Applied Sciences 18, 603–613
41. Schnack, E., Türke, K. (1995): Coupling of FEM and BEM for elastic structures. In this proceedings
42. Simon, H.D. (1991): Partitioning of unstructured problems for parallel processing. Comput. System in Eng. 2, 135–148
43. Steinbach, O., Wendland, W.L. (1995): Efficient preconditioners for boundary element methods and their use in domain decomposition. DFG-Schwerpunkt "Randelementmethoden", Report 95-19, University Stuttgart. Also in this proceedings.
44. Tong, C.H., Chan, T.F., Kuo, C.J. (1991): A domain decomposition preconditioner based on a change to a multilevel nodal basis. SIAM J. Sci. Stat. Comput. 12(6), 1486–1495

On the Use of a BEM Time-Stepping Procedure for Nonlinear and Unsteady Wave-Structure Interaction

O. Mahrenholtz, V. Schlegel, and C. Haack

1 Introduction

Computer aided analysis of fluid-structure interaction in ocean engineering problems requires efficient and reliable numerical methods. Due to free surface flow and gravity waves they have to consider: large and arbitrary shaped domains, free and moving surfaces, i.e. nonlinear and time-dependent boundary conditions. Assuming a two-dimensional potential flow, the direct boundary element formulation (BEM) is the most efficient solution method. Compared to other discretization methods it has minimum discretization expenses, a better representation of complex boundary geometries, and a simple approach to moving boundaries. This is achieved by coupling the BEM with a time-stepping procedure. For numerical simulation of time-dependent, nonlinear (overturning) gravity waves this approach, introduced by Longuet-Higgins and Cokelet [10], now is an optimized and reliable method in use, [6, 7, 12, 14]. Recently even a three-dimensional wave analysis is proposed: [1, 3, 15, 20]. The extension of this BEM time-stepping approach to wave-structure interaction is reasonable, [19], especially to wave channel simulations, [6, 18]. A direct BEM formulation allows for different kind of boundary conditions on each finite boundary element. Hence, this proposal pay regard to a wide range of different wave-structure interaction problems, ref. Fig. 1.1.

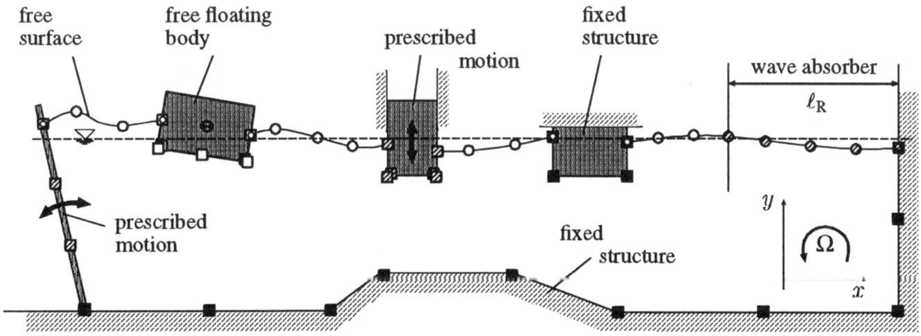

Fig. 1.1. Wave-structure interaction in wave tank applications

In this contribution some special aspects of BEM time-stepping procedures for the analysis of nonlinear and unsteady wave-structure interaction are discussed:

- space and time discretization (numerical stability, time-step control),
- fluid particles inside the domain,
- absorbing boundary condition,
- pressure formulations for freely and constrained floating structures.

The proposed numerical wave tank simulation obviously ask for proof of modeling and accuracy by wave tank experiments (typical laboratory facilities for the analysis of wave-structure interaction). Therefore, the aforementioned aspects are analyzed by direct comparison of experiment and BEM computation.

2 Wave-Structure Interaction

The interaction of gravity waves with fixed or floating structures is considerably complex, see Fig. 2.1. Within the time $[t_1, t_2]$ the geometry of the boundary Γ and the submerged surface of the structure is completely changing.

This is described by the kinematic and dynamic free surface condition and the equations of motion of structures, respectively. In case of closed domains (tanks, containers etc.) also a change of the mean water level (SWL) has to be considered.

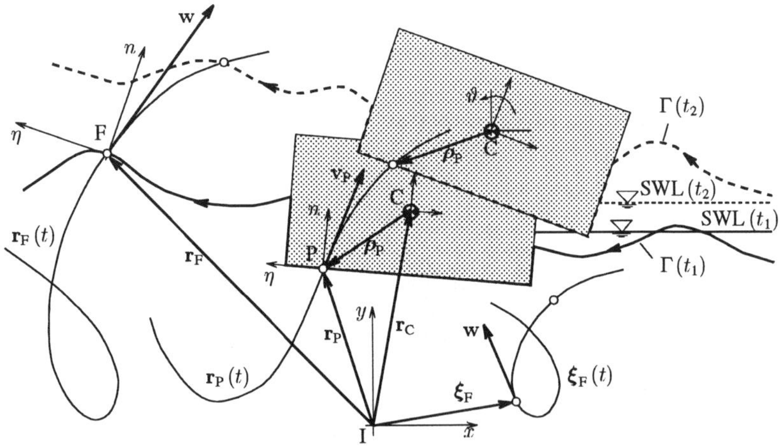

Fig. 2.1. Time-dependent wave-structure interaction

2.1 Free Surface Flow

If an irrotational flow and an incompressible fluid are assumed, the flow field $\mathbf{w}$ can be expressed by the gradient of a scalar velocity potential ϕ, i.e. $\mathbf{w} = \nabla\,\phi$. The conservation of mass now is described by Laplace's equation

$$\operatorname{div}\mathbf{w} = \operatorname{div}\nabla\,\phi = \nabla^2\phi = 0\,, \tag{2.1}$$

and the conservation of momentum reduces to the Euler equation (inviscid flow)

$$\varrho\,\frac{\mathrm{D}\mathbf{w}}{\mathrm{D}\,t} = \varrho\left(\frac{\partial\mathbf{w}}{\partial t} + \left(\mathbf{w}\nabla^{\mathrm{T}}\right)\mathbf{w}\right) = -\nabla\,p + \mathbf{k}\,, \tag{2.2}$$

with the fluid density ϱ, the pressure p and the specific volume forces $\mathbf{k}$. The integration of equation (2.2) with respect to spatial coordinates gives the scalar Bernoulli equation. For a point P, which is arbitrarily moved in the flow, it becomes the form

$$\left[\frac{\mathrm{d}\phi}{\mathrm{d}t}\right]_{\mathrm{P}} = \underbrace{-\frac{1}{\varrho}\,p - g\,y_{\mathrm{P}} - \frac{|\mathbf{w}|^2}{2}}_{\partial\phi/\partial t} + \mathbf{v}_{\mathrm{P}}^{\mathrm{T}}\,\mathbf{w}\,, \tag{2.3}$$

where g is the gravitational acceleration and y gives the vertical displacement. The nonlinear free surface flow is described by the kinematic and dynamic boundary condition of a fluid particle F on the free surface. From a Lagrangian formulation the velocity $\mathbf{v}_{\mathrm{F}}$ of the fluid particle is obtained as (kinematic free surface condition)

$$\mathbf{v}_{\mathrm{F}} = \frac{\mathrm{D}\mathbf{r}_{\mathrm{F}}}{\mathrm{D}t} = \frac{\partial\mathbf{r}_{\mathrm{F}}}{\partial t} + \left(\mathbf{w}\,\nabla^{\mathrm{T}}\right)\mathbf{r}_{\mathrm{F}} = \mathbf{w} = \nabla\,\phi\,, \tag{2.4}$$

where $\mathbf{r}_{\mathrm{F}}$ is the position vector of the fluid particle. With $\mathbf{v}_{\mathrm{F}} = \mathbf{w}$, and with the pressure at any position of the free surface is equal to the atmospheric pressure p_{amb}, equation (2.3) reads

$$\frac{\mathrm{D}\phi}{\mathrm{D}t} = \left[\frac{\mathrm{d}\phi}{\mathrm{d}t}\right]_{\mathrm{F}} = -\frac{1}{\varrho}\,p_{\mathrm{amb}} - g\,y_{\mathrm{F}} + \frac{|\mathbf{w}|^2}{2}\,, \tag{2.5}$$

and gives the time-dependent and nonlinear dynamic boundary condition of free surface flow.

2.2 Rigid Body Motion

Within the framework of potential theory and the herewith derived linear wave theories floating body dynamics are normally given by linearized equations of motion, [11]: i.e. small amplitude motions of waves and structures. In case of time-dependent nonlinear wave-structure interaction a more general formulation has to be used.

Therefore, the equations of motion of coupled or constrained floating structures are generated by the Newton-Euler formalism. Due to the couplings, the position vectors and the matrices of rotation of the considered bodies are no longer independent. The systems position now is described by (independent) generalized coordinates q_i of the vector $\mathbf{q}$. When introducing kinematical relations, the Newton-Euler equations are given by the applied $\boldsymbol{Q}_\mathrm{e}^*$ and constraint $\boldsymbol{Q}_\mathrm{z}^*$ forces and torques acting on the bodies

$$\mathbf{M}^* \, \ddot{\mathbf{q}} + \boldsymbol{K}^* = \boldsymbol{Q}_\mathrm{e}^* + \boldsymbol{Q}_\mathrm{z}^* \; . \tag{2.6}$$

Here, $\mathbf{M}^* = \bar{\mathbf{M}}\,\mathbf{J}$ denotes the product of the block diagonal mass matrix $\bar{\mathbf{M}}$ with the Jacobi matrix $\mathbf{J}$ of the system. All quantities independent of $\ddot{\mathbf{q}}$ are assigned to the vector of gyroscopic, centrifugal and Coriolis forces $\boldsymbol{K}^*$. The vector $\boldsymbol{Q}_\mathrm{e}^*$ describes the applied forces and torques, including hydrodynamic interaction, while the vector $\boldsymbol{Q}_\mathrm{z}^*$ denotes the constraint forces of the couplings. D'Alembert's principle is applied, in order to reduce the number of unknowns. Thus, the reduction of Newton-Euler equations (2.6) is achieved by left multiplication with the transposed Jacobi matrix $\mathbf{J}^\mathrm{T}$, and the constraint forces $\boldsymbol{Q}_\mathrm{z}^*$ are vanishing.

For an arbitrary point P (fixed, constrained or freely moving), see Fig. 2.1, the pressure is obtained by rearranging Bernoulli's equation (2.3) to the form

$$p\,(\mathbf{r}_\mathrm{P},\, t) = \varrho \left(-\,g\,(y_\mathrm{P} - y_\mathrm{SWL}) \;-\; \frac{|\mathbf{w}|^2}{2} \underbrace{+\,\mathbf{v}_\mathrm{P}^\mathrm{T}\,\mathbf{w} \;-\; \frac{\mathrm{d}\phi}{\mathrm{d}t}}_{-\,\partial\phi/\partial t} \right) . \tag{2.7}$$

Integrating the pressure field $p\,(\mathbf{x}_\mathrm{s}, t)$ on the submerged surface Γ_s of a structure gives the hydrodynamic force (see Fig. 2.2)

$$\boldsymbol{F}_\mathrm{h} = \int_{\Gamma_\mathrm{s}} p\,(\mathbf{x}_\mathrm{s}, t)\;\mathbf{n}(\mathbf{x}_\mathrm{s})\;\mathrm{d}\gamma \, , \tag{2.8}$$

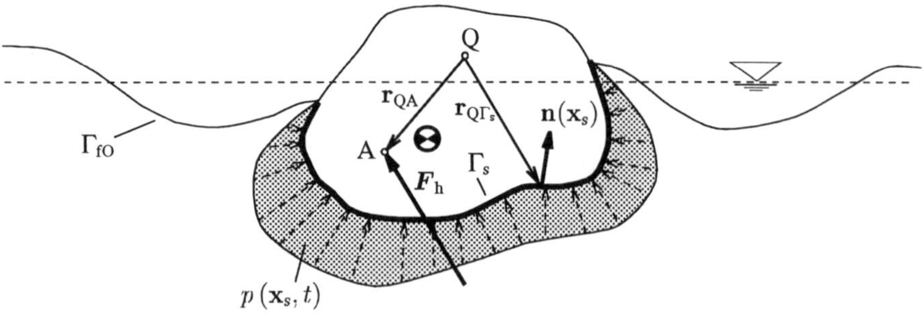

Fig. 2.2. Hydrodynamic pressure and forces

with the unit normal vector $\mathbf{n}(\mathbf{x}_s, t)$ directed out of the fluid domain. The position $\mathbf{x}_s$ of the submerged surface depends on the position $\mathbf{x}_{fO}(t)$ of the free surface (water sloshing, etc.), the position of the structure, and time. With respect to an arbitrary point Q, the resulting hydrodynamic torque reads

$$\boldsymbol{M}_{\mathrm{h}}^{\mathrm{Q}} = \int_{\Gamma_s} \mathbf{r}_{\mathrm{Q}\Gamma_s}(\mathbf{x}_s) \times \left(p(\mathbf{x}_s, t)\, \mathbf{n}(\mathbf{x}_s)\right) \mathrm{d}\gamma , \tag{2.9}$$

where $\mathbf{r}_{\mathrm{Q}\Gamma_s}(\mathbf{x}_s)$ is the time-depenent distance vector from Q to the submerged surface Γ_s, ref. Fig. 2.2. In case of two-dimensional applications the form

$$\boldsymbol{M}_{\mathrm{h}}^{\mathrm{Q}} = \mathbf{r}_{\mathrm{QA}} \times \boldsymbol{F}_{\mathrm{h}} \tag{2.10}$$

is more convenient. Here, the hydrodynamic forces are directly applied; however, the unknown time-dependent distance vector $\mathbf{r}_{\mathrm{QA}}$, connecting the two points Q and A, now has to be described. The equations (2.7)–(2.10) provide all relevant quantities of a hydrodynamic analysis.

3 BEM Time-Stepping Technique

The dynamics of wave-structure interaction is treated as an initial value problem given in state-space representation by the following system of coupled, nonlinear ordinary differential equations (ODEs)

$$\boldsymbol{X} = \begin{bmatrix} \boldsymbol{X}_1 = \mathbf{q} \\ \boldsymbol{X}_2 = \dfrac{\mathrm{d}\mathbf{q}}{\mathrm{d}t} \\ \boldsymbol{X}_3 = \mathbf{r}_{\mathrm{F}} \\ \boldsymbol{X}_4 = \phi(t) \end{bmatrix} , \quad \frac{\mathrm{d}\boldsymbol{X}}{\mathrm{d}t} = \begin{bmatrix} \boldsymbol{X}_2 \\ \mathrm{M}^{-1}\left(\boldsymbol{Q}_{\mathrm{e}} - \boldsymbol{K}\right) \\ \nabla \boldsymbol{X}_4 \\ -\dfrac{1}{\varrho} p_{\mathrm{amb}} - g\, y_{\mathrm{F}} + \dfrac{|\mathbf{w}|^2}{2} \end{bmatrix} . \tag{3.1}$$

Due to the pressure formulation (2.7), the vector of generalized forces $\boldsymbol{Q}_{\mathrm{e}}$ depends on ϕ and $\mathrm{d}\phi/\mathrm{d}t$. Therefore, the numerical solution procedure of wave-structure interaction has to consider both: the velocity potential ϕ and its time derivative.

By means of a BEM approach, see e.g. [2], Laplace's equation (2.1) is transformed to a boundary integral equation (BIE)

$$C(\boldsymbol{\xi})\, \phi(\boldsymbol{\xi}) = \int_{\Gamma(t)} w_n(\boldsymbol{x})\, G(\boldsymbol{x}, \boldsymbol{\xi}) - \phi(\boldsymbol{x}) \frac{\partial G(\boldsymbol{x}, \boldsymbol{\xi})}{\partial n} \, \mathrm{d}\Gamma(\boldsymbol{x}) , \tag{3.2}$$

where $\Gamma(t)$ is the boundary of the considered fluid domain, C is a geometry-dependent factor, G the fundamental solution of the problem, ϕ the velocity potential, $w_n = \partial\phi/\partial n$ its normal derivative and $\boldsymbol{x}$, $\boldsymbol{\xi}$ denote the position vectors for field and source points, respectively. This integral equation is discretized by NE finite boundary elements (Fig. 3.1) with the approximations $\tilde{\phi}$ and $\tilde{w}_n$ and results in

$$C_i\,\phi(\boldsymbol{\xi}_i) + \sum_{j=1}^{\mathrm{NE}} \fint_{\Gamma_j} \frac{\partial\,G(\boldsymbol{x},\boldsymbol{\xi}_i)}{\partial\,n}\,\tilde{\phi}(\boldsymbol{x})\;\mathrm{d}\,\Gamma = \sum_{j=1}^{\mathrm{NE}} \int_{\Gamma_j} \tilde{w}_n(\boldsymbol{x})\,G(\boldsymbol{x},\boldsymbol{\xi}_i)\;\mathrm{d}\,\Gamma, \quad (3.3)$$

with the nodes $\boldsymbol{\xi}_i$ of the boundary discretization.

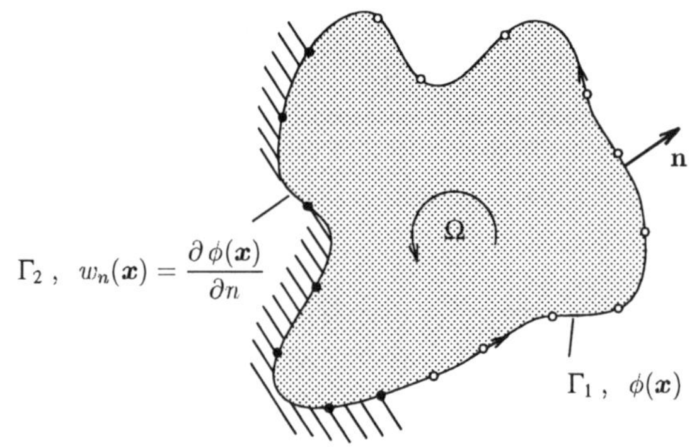

Fig. 3.1. Fluid domain Ω, boundary Γ, and boundary conditions ϕ and $\partial\phi/\partial n$

The main advantage of this approach is that on each boundary element either Dirichlet (ϕ) or Neumann ($\partial\phi/\partial n$) data can be given, see Fig. 3.1. Hence, it can be applied to a wide range of different problems, depending on the boundary conditions only. The approximation of the boundary (geometry and conditions) depends on the order of the Ansatz-functions or the type of boundary elements, respectively. For BEM simulations of nonlinear wave-structure interaction especially

- cubic spline-elements on the free surface and
- quadratic elements on all the other boundaries

show best results, [6], [8]. Depending on these element formulations the given boundary data ($\bar{\phi}(\boldsymbol{x})$, $\boldsymbol{x} \in \Gamma_1$ and $\bar{w}_n(\boldsymbol{x})$, $\boldsymbol{x} \in \Gamma_2$) is interpolated and the unknown boundary data ($\hat{\phi}(\boldsymbol{x})$, $\boldsymbol{x} \in \Gamma_2$ and $\hat{w}_n(\boldsymbol{x})$, $\boldsymbol{x} \in \Gamma_1$) is approximated. The BEM approach (3.3) now results in the more convenient matrix formulation

$$\underbrace{\left[\left(\hat{\mathbf{C}} + \hat{\mathbf{H}}\right) \quad -\hat{\mathbf{G}}\right]}_{\mathbf{A}_\Gamma} \underbrace{\begin{bmatrix} \hat{\varphi} \\ \hat{w}_n \end{bmatrix}}_{\mathbf{y}} = \underbrace{\left[\bar{\mathbf{G}} \quad -(\bar{\mathbf{C}} + \bar{\mathbf{H}})\right] \begin{bmatrix} \bar{w}_n \\ \bar{\varphi} \end{bmatrix}}_{\mathbf{b}}, \quad (3.4)$$

where the matrix of coefficients $\mathbf{A}_\Gamma$ typically is a dense and unsymmetric matrix. In this BEM proposal geometrical singularities of corners are treated by a double-node concept. The use of spline-elements asks for additional boundary data with respect to the local coordinate η of the finite boundary elements

$$\boldsymbol{\varphi} = \left[\varphi \quad \frac{\partial \varphi}{\partial \eta}\right]^{\mathrm{T}} \quad \text{and} \quad \boldsymbol{w}_n = \left[w_n \quad \frac{\partial w_n}{\partial \eta}\right]^{\mathrm{T}} . \tag{3.5}$$

This allows with the Jacobian matrix $\mathbf{J}$ direct computation of the kinematic boundary condition in (2.4) or (3.1), respectively

$$\frac{\mathsf{D}\mathbf{r}_\mathrm{F}}{\mathsf{D}t} = \mathbf{w} = \nabla\, \phi(\mathbf{r}_\mathrm{F},\, t) = \begin{bmatrix} \partial\phi/\partial x \\ \partial\phi/\partial y \end{bmatrix} = \mathbf{J}^{-1} \begin{bmatrix} \partial\phi/\partial\eta \\ \partial\phi/\partial n \end{bmatrix} . \tag{3.6}$$

Moreover, an additional smoothing (see e.g. [20]) is not necessary anymore. If using the double-node concept and spline-element formulations, a least-squares solution technique of the BIE (3.2) is reasonable

$$\mathbf{A}_\Gamma^\mathrm{T} \mathbf{Q} \mathbf{A}_\Gamma\, \mathbf{y} \;=\; \mathbf{A}_\Gamma^\mathrm{T} \mathbf{Q}\, \mathbf{b}\, . \tag{3.7}$$

All additional constraints are considered in the least-squares sense within the BEM solution procedure, with a now dense but symmetric and positive definite matrix of coefficients $\mathbf{A}_\Gamma^\mathrm{T}\mathbf{Q}\mathbf{A}_\Gamma$. Here, $\mathbf{Q}$ is a diagonal matrix of the integration weights multiplied by Jacobian's determinant.

3.1 Pressure Formulation

By means of the BEM solution and the state-space representation (3.1) all quantities of the pressure equation (2.7) are given now, except the time-derivative of the potential ϕ. Obviously there are two ways to describe the hydrodynamic pressure:

- to give an finite difference approximation of $\mathsf{d}\phi/\mathsf{d}t$ or
- to solve a BIE for $\nabla^2\, \partial\phi/\partial t = 0$ (though there is no physical interpretation).

Latter depends on the application. Besides ϕ and $\partial\phi/\partial n$, additional boundary conditions $\partial\phi/\partial t$ and $\partial^2\phi/\partial n\, \partial t$ are now required. In case of fixed structures [17], or linear dependencies this is possible [4, 19]. Considering nonlinear motion, it becomes most difficult, [5]. The additional BIE now depends on the problem under consideration.

Therefore, the more general finite difference approximation is used here:

$$\frac{\mathsf{d}\phi(\mathbf{r}_\mathrm{P}\,(t_j),\, t_j)}{\mathsf{d}t} = \frac{\phi\,(\mathbf{r}_\mathrm{P}\,(t_j + \Delta t_i),\, t_j + \Delta t_i) \;-\; \phi\,(\mathbf{r}_\mathrm{P}\,(t_j),\, t_j)}{\Delta t_i} . \tag{3.8}$$

This is easily implemented into a higher order time-stepping scheme, e.g. a 4th order Runge-Kutta method with equidistant time-levels (Newton's 3/8-rule). Moreover, equation (3.8) is independent of the problem and gives a local formulation. In the context of BEM: required data refers to single nodes on the boundary and is given directly by the BEM solution of (3.4). Once the hydrodynamic pressure is known, the hydrodynamic forces are computed and refering (3.1) they give the rigid body dynamics, i.e.

$$\nabla^2 \phi(t) \implies \phi(t) \implies p\,(\mathbf{r}_\mathrm{P},\, t) \implies \boldsymbol{F}_\mathrm{h}\,(\mathbf{x}_\mathrm{s}, t) \implies \ddot{\mathbf{q}},\, \dot{\mathbf{q}},\, \mathbf{q}\,. \tag{3.9}$$

Using basic kinematical relations the velocity of an arbitrary point P on the structure, ref. Fig. 2.1, is given in generalized coordinates $\mathbf{q}$ by:

$$\left[\mathbf{v}_\mathrm{P}\,(\dot{\mathbf{q}}, \mathbf{q}, t)\right]_\mathrm{I} = \left[\dot{\mathbf{r}}_\mathrm{C}\;(\dot{\mathbf{q}}, \mathbf{q}, t)\right]_\mathrm{I} \,+\, \boldsymbol{\omega}(\dot{\mathbf{q}}, \mathbf{q}, t) \,\times\, [\boldsymbol{\Omega}\,(\mathbf{q}, t)]_\mathrm{IC}\,[\boldsymbol{\rho}_\mathrm{P}]_\mathrm{C}\;, \tag{3.10}$$

with the position vector $\mathbf{r}_\mathrm{C}$ of the center of mass, the position vector ρ_P of the point P in a body-fixed frame, the vector of angular velocities $\boldsymbol{\omega}$, and the matrix of rotation $\boldsymbol{\Omega}$. The boundary condition on the submerged body surfaces now reads

$$w_n = \frac{\partial\phi(\mathbf{r}_\mathrm{P}, t)}{\partial n} = -\,\left[\mathbf{v}_\mathrm{P}^\mathrm{T}\,(\dot{\mathbf{q}}, \mathbf{q}, t)\right]_\mathrm{I}\;\mathbf{n}\,. \tag{3.11}$$

3.2 Space and Time Discretization

The solution of the proposed BEM time-stepping procedure depends on the order of approximation and the discretization in space and time. In order to improve accuracy and avoid (saw tooth) instabilities at the free surface higher order approximations are used, [6]:

- cubic spline-elements (BEM) and
- an explicit 4th order Runge-Kutta scheme.

Because the BEM is part of the operator of the initial value problem (3.1), the total error of the BEM domain solution has to be less than the total error of the Runge-Kutta method. In case of a smaller time-step, the space discretization has to be adapted, i.e. the number of boundary elements has to increase. Hence, the number of operations and CPU-time does not depend linearly on size and number of time-steps. For many practical applications a fraction $\Delta\boldsymbol{x}/\Delta t < 5$ of space/time discretization shows reliable results.

Another criterion that typically is used for these kind of problems, e.g. [13, 20], is the condition

$$\mathrm{CFL} = \left|\frac{\mathrm{d}\,\mathbf{r}(\boldsymbol{\xi})}{\mathrm{d}\,t}\right|\,\frac{\Delta t}{\Delta\boldsymbol{x}(t)} \;\leq\, 1\;, \tag{3.12}$$

of Courant-Friedrichs-Lewy (CFL). Thus, a node $\boldsymbol{\xi}(t)$ of the moving boundary $\Gamma(t)$ is allowed to move at most the distance $\Delta\boldsymbol{x}$ within the time Δt. During the numerical simulation the quantities $\Delta\boldsymbol{x}/\Delta t$ and CFL are computed on the critical, i.e. moving boundaries. They serve as a criterion for numerical stability as well as for time-step control.

4 Numerical Wave Tank Simulation

As mentioned typical laboratory facilities for wave-structure analysis are wave tanks. The main aspects of the proposed numerical method for the time-domain analysis of wave-structure interaction are directly compared with experimental data of the wave flume at hand: length $\ell = 12.3\,\text{m}$, width $b = 1.5\,\text{m}$, and water depth $d = 1\,\text{m}$. It is operated by a flap-type wave generator with or without a wave absorber at the end of the tank, ref. Fig. 1.1. The experimental set up gives e.g. wave elevations at different positions in the flume and the hydrodynamic pressure on the wave generator.

4.1 Space Discretization and Time-Step Control

The local representation of the time-dependent free surface wave elevation is depicted in Fig. 4.1. The solid line gives the experimental results of a wave probe and the symbols show the numerical results of the BEM time-stepping procedure, using 20 or 80 boundary elements on the free surface. Obviously the resolution of the gravity waves becomes more accurate the more elements are used. This test (both experiment and BEM) is carried out without any wave absorber in order to give the boundary conditions exactly at the end of the wave tank (no flux: $\partial\phi/\partial n = 0$). The time-stepping method uses a constant time-step $\Delta t \approx 10^{-2}\,\text{s}$, i.e. the real time simulation of $\approx 10\,\text{s}$ has exploited the BEM in the order of 10^3 times, without any instabilities.

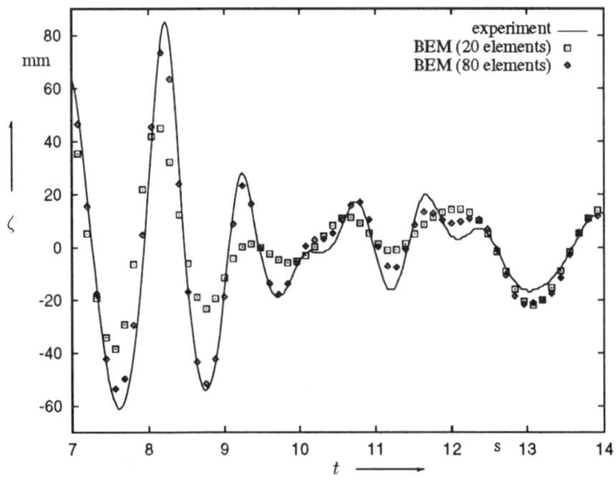

Fig. 4.1. Free surface wave elevation; comparison of experiment and BEM computation

While the results of Fig. 4.1 are achieved with a constant time-step, Fig. 4.2 compares the results with and without time-step control. Here the wave elevation is given for another test case. Reference (solid line) is the BEM time-stepping procedure with $\Delta t = 0.03\,\mathrm{s}$. The time-step is controlled by the CFL condition (3.12) and starts with $\Delta t = 0.12\,\mathrm{s}$. The wave elevation with time-step control shows some oscillation for $6\,\mathrm{s} < t < 8\,\mathrm{s}$ but still without instability, ref. Fig. 4.2. During computation the step size is changed, depending on CFL. Fig. 4.3 shows that CFL is larger with time-step control (dashed lines) compared to CFL with fixed time-step $\Delta t = 0.03\,\mathrm{s}$ (solid lines). As stability or control criterion CFL < 0.7 shows best results. Besides giving a reliable stability criterion, this enables a CPU-time reduction (speed-up $2 \div 3$). Space discretization is fixed to the finest grid.

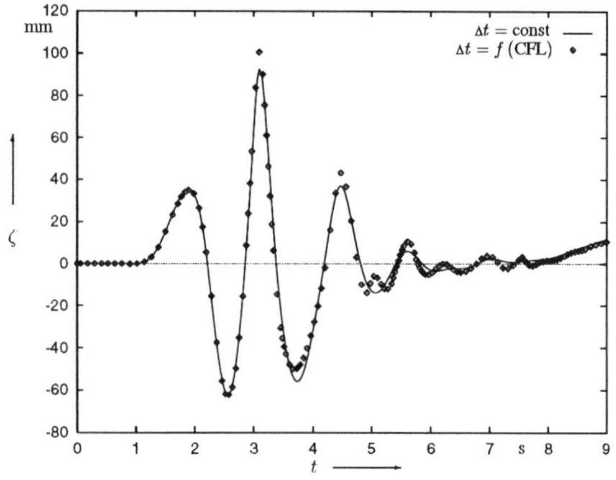

Fig. 4.2. Free surface wave elevation; computations with ($\Delta t = f(\mathrm{CFL}, t)$) and without ($\Delta t = const$) time-step control

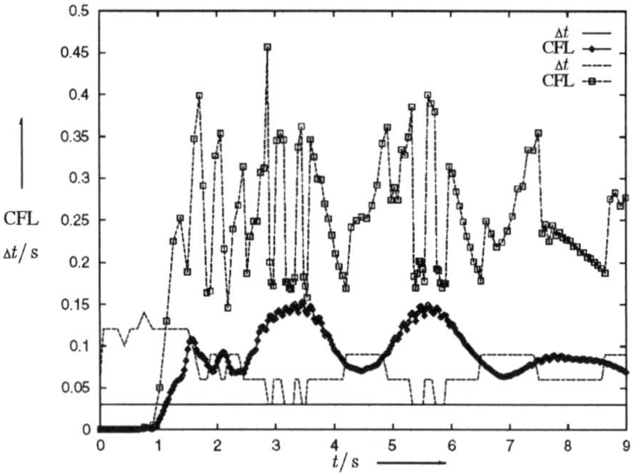

Fig. 4.3. Time histories of time-step and CFL condition with (dashed lines: $\Delta t = f(\mathrm{CFL}, t)$) and without (solid lines: $\Delta t = const$) step-size control

4.2 Trajectories of Fluid Particles Inside the Domain

The Lagrangian formulation (2.4) of the free surface kinematic boundary condition enables a simple approach for particle tracing on the free surface. A detailed analysis of fluid particle dynamics near fixed or floating structures is given by [9]. In some cases, [17], among free surface flow the fluid domain is of interest also — measurement becomes here most difficult. The described BEM time-stepping analysis of wave-structure interactions allows for particle tracing inside the domain, see Fig. 2.1

$$\boldsymbol{\xi}_{\mathrm{F}}(t) = \boldsymbol{\xi}_{\mathrm{F}}(t_0) + \int\limits_{t_0}^{t_{\mathrm{end}}} \mathbf{w} \, \mathrm{d}t \,. \tag{4.1}$$

Once all boundary data is given on the boundary $\Gamma(t)$ of the fluid domain Ω the flow field $\mathbf{w} = \nabla\phi$ can be expressed at any point $\boldsymbol{\xi}_{\mathrm{F}}$. The BIE (3.2) now becomes

$$\frac{\partial\, \phi(\boldsymbol{\xi}_{\mathrm{F}})}{\partial x_i} = \int\limits_{\Gamma} \frac{\partial \phi(\boldsymbol{x})}{\partial n} \frac{\partial G(\boldsymbol{x}, \boldsymbol{\xi}_{\mathrm{F}})}{\partial x_i} - \frac{\partial^2\, G(\boldsymbol{x}, \boldsymbol{\xi}_{\mathrm{F}})}{\partial n\, \partial x_i}\, \phi(\boldsymbol{x}) \; \mathrm{d}\,\Gamma(\boldsymbol{x}) \,, \tag{4.2}$$

with x_i the cartesian coordinates. The gradient ∇G of the fundamental solution is

$$\frac{\partial G}{\partial x_i} = \frac{-\, r_i}{2^{(\dim - 1)}\, \pi\, (\boldsymbol{r}^{\mathrm{T}} \boldsymbol{r})^{\dim/2}} \,, \tag{4.3}$$

where dim denotes the problems dimension, and r_i are the components of the distance vector $\boldsymbol{r} = \boldsymbol{x} - \boldsymbol{\xi}_{\mathrm{F}}$. The normal derivative of the gradient $\nabla(\nabla^{\mathrm{T}}\, G)\, \mathbf{n}$ reads

$$\frac{\partial^2 G}{\partial n\, \partial x_i} = \frac{\dim\, r_i\, \boldsymbol{r}^{\mathrm{T}} \mathbf{n}}{2^{(\dim - 1)}\, \pi\, (\boldsymbol{r}^{\mathrm{T}} \boldsymbol{r})^{(\dim/2+1)}} - \frac{n_i}{2^{(\dim - 1)}\, \pi\, (\boldsymbol{r}^{\mathrm{T}} \boldsymbol{r})^{\dim/2}} \,, \tag{4.4}$$

with n_i components of the normal vector $\mathbf{n}$.

The basic result of particle tracing inside the domain shows Fig. 4.4 for the simulation of linear gravity waves. In this case particle dynamics is described by wave theories. Numerical simulations with the BEM time-stepping procedure pays regard to three significant items of linear wave theory, see Fig. 4.4:

- in case of deep water, with $d/\lambda > 1/2$, particle trajectories are circles with decreasing diameters for greater water depth,
- down to the bottom of the wave tank an increasing horizontal movement in case of shallow water waves ($d/\lambda \ll 1$),
- in all cases the vertical component of the fluid particle orbits is vanishing near the bottom.

Note that Fig. 4.4 shows no circles due to different scaling of x, y-axes.

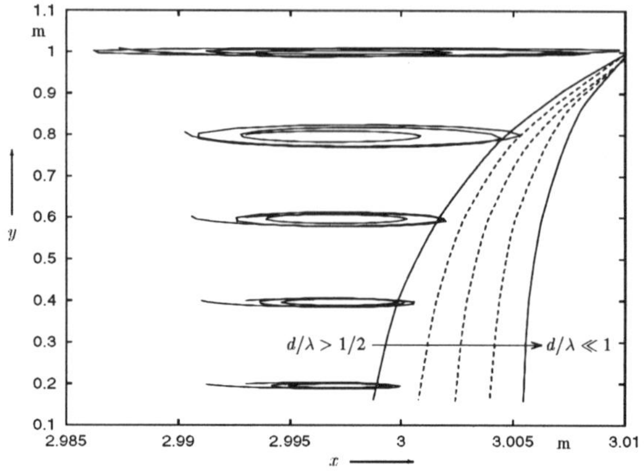

Fig. 4.4. Numerical results of particle-tracing; fluid particles inside linear gravity waves

4.3 Absorbing Boundary Condition

In order to avoid reflections linear wave tank analysis of wave-structure interaction makes use of wave absorbers at the end of the flume. Among many others, ref. [16], in case of a BEM time-stepping procedure a modified dynamic free surface boundary condition (2.5) can be used

$$\frac{\mathsf{D}\,\phi(\mathbf{r}_\mathrm{F},\,t)}{\mathsf{D}t} = -\frac{1}{\varrho}\,p_\mathrm{amb} - g\,y_\mathrm{F} + \frac{|\mathbf{w}|^2}{2} - \mu_\mathrm{R}\,\phi(\mathbf{r}_\mathrm{F},\,t)\,. \tag{4.5}$$

The added Rayleigh damping $\mu_\mathrm{R}\,\phi(\mathbf{r}_\mathrm{F},\,t)$ results in an irrotational, inviscid dissipation of energy. Compared to other approaches the main advantage again is in context of BEM: on certain nodes of the boundary discretization the boundary data is modified. This is easily implemented in the BEM time-stepping method.

Anyway, now the parameter μ_R of Rayleigh damping has to be identified. Best result is achieved with a linearly increasing distribution (up to $\mu_{\mathrm{R\,max}}$) on the damping zone, ref. Fig. 2.1. If the length ℓ_R of the damping zone is adapted to the wave length λ an optimal parameter value $\mu_{\mathrm{R\,max}} = 4\ \mathrm{s}^{-1}$ is found for a wide range of different applications (wave height, steepness, frequency).

Fig. 4.5 gives the damping of a wave group. Here the wave elevations at different positions in the flume are depicted. Experimental and numerical BEM data show very good agreement at these points. Hence, the Rayleigh damping boundary condition (4.5) pays regard for the absorption of the wave group in the wave tank. In addition Fig. 4.5 shows the numerical results in case of no artificial damping, i.e. no flux condition at the end of the flume. Large reflections are evident.

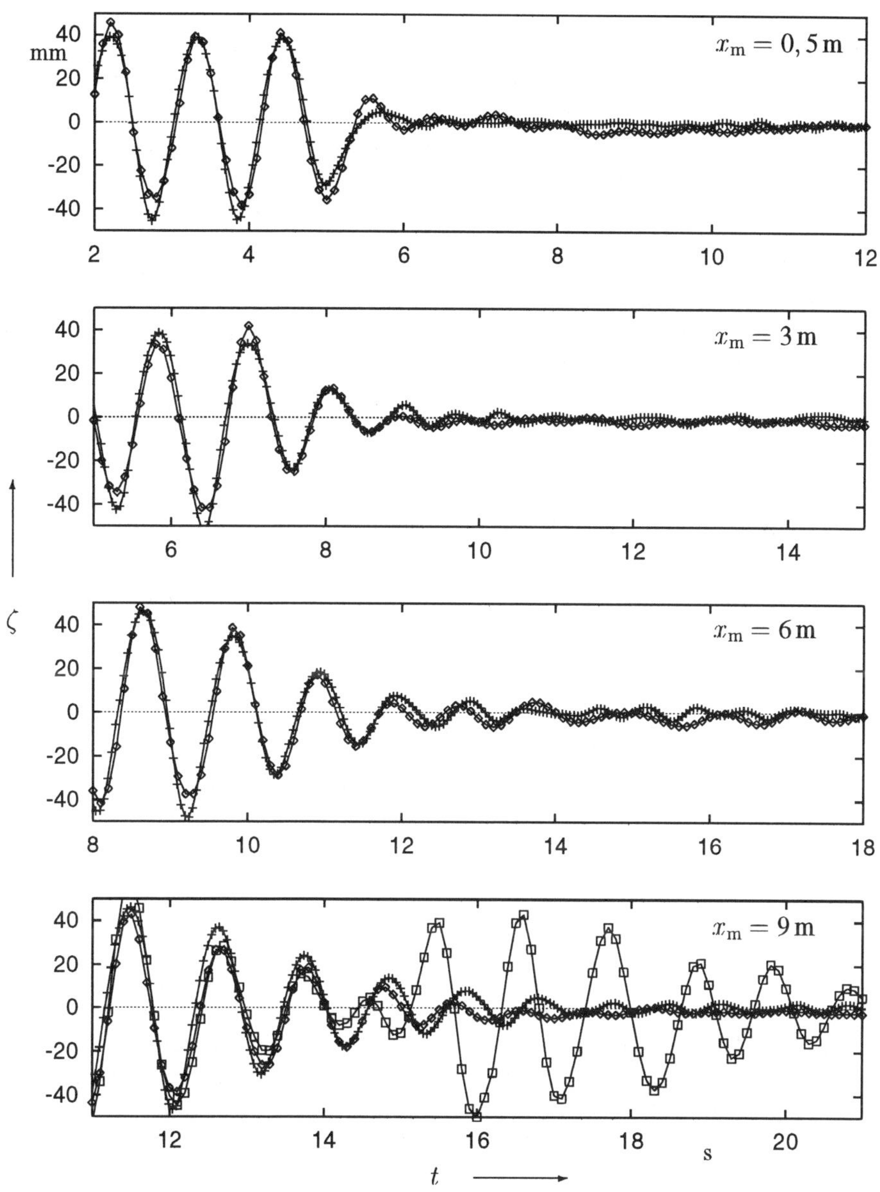

Fig. 4.5. Damping of free surface elevation at different positions x_m of the wave tank; + : experiment, ◊ : BEM with Rayleigh damping, □ : BEM without damping

4.4 Pressure Computation

The main aspect of wave-structure interaction is the hydrodynamic pressure in case of large amplitude gravity waves. This gives the dynamics of floating, and the loads on fixed or moved structures. The basic example of wave-structure interaction in wave tank experiments is the wave generator problem.

In order to proof modeling and accuracy of the proposed pressure computation the BEM time-stepping results are compared with experimental pressure data at hand. Fig. 4.6 shows the hydrodynamic pressure p_{dyn} (without the hydrostatic component) at different positions D1 ... D5 on the flap type wave generator. The positions (depth) with regard to SWL are: D1=0 mm, D2=25 mm, D3=75 mm, D4=175 mm, D5=375 mm. Boundary conditions on the wave generator are given as input for the BEM by means of additional measurement data (position and acceleration). Fig. 4.6 depicts the hydrodynamic pressure p_{dyn}. Solid lines give the results of experiments, and symbols the results of BEM computations, respectively.

The direct comparison of experiment and BEM computation shows that the proposed numerical methods gives wave-structure interaction in detail:

- shocks during starting procedures,
- large amplitudes of the hydrodynamic pressure due to sloshing effects,
- emerging of pressure transducers (D1),
- negative hydrodynamic components in greater water depth.

5 Constrained Floating Bodies

The interaction of gravity waves with fixed or floating structures is a highly developed subject and many mathematical models are available in case of small amplitudes, [11]. For either large amplitude waves or large or constrained motions of structures most of the commonly used models failed due to their restrictions. A typical example of freely and constrained floating bodies is depicted in Fig. 5.1.

Time-domain simulations of floating structures in free surface flow by means of equations (3.1) and (3.4) allow a more detailed analysis of these considerably complex wave-structure interactions. The described procedure can generally be applied to a wide range of different problems. Besides, the hydrodynamic loads acting on the structures, this formulation gives e.g. the moving boundaries, the velocity potential, and the flow field even in the case of large amplitude motions.

Fig. 5.2 shows the initial discretization of the aforementioned example in the area of the floating structures (Body 2, Body 3 ref. Fig. 5.1). The time evolution of the moving boundaries (free surfaces and submerged structures) is given in Fig. 5.3. Time histories of generalized coordinates $\mathbf{q} = [\vartheta_1(t) \quad \vartheta_2(t)]^{\mathrm{T}}$ are depicted in Fig. 5.4. With these results the position and velocity of arbitrary points on the rigid bodies (e.g. center of mass) can also be given by (time-dependent and nonlinear) kinematical relations.

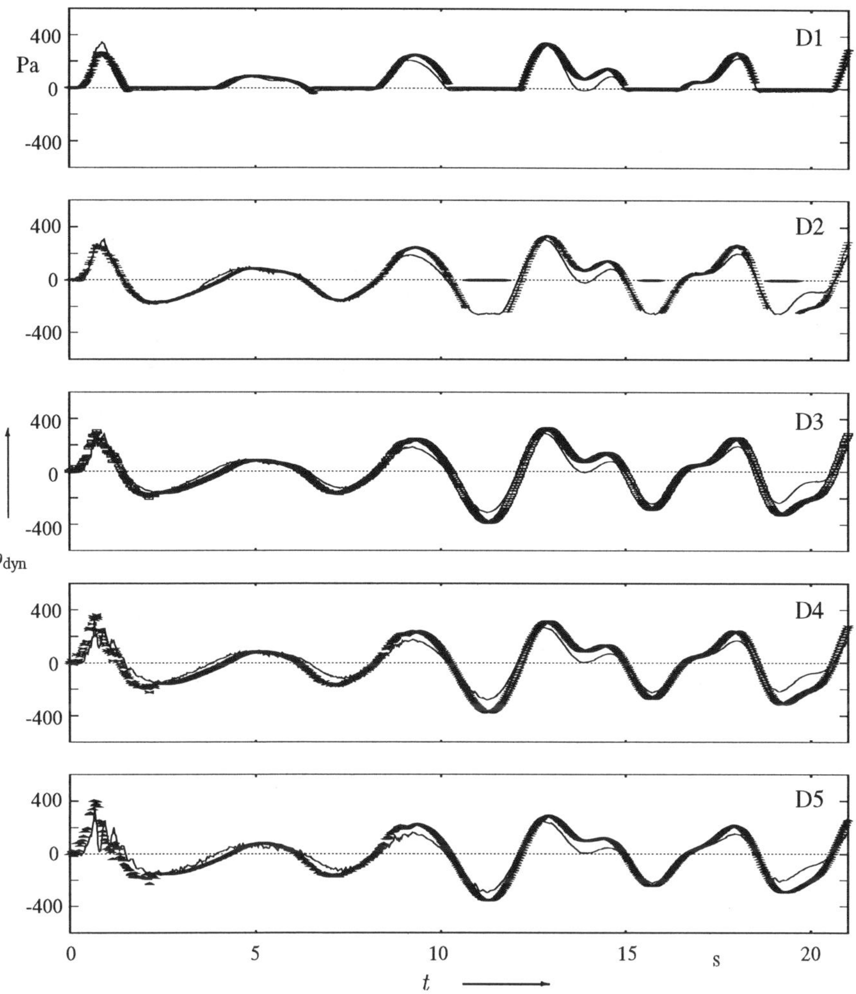

Fig. 4.6. Hydrodynamic pressure on a flap type wave generator; comparison of experiment (solid lines) and BEM computation (symbols) at the positions D1 ... D5

6 Conclusions

A boundary element formulation for the analysis of fluid-structure interaction problems considering nonlinear gravity waves is presented. Because all quantities of interest are located on the boundary itself and regarding recent developments in boundary element methods (BEM) this is a reliable, most efficient approach with minimal discretization expenses. The major advantage

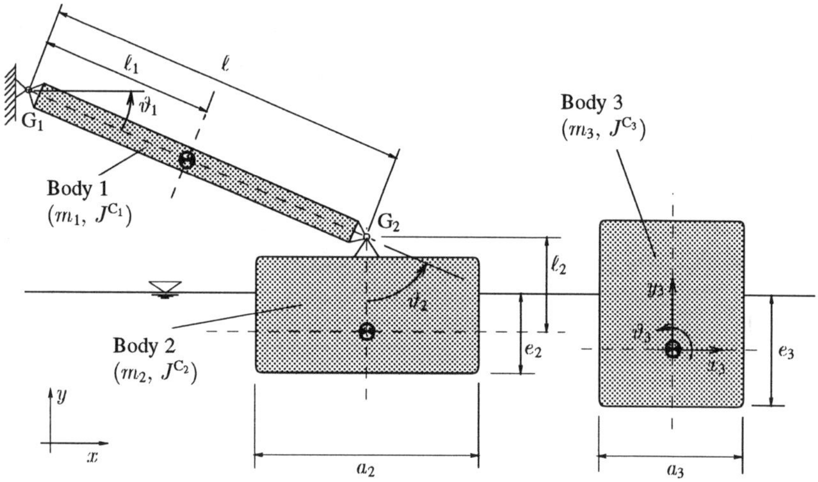

Fig. 5.1. Definition sketch of freely and constrained floating bodies

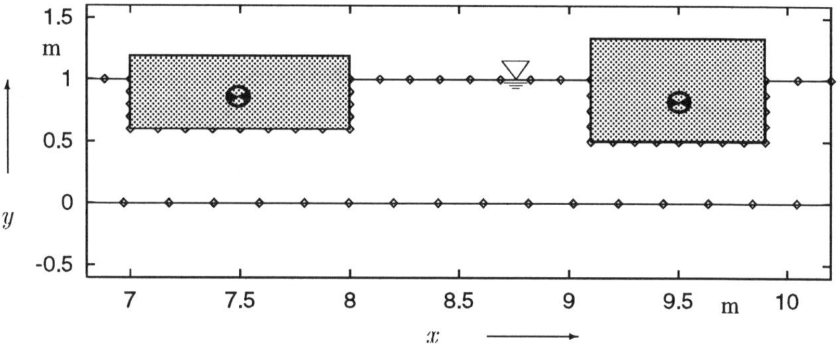

Fig. 5.2. Boundary discretization in the area of the floating structures (Body 2, Body 3)

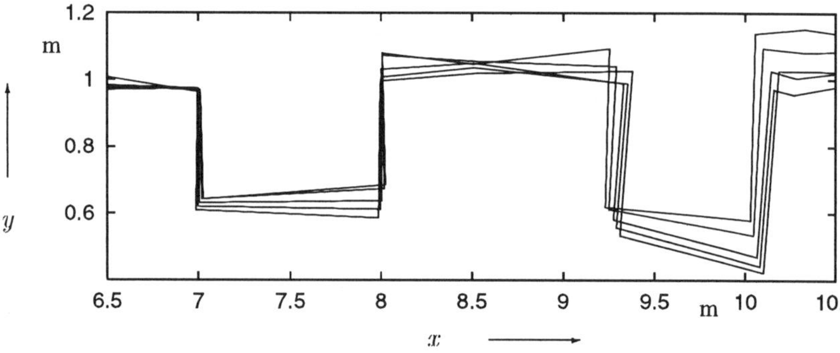

Fig. 5.3. Time evolution of the boundary in the area of the floating structures (Body 2, Body 3)

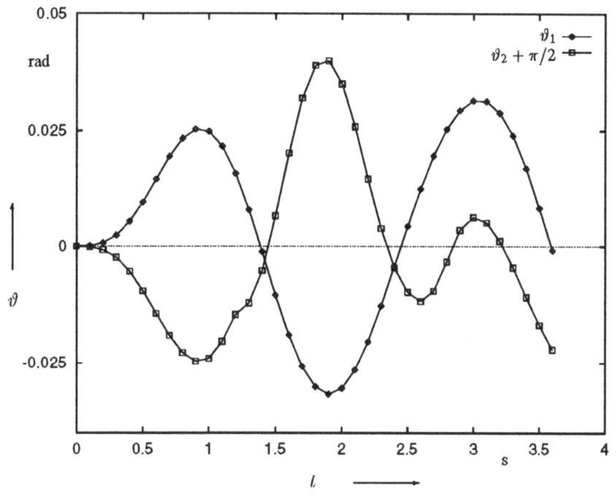

Fig. 5.4. Time histories of generalized coordinates $\mathbf{q}(t)$

of a BEM formulation is the direct numerical implementation of the equations of motion in an explicit form: for fluid particles on the free surface and for freely or constrained floating bodies. The coupling with a time-stepping procedure allows even for time-dependent and nonlinear conditions.

In particular this approach provides the extension to a numerical wave channel concept. The proposed method is applied to numerical simulations of wave tank experiments. Besides a direct comparison with the BEM computation, measurement data provide a more detailed analysis of the numerical scheme, e.g. with regard to space and time discretization, order of discretization, types of boundary elements, CPU-time requirement, stability criterions or pressure formulations.

A criterion is given that allows for time-step control of a higher order Runge-Kutta method, and pays regard to BEM discretization in case of moving boundaries. The BEM time-stepping procedure is a simple approach to trajectories of fluid particles on the free surface and inside the domain. Particle tracing shows good agreement with basic analytical results of linear wave theory. The presented pressure formulation is independent of the form of application. The computed hydrodynamic pressure takes hold of time-dependent, large amplitude motion (water sloshing). The main features of this approach for freely and constrained floating body (multi-body) dynamics are presented. In context of BEM: the analysis takes full advantage of physical data at discrete points (nodes) of the boundary only.

References

1. Boo, S.Y., Kim, C.H., Kim, M.H. (1994): A numerical wave tank for nonlinear irregular waves by 3-D higher order boundary element method. International Journal of Offshore and Polar Engineering 4, 265–272
2. Brebbia, C.A., Telles, J.C.F., Wrobel, L.C. (1984): Boundary element techniques. Springer, Berlin

3. Broeze, J. (1993): Numerical modelling of nonlinear free surface waves with a 3D panel method. PhD thesis, Enschede: University of Twente
4. Cointe, R., Geyer, P., King, B., Molin, B., Tramoni, M. (1991): Nonlinear and linear motions of a rectangular barge in a perfect fluid. In: Eighteenth Symposium on Naval Hydrodynamics, Washington, D.C.. National Academy Press, pp. 85–99
5. van Daalen, E.F.G. (1993) : Numerical and theoretical studies of water waves and floating bodies. PhD thesis, Enschede: University of Twente
6. Gravert, P. (1987): Numerische Simulation extremer Schwerewellen im Zeitbereich mit direkter Randelementmethode und Zeitschrittverfahren. Fortschr.-Ber. VDI Reihe 7, Nr. 132. VDI-Verlag, Düsseldorf
7. Grilli, S.T., Svendsen, I.A. (1990): Corner problems and global accuracy in the boundary element solution of nonlinear wave flows. Engineering Analysis with Boundary Elements 7, 178–195
8. Haack, C., Mahrenholtz, O., Schlegel, V. (1993): The interaction of nonlinear gravity waves with fixed and floating structures. In: Brebbia, C.A., Rencis, J.J. (eds.) Boundary Elements XV, Vol. 1: Fluid Flow and Computational Aspects. CMP, Southampton, pp. 249–262
9. Haack, C., Mahrenholtz, O., Schlegel, V. (1994): A direct BEM formulation for nonlinear free surface flow near fixed or floating structures. In: Wagner, S., Hirschel, E.H., Périaux, J., Piva, R. (eds.) Proc. 2nd European Computational Fluid Dynamics Conference (Stuttgart, Germany, 1994). Wiey, Chichester, pp. 873–879
10. Longuet-Higgins, M.S., Cokelet, E.D. (1976): The deformation of steep surface waves on water. A numerical method for computation. Proc. Roy. Soc. A 350, 1–26
11. Mei, C.C. (1989): The Applied Dynamics of Ocean Surface Waves. Advanced Series on Ocean Engineering, Vol. 1. World Scientific, Singapore
12. New, A.L., McIver, P., Peregrine, D.H. (1985): Computation of overturning waves. J. Fluid Mech. 150, 233–251
13. Pawlowski, J.S., Baddour, R.E. (1994): Recent advances in numerical modelling of non-linear propagation of water waves and their interaction with impermeable bodies. In: Rahman, M. (ed.) Ocean Waves Engineering. CMP, Southampton, pp. 165–200
14. Peregrine, D.H. (1987): Steep unsteady water waves and boundary integral methods. In: Cruse, Th.A. (ed.), Advanced Boundary Element Methods. Springer, Berlin, pp. 287–295
15. Romate, J.E. (1989): The numerical simulation of nonlinear gravity waves in three dimensions using a higher order panel method. PhD thesis, Enschede: University of Twente
16. Romate, J.E. (1992): Absorbing boundary conditions for free surface waves. J. of Comp. Physics 99, 135–145
17. Skourup, J., Jonsson, I.G. (1992): Computations of forces on, and particles around, horizontal cylinders under steep waves. Ocean Engng 19, 527–553
18. Skourup, J., Sterndorff, M.J., Hansen, E.A. (1992): Numerical modelling of wave-structure interaction by a three-dimensional non-linear boundary element method: A step towards the numerical wave tank. Ocean Engng 19, 437–460
19. Vinje, T., Brevig, P. (1981): Nonlinear ship motions. In: 3rd International Conference on Numerical Ship Hydrodynamics (Paris, France), pp. 257–266
20. Xü, H., Yue, D. (1994): Computations of fully-nonlinear three-dimensional water waves. In: Nineteenth Symposium on Naval Hydrodynamics. National Academy Press, Washington D.C., pp. 177–201

Integral Equations arising from Instationary Flows around Thin Wings

R. Hinder and E. Meister

1 Introduction

Within the scope of a three-dimensional linear theory we analyse a mixed screen boundary value problem for the Helmholtz equation $(\triangle + k^2)\Phi = 0$ where Φ is a perturbation velocity potential, generated by the presence of an oscillating wing in a basic flow.

Not all Cauchy data for Φ are given explicitly on the two sides of the screen. The missing Cauchy data depend on the wing circulation Γ and must be evaluated by the Kutta condition: $\nabla\Phi$ is finite near the trailing edge x_t of the wing. Using the boundary integral method we transform this mixed screen boundary value problem into the hypersingular boundary integral equation $D_L\widetilde{\Psi} = h(\Gamma)$, where D_L denotes the normal derivative of the double layer potential with respect to the projection L of the lifting surface on the screen, the unknown density $\widetilde{\Psi}$ and a right-hand side h depending on the unknown wing circulation Γ. To fulfil the Kutta condition in such a way that all appearing terms can be defined mathematically exactly and belong to spaces which are physically meaningful, we propose to fix Γ by the condition of vanishing stress intensity factors of Φ near x_t up to a certain order such that $\nabla\Phi|x_t \in W_2^\sigma(x_t) \subset L_2(x_t)$ for appropriate $\sigma > 0$.

In the two-dimensional case, and if L is the left half-plane in $\mathbf{R}^2$, we have an explicit formula to calculate Γ and we can control the regularity of Γ and Φ. In the case of a polygonal wing projection L we propose a strategy to calculate Γ by minimizing the stress intensity factors near x_t in the sense of least squares.

2 The problem

Our aim is to make a contribution to answer the question why aeroplanes can fly. Of course, one answer to this question is, that aeroplanes fly on account of a higher pressure on the lowerside than on the upperside of the wings of the aeroplane. But to give more precise answer to this question one needs a detailed knowledge of the distribution of the pressure, respectively of the velocity, at the sides of the wings.

One main aim for a mathematician working on an engineering problem is to prove the convergence of numerical algorithms for the mathematical problems. But in order to do this, one needs a detailed knowledge about solution spaces and their corresponding norms, i.e. one needs an existence,- uniqueness- and regularity theory for the specific problem. But such a strict mathematical theory is, up to now, not available for lifting surfaces, especially in the full three-dimensional situation. One even cannot define fundamental quantities appearing in the models without mathematical ambiguity.

Of course, a lot of models for calculating the lift, which have been realized on computers, can be found in the literature using modern numerical techniques, e.g. the evaluation of singular and hypersingular integrals. But the important fundamental problem of the correct solution spaces has not been solved.

Hence, one needs particularly existence,- uniqueness- and regularity results in such a way that the appearing quantities can be defined exactly and belong to physically reasonable spaces.

This philosophy is the starting point for our investigations. Of course, in these first attempts we have to admit a lot of idealizations, which are not approved from the practical point of view.

We now formulate under certain idealizing assumptions within the scope of a linear three-dimensional theory a model for the calculation of a perturbation velocity potential Φ, generated by the presence of an oscillating wing in a basic subsonic flow, which seems to be suitable for proving results like those mentioned above.

Starting point for our investigations is a fourthysix years old paper by Reissner [18], who formulated under similarly idealizing assumptions conditions which have to be fulfilled by Φ. These conditions have been reformulated by Meister [13] in the two-dimensional case as a more handable model and based on his ideas by Hebeker [8] in the three-dimensional situation. This model, which we shall recall shortly, has been studied in more details in the papers [9] and [11].

We assume that the gas flows with a constant subsonic velocity in the positive x-direction and is inviscid, barotropic, compressible, and weakly damping. Furthermore, we assume that the velocity of the flowing gas is small, compared with the speed of sound.

We assume as an obstacle a thin, weakly cambered, time harmonically oscillating wing into the region of the flowing gas which is divided into an upper and a lower part, respectively, by the (x,y)-plane.

Note, that the vorticity theorem of Helmholtz implies the existence of a region of vortices behind the wing, the wake, which we have to admit in the model.

On account of our assumptions (small velocity, thin and weakly cambered wing), the normal component of the velocity is prescribed on the wingsurfaces by its form ([1], [18] and others). Therefore, one can represent the vertical component of the velocity on the upper and lower side of the wing as functions on the projection L of the wing onto the (x,y)-plane.

Now one postulates within the scope of a linearized theory the existence of a perturbation velocity potential Φ, induced by the presence of the wing, such that

$$\text{Total potential} \;=\; \underbrace{\text{perturbation potential } \Phi}_{\text{"}small\text{"}} + \text{ potential without wing,}$$

for which one calculates the boundary conditions linearized in the (x,y)-plane [1, 18].

The functions, obtained from the requirement of a total vanishing normal velocity at the wingsurfaces, can be split into a symmetric and an antisymmetric part. It is known, that the symmetric part gives no contributions to the lift, so that we can assume a given Neumann data $g = \frac{\partial \Phi}{\partial n}$ on the projection L of the wings onto the (x,y)-plane [1, p. 210].

We agree upon the following notations and assumptions:

L denotes the projection of the wing
W denotes the projection of the region of shed vortices
} onto the (x,y)-plane.

We assume:

- L is a bounded, simply connected Lipschitz domain.
- The two straight lines bounding the strip I touch the boundary ∂L of L exactly in one point, the tips of the lifting surface. After removing these two points from ∂L, it is divided into a leading edge x_l and a trailing edge x_t, which we assume to be parametrisized by $x_l = x_l(y), x_t = x_t(y), y \in (y_{min}, y_{max})$. Every straight line in the strip I, and parallel to the x-axis, is assumed to intersect x_l and x_t in exactly one point [2].

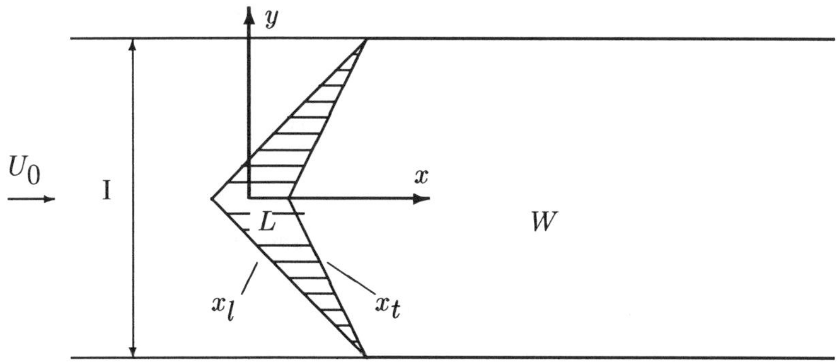

In the following we denote by S the open screen $L \cup x_t \cup W$ where $W := \{(x, y) : y \in (y_{min}, y_{max}), x > x_t(y)\}$.

Finally, we agree that the normal vector n on S is the outer normal of $\mathbf{R}^3_+ := \{(x,y,z) \in \mathbf{R}^3 : z > 0\}$ and that the potential jump on S is the "limit from $\mathbf{R}^3_-$ to S minus the limit from $\mathbf{R}^3_+$ to S" and denote this limit with $[\Phi]$, respectively, $\left[\frac{\partial \Phi}{\partial n}\right]$.

Now we can formulate the problem of calculating the perturbation velocity potential Φ explicitly.

The perturbation potential Φ (induced by a time harmonically oscillating wing) has to fulfil — after a reductionprocedure — the Helmholtz equation

$$(\Delta + k^2)\Phi = 0 \text{ in } \mathbf{R}^3 \backslash \overline{S}, \ \ \text{Re } k > 0\,, \ \ \text{Im } k > 0 \tag{2.1}$$

where k denotes the reduced wavenumber of the oscillating wing ([1], [13]). Re k, Im $k > 0$ hold on account of our assumption of a weakly damping gas.

The normal derivative $\frac{\partial \Phi}{\partial n}$ on L is determined by the assumption of a vanishing normal component of the total velocity at the wings:

$$\frac{\partial \Phi}{\partial n}(x,y) := g(x,y) \text{ on } L. \tag{2.2}$$

It is known ([1], [18]), that

$$\left[\frac{\partial \Phi}{\partial n}\right] = 0 \text{ on } W \tag{2.3}$$

and (2.2) implies

$$\left[\frac{\partial \Phi}{\partial n}\right] = 0 \text{ on } L. \tag{2.4}$$

In the case of a time harmonically oscillating wing the perturbation pressure p_S and the perturbation velocity potential Φ are connected by

$$p_S\,(x,y,z) = c\,(x,y) \left(\frac{\partial}{\partial x} - i\frac{k}{M}\right) \Phi\,(x,y,z) \tag{2.5}$$

with a continuous function c and the Machnumber M (here: $M < 1$).

It is well-known that the wake cannot carry a jump of the pressure:

$$[p_S] = 0 \text{ on } W. \tag{2.6}$$

Furthermore, we assume a condition of finite energy:

$$\Phi \in W_2^1(\mathbf{R}^3 \backslash \overline{S})\,, \tag{2.7}$$

where W_2^1 denotes the usual Sobolev-Slobodeckij space.

Finally, Φ has to fulfil the Kutta condition in a sense prescribed precisely in latter paragraphs. Classically, (e.g. [18, p. 826]) the Kutta condition is usually formulated in the form:

$$\left.\begin{array}{l}\text{The perturbation velocity } \nabla\Phi \text{ remains finite by an}\\ \text{(three-dimensional) approach}\\ \text{to the trailing edge } x_t \text{ of the wing projection } L\end{array}\right\}. \tag{2.8}$$

Following [13], [8] we reformulate some of these conditions in form of a more suitable model.

From (2.5) und (2.6) we obtain

$$0 = [ps]_W = c(\frac{\partial}{\partial x}[\Phi]_W - i\frac{k}{M}[\Phi]_W),$$

$$\text{i.e. } [\Phi]_W\,(x,y) = \alpha(y)exp\left(i\frac{k}{M}x\right) \text{ on W.}$$

$\alpha(y)$ can be expressed by the wing circulation Γ. Let

$$\Psi\,(x,y) := [\Phi]_L\,(x,y) \text{ on } L \tag{2.9}$$

denote the unknown jump across L. For $\Phi \in W_2^1(\mathbf{R}^3\backslash\overline{S})$ the jump $[\Phi]_S$ across S is definable in a senseful way, but $[\Phi]_S$ is in general not a continuous function in a neighbourhood of x_t. Our definition of the Kutta condition will rise the regularity of Φ in $\mathbf{R}^3_+ \cap U$ and $\mathbf{R}^3_- \cap U$ with an open subset $U \subset \mathbf{R}^3$, $x_t \subset U$, such that $[\Phi]$ is continuous on x_t. Hence, we obtain with $\Psi \equiv 0$ on the leading edge x_l

$$\Psi(x_t(y),y) = \alpha(y)exp\left(i\frac{k}{M}x_t(y)\right) = \int\limits_{x_l(y)}^{x_t(y)} \frac{\partial\Psi}{\partial x}dx =: \Gamma(y), \text{ i.e.}$$

$$[\Phi]_W\,(x,y) = \Gamma(y)exp\left(i\frac{k}{M}(x - x_t(y))\right) \text{ on } W. \tag{2.10}$$

Collecting these facts we can formulate the complete problem as follows: Find $\Phi \in W_2^1(\mathbf{R}^3\backslash\overline{S})$ with

$$\left.\begin{array}{rcl}(\Delta + k^2)\Phi &=& 0 \text{ in } \mathbf{R}^3\backslash\overline{S}\,,\ \text{Re } k > 0 \text{ and Im } k > 0\\ [\Phi]_L &=:& \Psi\\ [\Phi]_W &=& \Gamma(y)exp\left(i\frac{k}{M}(x - x_t(y))\right)\\ \left[\frac{\partial\Phi}{\partial n}\right]_L &=& \left[\frac{\partial\Phi}{\partial n}\right]_W = 0\\ \frac{\partial\Phi}{\partial n} &=& g \text{ on } L\\ |\nabla\Phi| < \infty \text{ on } x_t && \text{(Kutta condition)}\end{array}\right\}. \tag{2.11}$$

$$S := L \cup x_t \cup W$$

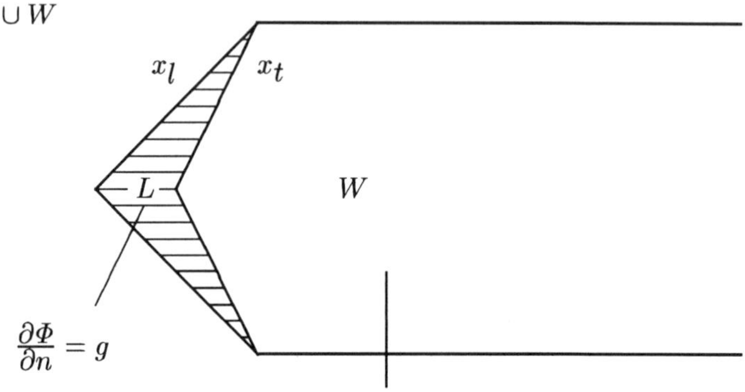

$$[\Phi]_L =: \Psi \qquad [\Phi]_W = \Gamma(y)\ \exp\left(i\frac{k}{M}(x - x_t(y))\right)$$

$$\left[\frac{\partial \Phi}{\partial n}\right]_L = 0 \qquad \left[\frac{\partial \Phi}{\partial n}\right]_W = 0$$

In (2.11) g, k, M and x_t are given, Ψ, Γ and, of course, Φ are unknown. Thus, our problem is a MIXED SCREEN – BOUNDARY – VALUE – PROBLEM with partly unknown Cauchy data.

3 Mathematical ansatz

In this section we reduce problem (2.11) to the calculation of the wing circulation Γ, where we cannot apply the high standard of accuracy involved usually in this paper by treating difficulties coming up near the tips of the wing. Concretely, we transform (2.11) into the hypersingular integral equation (3.14) on L. Then, we formulate regularity conditions for Γ, respectively for functions depending on Γ, such that Φ, constructed by using the solution of the boundary integral equation (3.14), solve (2.11) with (possible) exception of the Kutta condition, i.e. (2.11) is reduced to the calculation of Γ.

During the last years a lot of papers concerning screen problems with smooth and bounded edges have been published (i.e. [19], [20]). Contrary to these papers, our screens L and S are not as such smooth. Furthermore, the boundary ∂S of S is not bounded. But combining results of Costabel [3] who investigated elliptic equations on Lipschitz domains and of v. Petersdorff [15] who considered the Helmholtz equation under certain boundary conditions also for domains with unbounded boundaries and Stephan [19] who treated the Helmholtz equation in the exterior of a smooth and bounded screen under Dirichlet- and Neumann conditions we can formulate results of potential theory adapted to our situation which we present below (the proofs can be found in [9], [11]; There we took advantage that our screens L and S are plane and that S is with the possible exception of a bounded set a rectangle).

In order to formulate these results we recall briefly the definition of some well-known spaces. In the following we denote by $W_2^s(\Omega)$ for a domain $\Omega \subset \mathbf{R}^n$ the usual Sobolev-Slobodeckij spaces and by $H^s(\Omega)$ the Bessel potential spaces (e.g. [22]). Here the Fourier-transformed $\widehat{\varphi}$ of a tempered distribution φ is normed by $\widehat{\varphi}(u) := \int_{\mathbf{R}^n} \varphi(\eta) exp(i < \eta, u >) d\eta$. Furthermore, we need the spaces $\widetilde{H}^s(\Omega) := \{\varphi \in H^s(\mathbf{R}^n) : supp\ \varphi \subset \overline{\Omega}\}$ with the H^s-norm of the extension by 0 as norm.

For functions

$$\begin{aligned} \varphi \in H_P^1(\mathbf{R}_\pm^3) := \{\varphi \in H^1(\mathbf{R}_\pm^3) : \|\varphi\|^2_{H_P^1(\mathbf{R}_\pm^3)} \\ := \|\varphi\|^2_{H^1(\mathbf{R}_\pm^3)} + \|(\Delta + k^2)\varphi\|^2_{L_2(\mathbf{R}_\pm^3)} < \infty\} \end{aligned}$$

we can define the normal derivative $\gamma_1 := \frac{\partial}{\partial n}$ as usual (e.g. [3], [19], [9], [11]) by

$$\left. \begin{array}{rcl} \gamma_1 & : & H_P^1(\mathbf{R}_\pm^3) \longrightarrow H^{-1/2}(\mathbf{R}^2) \\ \varphi & \overset{\gamma_1}{\mapsto} & (f \mapsto < \gamma_1\varphi, f >_{L_2(\mathbf{R}^2)}) \end{array} \right\} \tag{3.1}$$

with

$$\left. \begin{array}{rcl} \gamma_1\varphi & : & \overline{H^{1/2}(\mathbf{R}^2)} \longrightarrow \mathbf{C} \\ < \gamma_1\varphi, f >_{L_2(\mathbf{R}^2)} & := & \pm \int_{\mathbf{R}_\pm^3} \overline{\gamma_0^- f} \cdot (\Delta + k^2)\varphi dV \\ & & \pm \int_{\mathbf{R}_\pm^3} \nabla\varphi \overline{\nabla\gamma_0^- f} dV \mp k^2 \int_{\mathbf{R}_\pm^3} \varphi \overline{\gamma_0^- f} dV \end{array} \right\}. \tag{3.2}$$

Here $\gamma_0^- : H^{1/2}(\mathbf{R}^2) \longrightarrow H^1(\mathbf{R}^3)$ denotes a right-inverse of the trace operator γ_0.

We need the following definitions of the potential jumps:

Definition 3.1. *For $\varphi \in L_2(\mathbf{R}^2)$ with $\varphi_- := \varphi |_{\mathbf{R}_-^3} \in H_P^1(\mathbf{R}_-^3)$ and $\varphi_+ := \varphi |_{\mathbf{R}_+^3} \in H_P^1(\mathbf{R}_+^3)$ the following jumps are well defined:*

$$\left. \begin{array}{rcl} [\varphi]_{\mathbf{R}^2} & := & (\gamma_0\varphi_-) - (\gamma_0\varphi_+) \\ \left[\frac{\partial\varphi}{\partial n}\right]_{\mathbf{R}^2} := [\gamma_1\varphi]_{\mathbf{R}^2} & := & (\gamma_1\varphi_-) - (\gamma_1\varphi_+) \end{array} \right\}. \tag{3.3}$$

The following theorem holds:

Theorem 3.1. *(representation formula)*
Let M denote one of the screens L or S and let $\Phi \in W_2^1(\mathbf{R}^3 \backslash \overline{M})$ be a weak solution of the Helmholtz equation in $\mathbf{R}^3 \backslash \overline{M}$ with $Re\, k > 0$ and $Im\, k > 0$. Then in $\mathbf{R}^3 \backslash \overline{M}$ the following representation formula holds:

$$\Phi(u) = -\int_M [\Phi](\xi) \frac{\partial}{\partial n_\xi} G(u,\xi) ds_\xi + \int_M \left[\frac{\partial \Phi}{\partial n}\right](\xi) G(u,\xi) ds_\xi, u \in \mathbf{R}^3 \backslash \overline{M} \tag{3.4}$$

with the fundamental solution

$$G(u,\xi) = \frac{exp(ik \mid u - \xi \mid)}{-4\pi \mid u - \xi \mid} \tag{3.5}$$

of the Helmholtz equation in the three-dimensional case.
For the densities we have:

$$[\Phi] \in \widetilde{H}^{1/2}(M), \left[\frac{\partial \Phi}{\partial n}\right] \in \widetilde{H}^{-1/2}(M). \tag{3.6}$$

The proof is presented in [9], respectively [11], where we used ideas from [19] and results from [15]. The main idea is to extend M to $\mathbf{R}^2$ and to use the well-known representation of Φ by a combination of the single layer and the double layer potential in the upper, respectively the lower half-plane of $\mathbf{R}^3$. Using that Φ is bounded away from M smoothly, the statements of the theorem follow.

Finally, we need the following

Lemma 3.1.
Let M denote one of the screens L, W or S. Then the operators of the double layer potential

$$\left.\begin{array}{rcl} K_M & : & \widetilde{H}^{1/2}(M) \longrightarrow H^1(\mathbf{R}^3_\pm), \text{ respectively,} \\ K_M & : & \widetilde{H}^{1/2}(M) \longrightarrow H^1_{loc}(\mathbf{R}^3 \backslash \overline{M}) \\ \varphi & \mapsto & -\int_M \varphi(\xi) \frac{\partial}{\partial n_\xi} G(u,\xi) ds_\xi \end{array}\right\} \tag{3.7}$$

and the operator "normal derivative of the double layer potential"

$$\left.\begin{array}{rcl} D_M & : & \widetilde{H}^{1/2}(M) \longrightarrow H^{-1/2}(M) \\ \varphi & \mapsto & -\frac{\partial}{\partial n_u} \int_M \varphi(\xi) \frac{\partial}{\partial n_\xi} G(u,\xi) ds_\xi \end{array}\right\} \tag{3.8}$$

are continuous. Furthermore, the following jump relations hold:

$$[\gamma_0 K_M \varphi] = \varphi, [\gamma_1 K_M \varphi] = 0 \text{ for } \varphi \in \widetilde{H}^{1/2}(M). \tag{3.9}$$

Remark 3.1.
In the literature screen problems for smooth and bounded screens are usually treated by extending the screen to a smooth, closed manifold $\partial\Omega$ such that $\partial\Omega$ is the boundary of a smoothly bounded domain Ω (e.g. Stephan [19]). In this case one is searching for solutions which fulfil the Sommerfeld radiation condition and one can admit real values for the reduced wave number k. Then, for $\varphi \in \widetilde{H}^{1/2}(M)$ one extends φ to $l\varphi \in H^{1/2}(\partial\Omega)$ by zero and obtains $(K_M l\varphi)|_\Omega \in H^1(\Omega)$ and $(K_M l\varphi)|_{\mathbf{R}^3\setminus\overline{\Omega}} \in H^1_{loc}(\mathbf{R}^3\setminus\overline{\Omega})$. Hence, one obtains solutions in the space $H^1_{loc}(\mathbf{R}^3\setminus\overline{M})$.

Because the screen S is unbounded, we have to require $\mathrm{Re}\, k > 0$ and $\mathrm{Im}\, k > 0$ to find an unique solution of the screen problem with S as screen in physically adequate spaces with a reasonable behaviour at infinity. We also used the idea outlined above: We have extended S to $\mathbf{R}^2 \times \{0\}$. Ω corresponds to the upper halfspace $\mathbf{R}^3_+$. (3.7) implies $(K_M\varphi)|_\Omega \in H^1(\Omega)$ and $(K_M\varphi)|_{\mathbf{R}^3\setminus\overline{\Omega}} \in H^1(\mathbf{R}^3\setminus\overline{\Omega})$. Nevertheless, the image $K_M(\widetilde{H}^{1/2}(M))$ is not contained in $H^1(\mathbf{R}^3\setminus\overline{M})$: The double layer potential K_M "jumps" by passing through the manifold M. So one cannot expect the existence of an extension of $(K_M)|_{\mathbf{R}^3_-}$ and $(K_M)|_{\mathbf{R}^3_+}$ into the space $H^1(\mathbf{R}^3)$. But for $\varphi \in \widetilde{H}^{1/2}(M)$, $K_M\varphi \in W^1_2(\mathbf{R}^3\setminus\overline{M})$, because $H^1(\mathbf{R}^3_\pm) = W^1_2(\mathbf{R}^3_\pm)$ and, as a solution of the homogeneous Helmholtz equation, $K_M\varphi$ is smooth away from M.

Under the assumption of a known circulation Γ we now derive a boundary integral equation for the unknown layer Ψ on L in order to find conditions for the calculation of Γ.

Because $\left[\frac{\partial\Phi}{\partial n}\right]_L = \left[\frac{\partial\Phi}{\partial n}\right]_W = 0$, the representation formula (3.4) implies for a solution $\Phi \in W^1_2(\mathbf{R}^3\setminus\overline{S})$ of the problem (2.11)

$$\Phi(u) = -\int_S [\Phi](\xi)\frac{\partial}{\partial n_\xi}G(u,\xi)ds_\xi. \tag{3.10}$$

Inserting the Cauchy data for $[\Phi]$ in (3.10), we obtain after application of the normal derivative γ_1

$$\begin{aligned}\frac{\partial\Phi}{\partial n}(u) &= -\frac{\partial}{\partial n u}\int_L \Psi(\xi)\frac{\partial}{\partial n_\xi}G(u,\xi)ds_\xi \\ &\quad -\frac{\partial}{\partial n u}\int_W \Gamma(\xi_2)exp\left(i\frac{k}{M}(\xi_1 - x_t(\xi_2))\right)\frac{\partial}{\partial n_\xi}G(u,\xi)ds_\xi.\end{aligned}$$

Passing to the limit $u = (u_1, u_2, u_3) \longrightarrow (u_1, u_2, 0) \cong: u$ we obtain together with $\frac{\partial\Phi}{\partial n} = g$ on L and (3.9) for $u \in L$ the boundary integral equation

$$D_L\Psi(u) = g(u) + \frac{\partial}{\partial n_u} \int\limits_W \Gamma(\xi_2) exp\left(i\frac{k}{M}(\xi_1 - x_t(\xi_2)) \right) \frac{\partial}{\partial n_\xi} G(u,\xi) ds_\xi \tag{3.11}$$

for Ψ.

Additionally, the boundary conditions

$$\left. \begin{array}{lll} \Psi \equiv 0 & \text{on} & x_l \\ \Psi \equiv \Gamma & \text{on} & x_t \end{array} \right\} \tag{3.12}$$

have to hold.

But it is well-known that boundary integral equations coming up from screen problems generate "zero boundary values" for the solution at the boundary of the screen. One has to expect $\Psi \in \widetilde{H}^{1/2}(L)$ (e.g. [19] or theorem 3.1), which is not compatible with the condition $\Psi \equiv \Gamma$ on the trailing edge x_t. The circulation Γ cannot vanish on x_t. Otherwise $[\Phi]_W \equiv 0$ would hold.

Therefore, following Hebeker [8], we introduce a smooth cut-off function χ which is identical to 1 on x_t and 0 on x_l with the exception of small neighbourhoods of the tips of L. The substitution

$$\widetilde{\Psi}(u_1, u_2) := \Psi(u_1, u_2) - \chi(u_1, u_2)\Gamma(u_2) \tag{3.13}$$

transforms (3.11) into the boundary integral equation

$$D_L\widetilde{\Psi}(u) = h(u), \; u \in L \tag{3.14}$$

with

$$\left. \begin{array}{rcl} h(u) & = & g(u) \\ & + & \frac{\partial}{\partial n_u} \int\limits_W \Gamma(\xi_2) exp\left(i\frac{k}{M}(\xi_1 - x_t(\xi_2)) \right) \frac{\partial}{\partial n_\xi} G(u,\xi) ds_\xi \\ & + & \frac{\partial}{\partial n_u} \int\limits_L \Gamma(\xi_2)\chi(\xi) \frac{\partial}{\partial n_\xi} G(u,\xi) ds_\xi \end{array} \right\} . \tag{3.15}$$

Then $\widetilde{\Psi}$ vanishes on ∂L up to small parts of ∂L.

The behaviour of $[\Phi]$ near the wing tips (side edges of the wing!) has to be analysed in more detail from the analytical as well as from the physical point of view. One needs a model which includes the shedding of vortices from the tips (side edges!) of the wings.

But we think that the error which results of the non-vanishing of $\widetilde{\Psi}$ on ∂L near the tips of the wings can be neglected within the scope of this model, because the wings are on account of our assumptions narrow near these points, i.e. we expect that $\Gamma(y)$, as a line integral along cross-sections of L, becomes small there. Results by engineers also support that Γ becomes small near the tips of the wings.

The following lemma holds:

Lemma 3.2.
The boundary integral equation (3.14) $D_L\widetilde{\Psi} = h$ *is for* $h \in H^{-1/2}(L)$ *uniquely solvable with the solution* $\widetilde{\Psi} \in \widetilde{H}^{1/2}(L)$.

The idea of proving lemma 3.2 is standard (e.g. [19]). The detailed proof, adapted to our situation, can be found in [9], [11]. Using that the operator "normal derivative of the double layer potential" is a pseudodifferentialoperator with symbol

$$D(\xi_1, \xi_2) = \frac{1}{2}\sqrt{\xi_1^2 + \xi_2^2 - k^2} \tag{3.16}$$

(e.g. [6]), i.e.

$$D_L\varphi = \frac{1}{2}p_L F^{-1}\sqrt{\xi_1^2 + \xi_2^2 - k^2} \cdot \widehat{\varphi} \;\forall \varphi \in \widetilde{H}^{1/2}(L), \tag{3.17}$$

where p_L denotes the operator of restriction onto L and F, respectively $\widehat{\cdot}$, the Fouriertransformation (branch-cut of the square root equal to $\mathbf{R}_-$), one shows in our situation of a plane screen by elementary calculations the coercivity estimate

$$| < D_L\varphi, \varphi >_{L_2(L)} | \geq c\,\|\varphi\|^2_{\widetilde{H}^{1/2}(L)} \quad \forall \varphi \in \widetilde{H}^{1/2}(L) \tag{3.18}$$

and that $a(\varphi, f) := < D_L\varphi, f >_{L_2(L)}$ is bounded on $\widetilde{H}^{1/2}(L) \times \widetilde{H}^{1/2}(L)$. Using lemma 3.1 one can prove the

Theorem 3.2.
Let Γ *denote a function defined on the trailing edge* x_t *of* L *with*

$$\left.\begin{array}{l} g \in H^{-1/2}(L) \\ \widetilde{h},\ [\Phi]_S \in \widetilde{H}^{1/2}(S) \end{array}\right\} \tag{3.19}$$

with

$$\widetilde{h}(\xi) = \left\{\begin{array}{ll} \Gamma(\xi_2)\cdot\chi(\xi) & for \quad \xi \in L \\ \Gamma(\xi_2) & for \quad \xi \in x_t \\ \Gamma(\xi_2) exp\left(i\frac{k}{M}(\xi_1 - x_t(\xi_2))\right) & for \quad \xi \in W \end{array}\right\}, \tag{3.20}$$

$$[\Phi]_S(\xi) = \left\{\begin{array}{ll} \Psi(\xi) & for \quad \xi \in L \\ \Gamma(\xi_2) & for \quad \xi \in x_t \\ \Gamma(\xi_2) exp\left(i\frac{k}{M}(\xi_1 - x_t(\xi_2))\right) & for \quad \xi \in W \end{array}\right\} \tag{3.21}$$

and

$$\Psi(\xi) = \widetilde{\Psi}(\xi) + \Gamma(\xi_2)\cdot\chi(\xi) \tag{3.22}$$

(compare (3.13)) and the solution $\widetilde{\Psi}$ *of (3.14).*

Then, the double layer potential (3.10) with the density $[\Phi]_S$

$$\Phi(u) := \left(K_S[\Phi]_S\right)(u) = -\int_S [\Phi]_S(\xi)\frac{\partial}{\partial n_\xi}G(u,\xi)ds_\xi \tag{3.23}$$

is a solution of (2.11) with (possible) exception of the Kutta condition.

Remark 3.2.
We saw in this paragraph that the main problem is the calculation of the wing circulation Γ in such a way, that the assumptions in theorem 3.2 and the usual Kutta condition are fulfilled. The Kutta condition has been postulated to fix Φ uniquely. This point of view is reflected in our model.

It turned out that it is a non-trivial problem to define the circulation Γ meaningfully. It is well-known that spaces like H^1_{loc} or, in our situation on account of a weakly damping gas, W^1_2 are canonically solution spaces for screen problems for the Helmholtz equation. This fact coincides with physical arguments: $\Phi \in W^1_2(\mathbf{R}^3\backslash\overline{S})$ means that Φ and the partial derivatives of Φ of order 1 (= velocity components) are square integrable, i.e. these spaces are adapted to a finite energy, respectively a locally finite energy. (2.11) shows that Γ is just the restriction $[\Phi]|_{x_t}$. But on account of the poor regularity of Φ it is not possible to define this restriction by twice applying the (sharp) trace theorem. Furthermore, $\nabla\Phi$ will blow up as x_t is approached.

We need more regularity near x_t. This additional regularity must be gained by the Kutta condition. Hence, we have to specify the classical definition of the Kutta condition for our purposes. This idea leads to the

Definition 3.2.
We say that the Kutta condition is fulfilled, iff $\Phi \in W_2^{2+\sigma}(U_\pm)$, where $U_\pm$ denotes the intersection of $\mathbf{R}^3_\pm$ with a small three-dimensional neighbourhood U of x_t and a positive number σ.

Of course, if Φ fulfills the Kutta condition in the sense of definition 3.2 one can define Γ by traces and $\nabla\Phi|_{x_t} \in W_2^\sigma(x_t)$.

In the next two paragraphs we demonstrate in the half-space situation, respectively in the pure two-dimensional case, where we can calculate Γ analytically, how the Kutta condition can be fulfilled.

4 The solution of problem (1.11) in the case of the wing projection L equal to the left half-plane

In this paragraph we constantly denote by L the "left half-plane" in $\mathbf{R}^2$, i.e. $L := \{(x,y) \in \mathbf{R}^2 : x < 0\}$. Hence, $W = \{(x,y) \in \mathbf{R}^2 : x > 0\}$, $S = \mathbf{R}^2$ and $x_t(y) = 0 \ \forall\ y \in \mathbf{R}$.

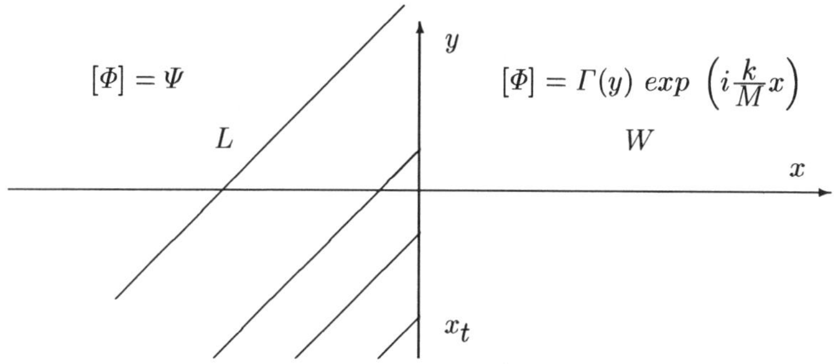

For calculating Γ we can choose the cut-off function χ in (3.13) in the half-space case independent of the second argument. More precisely, we choose χ as a twice continuously differentiable extension of $exp\left(i\frac{k}{M}\xi_1\right)$ from $\mathbf{R}_+$ to $\mathbf{R}$ with an appropriate behaviour at infinity, e.g.

$$\chi(\xi) = \chi(\xi_1) := \left\{ \begin{array}{ll} \exp\left(i\frac{k}{M}\xi_1\right) & \text{for} \quad \xi_1 \geq 0 \\ 6\exp\left(-i\frac{k}{M}\xi_1\right) - 32\exp\left(-i\frac{k}{2M}\xi_1\right) & \\ +27\exp\left(-i\frac{k}{3M}\xi_1\right) & \text{for} \quad \xi_1 < 0 \end{array} \right\}. \tag{4.1}$$

With the abbreviation

$$\widetilde{\Gamma}(\xi) := \Gamma(\xi_2)\chi(\xi_1) \tag{4.2}$$

we can write the boundary integral equation (3.14) as

$$D_L\widetilde{\Psi}(u) = g(u) + \frac{\partial}{\partial n_u}\int\limits_{\mathbf{R}^2} \widetilde{\Gamma}(\xi)\frac{\partial}{\partial n_\xi}G(u,\xi)ds_\xi \tag{4.3}$$

on L with the additional condition $\widetilde{\Psi} = 0$ on x_t.

Denoting by p_- the operator of restriction onto L and by lg an arbitrary extension of $g \in H^s(L)$ to $lg \in H^s(\mathbf{R}^2)$ with a real number s, which we have to concretize more precisely, we can interprete (4.3) as the pseudodifferential Wiener-Hopf equation

$$p_-D\widetilde{\Psi} = p_-(lg - D\widetilde{\Gamma}). \tag{4.4}$$

Here again D denotes the pseudodifferential operator "normal derivative of the double layer potential", generated by the symbol $D(\xi) = \frac{1}{2}\sqrt{\xi_1^2 + \xi_2^2 - k^2}$.

If the right hand-side of (4.4) belongs to the space $H^{-1/2}(L)$, we know from lemma 3.2 that (4.4) possesses an unique solution $\widetilde{\Psi} \in \widetilde{H}^{1/2}(L)$. But the proof of lemma 3.2 takes no advantage from the half-space geometry.

Using ideas and modified techniques of Èskin [6], who makes use of the half-space geometry, we can give an explicit representation of the solution

$\widetilde{\Psi}$ if $p_-(lg - D\widetilde{\Gamma}) \in H^{-1/2+\sigma}(L)$ where $0 < \sigma < \frac{1}{2}$. If $p_-(lg - D\widetilde{\Gamma}) \in H^{1/2+\sigma}(L)$ we can split $\widetilde{\Psi}$ additiv into a regular part $\widetilde{\Psi}_{reg}(\Gamma)$ and a singular part $\widetilde{\Psi}_{sing}(\Gamma)$, where the representation is explicit and especially explicit with respect to the unknown circulation Γ. The regularity result $\widetilde{\Psi}_{reg} \in \widetilde{H}^{3/2+\sigma}(L)$ holds then.

It is well-known that solutions of boundary integral equations like (4.3), respectively (4.4), behave like $\sqrt{r}$ near the boundary ∂L of the screen L where r denotes the distance to ∂L (e.g. [19]), i.e. $\nabla\widetilde{\Psi}$ behaves like $\frac{1}{\sqrt{r}}$. This asymptotic behaviour is not compatible with a finite perturbation velocity near the trailing edge x_t. Responsible for this poor regularity is the singular part $\widetilde{\Psi}_{sing}(\Gamma)$. Hence, we try to eliminate $\widetilde{\Psi}_{sing}(\Gamma)$ by using our last existing degree of freedom, i.e. we want to solve the equation

$$\widetilde{\Psi}_{sing}(\Gamma) = 0 \tag{4.5}$$

for the unknown circulation Γ. This equation can be realized by the condition

$$\left.\begin{aligned} &\int_{-\infty}^{\infty} \frac{1}{\sqrt{\xi_1 + i\sqrt{\xi_2^2 - k^2}}}\, \widehat{lg}(\xi)d\xi_1 \\ &= \frac{1}{2}\int_{-\infty}^{\infty} \sqrt{\xi_1 - i\sqrt{\xi_2^2 - k^2}}\, \widehat{\chi}(\xi_1)\widehat{\Gamma}(\xi_2)d\xi_1 \end{aligned}\right\} . \tag{4.6}$$

(The details are published in [9], respectively in [11]).

The right hand-side in (4.6) is of the form $\alpha(\xi_2)\cdot\widehat{\Gamma}(\xi_2)$. We can determine $\alpha(\xi_2)$ explicitly by calculating at first $\widehat{\chi}(\xi_1)$ with elementary methods and then $\alpha(\xi_2)$ by inserting this result into the right hand-side of (4.6) and making use of the residuum theorem. As a result we obtain from (4.6)

$$\widehat{\Gamma}(\xi_2) = \frac{\displaystyle\int_{-\infty}^{\infty} \frac{1}{\sqrt{\xi_1 + i\sqrt{\xi_2^2 - k^2}}}\, \widehat{lg}(\xi)d\xi_1}{\pi\sqrt{-\frac{k}{M} - i\sqrt{\xi_2^2 - k^2}}} \tag{4.7}$$

Especially, if $\widetilde{\Psi}_{sing}(\Gamma) = 0$ we have $\widetilde{\Psi}(\Gamma) = \widetilde{\Psi}_{reg}(\Gamma) \in \widetilde{H}^{3/2+\sigma}(L)$.

If we assume even the still higher regularity $g \in H^{1+\sigma}(L)$ and choose $lg \in H^{1+\sigma}(\mathbf{R}^2)$ we obtain from (4.7) $\Gamma \in H^{3/2+\sigma}(\{0\} \times \mathbf{R})$ and $D\widetilde{\Gamma} \in H^{1/2+\sigma}(\mathbf{R}^2)$, i.e. the assumptions for splitting $\widetilde{\Psi}(\Gamma)$ into $\widetilde{\Psi}_{reg}(\Gamma) + \widetilde{\Psi}_{sing}(\Gamma)$ are fulfilled.

Furthermore, the density $[\Phi]||_S$ consisting of the sum of the solution $\widetilde{\Psi}(\xi)$ and $\Gamma(\xi_2)\chi(\xi_1)$ (resulting from the resubstitution according to (3.13)) on L and $\Gamma(\xi_2)\cdot exp\left(i\frac{k}{M}\xi_1\right)$ on $\overline{W}$ belongs to $H^{3/2+\sigma}(\mathbf{R}^2)$. This leads to the

Theorem 4.1.
Let $g \in H^{1+\sigma}(L)$ with $0 < \sigma < \frac{1}{2}$. Let Γ be as in (4.7) with an arbitrary extension $lg \in H^{1+\sigma}(\mathbf{R}^2)$ of g. Then the double layer potential

$$\Phi(u) = -\int_S [\Phi]_S(\xi) \frac{\partial}{\partial n_\xi} G(u,\xi) ds_\xi \tag{4.8}$$

with the density

$$[\Phi]_S(\xi) = \left\{ \begin{array}{lll} \widetilde{\Psi}(\xi) + \Gamma(\xi_2)\chi(\xi_1) & for & \xi \in L \\ \Gamma(\xi_2) exp\left(i\frac{k}{M}\xi_1\right) & for & \xi \in \overline{W} \end{array} \right\} \tag{4.9}$$

solves problem (2.11).
Especially $\Phi|_{\mathbf{R}^3_+} \in H^{2+\sigma}(\mathbf{R}^3_+)$, $\Phi|_{\mathbf{R}^3_-} \in H^{2+\sigma}(\mathbf{R}^3_-)$ holds, i.e. the Kutta condition is fulfilled in the sense of definition 3.2.

5 The solution of problem (1.11) in the purely two-dimensional case

The techniques of proving the representation of $\widetilde{\Psi}_{sing}(\Gamma)$ in the last paragraph are based essentially on the Wiener-Hopf method which displays its power especially in the half-space case. If L is a more general screen, one needs a new idea in order to determine Γ because explicit representation formulas for $\widetilde{\Psi}$ or even $\widetilde{\Psi}_{sing}$ are not available.

But during the last years several authors proved results about the behaviour of solutions of special boundary value problems in two- and three-dimensional domains, respectively, near edges and corners of the boundary of the domain. We want to use these outcomes. Hence, we transform the boundary integral equation (3.14) into a screen problem with Neumann data prescribed on the screen L. Then, we interprete the exterior of the screen L as a degenerated Lipschitz domain, fix Γ by the requirement of a vanishing singular part of the solution of this screen problem and analyse the regularity of Φ constructed with Γ choosen in the described manner.

We present this idea here at first in the two-dimensional case, where we also can calculate Γ explicitly. In the two-dimensional case the wing projection L is an interval (x_l, x_t) and W is the interval (x_t, ∞). Especially Γ is then a complex number and not a function.

Starting from (2.11) in the two-dimensional case one obtains with the methods outlined in 2 correspondingly the hypersingular boundary integral equation

$$D_L \widetilde{\Psi}(u) = h(u) \tag{5.1}$$

with the normal derivative D_L of the double layer potential. Because x_t and Γ are scalars now, the right hand-side h is given by

$$\left.\begin{aligned} h(u) = g(u) \quad &+ \quad \Gamma exp\left(-i\frac{k}{M}x_t\right)\frac{\partial}{\partial n_u}\int\limits_W exp\left(i\frac{k}{M}\xi\right)\frac{\partial}{\partial n_\xi}G(u,\xi)ds_\xi \\ &+ \quad \Gamma\frac{\partial}{\partial n_u}\int\limits_L \chi(\xi)\frac{\partial}{\partial n_\xi}G(u,\xi)ds_\xi \end{aligned}\right\}, \tag{5.2}$$

where $G(u,\xi)$ denotes now the fundamental solution of the Helmholtz equation in $\mathbf{R}^2$ (Details can be found in [9] or [11]). One can show that the boundary integral equation (5.1) is equivalent to the following screen boundary value problem:

Find $\varphi \in W_2^1(\mathbf{R}^2\backslash\overline{L})$ with

$$\left.\begin{aligned} (\Delta + k^2)\varphi &= 0 \text{ in } \mathbf{R}^2\backslash\overline{L},\ \operatorname{Re} k > 0,\ \operatorname{Im} k > 0 \\ \frac{\partial\varphi}{\partial n} &= h \text{ on } L \end{aligned}\right\}. \tag{5.3}$$

For fixing Γ we assume, similar to the last paragraph, a higher regularity for h. More precisely, we require $h \in H^{1/2+\sigma}(L)$ where $0 < \sigma < \frac{1}{2}$. Then the unique solution $\varphi \in W_2^1(\mathbf{R}^2\backslash\overline{L})$ of (5.1) possesses the representation

$$\left.\begin{aligned} \varphi(\Gamma) &= \varphi_{reg}(\Gamma) + \varphi_{sing}(\Gamma) \text{ with} \\ \varphi_{sing}(\Gamma) &= c_l(\Gamma)\zeta_l\sqrt{\rho_l}\cos\frac{\theta_l}{2} + c_t(\Gamma)\zeta_t\sqrt{\rho_t}\cos\frac{\theta_t}{2} \end{aligned}\right\}, \tag{5.4}$$

where we indicated the dependence of the appearing terms on Γ explicitly.

Here $c_l(\Gamma)$ and $c_t(\Gamma)$ denote complex numbers independent of the spherical coordinates. ζ_l und ζ_t are smooth, two-dimensional cut-off functions centered in $(x_l, 0)$, respectively $(x_t, 0)$, whose supports are contained in disjoint domains Ω_l, respectively Ω_t. Finally, we denote by (ρ_l, θ_l), respectively (ρ_t, θ_t), in $(x_l, 0)$, respectively, $(x_t, 0)$ centered polar coordinates.

For the regular part of φ the following result holds:

$$\varphi_{reg}(\Gamma) \in W_2^{2+\sigma}(\mathbf{R}^2\backslash\overline{L}). \tag{5.5}$$

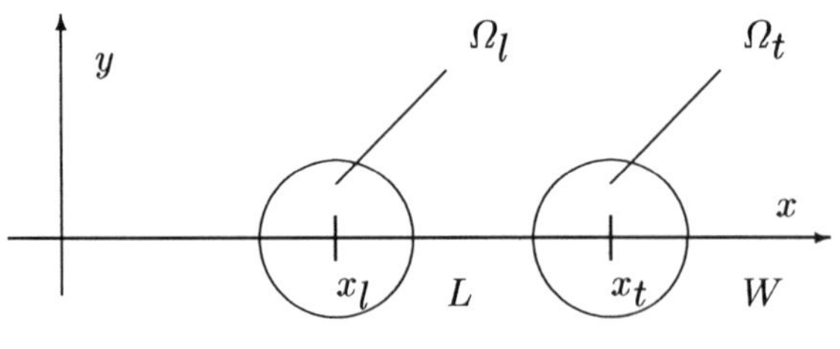

In order to fulfil the Kutta condition we try to fix the circulation Γ by claiming $\varphi_{sing}(\Gamma) = 0$ on Ω_t, i.e. we try to evaluate Γ by the condition $c_t(\Gamma) = 0$.

But, up to now, it is not possible to determine φ_{reg}, respectively φ_{sing}, i.e. we don't now the explicit dependence of c_t on Γ.

Thus, we try to choose Γ in a way that that part of $\varphi(\Gamma)$ is filtered out which behaves like $\sqrt{\rho_t}\cos\frac{\theta_t}{2}$.

To do this we augment – after a normalization – the singular function $\varphi_1 = \zeta_t\sqrt{\rho_t}\cos\frac{\theta_t}{2}$ by other functions $\varphi_2, \varphi_3, \cdots$ to a complete orthonormal system in $L_2(\Omega_t)$. With respect to this function system we expand $\varphi(\Gamma)$ in the form
$\varphi(\Gamma) = \sum\limits_{n=1}^{\infty} < \varphi(\Gamma), \varphi_n > \varphi_n$. We require then that the scalar factor corresponding to φ_1 in this expansion vanishes, i.e. we demand

$$\int\limits_{\Omega_t} \varphi(\Gamma)\,\zeta_t\sqrt{\rho_t}\cos\frac{\theta_t}{2}dV = 0. \tag{5.6}$$

The choice of the L_2-product in (5.6) seems to be unnatural but we can show that (5.6) is fulfilled iff

$$< \varphi(\Gamma), \zeta_t\sqrt{\rho_t}\cos\frac{\theta_t}{2} >_{W_2^1(\Omega_t\backslash\overline{L})} = 0 \tag{5.7}$$

holds.

Using $\varphi(\Gamma) = K_L D_L^{-1}h(\Gamma)$ with the double layer potential K_L and the solution $D_L^{-1}h(\Gamma)$ of (5.1) we can solve (5.6) on account of the affine linearity of h with respect to Γ explicitly. The result is formulated in

Theorem 5.1.
Let $0 < \sigma < \frac{1}{2}$ and $g \in H^{1/2+\sigma}(L)$. Let furthermore be

$$\left.\begin{aligned}
\Gamma &= \frac{Z}{N} \text{ with} \\
Z &= -\int\limits_{\Omega_t} K_L(D_L^{-1}g(u))\zeta_t\sqrt{\rho_t}\cos\frac{\theta_t}{2}dV \text{ and} \\
N &= \int\limits_{\Omega_t} K_L(D_L^{-1}(exp\left(-i\frac{k}{M}x_t\right) \times \\
&\times \frac{\partial}{\partial n u}\int\limits_{W} exp\left(i\frac{k}{M}\xi\right)\frac{\partial}{\partial n_\xi}G(u,\xi)ds_\xi \\
&+ \frac{\partial}{\partial n u}\int\limits_{L}\chi(\xi)\frac{\partial}{\partial n_\xi}G(u,\xi)ds_\xi))\zeta_t\sqrt{\rho_t}\cos\frac{\theta_t}{2}dV
\end{aligned}\right\}, \tag{5.8}$$

where we assume $N \neq 0$ ($Z = 0$ is physically unrealistic).

Then, the double layer potential $\Phi = K_S[\Phi]_S$ with

$$[\Phi]_S(\xi) = \left\{ \begin{array}{ll} [\varphi](\xi) + \Gamma\chi(\xi) & \text{for} \quad \xi \in L \\ \Gamma & \text{for} \quad \xi = x_t \\ \Gamma \exp\left(i\frac{k}{M}(\xi - x_t)\right) & \text{for} \quad \xi \in W \end{array} \right\} \tag{5.9}$$

solves the two-dimensional version of problem (2.1) where φ *denotes the solution of (5.3). For* $\Phi \in W_2^1(\mathbf{R}^2 \backslash \overline{S})$

$$\Phi|_{\Omega_t \pm} \in H^{2+\sigma}(\Omega_t \pm) \tag{5.10}$$

holds, i.e. the Kutta condition is fulfilled in the sense of definition 3.2.

6 A three-dimensional example

Up to now it is not possible to calculate Γ in the full three-dimensional case analytically, because neither explicit representation formulas for the singular part of solutions $\widetilde{\Psi}$ of the boundary integral equation (3.14) nor for solutions φ of the corresponding screen boundary value problem like (5.3) are available. Only structural explicit representations for φ_{sing} have been proven for screen boundary value problems with a polygonal screen L (e.g. [16],[17]). These representations are much more complicated as in the half-space case, respectively the two-dimensional case, because besides edge singularities also corner singularities appear. Hence, we can present only a numerical ansatz for calculating Γ within the scope of the least squares method by minimizing the stress intensity factor of φ with respect to x_t using the ideas of the last paragraph on cross-sections of L.

We demonstrate this idea here for the case of a screen L with a corner of its boundary of opening angle $\alpha = \frac{\pi}{2}$ at the trailing edge, but this idea also works in the case of arbitrary polygonal screens L.

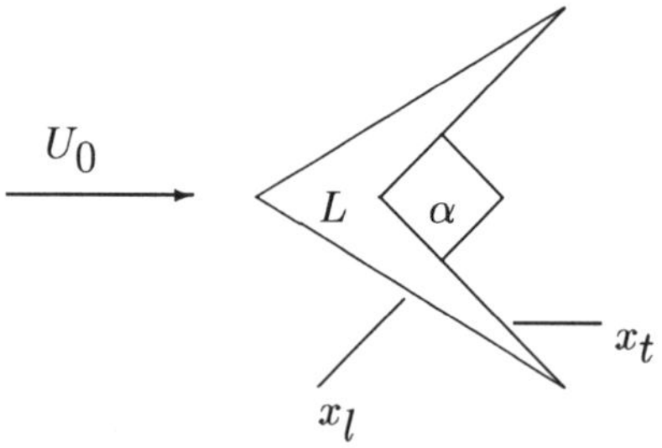

Our idea is to transform the boundary integral equation (3.14) again into a screen boundary value problem like (5.3) and to describe the singular part $\varphi_{sing}(\Gamma)$ by the structural asymptotic representation formulas presented in [16], respectively [17]. We recognized that the first edge singularity in the case of a polygonal screen is always stronger than the first corner singularity, i.e.

it is possible to subordinate the influence of the first corner singularity into the first edge singularity, where we describe the influence of the first corner singularity by numerical results of Morrison and Lewis [14]. This idea leads in the case of a polygonal screen with opening angle $\alpha = \frac{\pi}{2}$ to the following

Theorem 6.1.
Let be $h \in H^{2.3146}(L)$ and assume the existence of a function $F \in H^{3.8146}(\mathbf{R}^3)$ with $\frac{\partial F}{\partial n} = h$ on L. Assume furthermore, that $\lambda = 2.3146$ is not a corner exponent. Then, the boundary integral equation (3.14) is equivalent to the three-dimensional screen boundary value problem (5.3). The unique solution $\varphi \in W_2^1(\mathbf{R}^3 \backslash \overline{L})$ of this problem can be represented near x_t in the form

$$\left.\begin{array}{rcl} \varphi(\Gamma) & = & \varphi_{reg}(\Gamma) + \varphi_{sing}(\Gamma) \text{ with} \\ \varphi_{sing}(\Gamma) & = & \zeta_1(\theta_1) h^1_{1,0}(\Gamma)(r)\sqrt{\rho_1}\cos\frac{\eta_1}{2} \\ & + & \zeta_2(\theta_2) h^2_{1,0}(\Gamma)(r)\sqrt{\rho_2}\cos\frac{\eta_2}{2} \end{array}\right\} \tag{6.1}$$

with $\varphi_{reg}(\Gamma) \in W_2^{2.3146-\varepsilon}(U \backslash L)$ with an arbitrary $\varepsilon > 0$ and a three-dimensional neighbourhood U of x_t. Here, ζ_1, ζ_2 denote smooth cut-off functions centered in the both arcs of x_t, $h^j_{1,0}$ the stress intensity factors, ρ_j the distance to x_t, the distance r to the corner is a variable on the trailing edge and η_j, θ_j are angles resulting from polar-coordinates (compare [16], [17], [9]).

According to the ideas outlined in the last section, we try to fix Γ such that

$$< (KD_L^{-1}h(\Gamma)) \,|_{E_{j,r}}, \zeta_j(\theta_j)\sqrt{\rho_j}\cos\frac{\eta_j}{2} >_{L_2(E_{j,r})} = 0 \tag{6.2}$$

holds for almost all points on x_t with distance r to the corner. Here, $E_{j,r}$, $j = 1, 2$, denotes the intersection of planes which are orthogonal to the arcs of x_t and have the distance r to the corner with domains containing the supports of ζ_1, respectively ζ_2. For calculating Γ numerically, we propose to solve the problem

$$\sum_{j=1}^{2} \| < (KD_L^{-1}h(\Gamma))|_{E_{j,r}}, \zeta_j(\theta_j)\sqrt{\rho_j}\cos\frac{\eta_j}{2} >_{L_2(E_{j,r})} \|^2 \longrightarrow \min_{\Gamma} \tag{6.3}$$

in the $L_2(\{r : r \in (0, E_{j,\max})\})$-norm with the method of least squares by choosing Γ as a linear combination of $C_0^\infty(x_t)$-functions.

7 Concluding remarks

We note that an acceptable mathematical theory for calculating the perturbation velocity potential Φ, even within the scope of a linear theory, is up to now far away. From the mathematical point of view one needs a more detailed representation of the solution $\varphi(\Gamma)$ of the three-dimensional version of (5.3) near x_t in which the stress intensity factors should appear explicitly in dependence of the arc length such that one can formulate and solve a corresponding equation for Γ on x_t analytically.

New results from Russian scientists let us hope to find a suitable asymptotic representation formula in the case of smoothly bounded wing projections L. But for more practically purposes especially wing projections with corners are of interest. But in this case, for finding sufficiently explicit representation formulas for the solution $\varphi(\Gamma)$ of the screen problem near x_t, we think that we need a more detailed knowledge of higher eigenvalues and corresponding eigenfunctions of the Laplace-Beltrami operator (compare [16], [17]) in order to eliminate sufficiently many singular terms in the expansion of $\varphi(\Gamma)$ near x_t by choosing Γ in such a way to increase the regularity of Φ up to an order, such that $\nabla\Phi$ becomes square integrable on x_t. But we expect that these eigenvalues are not bounded far away from each other, such that besides analytical problems also numerical ones will appear.

In this manuscript we presented the philosophy to fix the perturbation velocity potential by conditions on the one-dimensional manifold x_t. In the case of a trailing edge with corners the calculation of Γ will depend essentially on the opening angles of the corners such that Φ has to be evaluated about four dimensions. At this point the question appears wether a more practical relevant model should be formulated in form of a mesh problem holding in four dimensions.

Another problem appears at the tips of the wings. Within the scope of the model presented in 2 it is not possible to allow edges for the points of maximal y-extension on account of further problems of modeling, because vortices would leave the wings there. Practical experiences show that Γ has to vanish near the tips of the wings. The circulation obtained by our model will not vanish there in general.

Hence, one has to treat three different kinds of points on the trailing edge x_t separately: Points where x_t is smooth, corner points of x_t, and boundary points of x_t. Here we found analytical solutions only for points where x_t is smooth and this happens only in the special situation that x_t consists only of one point, respectively x_t is the whole line $x_t = \{(x, y) \in \mathbf{R}^2 : x = 0\}$.

Furthermore, we assumed that the wake W is a semi-infinite strip with constant width. From the practical point of view we cannot support this assumption. W will grow in the y-direction with increasing x as a simple observation of an aeroplane or a ship shows, and the wake will not be of infinite length. Hence, one has to determine W in a more detailed, time depending model by solving a free boundary value problem.

Nevertheless, we want to present our results because, for the first time, we can touch the main problems of lifting surface theory which will be, as we think, also of interest for more directly, pratically developments. The calculation of Γ is not trivial and requires further investigations!

The problems outlined have to be solved in near future, because only after a solution of these problems essential progress in lifting surface theory could be expected. Also other topics which lead to similar problems, e.g. in elastodynamic fracture theory, can be solved only after a solution of the problems outlined.

Besides the questions quoted, we want to list some further problems in connection with lifting surface theory for which one needs a complete solution of (2.11):

a. The model has to be adapted to more practical purposes. The assumption of a thin, weakly cambered wing is not senseful from the practical point of view. Also regions where the flow separates from the wings have to be considered.

b. The assumption of a weakly damping gas is not reliable. One has to take into account real wave numbers, e.g. by introduction of weighted spaces or by using the limiting absorption principle. Furthermore, it is desirable to consider the wave number k as a parameter and to study Γ and Φ in dependence of k.

c. wings moving near the ground

d. correction formulas for a lifting surface in a wind-tunnel

e. arbitrary time dependence of the motion

f. proving of the convergence of numerical approximation schemes in finite dimensional spaces which take advantage of the trailing edge situation

g. programming of a theoretically assured algorithm on the base of graded meshes which uses modern techniques for evaluating singular and hypersingular integrals.

References

1. Bisplinghoff, R. L., Ashley, H., Halfman, R. L. (1957): Aeroelasticity. Addison Wesley, Massachusetts
2. Brakhage, H. (1971): Mathematische Tragflächentheorie. Unpublished
3. Costabel, M. (1988): Boundary integral operators on Lipschitz domains: Elementary results. SIAM J. Math. Anal., 19, 613-626
4. Costabel, M., Stephan, E. P. (1985): Boundary integral equations for mixed boundary value problems in polygonal domains and Galerkin approximation. Mathematical models and methods in mechanics 15, 175-251. Banach Center Publications, Warsaw
5. Dauge, M. (1988): Elliptic boundary value problems on corner domains. Lecture notes in mathematics 1341. Springer
6. Èskin, G. I. (1981): Boundary value problems for elliptic pseudodifferential equations. Translations of mathematical monographs 52. American Mathematical Society, Providence Rhode Island
7. Grisvard, P. (1985): Elliptic problems in nonsmooth domains. Pitman, Boston

8. Hebeker, F. C. (1988): Mathematische Theorie instationärer Flügelströmungen. Report for the VW-Project
9. Hinder, R. (1994): Zur Problematik der KUTTA-Bedingung. PhD thesis, Darmstadt
10. Hinder, R. (1995): About the Definition of the Kutta condition. In: Feistauer, M., Rannacher, R., Kozel, K. (eds.) Numerical Modelling in Continuum Mechanics. Proceedings of the 2nd Summer Conference held in Prague, 22-25 August,1994, pp.124-132
11. Hinder, R., Meister, E. (1995): Regarding some problems of the Kutta condition in lifting surface theory. To appear
12. Laschka, B. (1963): Zur Theorie der harmonisch schwingenden tragenden Fläche bei Unterschallströmung. Zeitschr. f. Flugwissensch. 11, 265-292
13. Meister, E. (1983): Randwertaufgaben der Funktionentheorie. Teubner, Stuttgart
14. Morrison, J. A., Lewis, J. A. (1976): Charge singularity at the corner of a flat plate. SIAM J. Appl. Math. 31, 233-250
15. von Petersdorff, T. (1989): Boundary integral equations for mixed Dirichlet, Neumann and transmission problems. Math. Methods Appl. Sci. 11, 185-213
16. von Petersdorff, T., Stephan, E. P. (1990): Decompositions in edge and corner singularities for the solution of the Dirichlet problem of the Laplacian in a polyhedron. Math. Nachr. 149, 71-104
17. von Petersdorff, T., Stephan, E. P. (1990): Regularity of mixed boundary value problems in R^3 and boundary element methods on graded meshes. Math. Meth. Appl. Sci. 12, 229-249
18. Reissner, E. (1949): Boundary value problems in aerodynamics of lifting surfaces in non-uniform motion. Bull. Amer. Math. Soc. 55, 825-850
19. Stephan, E. P. (1987): Boundary integral equations for screen problems in R^3. Integral Equations Operator Theory 10, 236-257
20. Stephan, E. P., Wendland, W. L. (1984): An augmented Galerkin procedure for the boundary integral method applied to two-dimensional screen and crack problems. Appl. Anal. 18, 183-220
21. Weissinger, J. (1963): Theorie des Tragflügels bei stationärer Bewegung in reibungslosen, inkompressiblen Medien. In: Flügge, S. (ed.) Handbuch der Physik, Strömungsmechanik II. Springer, pp.385-437
22. Wloka, J. (1987): Partial differential equations. Cambridge University Press

Acknowledgements: The authors would like to thank the members of the research group "Mathematische Methoden der Physik" at the TH Darmstadt and Prof. Dr. M. Costabel.

Multiscale Methods for the Solution of the Helmholtz and Laplace Equations

W. Dahmen, B. Kleemann, S. Prössdorf, and R. Schneider

1 Introduction

This paper presents some numerical results about applications of multiscale techniques to boundary integral equations. The numerical schemes developed here are to some extent based on the results of the papers [6]–[10]. Section 2 deals with a short description of the theory of generalized Petrov–Galerkin methods for elliptic periodic pseudodifferential equations in $\mathbb{R}^n$ covering classical Galerkin schemes, collocation, and other methods. A general setting of multiresolution analysis generated by periodized scaling functions as well as a general stability and convergence theory for such a framework is outlined. The key to the stability analysis is a local principle due to one of the authors. Its applicability relies here on a sufficiently general version of a so-called discrete commutator property of wavelet bases (see [6]). These results establish important prerequisites for developing and analysing methods for the fast solution of the resulting linear systems (Section 2.4). The crucial fact which is exploited by these methods is that the stiffness matrices relative to an appropriate wavelet basis can be approximated well by a sparse matrix while the solution to the perturbed problem still exhibits the same asymptotic accuracy as the solution to the full discrete problem. It can be shown (see [7]) that the amount of the overall computational work which is needed to realize a required accuracy is of the order $O(N(\log N)^b)$, where N is the number of unknowns and $b \geq 0$ is some real number.

We focus here on two problems which are solved by *fully discrete collocation* wavelet methods. Section 3 is devoted to various numerical experiments for the exterior Dirichlet problem for the Helmholtz equation. In this case the theoretical results for periodic problems apply and are confirmed by the numerical tests. In Section 4 we present some new results concerning a Dirichlet problem for the Laplace equation over three dimensional polyhedral domains. Linear systems with $\approx$ 100,000 unknowns are solved which corresponds to fully populated matrices of the same order. In this case our goal is not yet to present a fully developed scheme for such a complex three dimensional problem. In fact, one should note first that the general theory described in Section 2 is not directly applicable to this problem because it is not periodic and the boundary is not smooth. Therefore our main goal here is to explore to which extent the theoretical predictions from the model problem can still be confirmed under these less ideal circumstances. In particular, we focus on compression properties and the convergence behavior of the corresponding solutions of the perturbed problem. Therefore we content ourselves here still with a rather expensive way of computing the compressed stiffness matrices. Speeding up this portion of the solution process is less dependent on the topology of the domain and is meanwhile understood conceptually. Essentially following the analysis in [10], the practical realizations and numerical tests presented here do confirm a similar behavior as predicted by the theory for the periodic model problem.

2 Multiscale Methods

2.1 Periodic Pseudodifferential Equations

At this stage we focus on the model case of periodic pseudodifferential equations to exploit the full advantages of Fourier transform techniques in connection with appropriate representations for the class of operators under consideration. However, we do consider variable symbols and it should be mentioned that this class covers all the classical examples such as Hörmander's class, in particular, those operators arising in connection with boundary element methods. Moreover, most of the methods used here are of local nature and thus apply in essence to the case of non–periodic equations, as well (see [8]).

Consider the discrete Fourier transform by $\tilde{u}(k) := \int_{[0,1]} u(x) \mathrm{e}^{-2\pi \mathrm{i} x \cdot k} \mathrm{d}x$, $k \in \mathbb{Z}^n$, where $x \cdot k = \sum_{j=1}^n x_j k_j$. Let $\mathcal{T}^n$ denote the n–dimensional torus. Then a periodic pseudodifferential operator is defined by

$$\sigma(x, D) u(x) := \sum_{k \in \mathbb{Z}^n} \sigma(x,k) \tilde{u}(k) \mathrm{e}^{2\pi \mathrm{i} k \cdot x} .$$

The function $\sigma \in C^\infty(\mathcal{T}^n \times \mathbb{Z}^n)$, which is called the *symbol* of the operator $\sigma(x, D)$, is assumed to belong to a certain class $\sum^\mu$ for some $\mu \in \mathbb{C}$. Here $\sum^\mu$ is comprised of all symbols σ of the form $\sigma = \sigma_0 + \sigma_1$, where $\sigma_0 \in C^\infty(\mathcal{T}^n \times (\mathbb{R}^n/\{0\}))$ is homogeneous of degree $\mu \in \mathbb{C}$, i.e. $\sigma_0(x, 0) = 1$, $\sigma_0(x, \lambda k) = \lambda^\mu \sigma_0(x,k)$, for $\lambda > 0$, $k \neq 0$, and $|\partial^\beta_{(x)} \tau^\alpha_{(k)} \sigma_1(x,k)| \leq c_{\alpha,\beta}(1+|k|)^{r_1 - |\alpha|}$ for $x \in \mathcal{T}^n$, $k \in \mathbb{Z}^n$, and for some $r_1 < \operatorname{Re} \mu = r$. Here ∂ stands for the partial differential operator and τ for the partial forward difference operator. By $\mathcal{A}^r$ we denote the class of operators of the form $A = \sigma(\cdot, D) + K$ where $\sigma \in \sum^\mu$, $r = \operatorname{Re} \mu$ is the *order* of A, and $(Ku)(x) = \int_{\mathcal{T}^n} k(x,y) u(y) \mathrm{d}y$ with $k \in C^\infty(\mathcal{T}^n \times \mathbb{R}^n)$ is a smoothing operator. The operator $\sigma(x, D) \in \mathcal{A}^r$ is called *elliptic* if, for sufficiently large $|k|$, $|\sigma_0(x,k)| \geq c|k|^r$, $x \in \mathcal{T}^n$.

Any $A \in \mathcal{A}^r$ is a bounded linear operator $A : H^t \to H^{t-r}$, $t \in \mathbb{R}$. This operator is Fredholm if and only if it is elliptic. Here H^t denotes the classical periodic Sobolev space of order t, equipped with the norm $\|u\|_t := \left(\sum_{k \in \mathbb{Z}^n} (1+|k|)^{2t} |\tilde{u}(k)|^2\right)^{1/2}$.

2.2 Multiresolution and Wavelets

Multiresolution or Multiscale Analysis (MSA) is by now a well–studied notion [16]. Here we focus only on those variants which are useful for our purpose. MSA of $L^2(\mathbb{R})$ is a sequence of nested closed subspaces $\cdots \subset V_{-1} \subset V_0 \subset V_1 \subset \cdots \subset L^2(\mathbb{R})$ with

1. $\overline{\cup_j V_j} = L^2(\mathbb{R})$,
2. $\cap_j V_j = \{0\}$,
3. $f(x) \in V_j \Leftrightarrow f(2x) \in V_{j+1}$,
4. There is a function $\varphi \in L^2(\mathbb{R})$ such that the translates $\{\varphi(x-k)\}$ $(k \in \mathbb{Z})$ form an orthonormal basis of V_0.

The function φ is called *scaling function.* Obviously, the functions $\varphi_{j,k}(x) := 2^{j/2}\varphi(2^j x - k)$ form an orthonormal basis of V_j $(j \in \mathbb{Z})$ (see [11] for examples). Since $\varphi \in V_0 \subset V_1$ there exists a sequence $\{h_k\}_{k\in\mathbb{Z}}$ (which is called the mask or the filter of φ) such that φ satisfies the *scaling equation*

$$\varphi(x) = \sqrt{2} \sum_{k\in\mathbb{Z}} h_k \varphi(2x-k)\,, \; x \in \mathbb{R}. \tag{2.1}$$

Moreover, the mask $\{h_k\}$ is finite if and only if φ is compactly supported. (2.1) is the key to the constructions of orthogonal wavelet bases and of fast wavelet algorithms. The wavelet space W_j $(j \in \mathbb{Z})$ is defined as the orthogonal complement of V_j with respect to $V_{j+1} = V_j \oplus W_j$. Thus

$$L^2(\mathbb{R}) = \oplus_{j\in\mathbb{Z}} W_j = V_m \oplus \oplus_{j\geq m} W_j \quad (\text{for each } \; m \in \mathbb{Z}). \tag{2.2}$$

One of the main results of MSA reads as follows. Given φ, then there exists $\psi \in L^2(\mathbb{R})$ such that the functions $\psi_{j,k}(x) := 2^{j/2}\psi(2^j x - k)$, $k \in \mathbb{Z}$, form an orthonormal basis of W_j $(j \in \mathbb{Z})$. The function ψ is called the *mother wavelet* of MSA and is defined by

$$\psi(x) = \sqrt{2} \sum_{k\in\mathbb{Z}} g_k \varphi(2x-k), \quad g_k = (-1)^k h_{1-k}. \tag{2.3}$$

An important property of wavelets is that certain moments vanish, i.e.,

$$\int_{\mathbb{R}} x^l \psi(x) dx = 0, \quad 0 \leq l < d^*, \tag{2.4}$$

where in the case of orthogonal wavelets d^* is the order of polynomials which can be written as linear combinations of the translates $\varphi(\cdot - k), k \in \mathbb{Z}$.

More flexibility is offered by the concept of biorthogonal wavelets [11] which permits the employment of B–splines as scaling functions. In particular, the possibility of raising the order of moment conditions turns out to be essential for balancing compression and convergence rates.

In the following we need a *periodic* version of MSA. To this end, let $[\cdot]$ denote the periodization operator defined by $[f](x) := \sum_{k\in\mathbb{Z}} f(x+k)$ for any compactly supported function f. Given a compactly supported MSA $\varphi \in L^2(\mathbb{R})$, then the functions $[\varphi_{m,k}]$ and $[\psi_{m,k}]$ are 1–periodic. MSA of $L^2(\mathbb{R}^n)$ can be defined in a completely analogous manner. However, one needs a finite number of mother wavelets depending on the type of scaling.

2.3 Generalized Petrov–Galerkin Schemes

The spaces V_j will be used as trial spaces for the approximate solution of the equation

$$Au = f\,, \tag{2.5}$$

where $A \in \mathcal{A}^r$ and $f \in H^{t-r}$ is given. In the following we will fix one such t and assume that $\eta \in H^{r-t}(\mathbb{R}^n)$ is a fixed linear functional with support in some compact set $\Gamma \subset \mathbb{R}^n$. Defining the functionals $\eta_{j,k}$ by $\eta_{j,k}(f) := 2^{-nj/2}\eta(f(2^{-j}(\cdot+k)))$, $k \in \mathbb{Z}^{n,j}$, we seek for an element $u_j \in V_j$ satisfying

$$\eta_{j,k}(Au_j) = \eta_{j,k}(f)\,,\ k \in \mathbb{Z}^{n,j} := \mathbb{Z}^n/2^j\mathbb{Z}^n\,. \tag{2.6}$$

Clearly, $\eta = \varphi$ corresponds to a classical Galerkin scheme, while $\eta = \delta(\cdot - x_0)$ give rises to collocation at the points $2^{-j}(k + x_0)$, $k \in \mathbb{Z}^{n,j}$, $j \in \mathbb{N}_0$.

Our first objective is to study the solvability of (2.6), and if this is the case, the convergence of the solution u_j as $j \to \infty$. The key to this problem is a suitable stability concept. To this end, it is convenient to consider projectors of the form

$$Q_j f = \sum_{k\in\mathbb{Z}^{n,j}} \eta_{j,k}(f)\zeta_{j,k},$$

where the $\zeta_{j,k}$ are suitable basis functions satisfying $\eta_{j,k}(\zeta_{j,l}) = \delta_{l,k}$ (possibly spanning spaces different from V_j, see [6] Sect. 4). Then (2.6) can be rewritten as operator equation $Q_jAu_j = Q_jf$. The scheme (2.6) is called (t,r)–*stable* if

$$\|Q_jAv\|_{t-r} \geq c\|v\|_t \tag{2.7}$$

for all $v \in V_j$ uniformly in $j \in \mathbb{N}_0$. Clearly, (2.7) means that the finite dimensional operators $A_j := Q_jAP_j : H^t \to H^{t-r}$ have uniformly bounded inverses $A_j^{-1} : \mathrm{im}\ Q_j \to \mathrm{im}\ P_j$.

It turns out that the stability of (2.6) is equivalent to the ellipticity of a certain function λ which shares many features with the principal symbol σ_0 of the operator A and which is called the *numerical symbol* (more precisely, the "symbol of the Petrov–Galerkin scheme (2.6)"). The numerical symbol is defined by

$$\lambda(x,y) := \sum_{k\in\mathbb{Z}^n} \sigma_0(x, y+k)\hat{\varphi}(y+k)\overline{\hat{\eta}(y+k)} \tag{2.8}$$

for all $x, y \in \mathcal{T}^n$ (provided that the series on the right hand side of (2.8) converges absolutely for all $x, y \in \mathcal{T}^n$), where ^ stands for the Fourier transform.

Theorem 2.1 ([6]). *The scheme* (2.6) *is* (t, r)*–stable if and only if the numerical symbol* λ *is elliptic, i.e.* $|\lambda(x, y)| \geq c|\theta(y)|^r$ *holds uniformly in* $x \in \mathcal{T}^n$*, where* $\theta = (\theta_1, ..., \theta_n)^T$, $\theta_j(y) = e^{2\pi i y_j} - 1$.

Example 2.1. Consider the knot collocation (i.e. $x_0 = 0$) with tensor product splines of degree d. Then $\hat{\eta} = 1$ and $\hat{\varphi}(\xi) = \prod_{l=1}^n (\sin \pi\xi_l)^{d+1} / \ (\pi\xi_l)^{d+1}$. Clearly, if d is odd, then λ is elliptic if and only if A is strongly elliptic, i.e. Re $\sigma_0 \geq$ const > 0. Hence, in that case, Theorem 1 provides stability of the collocation with odd degree splines for strongly elliptic operators (cf. [1] for the case $n = 1$ and [6] for the multidimensional case). Notice that, in the case of the classical Galerkin scheme, the ellipticity of the numerical symbol for strongly elliptic operators is a consequence of the stability property of φ.

Applying Theorem 1 in combination with well known Galerkin techniques and approximation properties of the functions $[\varphi_{j,k}]$ one obtains optimal estimates for the error $\|u - u_j\|_t$ for a certain range of Sobolev norms (see [6]).

2.4 Matrix Compression and Fast Solution

This technique is essentially based on estimates for the asymptotic behavior of the entries of A_j ($j \to \infty$) in the wavelet representation. We will present such estimates only for the case of classical Petrov–Galerkin schemes, i.e., when η is actually an L^2–function.

Theorem 2.2 ([7, 17]). *Let* $2d^* + n + r > 0$ *where* d^* *denotes the number of vanishing moments of the wavelets. Then for* $A \in \mathcal{A}^r$ *the estimate*

$$|(A[\psi_{l,k}], [\psi_{l',k'}])| \leq c \, \frac{2^{-(l+l')(n/2+d^*)}}{\varrho^{n+r+2d^*+2}} \tag{2.9}$$

holds, where $\varrho =$ dist (supp $[\psi_{l,k}]$, supp $[\psi_{l',k'}]$) *and the constant* c *depends only on* r, n *and* d^*.

(2.9) are the crucial estimates which compression criteria rely on. Such criteria allow us to avoid the computation of the full stiffness matrix in the wavelet representation and tell us which entries must be computed in order to guarantee a required accuracy. Moreover, by realizing sufficiently high accuracy on lower scales, the asymptotical convergence rates of the solutions to the uncompressed systems can be preserved for those of the compressed system. Such a compression strategy reduces the computational work to the order $O(N(\log N)^b)$, where $N = 2^{jn}$ and b is a positive number ([7]).

3 Exterior Dirichlet problem for the HELMHOLTZ equation in 2D

3.1 Statement of the problem

The treatment of scattering of time–harmonic acoustic and electromagnetic waves on infinitely long cylindrical obstacles in $\mathbb{R}^3$ with simply connected cross section $D_- \subset \mathbb{R}^2$ and smooth boundary Γ leads to an exterior boundary value problem for the Helmholtz equation in $\mathbb{R}^2$.

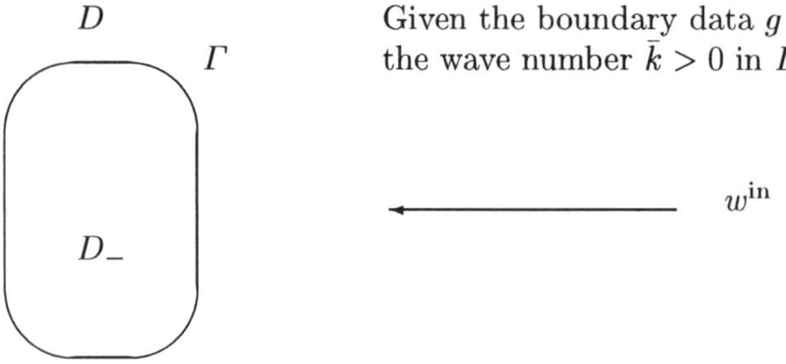

Given the boundary data g and the wave number $\bar{k} > 0$ in D, real.

We seek the solution w of the following problem

$$\Delta w(x) + \bar{k}^2 w(x) = 0 \quad \text{in} \quad D = \mathbb{R}^2 \backslash \overline{D_-}\,, \tag{3.1}$$

$$w(x) = g(x) \quad \text{on} \quad \Gamma\,, \tag{3.2}$$

$$\frac{\partial w(x)}{\partial r} - \mathrm{i}\bar{k} w(x) = o(r^{-1/2})\,,\ r = |x| \to \infty \tag{3.3}$$

uniformly in all directions (*Sommerfeld's radiation condition*). This problem is known to have a unique solution (cf. [5]).

Using a *single–layer approach* (indirect method), one seeks w in the form

$$w(x) = \int_\Gamma \phi(x,y)\xi(y)\mathrm{d}s(y)\,,\ x \in D\,, \tag{3.4}$$

where s is the arc length parametrization, and the fundamental solution corresponding to the Helmholtz equation is given by

$$\phi(x,y) := \frac{1}{2\mathrm{i}} H_0^{(1)}(\bar{k}|x-y|)\,, x \neq y\,.$$

Here $H_0^{(1)}$ means the Hankel function of order zero and of the first kind. Substituting (3.4) into (3.2) gives the boundary integral equation for the unknown density ξ:

$$\int_\Gamma \phi(x,y)\xi(y)\mathrm{d}s(y) = g(x)\,,\ x \in \Gamma\,. \tag{3.5}$$

Equation (3.5) is uniquely solvable in $H^s(\Gamma)$ provided that the homogeneous Dirichlet problem for the interior of Γ admits only the trivial solution.

Let $x(t) = (x_1(t), x_2(t))$, $0 \le t \le 2\pi$, be a 2π–periodic regular parametrization of Γ satisfying $[x_1'(t)]^2 + [x_2'(t)]^2 > 0$, $t \in \Gamma$. Setting in (3.5) $u(t) := \xi(x(t)) \left\{[x_1'(t)]^2 + [x_2'(t)]^2\right\}^{1/2}$ and $f(t) := g(x(t))$, the exterior Dirichlet problem for the Helmholtz equation leads to the following *logarithmic singular* integral equation of the first kind:

$$(Au)(t) := \frac{1}{2\pi}\int_0^{2\pi} K(t,\tau)u(\tau)\mathrm{d}\tau = f(t)\,,\ 0 \le t \le 2\pi, \tag{3.6}$$
$$K(t,\tau) := \frac{\pi}{\mathrm{i}} H_0^{(1)}(\bar{k}r(t,\tau))$$

where $r(t,\tau) := \{[x_1(t) - x_1(\tau)]^2 + [x_2(t) - x_2(\tau)]^2\}^{1/2}$.
In principle it is possible to separate the logarithmic part of the Hankel function and treat it separately. There exist several fast methods for the solution of the logarithmic kernel equation (also called Symm's equation). In particular, one can use the fast Fourier transform, apply the fast method [20] or the exponentially convergent method [14]. Here we do not separate the logarithmic part because we only consider it as a model problem for other cases where a separation is usually not applicable. A second related reason is to simplify the implementation by keeping it as independent as possible of further analytic investigations which have to be tuned to the particular application at hand.

3.2 Wavelet Discretization Method

For the numerical solution of (3.6) consider the knot collocation method on the following nested grids of $[0, 2\pi]$:
$\nabla^l = \{t_k^l | t_k^l := kh_l,\ k = 0, ..., N_l - 1,\ h_l = 2^{-l} \cdot 2\pi,\ N_l = 2^l\}$, $l = 0, 1, ..., j$.
The set of additional knots for passing from level $l-1$ to level l is denoted by $\Delta^{l-1} := \nabla^l \setminus \nabla^{l-1}$. For a corresponding Galerkin wavelet method the reader is referred to [17]. As trial functions we use here continuous piecewise linear functions spanned by the canonical periodized hat functions $[\varphi_{j,k}]$, i.e.,

$$\varphi(x) := \begin{cases} 1 + x & \text{if} \quad -1 \le x \le 0, \\ 1 - x & \text{if} \quad 0 \le x \le 1, \\ 0 & \quad\quad \text{else.} \end{cases}$$

In this case the mask in the two-scale relation (2.1) is

$$\{h_k\}_{k=-1}^{1} = \{\frac{1}{2}, 1, \frac{1}{2}\}. \tag{3.7}$$

Since the translates $\varphi(\cdot - k)$ are not pairwise orthogonal the simple recipe from (2.3) for forming orthogonal wavelets does no longer work. Nevertheless, one can resort to the literature for a long list of candidates $\psi(x)$ whose integer translates are stable and span the orthogonal or other complements of V_0 in V_1. In particular, for any desired order d^* of vanishing moments

$$\varphi(x) = \tfrac{1}{2}\varphi(2x+1) + \varphi(2x) + \tfrac{1}{2}\varphi(2x-1)$$

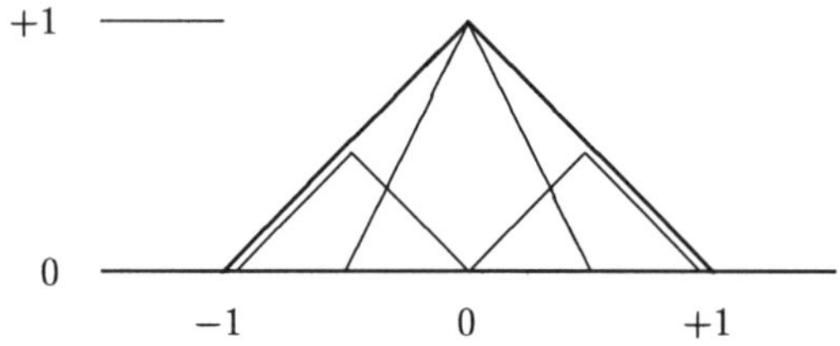

$$\psi(x) = \tfrac{1}{2}\varphi(2x+1) - \varphi(2x) + \tfrac{1}{2}\varphi(2x-1)$$

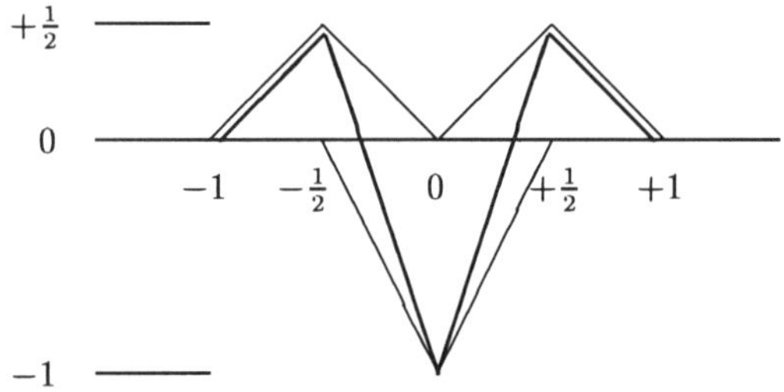

Fig. 3.1. Two-scale relations for our choice of generating function and mother wavelet.

biorthogonal wavelets have been constructed in [4] which all give rise to Riesz bases. Specifically, the wavelet for $d^* = 2$ from this family has support $[-1, 2]$ and the mask has the form

$$\{g_k\}_{k=-1}^{3} = \{\frac{1}{8}, \frac{1}{4}, -\frac{3}{4}, \frac{1}{4}, \frac{1}{8}\}. \tag{3.8}$$

To keep the support of ψ as small as possible we will focus here though on the following choice with also vanishing moments of order $d^* = 2$:

$$\{g_k\}_{k=-1}^{1} = \{\frac{1}{2}, -1, \frac{1}{2}\}, \tag{3.9}$$

which is depicted in Figure 3.1. These wavelets are likewise biorthogonal in the sense of [4] and therefore give rise to a Riesz basis as is shown in [18]. Since we are employing a collocation method the *Brandt/Lubrecht functionals* [2] are adequate "test wavelets" (because of pointwise evaluation) spanned by δ–distributions:

$$\eta_{l,k}(f) = f(t_{2k+1}^{l+1}) - \frac{1}{2}\Big[f(t_k^l) + f(t_{k+1}^l)\Big], \tag{3.10}$$

$$k = 0, ..., 2^{l-1} - 1, l = j-1, ..., 1; \; \eta_{0,k}(f) = f(t_k^1), k = 0, 1.$$

One easily verifies that the $\eta_{l,k}$ have also vanishing moments of order $d^* = 2$. Figure 3.2 illustrates for three levels l how the point values are transformed into the data $\eta_{l,k}(f)$. Obviously, the complexity of this transformation is of the order $\mathcal{O}(N_j)$ when j is as above the finest level. Given (3.10), the

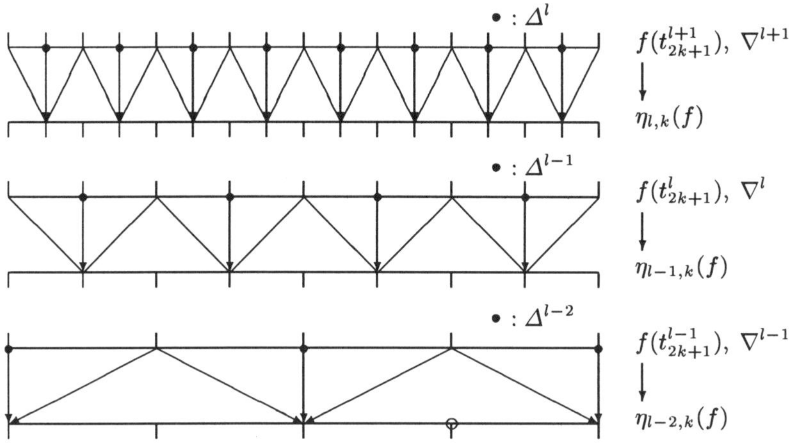

Fig. 3.2. Generation of the functionals of Brandt/Lubrecht. Open circles indicate those grid points whose values are to be calculated in the next step.

collocation stiffness matrix A_j on level j relative to the wavelet basis has entries of the form

$$(A_j)_{(l,k),(l',k')} := \eta_{l',k'}(A[\psi_{l,k}]), \tag{3.11}$$

$$k = 0, \ldots, 2^{l-1} - 1,\ k' = 0, \ldots, 2^{l'-1} - 1,\ l, l' < j\,.$$

It is easily seen that the integral operator A defined by (3.6) belongs to the class $\mathcal{A}^r$ (see Sect. 2) with $r = -1$ and $r_1 = -3$. Moreover, it is strongly elliptic and, hence, the considered collocation method is stable (see Theorem 2). Furthermore, the error estimate

$$\|u - u_j\|_s \le c2^{j(s-2)}\|u\|_2 \tag{3.12}$$

holds for all s, $-1 \le s < 3/2$, provided $f \in H^3$.

Using an `a-priori` *compression criterion* established in [7] and improved in [21] we only calculate and store approximations to those matrix entries of (3.11) obtained through quadrature for which

$$\operatorname{dist}(\Omega_{l,k}, \tilde{\Omega}_{l',k'}) \le \max\left\{a2^{-l}, a2^{-l'}, aj^{5/6} \cdot 2^{(2/3)j-(4/3)l-(2/3)l'}\right\}, \tag{3.13}$$

where $\Omega_{l,k}$ is the support of $[\psi_{l,k}]$, $\tilde{\Omega}_{l',k'}$ the support of $\eta_{l',k'}$, and a a constant that has to be chosen appropriately. Note that, according to the general theory, to realize an (asymptotically) almost optimal compression rate for the choice of piecewise linear wavelets with two vanishing moments it would be necessary to use Brandt/Lubrecht functionals with three vanishing moments. In this case also criterion (3.13) changes somewhat. Nevertheless, the present method turns out to be more efficient for the present scope of experiments because of the smaller support of the test functionals. Also our convergence studies confirm a sufficiently good nearly optimal performance of the present version in all our test problems (see Figs. 3.3,3.4,3.5).

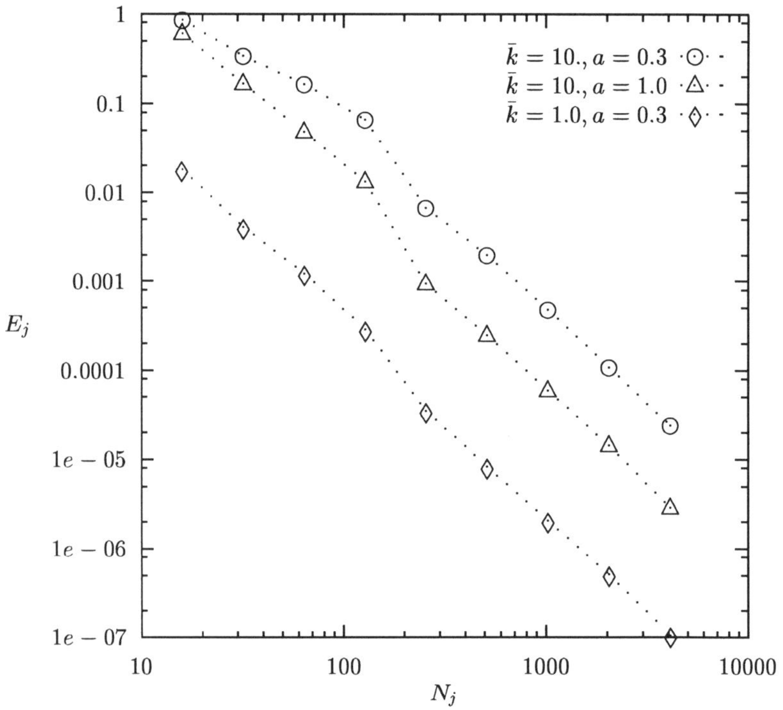

Fig. 3.3. Error E_j in logarithmic scale for various wave numbers $\bar{k}$ and constants a for the `a-priori` compression strategy. Γ = circle.

In order to get a fully discrete method we have to employ a suitable quadrature method to approximate the integrals. Because of the logarithmic singularity of the kernel function it is necessary to use an adapted rule for the quadrature to guarantee that the order of convergence is better than one. Therefore we choose the following graded quadrature points τ_q of the interval $[0, 2\pi]$ with grading parameter $\alpha \geq 1$. More precisely, the points are graded near the singular point $t \in (0, 2\pi)$ as follows:

$$\tau_q' = \pi \left| \frac{q}{\tilde{m}} \right|^\alpha \operatorname{sign} q + t, \quad q = -\tilde{m}, ..., \tilde{m} - 1.$$

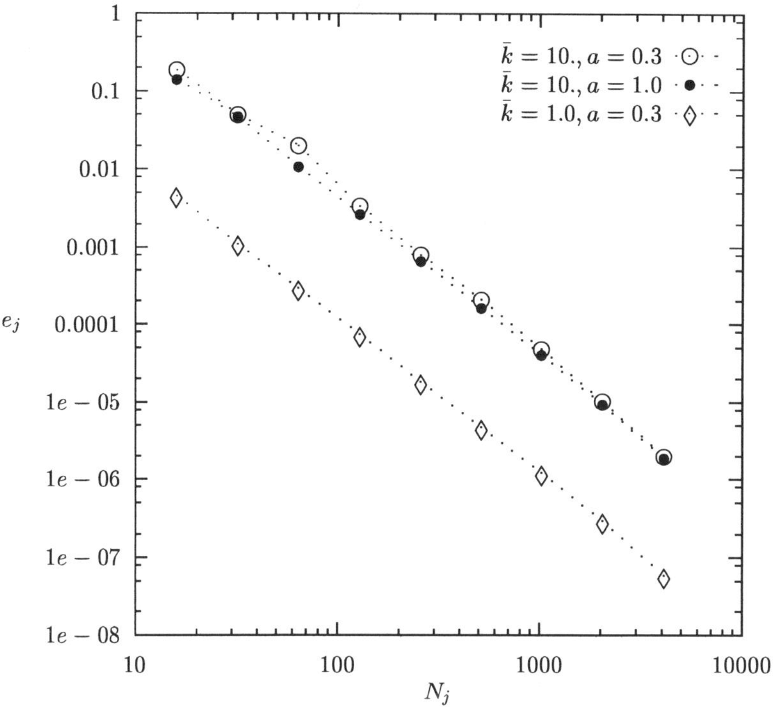

Fig. 3.4. Error e_j in logarithmic scale for various wave numbers $\bar{k}$ and constants a for the `a-priori` compression strategy. Γ = circle.

The integer $\tilde{m}$ has to be chosen proportional to N_j. To make the quadrature exact for the ansatz functions these graded points are united with equidistant ones: $\{\tau^{\mathrm{e}}\}_0^{N_j-1} = \{t_k^j\}$. Let $\tau_q, q = 0, \ldots, m-1$, $m = 2\tilde{m} + N_j$ be the quadrature points obtained by periodizing $\{\tau_q'\} \cup \{\tau^{\mathrm{e}}\}$. In our numerical examples we choose a grading parameter $\alpha = 2.5$ together with the trapezoidal rule. The latter is exact for linear polynomials, and hence sufficient to obtain second order convergence in L^2 (cf. (3.12)).

3.3 Assembling the compressed stiffness matrix

Let

$$(\tilde{A}_j[\psi_{l,k}])(t_i) = \sum_{\substack{q=0 \\ \tau_q \neq t}}^{m-1} K(t_i, \tau_q)\omega_q \psi_{l,k}(\tau_q), \quad k = 0, \ldots, 2^l - 1, \ l < j,$$

be the approximation of $(A\psi_{l,k})(t_i)$ obtained by quadrature. Here ω_q are quadrature weights.

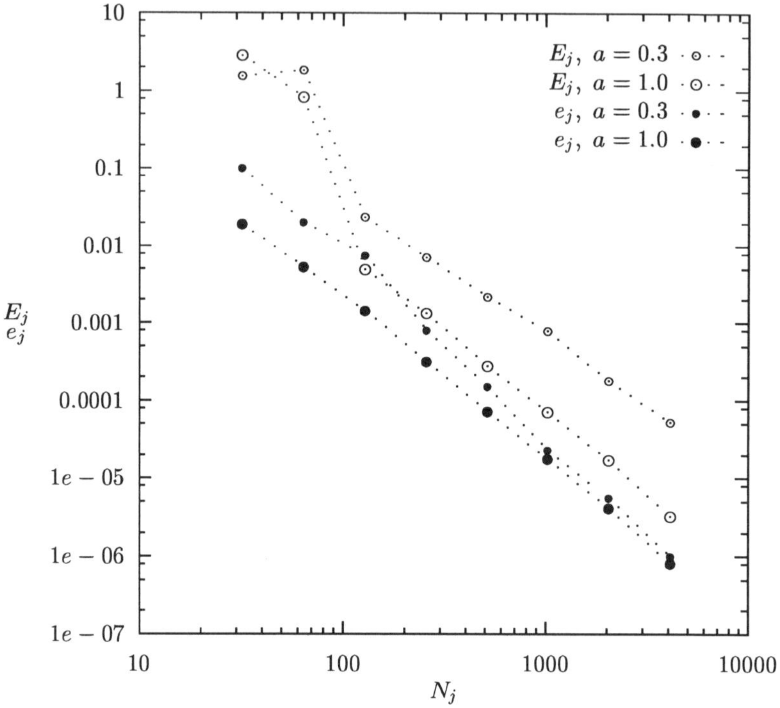

Fig. 3.5. Errors E_j, e_j in logarithmic scale for an elliptical cross section and fixed wave number $\bar{k} = 10$ for two different constants a for the `a-priori` compression strategy.

Keeping in mind that

$$\eta_{l',k'}(A[\psi_{l,k}]) = (A[\psi_{l,k}])(t^{l'+1}_{2k'+1}) - 1/2\{(A[\psi_{l,k}])(t^{l'}_{k'}) + (A[\psi_{l,k}])(t^{l'}_{k'+1})\}$$

and that the vanishing moments of the test wavelets $\eta_{l,k}$ imply that these quantities decay, their approximations obtained through quadrature are expected to exhibit the same behavior. Thus we have to compute the following expressions:

$$\eta_{l',i}(\tilde{A}_j[\psi_{l,k}]) = \sum_{\substack{q=0 \\ \tau_q \neq t^{l'+1}_{2i+1}}}^{m-1} K(t^{l'+1}_{2i+1}, \tau_q)\omega_q[\psi_{l,k}](\tau_q) - \frac{1}{2} \sum_{\substack{q=0 \\ \tau_q \neq t^{l'}_{i}}}^{m-1} K(t^{l'}_{i}, \tau_q)\omega_q[\psi_{l,k}](\tau_q) - \tag{3.14}$$

$$\frac{1}{2} \sum_{\substack{q=0 \\ \tau_q \neq t_{i+1}^{l'}}}^{m-1} K(t_{i+1}^{l'}, \tau_q)\omega_q[\psi_{l,k}](\tau_q),$$

$$i = 0, \ldots, N_{l'} - 1, \quad k = 0, \ldots, N_l - 1, \quad l, l' = 1, \ldots, j-1,$$

$$\begin{aligned} \eta_{l',i}(\tilde{A}_j[\varphi_{1,k}]) = & \sum_{\substack{q=0 \\ \tau_q \neq t_{2i+1}^{l'+1}}}^{m-1} K(t_{2i+1}^{l'+1}, \tau_q)\omega_q[\varphi_{1,k}](\tau_q) - \\ & \frac{1}{2} \sum_{\substack{q=0 \\ \tau_q \neq t_i^{l'}}}^{m-1} K(t_i^{l'}, \tau_q)\omega_q[\varphi_{1,k}](\tau_q) - \\ & \frac{1}{2} \sum_{\substack{q=0 \\ \tau_q \neq t_{i+1}^{l'}}}^{m-1} K(t_{i+1}^{l'}, \tau_q)\omega_q[\varphi_{1,k}](\tau_q), \end{aligned} \tag{3.15}$$

$$i = 0, \ldots, N_{l'} - 1, \quad k = 0, 1, \quad l' = 1, \ldots, j-1,$$

$$\eta_{0,i}(\tilde{A}_j[\psi_{l,k}]) = \sum_{\substack{q=0 \\ \tau_q \neq t_i^1}}^{m-1} K(t_i^1, \tau_q)\omega_q[\psi_{l,k}](\tau_q), \tag{3.16}$$

$$i = 0, 1, \quad k = 0, \ldots, N_l - 1, \quad l = 1, \ldots, j-1,$$

$$\eta_{0,i}(\tilde{A}_j[\varphi_{1,k}]) = \sum_{\substack{q=0 \\ \tau_q \neq t_i^1}}^{m-1} K(t_i^1, \tau_q)\omega_q[\varphi_{1,k}](\tau_q), \tag{3.17}$$

$$i = 0, 1, \quad k = 0, 1.$$

Here it is important to note that, according to our above remarks, these calculations are only performed for those pairs $(l', k'), (l, k)$ which satisfy condition (3.13). Thus storage is only required, and hence is directly reduced, for the significant entries of the matrix A_j^ε defined by

$$(A_j^\varepsilon)_{(l',k'),(l,k)} := \begin{cases} 0 & \text{if} \quad (3.13) \text{ does not hold,} \\ \eta_{l',k'}(\tilde{A}_j[\psi_{l,k}]) & \text{if} \quad (3.13) \text{ holds,} \end{cases}$$

where $k' = 0, \ldots N_{l'} - 1$, $k = 0, \ldots, N_l - 1$, $l, l' = 0, \ldots, j-1$ and where we have set $\psi_{0,k} := \varphi_{1,k}$.

Next let us denote by B_j the transformation (3.10) which takes the array of data $f(t_k^j)$, $k = 0, \ldots, N_j - 1$, into the array $\eta_{l,k}(f)$. One can check that A_j^ε is the compressed version of the matrix $B_j\tilde{A}_j$. Hence we end up with the following sparse matrix equation

$$A_j^\varepsilon u_j = B_j f_j, \tag{3.18}$$

where $f_j := (f(t_0^j), f(t_1^j), \ldots, f(t_{N_j-1}^j))^T$.

Note that it is possible to use the same quadrature formula as for a standard collocation method not based on wavelets for all expressions in (3.14)–(3.17). It is only necessary to shift the grading for the singularity to the actual singular point.

Clearly, the solution u_j of (3.18) consists of the coefficients of the approximate solution relative to the wavelet basis. If one wishes to represent the solution in terms of the nodal basis, a further transformation

$$u_j^\varphi = T_j u_j$$

has to be performed. This transformation also has the following familiar pyramid structure

$$\begin{array}{ccccccccc} w_{1,\cdot} & & w_{2,\cdot} & & w_{3,\cdot} & & \cdots & w_{j-1,\cdot} & \\ & \searrow & C_1 & \searrow & C_2 & & & \searrow & C_{j-1} \\ u_{1,\cdot} & \rightarrow & u_{2,\cdot} & \rightarrow & u_{3,\cdot} & \cdots & u_{j-1,\cdot} & \rightarrow & u_j^\varphi \\ & M_1 & & M_2 & & & & M_{j-1} & \end{array}, \tag{3.19}$$

where the matrices $M_i, C_i,\ \ i = 1, \ldots, j-1$ are sparse and contain the masks from (3.7) and (3.9). For a description of the algorithm in pseudocode see [10]. Thus T_j takes coefficients $u_j := \{u_{1,k}, w_{l,k}\}$ of the wavelet representation $\sum_{k=0,1} u_{1,k}\varphi_{1,k} + \sum_{l=1}^{j-1}\sum_{k\in\Delta^l} w_{l,k}\psi_{l,k}$ into the coefficients $u_j^\varphi := \{u_{j,k}\}$ of the corresponding nodal representation $\sum_{k\in\nabla^j} u_{j,k}\varphi_{j,k}$. Also the application of the transformation T_j is easily seen to require only $\mathcal{O}(N_j)$ operations.

3.4 Preconditioning and solution

The linear system (3.18) is solved with the aid of the iterative Krylov subspace method GMRES (cf. [19],[24]). This method has been proved to be efficient for this kind of systems (cf. e.g. [23],[13],[10]). Since $r = -1$, the condition numbers $\kappa(A_j^\varepsilon)$ grow like N_j so that the number of iterations which is necessary to accomplish a desired accuracy increases when N_j becomes large. The theory in [21] suggests multiplying A_j^ε with the $N_j \times N_j$ diagonal matrix D from the right where $D_{(l,k),(l',k')} = (\bar{k}^2 + 2^{2l})^{\frac{1}{2}} \cdot \delta_{(l,k),(l',k')}$. After solving then the system $A_j^\varepsilon D v_j = B_j f_j$ one has to transform the resulting vector according to $u_j = D v_j$ to obtain the wavelet coefficients of the solution.

We have also tested the following alternative preconditioning strategies based on diagonal scaling. Let $\bar{D}_{(l,k),(l',k')} = 2^l \cdot \delta_{(l,k),(l',k')}$.

1. Solve $\bar{D} A_j^\varepsilon u_j = \bar{D} B_j f_j$ or
2. solve $\bar{D}^{\frac{1}{2}} A_j^\varepsilon \bar{D}^{\frac{1}{2}} v_j = \bar{D}^{\frac{1}{2}} B_j f_j$ followed by corresponding diagonal back transformations.

The results of the tests for a fixed example are given in Table 3.1. The best preconditioner is the one suggested by the theory which gives a constant iteration count. May be that with higher order Brandt/Lubrecht functionals the iteration count can further be reduced. The corresponding tools are available and, in principle, easily implemented.

Table 3.1. Number of GMRES iterations without preconditioning and with different preconditioners for the following example: $\Gamma=$ ellipse, $\bar{k}=10$, $a=1.0$.

Preconditioning											
j	2	3	4	5	6	7	8	9	10	11	12
N_j	4	8	16	32	64	128	256	512	1024	2048	4096
without	3	5	9	14	18	21	29	33	37	40	44
$\bar{D}A_j^\varepsilon$	3	5	9	15	18	20	25	27	28	28	29
$\bar{D}^{\frac{1}{2}}A_j^\varepsilon\bar{D}^{\frac{1}{2}}$	3	5	9	15	17	19	24	27	26	26	27
$A_j^\varepsilon D$	3	5	9	14	**15**	**16**	**21**	**23**	**23**	**22**	**22**

The main steps of the solution procedure may be summarized now as follows:

1. Compute the matrix A_j^ε directly in the wavelet representation as described by (3.14) – (3.17) and apply the `a-priori` compression criterion (3.13) yielding a sparse matrix;
2. apply the transformation B_j (3.10) to the right-hand side: $B_j: \{f(t_k^j)\} \mapsto \{\eta_{j,k}(f)\},\ \ k=0,\ldots,N_j-1: B_jf_j$, which requires only $\mathcal{O}(N_j)$ operations;
3. solve the sparse matrix equation $A_j^\varepsilon Dv_j = B_jf_j$ by an iterative solver (e.g. GMRES). The complexity of this part is nearly $\mathcal{O}(N_j(\log N_j)^b)$;
4. back transform the solution of this linear system to $u_j = Dv_j$ ($= N_j$ operations), which provides the coefficient vector u_j relative to the wavelet basis and
5. (optionally) apply the transformation T_j which yields the coefficients $u_j^\varphi = T_ju_j$ for the nodal basis representation. Again this only requires $\mathcal{O}(N_j)$ operations.

Clearly, the multiplications with the diagonal matrices are applied to the iteration vector rather than to the matrix A_j^ε: $A_j^\varepsilon(Du_j^{(i)})$, with $u_j^{(i)}$ the actual iteration vector of u_j during the GMRES-iteration. This only requires $2N_j$ additional operations for the diagonal matrix times vector products in each iteration step.

Summing up the operation count in steps 2.-5. the overall complexity is of order $\mathcal{O}(N_j(\log N_j)^b)$ which gives rise to a fast solution of the linear system. Unfortunately, step 1., the matrix assemblation, is still of the order $\mathcal{O}(N_j^2)$ although only (nearly) $\mathcal{O}(N_j(\log N_j)^b)$ matrix entries are calculated. This is due to the poor quadrature scheme which does not take the distance between the supports $\Omega_{l,k}$ and $\Omega_{l',k'}$ into account. Meanwhile it has been shown in [21] that such a refined quadrature strategy reduces the computational work for assembling the matrix to the order of nonvanishing matrix entries.

After the evaluation of the solution, the relative L^2 error E_j of the solution and the convergence rate β_1 are calculated:

$$E_j = \frac{\sum_{i=0}^{N_j-1} |u_j(t_i^j) - u(t_i^j)|^2}{\sum_{i=0}^{N_j-1} |u(t_i^j)|^2}, \quad \beta_1 = -\frac{\log(E_j/E_{j-1})}{\log 2}.$$

3.5 Numerical examples for scattering

We consider the scattering of an E-polarized electromagnetic plane wave u^{in} by a perfectly conducting cylinder. In acoustics this problem is equivalent to the scattering of a plane wave by an impenetrable, sound-soft cylinder with smooth cross section. Let the incident wave be given by $u^{\mathrm{in}}(x) = \exp(\mathrm{i}\bar{k}\tilde{e}_{\mathrm{in}}x)$, with $\tilde{e}_{\mathrm{in}}$ a unit vector describing the direction of propagation of the incident wave, and let u^{s} be the scattered wave. Then the exterior wave (total field) $u^{\mathrm{e}} = u^{\mathrm{in}} + u^{\mathrm{s}}$ is also a solution of the Helmholtz equation $\Delta u^{\mathrm{e}} + \bar{k}^2 u^{\mathrm{e}} = 0$ in D and satisfies the boundary condition $u^{\mathrm{e}} = 0$ on Γ. So we have $u^{\mathrm{s}} = -u^{\mathrm{in}}$ on Γ and u^{s} also satisfies the conditions (3.1) and (3.3). Thus in terms of our initial unknowns we have:

- $w := u^{\mathrm{s}}$,
- $g := u^{\mathrm{s}}|_\Gamma = -\exp(\mathrm{i}\bar{k}\tilde{e}_{\mathrm{in}}x)|_\Gamma$ and if we set for the incident direction $\tilde{e}_{\mathrm{in}} = (1,0)$, we obtain

$$f(x(t)) := -(\cos(\bar{k}x_1(t)) + \mathrm{i}\sin(\bar{k}x_1(t))).$$

With these data one can then solve the integral equation (3.6).

The solution u^{s}, the backscattered electric field, exhibits the following asymptotic behavior which follows from the asymptotics for the Hankel function for large argument:

$$u^{\mathrm{s}}(x) = \frac{\exp(\mathrm{i}\bar{k}|x|)}{\sqrt{|x|}} \left\{ u^\infty(\tilde{x}) + \mathcal{O}\left\{\frac{1}{|x|}\right\}\right\}, \; |x| \longrightarrow \infty,$$

$$u^\infty(\tilde{x}) = \frac{\exp(\mathrm{i}\frac{\pi}{4})}{\sqrt{8\pi\bar{k}}} \int_\Gamma \xi(y)\exp(-\mathrm{i}\bar{k}\tilde{x}y)ds(y).$$

uniformly for all directions $\tilde{x} := x/|x|$. The function u^∞ is known as *far field pattern* or scattering amplitude of u^{s}. Substituting for $\xi(y)$ the solution $u(t)$ of our integral equation, we obtain

$$u^\infty(\tilde{x}) = \frac{\exp(\mathrm{i}\frac{\pi}{4})}{\sqrt{8\pi\bar{k}}} \int_\Gamma u(t)\exp(-\mathrm{i}\bar{k}\tilde{e}_{\mathrm{ob}}x(t))dt. \tag{3.20}$$

A more interesting quantity is the *radar cross section* (RCS). It is defined by $\sigma^{\mathrm{c}}(\theta') = 2\pi \lim_{|x|\to\infty} x\frac{|u^{\mathrm{s}}|^2}{|u^{\mathrm{in}}|^2}$, with $0° \leq \theta' \leq 360°$, the backscatter angle. With the aid of the asymptotics of u^{s} we obtain the following approximation $\sigma^{\mathrm{c}}(\theta') = 2\pi|u^\infty(\tilde{x})|^2$ for the RCS where $\tilde{x}$ is the direction given by θ'. When plotting the RCS as a function of the observation angle, the quantity

$$\sigma = 10\log_{10}(\sigma^{\mathrm{c}}/\lambda), \ \lambda = \text{ wavelength} \tag{3.21}$$

are shown in the figures rather than σ^{c} which is common in the engineering literature.

After the solution of the integral equation by the wavelet method, the integral (3.20) is calculated which is a functional of the solution. This is realized by the rectangular rule which in the periodic case is equivalent to the trapezoidal rule. Let the approximation of (3.20) be u_j^∞. For the determination of u^∞ and u_j^∞ we fix the direction of observation. For a sequence of discretizations we determine the error e_j and the convergence rate β_2 for the functionals u_j^∞:

$$e_j = \frac{|u_j^\infty - u^\infty|}{|u^\infty|}, \quad \beta_2 = -\frac{\log(e_j/e_{j-1})}{\log 2}.$$

For all numerical examples we have selected $\alpha = 2.5$ as an optimal grading parameter for the quadrature points and employing the `a-priori` compression criterion, we have examined the convergence and compression behavior for various wave numbers $\bar{k}$ and constants a. In the first example we consider a disc of radius $r = 0.6$ as cross section of a scattering obstacle (cf. e.g. [3]) and we set $\tilde{e}_{\mathrm{ob}} := (1,0)$ while the second considered obstacle is assumed to have an elliptical cross section with $a_x = 2b_y = 0.6$.

Figures 3.3 and 3.4 present the behavior of the errors E_j, e_j for the disc. Figure 3.5 shows results for the case of an ellipse. Figures 3.3,3.4,3.5 clearly show the predicted behavior for the convergence rate of E_j, e_j for the investigated boundaries if the constant a for the `a-priori` compression criterion is appropriately chosen. The convergence rates of ≈ 2 coincide with the expected convergence rate of the solutions to the full systems when using the same discretization method. Because of (3.12) for e_j a convergence rate of 3 is possible when using a quadrature of higher exactness.

The constant a from (3.13) in practical applications should be chosen so that the supports of $\psi_{j,k}, \eta_{j,k'}$ do not overlap on the finest level j. This leads to a fixed bandwidth for the diagonal bandmatrix on the finest levels j and determines all other bandwidths of the appearing block band matrices. For our special choice of $\psi_{l,k}, \eta_{l,k}$ we recommend a bandwidth $\bar{b}$ of $\bar{b} \geq 4$. For greater values of $\bar{k}$ or more complex domains D_- it may be necessary to increase this value by small amounts.

Obviously, the error e_j for the functional of the solution is not as sensitive to different compression constants a as the error E_j. This is due to the fact that a quadrature rule of higher exactness would yield a better compression rate for this error.

Figures 3.6 and 3.7 show the number nze of nonzero elements of the compressed matrix A_j^ε for some constants a and the time t_{GM} for the iterative solution of the linear system by a preconditioned GMRES. One clearly observes that the time depends on the wave number $\bar{k}$ as well as on the constant a for the compression criterion.

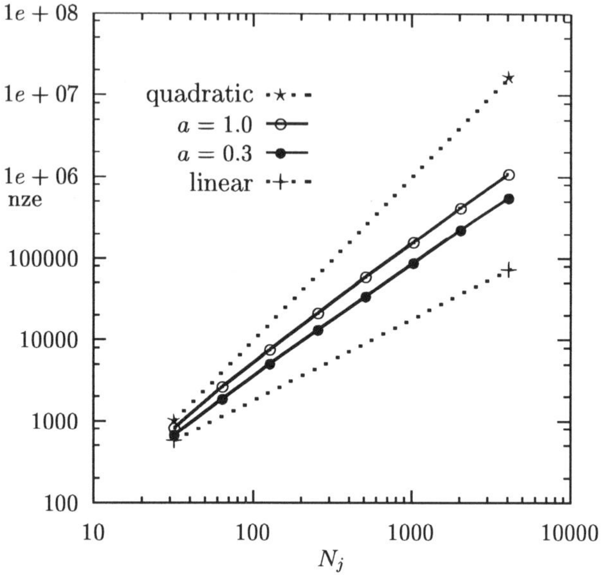

Fig. 3.6. Number nze of nonzero elements of the compressed matrix for two constants a for different wave numbers $\bar{k}$.

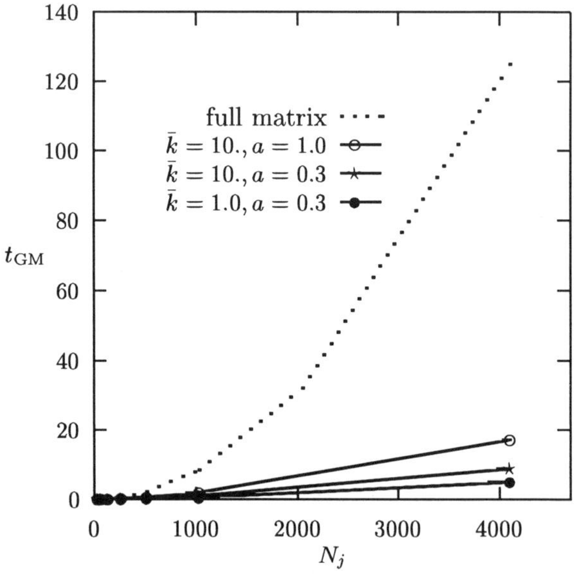

Fig. 3.7. CPU-time t_{GM} in seconds on a DEC 3000 AXP 500 α-processor workstation for the iterative solution of the arising linear system by GMRES for different wave numbers $\bar{k}$ and constants a.

Finally, Figs. 3.8,3.9,3.10 present some results for the RCS for two domains and two different wave numbers $\bar{k}$. The results agree with those obtained with a conventional collocation method.

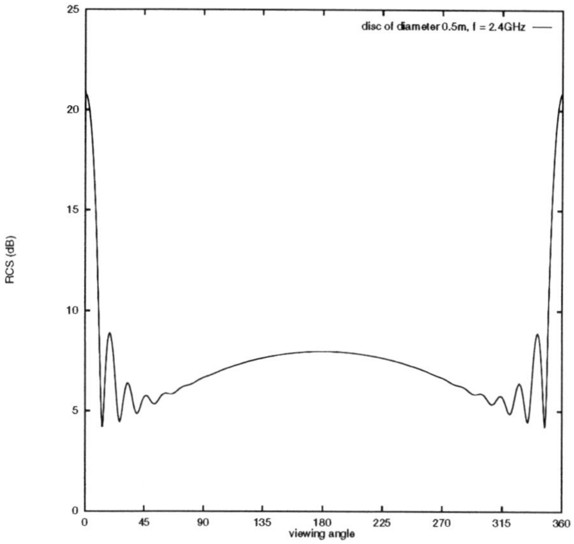

Fig. 3.8. Radar cross section (RCS) σ versus viewing angle for the scattering of a frequency of 2.4 GHz implying a wavelength of $\lambda = 0.125$ m at a disc with radius $r = 0.25$ m. The quantity $\bar{k} \cdot 2r$ determining the Hankel function values is equal to 25.13.

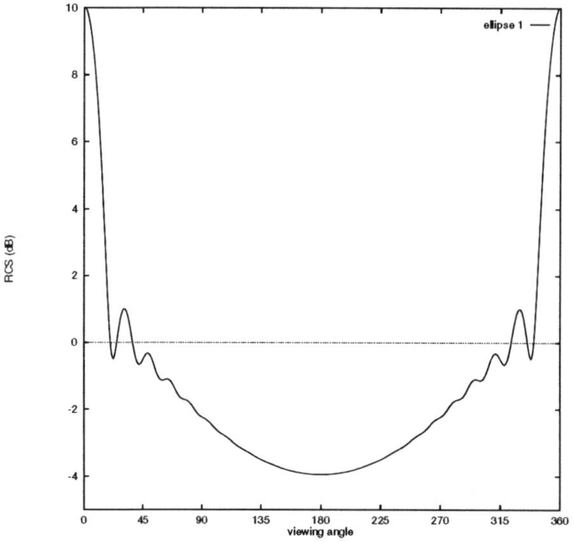

Fig. 3.9. Radar cross section (RCS) σ versus viewing angle for the scattering of a frequency of 10 GHz implying a wavelength of $\lambda = 30$ mm at an ellipse with $a_x =$ 60 mm, $b_y = 30$ mm. The quantity $\bar{k} \cdot 2a$ determining the Hankel function values is equal to 25.13.

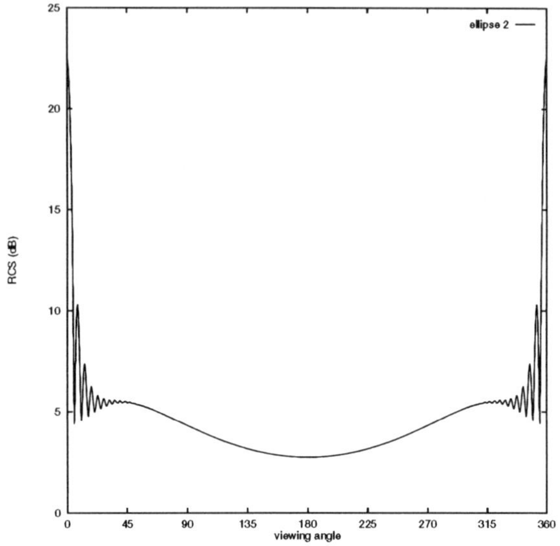

Fig. 3.10. Radar cross section (RCS) σ versus viewing angle for the scattering of a frequency of 10 GHz implying a wavelength of $\lambda = 30$ mm at an ellipse with $a_x = 300$ mm, $b_y = 150$ mm. The quantity $\bar{k} \cdot 2a$ determining the Hankel function values is equal to 125.66.

4 Dirichlet problem for the LAPLACE equation in 3D

In this section new results for the multiscale method proposed in [10] are presented for a large number of degrees of freedom N_j. Note that linear systems with nearly 100,000 unknowns are solved. Since the thresholding procedure is chosen so that the optimal asymptotic accuracy of the discretization error is preserved this corresponds to solving fully populated linear systems of the same size arising from non-wavelet approaches.

We start briefly recalling the main ingredients of the method and refer to [10] for further details. Note that the theory outlined in Section 2 does not apply directly because we are dealing now with a non-periodic problem and the boundary of the considered polyhedral domain is not smooth. Nevertheless, the decay of the matrix entries in the wavelet representation on a face of the polyhedron [10] is of the same form as for the periodic case [7]. Accordingly, the results are also similar to those predicted in the model problem.

It is a well known fact that the Dirichlet problem for the Laplace equation in some domain $\mathcal{P} \subset \mathbb{R}^3$ can be transformed into a second kind integral equation over its boundary [15]. For a polyhedral domain $\mathcal{P}$ this equation over the boundary $\Omega = \partial\mathcal{P}$ reads

$$Au := (I + 2W)u = f, \tag{4.1}$$

$$Wu(x) := [1/2 - \theta_\Omega(x)]u(x) + \frac{1}{4\pi}\int_\Omega \frac{n_y \cdot (x-y)}{|y-x|^3} u(y) \mathrm{d}_y\Omega,$$

where $\theta_\Omega(x)$ is the inner solid angle of Ω at $x \in \Omega$ and n_y the unit vector of the interior normal to $\mathcal{P}$ at y. W is the double layer potential operator. The kernel function $k(x,y) := \frac{1}{4\pi} n_y \cdot (x-y)|y-x|^{-3}$ vanishes if x and y are located on the same face of Ω.

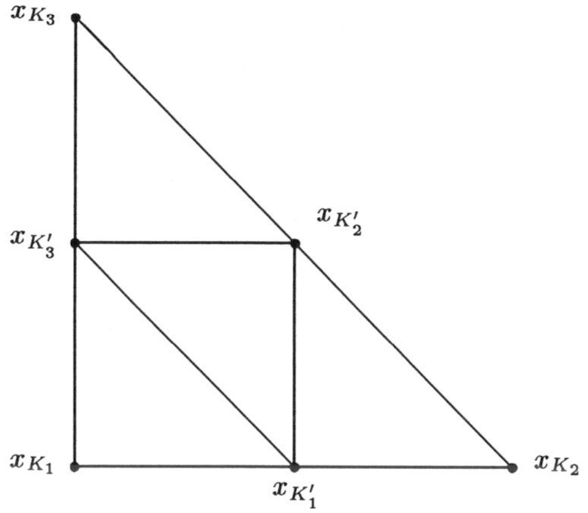

Fig. 4.1. Triangle $[x_{K_1}, x_{K_2}, x_{K_3}]$ of a face of a polyhedral boundary and its refinement. $\{K_1, K_2, K_3\} \subset \nabla^j \Rightarrow \{K_1, K_2, K_3, K'_1, K'_2, K'_3\} \subset \nabla^{j+1}$, $\{K'_1, K'_2, K'_3\} \subset \Delta^j$.

4.1 Triangulation and discretization

As mesh for the boundary Ω a regular triangulation is used. Given some initial partition (mesh Ω^0) of Ω into triangles, subsequent subdivision of each triangle of Ω^0 into four congruent subtriangles leads to Ω^1. This process generates a sequence of nested meshes Ω^j of *depth* or *level* $j \in \mathbb{N}_0$. Clearly, the *meshsize* h on the level j is proportional to 2^{-j}. The vertices x_K of the triangles are the *knot points* of the mesh. The indices K on a given level define a *grid* $\nabla^j = \{K : x_K \text{ is a knot in } \Omega^j\}$. Again, the set of additional knots added when passing from level $j-1$ to level j is $\Delta^{j-1} := \nabla^j \backslash \nabla^{j-1}$.

As trial functions, we employ properly normalized *Courant hat functions* $\varphi_{j,K}$ which are frequently used in finite element or boundary element methods. More precisely, each basis function $\varphi_{j,K}$ is a continuous function on Ω whose restriction to any triangle τ in Ω^j is affine and satisfies the nodal conditions $\varphi_{j,K}(x_{K'}) = 2^j \delta_{K,K'}$, for all $K, K' \in \nabla^j|_\tau$. The spaces $V_j := \text{span}\,\{\varphi_{j,K} : K \in \nabla^j\}$ of piecewise linear continuous functions on Ω are by construction nested $V_0 \subset V_1 \subset \cdots \subset V_{j-1} \subset V_j$. By considering the restriction to the planar faces of $\mathcal{P}$ one readily confirms second order accuracy of the trial spaces.

Consider knot collocation on the finest grid ∇^j, i.e., we wish to determine a piecewise linear and continuous function $u_j^\varphi \in V_j$ such that

$$Au_j^\varphi(x_K) = f(x_K), \quad K \in \nabla^j . \tag{4.2}$$

Thus the entries of the collocation stiffness matrix are given by

$$(A\varphi_{j,K'})(x_K) = \frac{1}{2\pi}\int_{\text{supp}\,\varphi_{j,K'}} \frac{n_y \cdot (x_K - y)}{|y - x_K|^3}\varphi_{j,K'}(y)\mathrm{d}_y\Omega, \tag{4.3}$$

$$K, K' \in \nabla^j, K \neq K'.$$

$\text{supp}\varphi_{j,K'}$ consists of the six triangles τ, for which knot K' is a common vertex. In order to get a fully discrete method quadrature which is known to be exact for polynomials p of degree at most two is used for approximating the integrals (e.g. [22]). To describe this, let $\tau = [x_{K_1}, x_{K_2}, x_{K_3}] \in \Omega^j$ and let $x_{K_1'}, x_{K_2'}, x_{K_3'}$ denote the midpoints of its edges $[x_{K_1}, x_{K_2}]$, $[x_{K_2}, x_{K_3}]$, $[x_{K_3}, x_{K_1}]$, respectively (Fig. 4.1). The quadrature formula for a function $v(x)$ then reads

$$\int_\tau v(x)\mathrm{d}x \simeq \frac{1}{3}\sum_{i=1}^{3} v(x_{K_i'})|\tau|. \tag{4.4}$$

Moreover, to treat the singularities in a proper way, usually a regularization technique is employed (cf. [13]) based on rewriting $Au = f$ as

$$2u(x) + \frac{1}{2\pi}\int_\Omega \frac{n_y \cdot (x - y)}{|y - x|^3}[u(y) - u(x)]\mathrm{d}y = f(x), \quad x \in \Omega. \tag{4.5}$$

Note that some calculations presented in Tables 4.1 and 4.2 are performed without singularity subtraction.

In summary one obtains the following linear system with the discrete collocation matrix

$$A_j^\varphi u_j^\varphi = f_j, \quad A_j^\varphi = (a_{K,K'})_{K,K' \in \nabla^j}. \tag{4.6}$$

4.2 Decomposition of Function Spaces

Our approach is based on a multiscale decomposition of the given trial spaces induced by the following two scale relations. The refinement equation (2.1) now takes the form

$$\varphi_{j,K} = \sum_{K' \in \nabla^{j+1}} m_{j,K',K}\varphi_{j+1,K'}, \; K \in \nabla^j. \tag{4.7}$$

The coefficients $m_{j,K',K}$ are called *mask* or *filter* coefficients. On those grid points located in the interior of the triangles $\tau \in \omega^0$ the entries of the sparse

$N_{j+1} \times N_j$ matrix $(m_{j,K',K})_{K' \in \nabla^{j+1}, K \in \nabla^j}$ can be represented by the following 7 point stencil

$$\begin{pmatrix} 0 & 1 & 1 \\ 1 & \mathbf{2} & 1 \\ 1 & 1 & 0 \end{pmatrix}. \tag{4.8}$$

Here the bold value indicates a position K in the coarse grid ∇^j. Relation (2.3) which complements the basis in V_j to one for V_{j+1}, now reads

$$\psi_{j,K} := \sum_{K' \in \nabla^{j+1}} c_{j,K',K} \varphi_{j+1,K'}, \quad K \in \Delta^j = \nabla^{j+1} \setminus \nabla^j, \tag{4.9}$$

and in the interior of each $\tau \in \Omega^0$, $c_{j,K',K}$ corresponds to one of the following 3 point stencils

$$\begin{pmatrix} 0 & 0 & 0 \\ -1 & \mathbf{2} & -1 \\ 0 & 0 & 0 \end{pmatrix} \frac{1}{2}, \begin{pmatrix} 0 & -1 & 0 \\ 0 & \mathbf{2} & 0 \\ 0 & -1 & 0 \end{pmatrix} \frac{1}{2}, \begin{pmatrix} 0 & 0 & -1 \\ 0 & \mathbf{2} & 0 \\ -1 & 0 & 0 \end{pmatrix} \frac{1}{2}, \tag{4.10}$$

depending on the direction of the edge containing the knot x_K, $K \in \Delta^j$. Here the boldface values correspond to indices $K \in \Delta^j$.

Now let as before T_j denote the transformation that takes the coefficient vector $\tilde{u}_j$ of some element of V_j relative to the *multiscale basis* $\{\varphi_{0,K} : K \in \nabla^0\} \cup \{\psi_{l,K} : K \in \Delta^l,\ l = 0, \ldots, j-1\}$ into the coefficient vector u_j^φ relative to the *nodal basis* $\{\varphi_{j,K} : K \in \nabla^j\}$. The linear system (4.6) is equivalent to

$$A_j \tilde{u}_j = T_j^* f_j, \quad f_j = f(x_K),\ K \in \nabla^j,$$

where $A_j := T_j^* A_j^\varphi T_j$ is the collocation stiffness matrix relative to the multiscale basis.

The transformation T_j has the same pyramid structure as described in Section 3. Due to the finite supports of the masks its application requires the order of $\mathcal{O}(N_j)$ operations.

The essential steps of the present preliminary version of the algorithm may now be described as follows:

1. The discretized collocation method in the nodal basis gives rise to a linear system (4.6). The complexity of assembling the matrix is $\mathcal{O}(N_j^2)$.
2. The transformations T_j, T_j^* lead to the matrix $A_j := T_j^* A_j^\varphi T_j$, so that the system is equivalent to $A_j \tilde{u}_j = T_j^* f_j$, and $u_j^\varphi = T_j \tilde{u}_j$. Since A_j^φ is fully populated the complexity of this part is still $\mathcal{O}(N_j^2)$. In view of the order of our quadrature rule, the estimates in [10] predict a significantly faster decay of the entries in A_j than of those in A_j^φ. Replacing all those entries in A_j by zero whose modulus stays below a given threshold th yields the *compressed* system $A_j^\varepsilon u_j = T_j^* f_j$. To improve the scheme one has to compute A_j^ε directly without going through A_j^φ. The theoretical concepts needed to perform this step in an efficient way have now become available [21].

3. This latter system is solved with the aid of a sparse iterative solver (e.g. GMRES) without preconditioning since the order of the operator is zero in this case.

The decay estimates (see [10]) allow us to estimate the deviation of the solution of the compressed system from the solution to the full system and, in some cases, also from the exact solution of $Au = f$. As pointed out above at this stage this method has still an overall complexity of $\mathcal{O}(N_j^2)$. Only the last step, after the thresholding procedure, has a complexity which is nearly linear and therefore leads to a fast solution of the linear system. As one can see in Tables 4.1 and 4.2, quadrupling the number N_j of unknowns leads to quadrupled nze only. In contrast to the method of Section 3 it does not save yet memory because in the first two steps, N_j^2 matrix entries are handled. Nevertheless, so far the primary objective of these test has been to confirm the performance of such techniques with regard to compression and convergence behavior in situations where not all the assumptions of the model case are fulfilled. The results show that efficient compression is possible without deteriorating the resulting accuracy of the discrete solutions significantly.

4.3 Numerical examples

For our numerical experiments we consider a Dirichlet problem for Laplace's equation $\Delta U(x) = 0, x \in \mathcal{P}$, and smooth Dirichlet data, $U(x)|_\Omega = f(x),\ \ x = ({}_1x, {}_2x, {}_3x) \in \Omega$. In spite of the smoothness of the Dirichlet data the solution will generally not be smooth because the boundary is not smooth.

Here we mainly present numerical experiments for two different polyhedra, a cube, and a pyramid. For a nonconvex domain (bench) and further more detailed results, see [10].

Recall that the solution U of the above Dirichlet problem with boundary data $f(x),\ x \in \Omega$ has the representation

$$U(x) = \frac{1}{4\pi} \int_\Omega \frac{n_y \cdot (x-y)}{|x-y|^3} u(y) \mathrm{d}_y \Omega, \quad x \in \mathcal{P}, \tag{4.11}$$

where u is the solution of the double layer potential equation $Au = f$. Specifically, for our tests we choose $f(x) := U(x)|_\Omega$ with the harmonic function $U(x) := \sqrt{\frac{11}{4}}(({}_1x+1)^2 + {}_2x^2 + {}_3x^2)^{-\frac{1}{2}}, x \in \mathcal{P}$.

In order to control our compression error, we have to monitor the error between the exact solution u, and the approximate solution u_j^φ obtained by the given (uncompressed) discretized collocation method on a grid ∇^j with $N = N_j$ knots. Since we are often interested in the potential U, or in some functionals of the boundary data u, rather than in the values of u, we directly go ahead and determine first an approximation of U as follows. Inserting u_j^φ into the representation formula above and computing the integral via the same quadrature rule used for the computation of the entries of the stiffness matrix, we derive the following formula for the approximate solution of the boundary value problem (cf. Fig.4.1):

$$U_j^{\varphi}(x) := \frac{1}{4\pi} \sum_{\tau \in \Omega^j} \frac{1}{3} \sum_{i=1}^{3} \frac{n_\tau \cdot (x - y^\tau_{K'_i})}{|x - y^\tau_{K'_i}|^3} u_j^{\varphi}(y^\tau_{K'_i}) |\tau|, \tag{4.12}$$

where for $\tau = [y_{K_1}, y_{K_2}, y_{K_3}]$ the points $y^\tau_{K'_i}$ are defined by $y^\tau_{K'_i} = \frac{1}{2}(y_{K_i} + y_{K_k}) \in \Delta^j, i = 1,2,3$ with $k = i+1$, for $i = 1,2$ and with $k = 1$ if $i = 3$, are the midpoints of the edges of the triangle τ and $u_j^{\varphi}(y^\tau_{K'_i}) := \frac{1}{2}(u_j^{\varphi}(y^\tau_{K_i}) + u_j^{\varphi}(y^\tau_{K_k}))$.

As a further control we replace the exact solution u of (4.5) by the numerical solution u_J on a very fine grid and compute $\mathrm{ERR}_j^{\varphi} = \frac{\|u_j^{\varphi} - u_J\|_0}{\|u_J\|_0}$ where here $\|u\|_0 := (\sum_{K \in \square^j} |u(x_K)|^2)^{1/2}$. This error is equivalent to the relative L^2-error (see [10]).

In the multiscale representation A_j of the operator we discard those elements which are below a given threshold th and end up with a compressed or sparsified matrix A_j^{ε}. The ratio between the total number N^2 of matrix elements of the full matrix and the number nze of nonzero elements after thresholding defines the *compression rate*= cpr $:= \frac{N^2}{\mathrm{nze}}$. By compression we obtain a perturbed system. The solution of this perturbed system yields the approximate solution u_j (which at this stage is given in its multiscale representation). Evaluating u_j at the points $y^\tau_{K'_i}$ and substituting these values into the discrete representation formula (4.12) for u_j^{φ}, provides $U_j(x)$. The evaluation is most efficiently performed by transforming the solution of the compressed system into nodal basis coefficients with the aid of T_j and exploiting then the localness of the basis functions $\varphi_{j,K}$.

Of course, the compression causes an additional error. An acceptable compression should have only a negligable influence on the precision of the final approximate solution. To monitor u_j we again replace the exact solution u of (4.5) by the numerical solution u_J on a very fine level J and compute $\mathrm{ERR}_j = \frac{\|u_j - u_J\|_0}{\|u_J\|_0}$. In order to determine which threshold results in a suitable compression, we compare this error with the error ERR_j^{φ} of the discretization scheme. Similarly we compare ERR_{x^i} with the corresponding error $|U(x^i) - U_j(x^i)|$. To estimate this error we compute the maximum $\mathrm{MERR}_x := \max_{i=1,2,3}|U(x^i) - U_j(x^i)|$. This is the second quantity that should help determining which compression rate still has a negligable effect on the precision of the approximate solution. Finally, we compute the additional error arising from the compression $\mathrm{ERR} := \frac{\|u_j^{\varphi} - u_j\|_0}{\|u_j^{\varphi}\|_0}$, which is also displayed in most of the tables.

The main results for the geometries we have tested are presented in Tables 4.1, and 4.2. The bold quantities refer to the largest threshold th and the corresponding solution u_j for which $\mathrm{ERR}_j \approx \mathrm{ERR}_j^{\varphi}$ on the same level j. In this case, thresholding apparently has a *negligable* influence on the solution

u_j. Furthermore for the bold quantities one has $\mathrm{MERR}_x \approx \mathrm{ERR}_{x^i}, i = 1,2,3$ and for $j > 4$ even $\mathrm{MERR}_x \leq \mathrm{ERR}_{x^i}, i = 1,2,3$. We observe that we can choose a larger threshold, and this results in a better compression for larger N, satisfying $\mathrm{MERR}_x \approx \mathrm{inf}_i \mathrm{ERR}_{x^i}$, $i = 1,2,3$.

Table 4.1. Number nze of nonzero elements, compression rate cpr, errors ERR and MERR_x for the multiscale algorithm for the Laplace equation on the cube for several thresholds th and levels j.

Cube						
j	N_j	th	nze	cpr	ERR	MERR_x
1	26	$3 \cdot 10^{-3}$	568	1.2	$2.0 \cdot 10^{-4}$	$2.9 \cdot 10^{-2}$
		$1 \cdot 10^{-2}$	**466**	**1.4**	$4.3 \cdot 10^{-3}$	$2.9 \cdot 10^{-2}$
		$3 \cdot 10^{-2}$	274	2.5	$8.5 \cdot 10^{-2}$	$3.0 \cdot 10^{-2}$
2	98	$1 \cdot 10^{-3}$	5176	1.8	$2.3 \cdot 10^{-3}$	$4.6 \cdot 10^{-3}$
		$3 \cdot 10^{-3}$	**3346**	**2.9**	$5.5 \cdot 10^{-3}$	$4.4 \cdot 10^{-3}$
		$1 \cdot 10^{-2}$	1990	4.8	$2.5 \cdot 10^{-3}$	$2.9 \cdot 10^{-3}$
3	386	$3 \cdot 10^{-4}$	36,484	4.1	$1.0 \cdot 10^{-3}$	$6.7 \cdot 10^{-4}$
		$1 \cdot 10^{-3}$	**21,124**	**7.0**	$4.4 \cdot 10^{-3}$	$5.2 \cdot 10^{-4}$
		$3 \cdot 10^{-3}$	12,646	11.8	$2.3 \cdot 10^{-2}$	$1.9 \cdot 10^{-3}$
4	1,538	$1 \cdot 10^{-4}$	401,272	5.9	$4.7 \cdot 10^{-4}$	$1.3 \cdot 10^{-4}$
		$3 \cdot 10^{-4}$	**235,588**	**10.3**	$1.3 \cdot 10^{-3}$	$1.2 \cdot 10^{-4}$
		$1 \cdot 10^{-3}$	125,524	18.8	$3.8 \cdot 10^{-3}$	$1.8 \cdot 10^{-4}$
5	6,146	$3 \cdot 10^{-5}$	2,098,108	18.0	$2.9 \cdot 10^{-4}$	$9.2 \cdot 10^{-6}$
		$1 \cdot 10^{-4}$	**1,118,296**	**33.8**	$8.4 \cdot 10^{-4}$	$4.0 \cdot 10^{-5}$
		$3 \cdot 10^{-4}$	647,032	58.4	$2.1 \cdot 10^{-3}$	$2.4 \cdot 10^{-5}$
6[a]	24,578	$3 \cdot 10^{-6}$	6,755,512	89.4	-	$5.0 \cdot 10^{-6}$
		$1 \cdot 10^{-5}$	**3,953,114**	**152.8**	-	$5.5 \cdot 10^{-6}$
		$3 \cdot 10^{-5}$	2,478,484	243.7	-	$2.1 \cdot 10^{-5}$
7[b]	98,306	$3 \cdot 10^{-6}$	**17,857,936**	**541.0**	-	$2.0 \cdot 10^{-6}$
		$3 \cdot 10^{-5}$	7,097,044	1361.0	-	$2.2 \cdot 10^{-5}$

[a] Calculations without singularity subtraction. Coarsest level is set to $l = 2$.
[b] Calculations without singularity subtraction. Coarsest level is set to $l = 3$.

In Figs. 4.2, and 4.3 the number nze of nonzero elements of the compressed matrices A_j^ε, for which we observed acceptable precision (bold quantities in Tables 4.1, and 4.2) versus the number of knot points $N = \#\nabla^j$ are plotted and compared with the N^2 elements of the dense matrix. Figure 4.3 exhibits a nearly linear increase of nze for greater N. The compression reduces storage significantly to $1/\mathrm{cpr} \cdot N^2$. Likewise the CPU time for the matrix-vector multiplications during the iterative solution of the linear system decreases substantially (see Fig.4.4).

For the solution of the discrete and compressed scheme we again use the iterative method GMRES. For the estimation of the condition numbers $\kappa(A_j^\varphi)$ and $\kappa(A_j^\varepsilon)$ we apply a direct solver. This method is an expert driver of the well known LA-package [12]. The number of iteration steps, the estimated condition numbers and the CPU time t_{GM} for the GMRES solution on a DEC 3000 AXP 500 workstation are presented in Table 4.3. The termination bound for the iteration process is chosen to be about $\mathrm{ERR}_j^{\varphi}/100$. Note that since the operator has order zero in this case and since we have not employed an orthogonal basis the condition of the original matrix should be better than that of the transformed matrix. Furthermore, compression should have a minor influence on the condition numbers which is also confirmed by our experiments.

Table 4.2. Number nze of nonzero elements, compression rate cpr, errors ERR and MERR_x for the multiscale algorithm for the Laplace equation on the pyramid for several thresholds th and levels j.

Pyramid						
j	N_j	th	nze	cpr	ERR	MERR_x
1	14	$3 \cdot 10^{-3}$	188	1.0	$1.1 \cdot 10^{-4}$	$2.2 \cdot 10^{-2}$
		$1 \cdot 10^{-2}$	**172**	**1.1**	$1.8 \cdot 10^{-4}$	$2.2 \cdot 10^{-2}$
		$3 \cdot 10^{-2}$	122	1.6	$5.6 \cdot 10^{-2}$	$3.7 \cdot 10^{-2}$
2	50	$1 \cdot 10^{-3}$	1948	1.3	$7.5 \cdot 10^{-4}$	$5.2 \cdot 10^{-3}$
		$3 \cdot 10^{-3}$	**1494**	**1.7**	$1.5 \cdot 10^{-3}$	$5.3 \cdot 10^{-3}$
		$1 \cdot 10^{-2}$	1016	2.5	$1.5 \cdot 10^{-2}$	$6.3 \cdot 10^{-3}$
3	194	$3 \cdot 10^{-4}$	22,546	1.7	$2.2 \cdot 10^{-4}$	$2.2 \cdot 10^{-4}$
		$1 \cdot 10^{-3}$	**15,108**	**2.5**	$1.6 \cdot 10^{-3}$	$1.8 \cdot 10^{-4}$
		$3 \cdot 10^{-3}$	9,736	3.9	$5.8 \cdot 10^{-3}$	$1.5 \cdot 10^{-4}$
4	770	$1 \cdot 10^{-4}$	159,022	3.7	$3.1 \cdot 10^{-4}$	$1.2 \cdot 10^{-4}$
		$3 \cdot 10^{-4}$	**96,244**	**6.2**	$1.1 \cdot 10^{-3}$	$1.2 \cdot 10^{-4}$
		$1 \cdot 10^{-3}$	56,668	10.5	$3.0 \cdot 10^{-3}$	$2.3 \cdot 10^{-4}$
5	3,074	$3 \cdot 10^{-5}$	1,179,686	8.0	$1.8 \cdot 10^{-4}$	$2.9 \cdot 10^{-5}$
		$1 \cdot 10^{-4}$	**660,124**	**14.3**	$5.1 \cdot 10^{-4}$	$2.8 \cdot 10^{-5}$
		$3 \cdot 10^{-4}$	389,018	24.3	$1.4 \cdot 10^{-3}$	$2.5 \cdot 10^{-5}$
6[a]	12,290	$3 \cdot 10^{-6}$	4,071,856	37.1	-	$9.0 \cdot 10^{-6}$
		$1 \cdot 10^{-5}$	**2,401,236**	**62.9**	-	$7.3 \cdot 10^{-6}$
		$3 \cdot 10^{-5}$	1,486,256	101.6	-	$9.2 \cdot 10^{-6}$
7[b]	49,154	$3 \cdot 10^{-6}$	**10,776,400**	**224.4**	-	$1.6 \cdot 10^{-6}$

[a] Calculations without singularity subtraction. Coarsest level is set to $l = 2$.
[b] Calculations without singularity subtraction. Coarsest level is set to $l = 3$.

Our aim was to show the nearly linear asymptotic behavior of the necessary number of nonzero elements nze also in the case of a large number of degrees of freedom. So quadrupling the number N_j of unknowns leads to quadrupled nze only. In spite of a relatively strong compression one observes acceptable accuracy. For $N \approx 100,000$ such a compression rate is about 500 and we observed this rate to be more or less independent of the underlying geometry. Therefore in CPU–time solving the linear system by an iterative scheme the speed up factor was dramatic – solving a system corresponding to $100,000$ degrees of freedom, only takes 3 minutes on a sequential workstation and the whole matrix goes into a main memory of $\approx$ 200 MB. Our method damps the coefficients away from the singularity. Thus the double layer potential operator for the Laplacian can be well compressed. We expect the same to persist for the double layer potential for the Stokes system.

Table 4.3. Laplace equation: Number of iterations it for the solution of the linear systems $A_j^\varphi u_j^\varphi = f_j$ and $A_j^\epsilon u_j = (T_j)^* f_j$, respectively, with GMRES; CPU-times t in seconds on a DEC 3000 AXP 500 α–processor workstation. Condition numbers κ of the corresponding matrices.

j	N_j	$\kappa(A_j^\varphi)$	it(A_j^φ)	$t(A_j^\varphi)$	cpr	$\kappa(A_j^\epsilon)$	it(A_j^ϵ)	$t(A_j^\epsilon)$
Cube:								
3	386	2.4	6	0.2	7.0	38.9	14	**0.1**
4	1,538	2.8	8	3.9	10.3	133.0	16	**0.9**
5	6,146	3.2	11	52.0	33.8	215.7	23	**6.0**
6	24,578	-	-	-	152.8[a]	-	29	**33.8**
7	98,306	-	-	-	541.0[b]	-	32	**178.0**
Pyramid:								
3	194	2.3	7	0.06	3.9	31.1	13	**0.03**
4	770	5.3	8	0.9	6.2	63.4	15	**0.3**
5	3,074	8.6	10	17.3	14.3	118.2	19	**2.6**
6	12,290	-	-	-	62.9[a]	-	28	**19.9**
7	49,154	-	-	-	224.4[b]	-	31	**74.2**

[a] Calculations without singularity subtraction. Coarsest level is set to $l = 2$.
[b] Calculations without singularity subtraction. Coarsest level is set to $l = 3$.

One advantage of the present multiscale technique applied to boundary integral equations is that we have additionally a simple `a posteriori` criterion to decide which coefficients are essentially required. Applying the present thresholding leads often to much better compression than the `a priori` choice of coefficients. This seems to apply to our double layer potential operator.

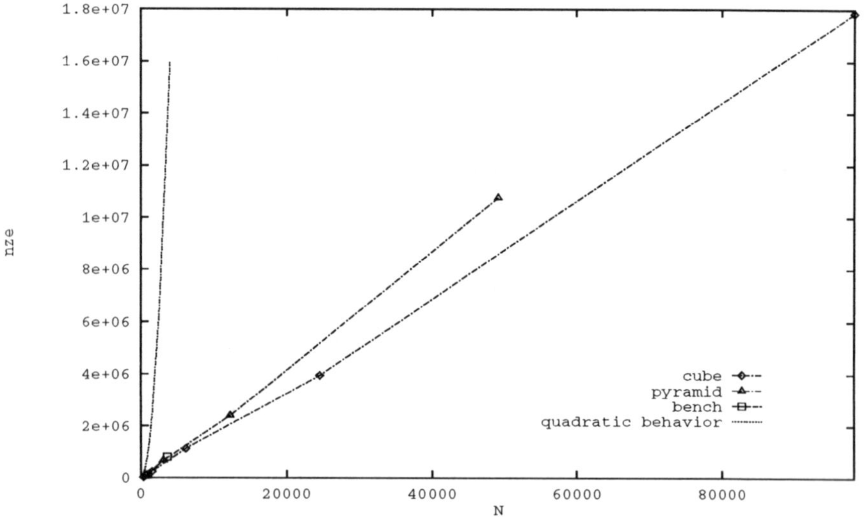

Fig. 4.2. Number nze of nonzero elements of the transformed matrix A_j^ϵ for the Laplace equation after thresholding in dependence of N_j. Dotted line for comparison: quadratic behavior of the number of elements of the original stiffness matrix.

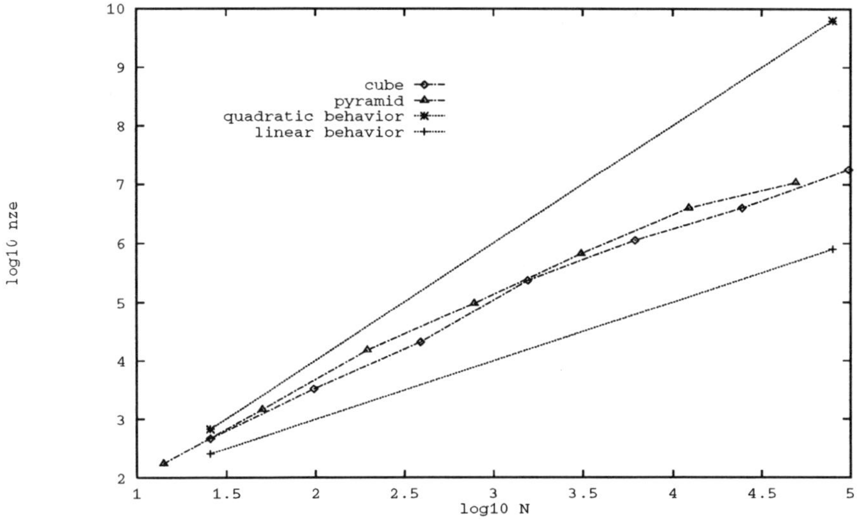

Fig. 4.3. Number nze of nonzero elements of the transformed matrix A_j^ϵ for the Laplace equation after thresholding in dependence of N_j in logarithmic scale. Dotted lines for comparison: quadratic behavior and linear behavior.

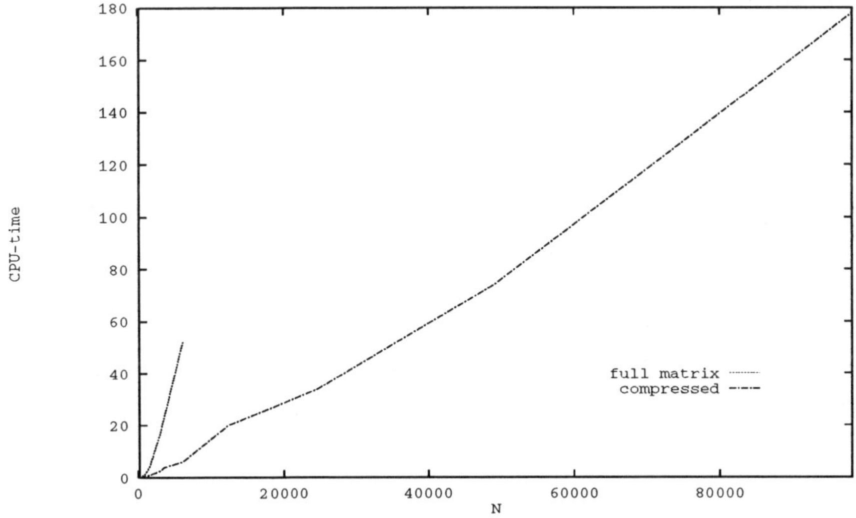

Fig. 4.4. CPU-time in seconds on a DEC 3000 AXP 500 α–processor workstation for the iterative solution of the linear systems for the Laplace equation with uncompressed (dotted line) and compressed matrix, respectively, with GMRES.

5 Concluding Remarks

We have outlined some theoretical foundations of multiscale methods and have carried out corresponding numerical experiments for two types of boundary integral equations. Galerkin methods being better understood we have concentrated here on collocation. The first case study concerning the Helmholtz equation is covered by the existing theory and shows that aside from the natural effects of large wave numbers also operators of order different from zero can be handled efficiently. The experiments are of significant help in clarifying the quantitative effects of the various ingredients of the scheme. The second example is a more complex three dimensional problem which is not fully covered by the analysis of the model problems. It could be shown though that nevertheless even for large problem sizes the basic compression and convergence properties predicted by the theory for the periodic case still persist to be valid.

Acknowledgement

We would like to thank A.Rathsfeld for helpful discussions and valuable suggestions.

References

1. Arnold, D.N., Wendland, W.L. (1985): The convergence of spline collocation for strongly elliptic equations on curves. Numer. Math. 47, 317–431
2. Brandt, A., Lubrecht, A.A. (1991): Multilevel matrix multiplication and fast solution of integral equations. J. Comp. Phys. 90, 348–370
3. Bristeau, M.O., Glowinski, R., Périaux, J. (1993): Using exact controllability to solve the Helmholtz equation at high wave numbers. In: Kleinman, R. et al. (eds.) Mathematical and Numerical aspects of wave propagation. SIAM Philadelphia
4. Cohen, A., Daubechies, I., Feauveau, J.-C. (1992): Biorthogonal bases of compactly supported wavelets. Comm. Pure and Appl. Math. 45, 485-560
5. Colton, D., Kress, R. (1983): Integral equation methods in scattering theory. Springer, Berlin Heidelberg New York
6. Dahmen, W., Prössdorf, S., Schneider, R. (1994): Wavelet approximation methods for pseudodifferential equations I: Stability and convergence. Math. Zeitschrift 215, 583–620
7. Dahmen, W., Prössdorf, S., Schneider, R. (1993): Wavelet approximation methods for pseudodifferential equations II: Matrix compression and fast solution. Advances in Computational Mathematics, 2nd Issue, 259–335
8. Dahmen, W., Prössdorf, S., Schneider, R. (1994): Multiscale methods for pseudodifferential equations on smooth closed manifolds. In: Chui, C.K., Montefusco, L., Puccio, L. (eds). Wavelets: Theory, Algorithms, and Applications. Academic Press, pp. 385–424

9. Dahmen, W., Prössdorf, S., Schneider, R. (1994): Multiscale methods for pseudodifferential equations. In: Schumaker, L.L., Webb, G. (eds.) Recent Advances in Wavelet Analysis. Academic Press, pp. 191–235
10. Dahmen, W., Kleemann, B., Prössdorf, S., Schneider, R. (1994): A multiscale method for the double layer potential equation on a polyhedron. In: Dikshit, H.P., Micchelli, C.A. (eds.) Advances in Computational Mathematics. World Scientific, Singapore New Jersey London Hong Kong, pp. 15–57
11. Daubechies, I. (1992): Ten Lectures on Wavelets. SIAM, Philadelphia, PA
12. Dongarra, J., Duff I., Sorensen, D., H.van der Vorst. (1991): Solving Linear Systems on Vector and Shared Memory Computers. The Society for Industrial and Applied Mathematics, University City Science Center, Philadelphia
13. Kleemann, B., Rathsfeld, A. (1993): Nyström's method and iterative solvers for the solution of the double layer potential equation over polyhedral boundaries. IAAS-Berlin, Preprint No. 36
14. Kress, R., Sloan, I.H. (1993): On the numerical solution of a logarithmic integral equation of the first kind for the Helmholtz equation. Numer. Math. 66, 199–214
15. Maz'ya, V.G. (1991): Boundary integral equations. In: Maz'ya, V.G., Nikol'skiĭ, S.M. (eds.) Encyclopaedia of Math. Sciences, vol. 27, Analysis IV. Springer, Berlin Heidelberg New York
16. Meyer, Y. (1992): Wavelets and Operators. Cambridge University Press
17. von Petersdorff, T., Schwab, C. (1994): Wavelet approximation for first kind boundary integral equations on polygons. Techn. Note BN–1157, Institute for Physical Science and Technology, University of Maryland at College Park
18. Rathsfeld, A. (1995): A wavelet algorithm for the solution of the double layer potential equation over polygonal boundaries. J. Integral Equations Appl. 7, No. 1, 47–98
19. Saad, Y., Schultz, M.H. (1986): GMRES: A generalized minimal residual algorithm for solving nonsymmetric linear systems. SIAM J. Sci. Stat. Comput. 7, No. 3, 856–869
20. Saranen, J., Vainikko, G. (1994): Fast solvers of integral and pseudodifferential equations on closed curves. To appear in Mathematics of Computation
21. Schneider, R. (1995): Multiskalen- und Wavelet-Matrixkompression: Analysisbasierte Methoden zur effizienten Lösung großer vollbesetzter Gleichungssysteme. Habilitationsschrift, TH Darmstadt
22. Stephan, E.P., Wendland, W.L. (1990): A hypersingular boundary integral method for two-dimensional screen and crack problems. Arch. Rational Mech. Anal. 112, 363–390
23. Vavasis, S.A. (1992): Preconditioning for boundary integral equations. SIAM J. Matrix Anal. Appl. 13, Vol. 3, 905–925
24. Walker, H.F. (1988): Implementation of the GMRES method using Householder transformations. SIAM J. Sci. Stat. Comput. 9, No. 1, 152–163

Galerkin-Type Boundary Element Analysis for 3D Elasticity Problems

H. Andrä and E. Schnack

1 Introduction

In this paper, we introduce a formulation that is based on the GALERKIN-type approach, leading to double integrations. In this formulation, the use of both singular and hypersingular boundary integral equation leads to a fully symmetric-definite equation system, a property of considerable advantage. The GALERKIN-type approximation of the hypersingular boundary integral equation does not lead to complicated point functionals like in the collocation method and is well suitable for non-overlapping domain-decomposition algorithms [4].

The requirement of performing double integrations of (hyper-)singular kernel functions has limited the implementations of symmetric GALERKIN boundary element method (GBEM) to those involving straight elements [11, 17, 18], to two-dimensional problems [11, 10], or to only potential and similar problems in three dimensions [21, 6].

Since analytical integration is not possible for general curved elements, one has to rely on numerical integration; but due to the singularities in the kernels ordinary numerical quadrature schemes fail to produce an accurate solution or need thousands of integration points. Many proposed methods approximate the curved elements by flat elements of approximately the same proportions tangential to the curved element at a projection point (see e.g. [8, 21, 6, 5]). Then integration is performed with respect to this special flat element or further mappings of the flat elements are used. Whereas, in this work the numerical quadrature is referred to a standard reference element like in the isoparametric finite element method.

In Section 2 we formulate the boundary integral equations for the mixed boundary value problem in linear elasticity and introduce the GALERKIN scheme. Section 3 is devoted to the integration method for the singular and regular integrals. By using special variable substitutions we obtain integrands, which have no singularities and can be treated numerically with standard GAUSSIAN quadrature. We give estimates of the quadrature errors by extending the estimation method of SCHWAB/ WENDLAND [23] from the two-dimensional case occuring in the collocation BEA to the four-dimensional case (see Theorem 3.2). In Section 4 we present some numerical tests.

2 Boundary integral equations in elasticity and the Galerkin boundary element analysis

2.1 Boundary value problem

We consider a solid body or structure occupying the (finite, simply connected, open) domain $\Omega \subset \mathbb{R}^3$, with a LIPSCHITZ boundary surface $\Gamma = \Gamma_1 \cup \Gamma_2$, $\Gamma_1 \cap \Gamma_2 = \emptyset$, mes $\Gamma_1 > 0$, with given DIRICHLET data on Γ_1, and NEUMANN data on Γ_2. The body is referred to a Cartesian orthogonal coordinate system $x = (x_1, x_2, x_3)$. We present a solution procedure via a system of boundary integral equations for the mixed boundary value problem of linear elasticity:

Problem 1. *(Boundary Value Problem)*
For given displacements u^* on Γ_1 and given tractions t^* on Γ_2 find the displacements $u = (u_1, u_2, u_3)$ in Ω satisfying the NAVIER *differential equations*

$$\Delta^* u = \mu \Delta u + (\lambda + \mu)\ grad\ \ div\ u = 0, \qquad \forall x \in \Omega,$$

with mixed boundary conditions

$$u_i = u_i^*, \qquad \forall x \in \Gamma_1, \tag{2.1}$$

$$t_i = \sum_{j=1}^{3} \sigma_{ji}(u)\nu_j = t_i^*, \forall x \in \Gamma_2 \quad (i = 1, 2, 3), \tag{2.2}$$

where σ_{ij} is the CAUCHY *stress tensor, t_i are the boundary tractions, ν_j is the unit outward normal, and $\mu > 0, \lambda > -\frac{2}{3}\mu$ are Lamé's elastic constants.*

2.2 Boundary integral equations

For the sake of simplicity we have restricted ourselves to the case of non body forces.
The operator Δ^* has the fundamental solution U (see [16, 3]) with

$$U_{ij}(x, y - x) = \frac{1}{16\pi G(1 - \nu)} \left(\frac{1}{r}\right) [(3 - 4\nu)\delta_{ij} + r_{,i} r_{,j}] \ (i, j = 1, 2, 3), \tag{2.3}$$

where $r_i = y_i - x_i, r = \sqrt{\left(\sum_{i=1}^{3} r_i^2\right)}, r_{,i} = \frac{\partial r}{\partial y_i} = \frac{r_i}{r}$.
We obtain fundamental solutions with stronger singularities from (2.3) by taking certain derivatives. We apply the stress operator [3] on $U(y, x)$ with respect to y to obtain the strongly singular fundamental solutions

$$T_{ij}(x, y - x) = \sum_{k=1}^{3} \sigma_{kj}(\nabla_y)([U_{il}]_{l=1}^3(y, x)) n_k(y) \qquad (i, j = 1, 2, 3), \tag{2.4}$$

where $n(y) = (n_i(y))_{i=1,3}$ is an arbitrary unit vector. If we apply the stress operator once more with respect to x, we obtain the hypersingular fundamental solution

$$S_{ij}(x, y-x) = \sum_{k=1}^{3} \sigma_{ki}(\nabla_x)([T_{lj}]_{l=1}^{3}(y,x))n_k(x) \qquad (i,j=1,2,3). \quad (2.5)$$

Furthermore, we introduce the trace mappings

$$\psi_i = \gamma_0^i \psi := \psi|_{\Gamma_i}, \qquad \varphi_i = \gamma_1^i \varphi := [\sigma_{lj}(\varphi)\nu_j|_{\Gamma_i}]_{l=1}^{3}$$

Now the boundary integral operators, which are associated with the fundamental solutions (2.3), (2.4), and (2.5), are defined by

$$\begin{aligned}
V_{ij}\psi_i &:= 2\gamma_{0,x}^j \int_{\Gamma_i} \gamma_{0,y}^i U(x,y-x)\psi_i(y)ds(y),\\
K_{ij}\varphi_i &:= 2 \text{ v.p. } \gamma_{0,x}^j \int_{\Gamma_i} \gamma_{0,y}^i T(x,y-x)\varphi_i(y)ds(y),\\
K'_{ij}\psi_i &:= 2 \text{ v.p. } \gamma_{0,x}^j \int_{\Gamma_i} \gamma_{0,y}^i T^T(y,x-y)\psi_i(y)ds(y),\\
D_{ij}\varphi_i &:= -2 \text{ p.f. } \gamma_{0,x}^j \int_{\Gamma_i\setminus B_\varepsilon(x)} \gamma_{0,y}^i S(x,y-x)\varphi_i(y)ds(y),
\end{aligned}$$

where $B_\varepsilon(x) = \{y|\ |y-x| < \varepsilon(\varepsilon > 0)\}$ is the ε-ball about $x \in \mathbb{R}^3$. K_{ij}, K'_{ij} are defined by CAUCHY singular integrals, and D_{ij} is defined by a hypersingular finite part integral in the sense of HADAMARD. The weak solution of the boundary value problem (Problem 1) with $u \in \mathbb{H}^1(\Omega) := (H^1(\Omega))^3$ is equivalent to finding a solution of the following problem:

Problem 2 (Boundary Integral Equation).
Let be given $\binom{u^*}{t^*} \in \mathbb{H}^{\frac{1}{2}}(\Gamma_1) \times \mathbb{H}^{-\frac{1}{2}}(\Gamma_2)$.
Find a boundary function $\binom{v_2}{t_1} \in V_\Gamma := \mathbb{H}^{\frac{1}{2}}(\Gamma_2) \times \mathbb{H}^{-\frac{1}{2}}(\Gamma_1)$ *satisfying the system of boundary integral equation*

$$D_{22}v_2 + K'_{12}t_1 = -D_{12}u^* + (I - K'_{22})t^* \quad \text{on } \Gamma_2, \qquad (2.6)$$

$$-K_{21}v_2 + V_{11}t_1 = (I + K_{11})u^* - V_{12}t^* \quad \text{on } \Gamma_1 \qquad (2.7)$$

in V'_Γ *(in the weak sense), i. e.*

$$\left\langle \begin{pmatrix} D_{22} & K'_{12} \\ -K_{21} & V_{11} \end{pmatrix} \binom{v_2}{t_1}, \binom{\varphi_2}{\psi_1} \right\rangle = \left\langle \begin{pmatrix} -D_{12} & I - K'_{22} \\ I + K_{11} & -V_{12} \end{pmatrix} \binom{u^*}{t^*}, \binom{\varphi_2}{\psi_1} \right\rangle$$

$$\forall \binom{\varphi_2}{\psi_1} \in V_\Gamma.$$

Under these assumptions, there exists a unique solution of Problem 2 [25].

2.3 Galerkin's Method for the Boundary Integral Equation

Now we formulate the GALERKIN's method for this boundary integral equation. We specify the conditions on the boundary Γ and its triangulations in a standard manner (see [23]). Let the boundary be a closed strong LIPSCHITZ surface and let Γ be partitioned into finitely many pieces S_l with $\Gamma = \bigcup_{l=1}^{L} \bar{S}_l$ where each of the S_l is the image of an analytic application $y = F^{(l)}(\eta)$ defined on a polygonal domain $V_l \in \mathbb{R}^2$. The $F^{(l)}$ are supposed to have analytic continuations beyond $\bar{V}_l$. The triangulations on Γ are obtained as images of regular quasiuniform triangulations of the parameter region V_l. All boundary elements of one surface piece S_l can be obtained by the composition of $F^{(l)}$ and one linear mapping $A^{(l)}$ together with translations a_i applied to a reference element

$$\hat{\pi} = \{z \in \mathbb{R}^2 : 0 < z_1 < 1, 0 < z_2 < z_1\}.$$

Then

$$\pi_i^h = \chi^{(i)}(\hat{\pi}) = F^{(l)} \circ A^{(l)}(\hat{\pi} + a_i)$$

defines the in general curved elements of S_l. Each element π_i^h belongs to one and only one surface piece S_l. Thus we omit the index l at $\pi_i^h = \chi^{(i)}(\hat{\pi})$. The approximating boundary element trial spaces are now defined by lifting finite element functions from the parameter partitions of $\bar{V}_l$ onto Γ via $(\chi^{(i)})^{-1}$. As

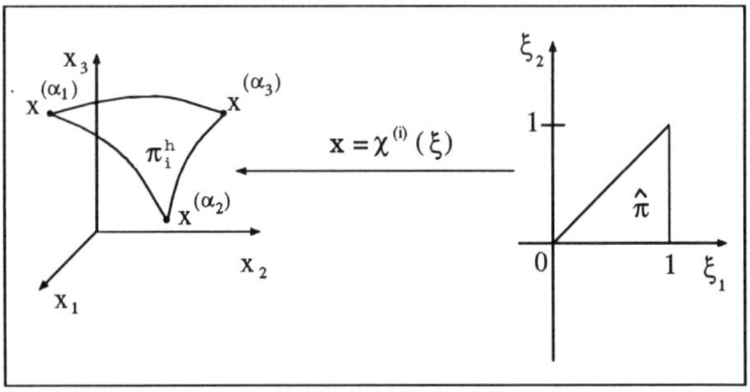

Fig. 2.1. Mapping of the reference element onto the original element

approximating subspace V_h of V_Γ we take piecewise polynomials of different degree for the approximation of v_2 or t_1. For the definition of the nodal basis of V_h we select on Γ a number of nodes, $\{x_\Psi^{(n)} : N = 1, \cdots, n_\Psi\}$, for approximating the boundary tractions and a second number of nodes, $\{x_\Phi^{(n)} : N = 1, \cdots, n_\Phi\}$, for approximating the boundary displacements and assume that the first $n_{\Psi,1}$ ($n_{\Phi,1}$) nodes belong to Γ_1. Suppose that there is a (scalar) standard local basis function $\Psi^{(i)}$ ($\Phi^{(i)}$) associated with each node $x_\Psi^{(i)}$ ($x_\Phi^{(i)}$). Since the transformation $\chi^{(i)}$ is invertible, local basis functions on π_i can be defined by

$$\tilde{\Psi}^{(i)} = \Psi^{(i)} \circ (\chi^{(i)})^{-1}, \quad \tilde{\Phi}^{(i)} = \Phi^{(i)} \circ (\chi^{(i)})^{-1}, \quad \forall x \in \pi_i.$$

Thus the boundary element bases are defined by

$$\tilde{\Psi} = \left[\begin{pmatrix} \tilde{\psi}^{(1)} \\ \tilde{\psi}^{(1)} \\ \tilde{\psi}^{(1)} \end{pmatrix}, \ldots, \begin{pmatrix} \tilde{\psi}^{(n_{\tilde{\Psi},1})} \\ \tilde{\psi}^{(n_{\tilde{\Psi},1})} \\ \tilde{\psi}^{(n_{\tilde{\Psi},1})} \end{pmatrix} \right], \quad \tilde{\Phi} = \left[\begin{pmatrix} \tilde{\phi}^{(n_{\tilde{\Phi},1}+1)} \\ \tilde{\phi}^{(n_{\tilde{\Phi},1}+1)} \\ \tilde{\phi}^{(n_{\tilde{\Phi},1}+1)} \end{pmatrix}, \ldots, \begin{pmatrix} \tilde{\phi}^{(n_{\tilde{\Phi}})} \\ \tilde{\phi}^{(n_{\tilde{\Phi}})} \\ \tilde{\phi}^{(n_{\tilde{\Phi}})} \end{pmatrix} \right],$$

and

$$V_h = \text{ span } \tilde{\Phi} \times \text{ span } \tilde{\Psi}$$

is taken as approximating subspace.

Problem 3 (Galerkin Scheme (exact integration)).
Find $\binom{v_{2,h}}{t_{1,h}} \in V_h$ *such that*

$$\left\langle \begin{pmatrix} D_{22} & K'_{12} \\ -K_{21} & V_{11} \end{pmatrix} \begin{pmatrix} v_{2,h} \\ t_{1,h} \end{pmatrix}, \begin{pmatrix} \varphi_h \\ \psi_h \end{pmatrix} \right\rangle$$
$$= \left\langle \begin{pmatrix} -D_{12} & I - K'_{22} \\ I + K_{11} & -V_{12} \end{pmatrix} \begin{pmatrix} u^* \\ t^* \end{pmatrix}, \begin{pmatrix} \varphi_h \\ \psi_h \end{pmatrix} \right\rangle \quad \forall \begin{pmatrix} \varphi_h \\ \psi_h \end{pmatrix} \in V_h. \tag{2.8}$$

Lemma 2.1. *[26]*
For the exact solution $\binom{v_2}{t_1} \in \mathbb{H}^{\alpha}(\Gamma_2) \times \mathbb{H}^{\alpha-1}(\Gamma_1)$ $(\alpha < 1)$ *the* GALERKIN *approximation* **with exact integration** $\binom{v_h}{t_h} \in V_h$ *satisfies the error estimate*

$$\left\| \begin{pmatrix} v_2 \\ t_1 \end{pmatrix} - \begin{pmatrix} v_{2,h} \\ t_{1,h} \end{pmatrix} \right\|_{\mathbb{H}^t(\Gamma_2)\times\mathbb{H}^{t-1}(\Gamma_1)} \leq ch^{s-t} \left\| \begin{pmatrix} v_2 \\ t_1 \end{pmatrix} \right\|_{\mathbb{H}^s(\Gamma_2)\times\mathbb{H}^{s-1}(\Gamma_1)}$$

if $t \leq 1, -1 \leq t \leq s \leq \alpha$. *The highest order of convergence is* $ch^{1+\alpha}$.

Next we write down the GALERKIN scheme in the form of a linear system of algebraic equations for the unknown coefficients $\binom{X_1}{X_2}$ in the spline basis (where X_1 is the vector of nodal tractions and X_2 is the vector of nodal displacements)

$$\begin{pmatrix} \mathbf{A_{11}} & -\mathbf{A_{12}} \\ \mathbf{A_{21}} & \mathbf{A_{22}} \end{pmatrix} \begin{pmatrix} X_1 \\ X_2 \end{pmatrix} = \begin{pmatrix} F_1 \\ F_2 \end{pmatrix}.$$

We note that $\mathbf{A_{11}}$ and $\mathbf{A_{22}}$ are both symmetric and positive definite and that $\mathbf{A_{12}} = \mathbf{A_{21}^T}$ [19]. This system can be solved by elimination of X_1 to build the Schur complement of the system matrix. The resulting reduced matrix is symmetric and positive definite.

3 Integration of singular and regular integrals

One drawback of the symmetric GALERKIN BEM is the necessity of calculating weak singular, singular and hypersingular integrals. However, the combined analytical and numerical integration of such integrals removes this drawback (see, e. g., [10, 21, 6]). The combined numerical and analytical integration method for four-dimensional CAUCHY singular integrals is based on the fundamental work from SAUTER [21]. The introduction of relative coordinates with respect to the original coordinate system leads to complicated integration bounds and a lot of technical effort (see [21, 6]). Therefore, the implementation of the proposed method for the direct computation of CAUCHY singular integrals, which seems to be more efficient than other methods based on regularizations [11, 14], which lead to weak singular integrals, is for the kernels of elasticity rather expensive. These disadvantages are avoided in this work by introducing relative coordinates in the parameter plane for the reference element.

For generating the GALERKIN system, the following finite part integrals have to be evaluated

$$\langle D_{22}\varphi_h^{(l)}, \varphi_h^{(k)}\rangle_{\mathbb{H}^{-\frac{1}{2}}(\Gamma_2)\times\mathbb{H}^{\frac{1}{2}}(\Gamma_2)} = \int_{\Gamma_2} \text{p.f.} \int_{\Gamma_2} [\varphi_h^{(l)}(x)]^T S(x, y-x)\varphi_h^{(k)}(y)\, ds(y)\, ds(x).$$

Remark 3.1. The shape functions $\varphi_h^{(.)}(.)$ have to be linear functions or polynomials of higher degree.

This integral can be regularized by using the identity

$$\text{p.f.} \oint_{\Gamma} S(x, y-x)\, ds(y) = 0 \qquad \forall x \in \Gamma.$$

The result is a sum of integrals with a reduced order of singularity

$$\begin{aligned}
&\int_{\Gamma_2} \text{p.f.} \int_{\Gamma_2} [\varphi_h^{(l)}(x)]^T S(y,x)\varphi_h^{(k)}(y)\, ds(y)\, ds(x) \qquad (3.1)\\
&= \int_{\Gamma_2} \text{p.f.} \int_{\Gamma} [\varphi_h^{(l)}(x)]^T S(y,x)[\varphi_h^{(k)}(y) - \varphi_h^{(k)}(x)]\, ds(y)\, ds(x)\\
&= \int_{\text{supp } \varphi_h^{(l)}} \text{p.f.} \int_{\text{supp } \varphi_h^{(k)}} [\varphi_h^{(l)}(x)]^T S(y,x)[\varphi_h^{(k)}(y) - \varphi^{(k)}(x)]\, ds(y)\, ds(x)\\
&- \int_{\text{supp } \varphi_h^{(l)} \cap \text{supp } \varphi_h^{(k)}} \text{p.f.} \int_{\Gamma \setminus \text{supp } \varphi_h^{(k)}} [\varphi_h^{(l)}(x)]^T S(y,x)\varphi_h^{(k)}(x)\, ds(y)\, ds(x).
\end{aligned}$$

Therefore, we have to compute integrals of the type

$$I = \int_{\pi_i^h} \text{p.f. } I_\varepsilon^h(x) ds(x), \qquad (3.2)$$

where

$$I_\varepsilon^h(x) := \int_{\pi_j^h \setminus B_\varepsilon(x)} k_n(x, y-x) \Phi((\chi^{(i)})^{-1}(x), (\chi^{(j)})^{-1}(y)) ds(y) \tag{3.3}$$

and

$$\Phi(\xi,\eta) = \sum_{|\alpha| \geq \max\{0,n\}} \frac{1}{\alpha!}(D_2^\alpha \Phi)(\xi,\xi)(\eta-\xi)^\alpha, \tag{3.4}$$

where $D_2^\alpha \Phi(\xi,\eta) := \sum_{\lambda=0}^{|\alpha|} \frac{\partial^{|\alpha|}\Phi(\xi,\eta)}{\partial\eta_1^\lambda \partial\eta_2^{|\alpha|-\lambda}}$ denote the derivatives with the multi-index $\alpha \in \mathbb{R}^2$ with respect to the second argument of $\Phi(.,.)$ and $B_\varepsilon(x) := \{y : |y-x| < \varepsilon\}$ for $\varepsilon > 0$. For $n = 1$ (hypersingular operator) the regularization (3.1) leads to $|\alpha| \geq 1$ in (3.4). We denote by $k_n(x, y-x)$ a component of the kernel functions U $(n = -1), T$ $(n = 0)$, or S $(n = 1)$ with the order of singularity $-2-n$. The components k_n of the kernel functions U, T, S (and also other kernel functions, which arise from any integral equation formulation of a strongly elliptic boundary value problem) can be written in the form

$$k_n(x, y-x) = \sum_{\kappa=0}^{\kappa_{max}} \varrho^{-2-n-\kappa} \sum_{|\alpha| \geq \kappa} c_\alpha(x)(y-x)^\alpha, \tag{3.5}$$

where $2 + n + \kappa$ is an odd number and

$$\varrho^2 = (y-x)\cdot(y-x) = \sum_{i=1}^{3}(y_i - x_i)^2.$$

The coefficients $c_\alpha(x)$ are supposed to be piecewise analytic. The shape functions Φ in (3.3) are products of elements of the matrices N_1, N_2. The cases of

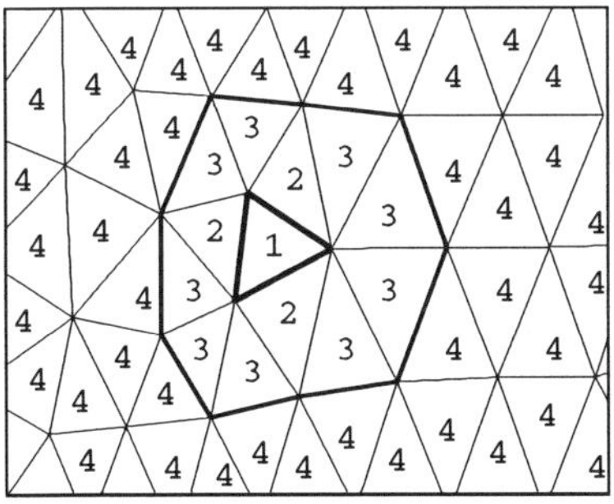

Fig. 3.1. 1: coincident, 2: edge-adj., 3: vertex-adj., 4: regular case

(1) coincident elements, (2) edge-adjacent elements, and (3) vertex-adjacent elements are potentially singular (see Fig. 3.1). In all of these cases the integration method for (3.2) consists of the following steps:

1. Introduction of relative coordinates.
2. Representation of the integration domains such that the integration over the relative coordinates is the outer integration [21, 6].
3. Consideration of the union of symmetric integration domains and exploitation of the parity conditions [12] $k_n(x, y-x) = (-1)^{1+n} k_n(x, x-y)$ of the kernels.
4. Introduction of DUFFY'S triangular coordinates $u = \binom{w_1 w_2}{w_2}$ for the remaining weak singular integrals.
5. Computation of the outer regular integral by using standard GAUSSIAN quadrature rules.

We describe these steps in detail for the case of coincident elements and give the results for the other singular cases in the next subsections. In the last subsection we describe the adaptive numerical quadrature in the regular case.

3.1 Coincident elements

First we describe the case of coincident elements, i. e. $\pi_i^h = \pi_j^h$. We suppose (for simplicity) that the mapping $\chi \equiv \chi^{(i)} \equiv \chi^{(j)}$ is analytic. Then we have

$$\chi_i(\eta) = \sum_{\alpha \in \mathbb{N}_0^2} \frac{1}{\alpha!} (D^\alpha \chi_i)(\xi)(\eta - \xi)^\alpha, \qquad i = 1, 2, 3. \tag{3.6}$$

Let us now write the finite part of the inner integral in (3.2) in local coordinates for an arbitrary $x \in interior(\pi_i^h)$.

Lemma 3.1. *If $x \in interior(\pi_i^h)$, or $\xi \in interior(\hat{\pi})$, then*

$$p.f.\ I_\varepsilon^h(x) = \ p.f.\ J_\varepsilon^h(\xi) = \ v.p.\ J_0^h(\xi) \le C_1 |\ln(\text{dist}(\xi, \partial\hat{\pi}))| + C_2$$

holds, where

$$J_\varepsilon^h(\xi) := \int_{\hat{\pi} \setminus B_\varepsilon(\xi)} \tilde{k}_n(\xi, \eta - \xi) \tilde{\Phi}(\xi, \eta) d\eta,$$

$$\tilde{k}_n(\xi, \eta - \xi) = k_n(\chi(\xi), \chi(\eta) - \chi(\xi)), \tag{3.7}$$

$$\tilde{\Phi}(\xi, \eta) := |J(\chi)(\xi)||J(\chi)(\eta)| \begin{cases} \Phi(\xi, \eta), & n \le 0 \\ \Phi(\xi, \eta) - \Phi(\xi, \xi), & n = 1 \end{cases}, \tag{3.8}$$

where J denotes the JACOBIAN *determinant. The positive constants C_1, C_2 do not depend on ε.*

Proof. In general, we have [23, 22, 13]

$$\text{p.f. } I_\varepsilon^h(x) = \text{ p.f. } J_\varepsilon^h(\xi) + \sum_{0 \le |\alpha| \le n} \tilde{c}_\alpha(\xi)(D_2^\alpha \tilde{\Phi})(\xi, \xi).$$

If k_n arises from any integral equation formulation of a strongly elliptic boundary value problem and $x \in \text{interior}(\pi_i^h)$, then $\tilde{c}_\alpha(\xi) = 0$ holds. We introduce in $\hat{\pi}$ polar coordinates about ξ:

$$\eta - \xi = r\begin{pmatrix}\cos\theta \\ \sin\theta\end{pmatrix}. \tag{3.9}$$

The transformed kernel $\tilde{k}_n$ is pseudo-homogeneous [13], i. e. for every $N \in \mathbb{N}$

$$\tilde{k}_n(\xi, \eta - \xi) = \frac{1}{r^{n+2}} \sum_{j=0}^{N} f_{n-j}(\xi, \theta) r^j + R_N(\xi, \eta - \xi), \tag{3.10}$$

where

$$|R_N(\xi, \eta - \xi)| \le C_N r^{N-n-2} \tag{3.11}$$

and $f_\kappa(\xi, \theta)$ are analytic 2π periodic functions in θ, which satisfy for $\kappa \le -2$ and $\kappa \ge 0$ the compatibility conditions

$$\int_0^{2\pi} f_\kappa(\xi, \theta) \cos^{\alpha_1}\theta \sin^{\alpha_2}\theta d\theta = 0 \text{ for } \begin{cases} |\alpha| = \kappa & \text{if } \kappa \ge 0 \\ |\alpha| = -\kappa - 2 & \text{if } \kappa \le -2. \end{cases} \tag{3.12}$$

By using the parameterization $r = R(\theta)$ of the boundary $\partial\hat{\pi}$ with respect to the variable θ, we obtain

$$\begin{aligned} J_\varepsilon^h(\xi) &= \int_0^{2\pi} \int_\varepsilon^{R(\theta)} \sum_{j=0}^{N} r^{-2-n+j} f_{n-j}(\xi, \theta) \\ &\quad \times \sum_{|\alpha| \ge \max\{0,n\}} \frac{1}{\alpha!} (D_2^\alpha \tilde{\Phi})(\xi, \xi) r^{|\alpha|} \cos^{\alpha_1}\theta \sin^{\alpha_2}\theta r dr d\theta \\ &\quad + \int_{\hat{\pi}\setminus B_\varepsilon(\xi)} R_N(\xi, \eta - \xi) \tilde{\Phi}(\xi, \eta) d\eta \\ &= \sum_{|\alpha| \ge \max\{0,n\}} \frac{1}{\alpha!} (D_2^\alpha \tilde{\Phi})(\xi, \xi) \int_0^{2\pi} f_{n-j}(\xi, \theta) \cos^{\alpha_1}\theta \sin^{\alpha_2}\theta \\ &\quad \times \begin{cases} \ln R(\theta) - \ln\varepsilon & \text{if } |\alpha| = n - j \\ (|\alpha| - n + j)^{-1} (R^{|\alpha|-n+j} - \varepsilon^{|\alpha|-n+j}) & \text{otherwise} \end{cases} \\ &\quad + \int_{\hat{\pi}\setminus B_\varepsilon(\xi)} R_N(\xi, \eta - \xi) \tilde{\Phi}(\xi, \eta) d\eta \end{aligned} \tag{3.13}$$

From (3.12), (3.11), it follows

$$J_\varepsilon^h(\xi) \le C_1 |\ln(\min_{0 \le \theta < 2\pi} R(\theta))| + C_2, \tag{3.14}$$

where the positive constants C_1 and C_2 do not depend on ε.
Then $\lim_{\varepsilon \to 0} J_\varepsilon^h(\xi) =:$ v.p. $J_0^h(\xi)$ exists and is called CAUCHY principal value and the assertion is proved.

Remark 3.2. In general there is an explicit dependence of strong ($n = 0$) and hypersingular ($n = 1$) integral operators upon the shape of ε-neighbourhoods used to define the finite part integrals, even if the integral exists as a CAUCHY principal value. KIESER shows in [12] that for operators defined by boundary potential operators and a sufficiently smooth diffeomorphism χ transforming

$$B_\varepsilon^{\tilde{r}}(\xi) = \{\eta \in \mathbb{R}^2 : \|\eta - \xi\| \leq \tilde{r}(\varepsilon, \theta), 0 < \theta < 2\pi\}$$

onto

$$B_\varepsilon^{\tilde{\varrho}}(\xi) = \{\eta \in \mathbb{R}^2 : \|\eta - \xi\| \leq \tilde{\varrho}(\varepsilon, \theta), 0 < \theta < 2\pi\}$$

for all $0 < \varepsilon \leq \varepsilon_0$ with some $\varepsilon_0 > 0$, where local polar coordinates in the parameter plane are used, one finds that

$$\text{p.f.} \int_{\hat{\pi} \setminus B_\varepsilon^{\tilde{r}}(\xi)} \tilde{k}_n(\xi, \eta - \xi)\, \tilde{\Phi}(\xi, \eta)\, d\eta = \text{p.f.} \int_{\hat{\pi} \setminus B_\varepsilon^{\tilde{\varrho}}(x)} \tilde{k}_n(\xi, \eta - \xi)\, \tilde{\Phi}(\xi, \eta)\, d\eta.$$

By using (3.6) we obtain the expansion

$$\tilde{k}_n(\xi, \eta - \xi) = \sum_{|\alpha| \geq \kappa} \varrho^{-2-n-\kappa} a_\alpha(\xi)(\eta - \xi)^\alpha \tag{3.15}$$

for the transformed kernel.

Theorem 3.1. *Let the following suppositions be fulfilled:*

1. $\pi_i^h = \pi_j^h = \chi(\hat{\pi})$, *where* χ *is analytic,*
2. *the kernel function* k_n *is of type (3.5), and* $\tilde{k}_n$ *is defined by (3.7),*
3. *the shape functions* $\tilde{\Phi}$ *are piecewise polynomials, i. e.*

$$\tilde{\Phi}(\eta) = \sum_{|\alpha| \geq \max\{0,n\}} \frac{1}{\alpha!}(D_2^\alpha \tilde{\Phi})(\xi, \xi)(\eta - \xi)^\alpha. \tag{3.16}$$

Then the singular integral (3.2) is equal to the following regular *integral (which has bounded integrands)*

$$\begin{aligned} I = \int_0^1 \int_0^1 \Big\{ &\int_0^{1-w_1} \int_0^{\xi_1} (\tilde{K}_n \tilde{\Phi}) \left(\xi, w_1 \tbinom{1}{w_2}\right) d\xi_2 d\xi_1 \\ &+ \int_{w_1(1-w_2)}^{1-w_1 w_2} \int_{w_1(1-w_2)}^{\xi_1} (\tilde{K}_n \tilde{\Phi}) \left(\xi, w_1 \tbinom{w_2}{w_2 - 1}\right) d\xi_2 d\xi_1 \\ &+ \int_{w_1(1-w_2)}^{1-w_1 w_2} \int_0^{\xi_1 + w_1(w_2-1)} (\tilde{K}_n \tilde{\Phi}) \left(\xi, w_1 \tbinom{w_2}{1}\right) d\xi_2 d\xi_1 \Big\} w_1 dw_1 dw_2 \end{aligned} \tag{3.17}$$

$$\begin{aligned} = \int_0^1 \int_0^1 \int_0^1 \int_0^1 \Big\{ &(\tilde{K}_n \tilde{\Phi}) \left((1 - w_1)\tilde{\xi}_1 \tbinom{1}{\tilde{\xi}_2}, w_1 \tbinom{1}{w_2}\right) \\ &+ (\tilde{K}_n \tilde{\Phi}) \left((1 - w_1)\tilde{\xi}_1 \tbinom{1}{\tilde{\xi}_2} + w_1(1 - w_2)\tbinom{1}{1}, w_1 \tbinom{w_2}{w_2 - 1}\right) \\ &+ (\tilde{K}_n \tilde{\Phi}) \left((1 - w_1)\tilde{\xi}_1 \tbinom{1}{\tilde{\xi}_2} + w_1(1 - w_2)\tbinom{1}{0}, w_1 \tbinom{w_2}{1}\right) \Big\} \\ &\times (1 - w_1)^2 w_1 \tilde{\xi}_1 d\tilde{\xi}_2 d\tilde{\xi}_1 dw_1 dw_2, \end{aligned} \tag{3.18}$$

where

$$(\tilde{K}_n\tilde{\Phi})(\xi,\eta) := \tilde{k}_n(\xi,\eta)\tilde{\Phi}(\xi,\xi+\eta) + \tilde{k}_n(\xi+\eta,-\eta)\tilde{\Phi}(\xi+\eta,\xi) \qquad (3.19)$$

Proof. By using Lemma 3.1 we can write the singular integral (3.2) in the form

$$\begin{aligned} I &= \int_{\hat{\pi}} \text{v.p. } J_0^h(\xi)d\xi \\ &= \int_{\hat{\pi}} \lim_{\varepsilon\to 0} \int_{\hat{\pi}\setminus B_\varepsilon^\varrho(\xi)} \tilde{k}_n(\xi,\eta-\xi)\ \tilde{\Phi}(\xi,\eta)d\eta d\xi \\ &= \lim_{\varepsilon\to 0} \iint_{(\hat{\pi}\times\hat{\pi})\setminus B_\varepsilon^\varrho(\xi)} \tilde{k}_n(\xi,\eta-\xi)\ \tilde{\Phi}(\xi,\eta)d\eta d\xi \end{aligned}$$

We introduce the relative coordinates $u = \eta - \xi$ to obtain

$$I = \lim_{\varepsilon\to 0} \int_E \int_{G(\xi)\setminus B_\varepsilon(0)} \tilde{k}_n(\xi,u)\tilde{\Phi}(\xi,\xi+u)dud\xi, \qquad (3.20)$$

where

$$\begin{aligned} E &= \{\xi \in \mathbb{R}^2 : 0 < \xi_1 < 1, 0 < \xi_2 < \xi_1\}, \\ G(\xi) &= \{u \in \mathbb{R}^2 : -\xi_1 < u_1 < 1-\xi_1, -\xi_2 < u_2 < u_1 + \xi_1 - \xi_2\}, \end{aligned}$$

and

$$B_\varepsilon(\zeta) = \{u \in \mathbb{R}^2 : \|u-\zeta\|_2 < \varepsilon\}.$$

Next we describe the integration domain, such that the integration over the relative coordinates u is the outer integration. (This step was performed for the original element in [21, 6], but the representation of the integration domains with respect to the parameter plane (reference elements) is much easier.) We decompose $E \times G(\xi)$ into the domains

$$\begin{aligned} E_1 &= \{-1 < u_1 < 0, u_1 < u_2 < 0, -u_1 < \xi_1 < 1, -u_2 < \xi_2 < \xi_1 - \Delta u\}, \\ E_3 &= \{-1 < u_1 < 0, 0 < u_2 < u_1 + 1, \Delta u < \xi_1 < 1, 0 < \xi_2 < \xi_1 - \Delta u\}, \\ E_5 &= \{-1 < u_1 < 0, -1 < u_2 < u_1, -u_2 < \xi_1 < 1, -u_2 < \xi_2 < \xi_1\}, \\ E_2 &= \{0 < u_1 < 1, 0 < u_2 < u_1, 0 < \xi_1 < 1-u_1, 0 < \xi_2 < \xi_1\}, \\ E_4 &= \{0 < u_1 < 1, u_1 - 1 < u_2 < 0, -u_2 < \xi_1 < 1-u_1, -u_2 < \xi_2 < \xi_1\}, \\ E_6 &= \{0 < u_1 < 1, u_1 < u_2 < 1, \Delta u < \xi_1 < 1-u_1, 0 < \xi_2 < \xi_1 - \Delta u\}. \end{aligned}$$

($\Delta u := u_2 - u_1$.) The CAUCHY principal value, i. e. the limit in (3.20), exists only for the whole integration domain $E \times G(\xi)$, but does not exist for arbitrary parts of the integration domain. Nevertheless the limit

$$\lim_{\varepsilon\to 0} \tilde{J}_\varepsilon^{(i)} := \lim_{\varepsilon\to 0} \iint_{(E_{2i-1}\cup E_{2i})\setminus B_\varepsilon(0)} \tilde{k}_n(\xi,u)\tilde{\Phi}(\xi,\xi+u)d\xi du \qquad (i = 1,2,3)$$

exists because of the symmetry of the domain E_i and E_{i+1} $(i = 1, 2, 3)$. For proving this assertion we estimate the regular integral $\tilde{J}_\varepsilon^{(i)}$ $(\varepsilon > 0)$:

$$\tilde{J}_\varepsilon^{(i)} = \iint_{E_{2i}\setminus B_\varepsilon(0)} \tilde{k}_n(\xi, u)\tilde{\Phi}(\xi, \xi + u) + \tilde{k}_n(\xi + u, -u)\tilde{\Phi}(\xi + u, \xi)d\xi du.$$

By taking the suppositions 1. – 3. into account, we obtain for the integrand functions

$$\tilde{\Phi}(\xi, \xi + u) = \begin{cases} \tilde{\Phi}(\xi, \xi) + \sum_{\lambda \geq 1} p_\lambda^{(1)}(\xi, u), & n \leq 0 \\ p_1^{(3)}(\xi, u) + \sum_{\lambda \geq 2} p_\lambda^{(3)}(\xi, u), & n = 1 \end{cases},$$

$$\tilde{\Phi}(\xi + u, \xi) = \begin{cases} \tilde{\Phi}(\xi, \xi) + \sum_{\lambda \geq 1} p_\lambda^{(2)}(\xi, u), & n \leq 0 \\ -p_1^{(3)}(\xi, u) + \sum_{\lambda \geq 2} p_\lambda^{(4)}(\xi, u), & n = 1 \end{cases}, \tag{3.21}$$

$$\tilde{k}_n(\xi, u) = \sum_{\kappa=0}^{\kappa_{max}} \varrho^{-2-n-\kappa} \sum_{|\alpha| \geq \kappa} c_\alpha(\chi(\xi))(\chi(\xi + u) - \chi(\xi))^\alpha,$$

$$\tilde{k}_n(\xi + u, -u) = \sum_{\kappa=0}^{\kappa_{max}} \varrho^{-2-n-\kappa} \sum_{|\alpha| \geq \kappa} (c_\alpha(\chi(\xi)) + \tag{3.22}$$

$$+ \sum_{\lambda \geq 1} p_\lambda^{(5)}(\xi, u))(\chi(\xi) - \chi(\xi + u))^\alpha,$$

where

$$p_\lambda^{(i)}(\xi, u) := \sum_{j=0}^{\lambda} g_{ij}(\xi) u_1^j u_2^{\lambda - j} \qquad (i = 1, ..., 5).$$

In the case $n = 0$, the summands

$$S_0 = \varrho^{-2-\kappa} \sum_{|\alpha| = \kappa} c_\alpha(\chi(\xi))[(\chi(\xi + u) - \chi(\xi))^\alpha + (\chi(\xi) - \chi(\xi + u))^\alpha]\tilde{\Phi}(\xi, \xi)$$
$$= 0$$

vanish because κ is an odd number. In the case $n = 1$, the summands

$$S_1 = \varrho^{-3-\kappa} \sum_{|\alpha| = \kappa} c_\alpha(\chi(\xi))[(\chi(\xi + u) - \chi(\xi))^\alpha -$$
$$-(\chi(\xi) - \chi(\xi + u))^\alpha] p_1^{(3)}(\xi, u) = 0$$

vanish because κ is an even number. Because of (3.21) and

$$|u_i| \leq C h^{-1} \varrho \qquad (i = 1, 2),$$
$$|(\chi(\xi + u) - \chi(\xi))^\alpha| \leq C \varrho^{|\alpha|} \qquad (C = const. \neq C(\varrho, h))$$

all of the remaining summands of $\tilde{J}_\varepsilon^{(i)}$ are obviously weakly singular. By introducing the DUFFY'S triangular coordinates $u = \binom{w_1}{w_1 w_2}$ for $\tilde{J}_0^{(1)}$, $u = \binom{w_1 w_2}{w_1(w_2-1)}$ for $\tilde{J}_0^{(2)}$, $u = \binom{w_1 w_2}{w_1}$ for $\tilde{J}_0^{(3)}$, we remove the weak singularity and obtain the regular integral (3.17) by summing up. Equation (3.19) follows from (3.17) by the following variable substitutions with respect to the three summands in (3.17)

$$\begin{aligned} \xi_1 &= (1-w_1)\tilde{\xi}_1, & \xi_2 &= (1-w_1)\tilde{\xi}_1\tilde{\xi}_2; \\ \xi_1 &= (1-w_1)\tilde{\xi}_1 + w_1(1-w_2), & \xi_2 &= (1-w_1)\tilde{\xi}_1\tilde{\xi}_2 + w_1(1-w_2); \\ \xi_1 &= (1-w_1)\tilde{\xi}_1 + w_1(1-w_2), & \xi_2 &= (1-w_1)\tilde{\xi}_1\tilde{\xi}_2. \end{aligned}$$

Now all assertions are proved.

Lemma 3.2. *Let the mapping χ be linear, and let the suppositions of Theorem 3.1 be fulfilled. Then the integrands of (3.17) are polynomials in ξ_1, ξ_2, w_1 and the integrations of (3.17) can be performed fully analytically.*

Proof. The shape functions $\tilde{\Phi}(\cdot,\cdot)$ are obviously polynomials in ξ_1, ξ_2, w_1, w_2. Therefore we have to consider kernel functions of the type $\tilde{k}_n(\xi + w_1\zeta, w_1\eta)$. By using the linearity of χ we obtain

$$\chi(\xi + w_1(\zeta+\eta)) - \chi(\xi + w_1\zeta) = w_1\chi(\eta)$$

and

$$\varrho^2 = \|\chi(\xi + w_1(\zeta+\eta)) - \chi(\xi + w_1\zeta)\|_2^2 = w_1^2\|\chi(\eta\|_2^2 =: w_1^2\tilde{\varrho}^2(\eta).$$

Hence

$$\tilde{k}_n(\xi + w_1\zeta, w_1\eta) = \sum_{\kappa=0}^{\kappa_{max}} (\tilde{\varrho}(\eta))^{-2-n-\kappa} \sum_{|\alpha|\geq\kappa} w_1^{-2-n-\kappa+|\alpha|} c_\alpha(\xi + w_1\zeta)(\chi(\eta))^\alpha,$$

where $c_\alpha(.)$ is analytic, is a polynomial in $\xi = (\xi_1, \xi_2)^T$. Thus the integrands of (3.17) are polynomials in ξ_1, ξ_2. In (3.17) the terms, which have negative exponents of w_1, vanish, because the integrands of (3.17) are regular (see Proof of Theorem 3.1), such that the integration over ξ_1, ξ_2, w_1 can be performed analytically. The remaining outer integrals over w_2 are of the type

$$\int_0^1 \frac{q_n(w_2)}{\sqrt{a^2w_2^2 + bw_2 + c^2}^m} dw_2, \tag{3.23}$$

where q_n is a polynomial of degree n and a, b, c are constants. This integral (3.23) can be evaluated by standard variable substitutions [1].

Now we propose a GAUSSIAN quadrature scheme suitable for integral (3.19). In ξ_1 and w_1 we apply the standard error estimates for GAUSSIAN quadrature formulae in terms of the derivatives of the integrand. The error for the ξ_2- and w_2-integration will be estimated by using derivative-free estimates based on the domain of analytic continuation of the integrand, i. e. we use ideas of SCHWAB/ WENDLAND [23] and RABINOWITZ / RICHTER [20]. But the application of derivative-free estimates in the case of four-dimensional Galerkin integrals seems to be new.

We substitute $\tilde{\xi}_1 = h^{-1}\zeta_1, \quad \tilde{\xi}_2 = \zeta_2, \quad w_1 = h^{-1}\zeta_3, \quad w_2 = \zeta_4$ in (3.19) to obtain

$$I = \int_0^h \int_0^1 \int_0^h \int_0^1 f(\zeta_1, \zeta_2, \zeta_3, \zeta_4) d\zeta_2 d\zeta_1 d\zeta_4 d\zeta_3,$$

where is independent of h.

Theorem 3.2. *Denote by $Q_\zeta^N[f] = \sum_{j=1}^N \omega_j^N f(\zeta_j^N)$ the* GAUSSIAN *quadrature formula of order N in ξ for the interval $[0,1]$. Then there exist constants $\delta > 0$ and $C > 0$ independent of h, such that*

$$\left|(Q_{w_2}^M Q_{w_1}^N Q_{\tilde{\xi}_1}^N Q_{\tilde{\xi}_2}^M - I)[h^2 f(h\tilde{\xi}_1, \tilde{\xi}_2, hw_1, w_2)]\right| \leq Ch^2(h^{2N} + e^{-(2\ln\delta)M}) \tag{3.24}$$

where I is defined in (3.17) or (3.19).

Proof. With I_ζ denoting the integration with respect to ζ, the error E of numerical integration is

$$\begin{aligned} &|E[h^2 f(h\tilde{\xi}_1, \tilde{\xi}_2, hw_1, w_2]| \\ &= h^2|(Q_{w_2}^M Q_{w_1}^N Q_{\tilde{\xi}_1}^N Q_{\tilde{\xi}_2}^M - I_{w_2} I_{w_1} I_{\tilde{\xi}_1} I_{\tilde{\xi}_2})[f(h\tilde{\xi}_1, \tilde{\xi}_2, hw_1, w_2)]| \end{aligned} \tag{3.25}$$

We reduce the estimate of (3.1) to four one-dimensional estimates of the errors $E_{\tilde{\xi}_1} := Q_{\tilde{\xi}_1}^N - I_{\tilde{\xi}_1}$ etc., as usual:

$$\begin{aligned} &|E[h^2 f]| \\ &= h^2|(I_{w_2} + E_{w_2}) \circ (I_{w_1} + E_{w_1}) \circ (I_{\tilde{\xi}_1} + E_{\tilde{\xi}_1}) \circ (I_{\tilde{\xi}_2} + E_{\tilde{\xi}_2})[h^2 f] - I[h^2 f]| \\ &\leq Ch^2 \sum_{i=1}^4 R_i[f], \end{aligned}$$

where

$$R_1[f] = \sup_{\tilde{\xi}_2, w_1, w_2 \in [0,1]} |E_{\tilde{\xi}_1}[f]|, \qquad R_2[f] = \sup_{\tilde{\xi}_1, w_1, w_2 \in [0,1]} |E_{\tilde{\xi}_2}[f]|,$$

$$R_3[f] = \sup_{\tilde{\xi}_1, \tilde{\xi}_2, w_2 \in [0,1]} |E_{w_1}[f]|, \quad R_4[f] = \sup_{\tilde{\xi}_1, \tilde{\xi}_2, w_1 \in [0,1]} |E_{w_2}[f]|.$$

The errors R_1, R_3 can be estimated via the TAYLOR expansion by using the degree of exactness:

$$R_1[f] \le C_N \sup_{\tilde{\xi}_1,\tilde{\xi}_2,w_1,w_2\in[0,1]} \left| \left(\frac{\partial}{\partial \tilde{\xi}_1}\right)^{2N} f(h\tilde{\xi}_1,\tilde{\xi}_2,hw_1,w_2)\right|$$

$$\le C_N h^{2N} \left\| \left(\frac{\partial}{\partial \zeta_1}\right)^{2N} f\right\|_{L_\infty(G)}, \tag{3.26}$$

$$R_3[f] \le C_N h^{2N} \left\| \left(\frac{\partial}{\partial \zeta_3}\right)^{2N} f\right\|_{L_\infty(G)}, \tag{3.27}$$

where $G = [0,h]\times[0,1]\times[0,h]\times[0,1]$. To estimate the terms R_2, R_4, we apply in this variable derivative-free error estimates (see [20]). Therefore we have to estimate ellipsoids of analyticity for f. Since $\tilde{\Phi}$ is a polynomial, it suffices to consider $\tilde{k}_n$. In local polar coordinates (3.9), the domain of analyticity of

$$\tilde{k}_n\left(\xi, r\binom{\cos\theta}{\sin\theta}\right) \qquad \text{and} \qquad \tilde{k}_n\left(\xi+u, -r\binom{\cos\theta}{\sin\theta}\right)$$

will be restricted by the zeros of r^2 and of

$$l_2(\theta) = c\{\lambda\cos^2\theta + 2\cos\gamma\cos\theta\sin\theta + \lambda^{-1}\sin^2\theta\}$$

corresponding to the first fundamental form of the surface Γ with respect to ξ or $\xi+u$. From (3.19) we have the following three cases correspondent to the three summands of f:

1. $r^2 = w_1^2(1+w_2^2) = 0 \quad \Longrightarrow \quad (w_2)_{1,2} = \pm i$
   $l_2(\theta) = cw_1^2 r^{-2}\{\lambda + 2\cos\gamma w_2 + \lambda^{-1}w_2^2\} = 0 \Longrightarrow (w_2)_{3,4} = -\lambda e^{\pm i\gamma}$,
2. $r^2 = w_1^2(2w_2^2 - 2w_2 + 1) = 0 \quad \Longrightarrow \quad (w_2)_{5,6} = \frac{1}{2}(1 + \pm i)$
   $l_2(\theta) = cw_1^2 r^{-2}\{\lambda w_2^2 + 2\cos\gamma w_2(w_2-1) + \lambda^{-1}(w_2-1)^2\} = 0$
   $\Longrightarrow (w_2)_{7,8} = \frac{1+\lambda e^{\pm i\gamma}}{\lambda^2+2\lambda\cos\gamma+1}$,
3. $r^2 = w_1^2(1+w_2^2) = 0 \quad \Longrightarrow \quad (w_2)_{9,10} = \pm i = (w_2)_{1,2}$
   $l_2(\theta) = cw_1^2 r^{-2}\{\lambda w_2^2 + 2\cos\gamma w_2 + \lambda^{-1}\} = 0 \Longrightarrow (w_2)_{11,12} = -\lambda^{-1}e^{\pm i\gamma}$.

Next we map $[0,1]$ onto $[-1,1]$ via $\zeta = 2w_2 - 1$:

$$\zeta_{1,2} = \zeta_{9,10} = \pm 2i - 1, \qquad \zeta_{3,4} = -2\lambda e^{\pm i\gamma} - 1, \qquad \zeta_{5,6} = \pm i,$$

$$\zeta_{7,8} = \frac{\pm i(2\lambda\sin\gamma) - \lambda^2 + 1}{\lambda^2 + 2\lambda\cos\gamma + 1}, 7 \qquad \zeta_{11,12} = -2\lambda^{-1}e^{\pm i\gamma} - 1.$$

In the same manner as in [23] we can construct ellipses $\mathcal{E}_{\delta_i^*}$ in the complex plane having foci ± 1, such that $\zeta_i \in \partial\mathcal{E}_{\delta_i^*}$, where δ_i^* is the sum of the semi-major and semi-minor axis of $\mathcal{E}_{\delta_i^*}$. We choose $\delta = \min_i \delta_i^*$, such that f is analytic in $\mathcal{E}_\delta \supset [-1,1]$. Then

$$R_4[f] \le (2\pi)^{\frac{1}{2}} c_4 e^{-(2\ln\delta)M} \sup_{2w_2-1\in\partial\mathcal{E}_\delta} |f|. \tag{3.28}$$

The function f is analytic in $\tilde{\xi}_2$, such that

$$R_2[f] \leq C_\delta e^{-(2\ln\delta)M} \tag{3.29}$$

holds for any $\delta > 1$. The assertion follows from (3.26), (3.29), (3.27), and (3.28).

Theorem 3.2 gives the orders N and M of the GAUSSIAN quadrature to obtain an error of a prescribed consistent order of h. In order to satisfy the consistent estimate $|E| \leq Ch^a$ for given $a > 0$, one should choose [23]

$$N \geq (a-2)/2 \qquad \text{and} \qquad M \geq \frac{|\ln h|}{ln\delta} N.$$

3.2 Edge-adjacent and vertex-adjacent cases

For the "weaker singular" edge-adjacent and vertex-adjacent cases a similar integration method is proposed. We give formulae with regular integrands for the computation of the singular integrals. In [21] similar results are obtained for polygonal bounded domains.

Theorem 3.3. *Let $x \in \pi_i^h$ and $y \in \pi_j^h$ be two elements with adjacent edges, such that*

$$\chi^{(i)}(0) = \chi^{(j)}(0) \text{ and } \chi^{(i)}(\tbinom{1}{1}) = \chi^{(j)}(\tbinom{1}{0}),$$

and let all other suppositions of Theorem 3.1 be fulfilled. Then the singular integral (3.2) is equal to the following regular *integral (which has bounded integrands)*

$$I = \int_0^1 \int_0^1 \int_0^1 \int_{v_2}^1 v_2^2 \Big\{ u_1 \Big[(\tilde{k}_n \tilde{\Phi})(\xi^{(1)}, \eta^{(1)}) + (\tilde{k}_n \tilde{\Phi})(\eta^{(1)}, \xi^{(1)}) \Big] \\ + (\tilde{k}_n \tilde{\Phi})(\xi^{(2)}, \eta^{(2)}) + (\tilde{k}_n \tilde{\Phi})(\eta^{(2)}, \xi^{(2)}) \Big\} d\xi_1 dv_2 du_1 du_2 \tag{3.30}$$

where $\xi^{(1)} = \binom{\xi_1}{v_2}$, $\eta^{(1)} = \binom{\xi_1 + v_2 u_1 (u_2 - 1)}{v_2 u_1 u_2}$, $\xi^{(2)} = \binom{\xi_1}{v_2(1-u_2)}$, $\eta^{(2)} = \binom{\xi_1 + v_2(u_1 - 1)}{v_2 u_1}$, *and* $(\tilde{k}_n \tilde{\Phi})(\xi, \eta) := \tilde{k}_n(\xi, \eta - \xi)\tilde{\Phi}(\xi, \eta)$ *with* $\chi(\xi) = \chi^{(i)}(\xi)$, $\chi(\eta) = \chi^{(j)}(\eta)$ *(see (3.7), (3.8)).*

Proof. By using the existence of p.f. $I_\varepsilon^h(x)$ $\forall x \in \pi_i^h$ (see (3.3)) we have

$$I = \int_{\pi_j^h} \text{p.f. } I_\varepsilon^h(x) ds(x) = \lim_{\tilde{\varepsilon} \to 0} \int_{\pi_i^h \setminus M_x(\tilde{\varepsilon})} \text{p.f. } I_\varepsilon^h(x) ds(x), \tag{3.31}$$

where $M_x(\tilde{\varepsilon}) := \{x \in \pi_i^h : \text{dist}(x, \partial\pi_j^h) < \tilde{\varepsilon}\}$. In a similar way as in the proof of Lemma 3.1 (see (3.13)), we obtain

$$\text{p.f. } I_\varepsilon^h(x) = I_0^h(x) \leq C_1 |\ln(\text{dist}(x, \partial\pi_j^h)| + C_2 \quad \forall x \notin \pi_j^h,$$

where the positive constants C_1, C_2 do not depend on ε. Hence

$$I = \lim_{\tilde{\varepsilon} \to 0} \int_{\pi_i^h \setminus M_x(\tilde{\varepsilon})} I_0^h(x) ds(x),$$

where the integrand $I_0^h(x)$ is weak singular, such that

$$I = \lim_{\varepsilon \to 0} \iint_{(\hat{\pi} \times \hat{\pi}) \setminus M_\xi(\varepsilon)} (\tilde{k}_n \tilde{\Phi})(\xi, \eta) d\eta d\xi, \tag{3.32}$$

where

$$M_\xi(\varepsilon) := \{(\xi, \eta) \in \hat{\pi} \times \hat{\pi} : \xi_2 < \varepsilon\}.$$

Now the goal is to remove the weak singularity by applying three-dimensional triangular coordinates. This requires to split up the integration domain into generalized pyramids. After decomposition of $\hat{\pi} \times \hat{\pi}$ in

$$E_1 = \{0 < \xi_1 < 1, \;\; 0 < \xi_2 < \xi_1, \;\; 0 < \eta_1 < \xi_1, \;\; 0 < \eta_2 < \eta_1\}, \tag{3.33}$$

$$E_2 = \{0 < \eta_1 < 1, \;\; 0 < \eta_2 < \eta_1, \;\; 0 < \xi_2 < \eta_1, \;\; 0 < \xi_2 < \xi_1\}, \tag{3.34}$$

we introduce the relative coordinate $u_1 = \xi_1 - \eta_1$ for the integration over E_1. The domain of integration can be described by $\tilde{E}_1 = \tilde{E}_{11} \cup \tilde{E}_{12}$, where

$$\begin{aligned}
\tilde{E}_{11} &= \{0 < \xi_2 < 1, 0 < u_1 < \xi_2, 0 < \eta_2 < \xi_2 - u_1, \xi_2 < \xi_1 < 1\}, \\
\tilde{E}_{12} &= \{0 < \xi_2 < 1, 0 < u_1 < 1, \\
&\qquad \max\{0, \xi_2 - u_1\} < \eta_2 < 1 - u_1, u_1 + \eta_2 < \xi_1 < 1\}.
\end{aligned}$$

are generalized pyramids. The variable substitutions $w_1 = u_1 + \eta_2$ in $\tilde{E}_{11}$ and $w_1 = u_1 + \eta_2$, $w_2 = u_1 + \eta_2 - \xi_2$ in $\tilde{E}_{12}$ lead to

$$\begin{aligned}
I = \lim_{\varepsilon \to 0} \Big\{ &\int_\varepsilon^1 \int_0^{\xi_2} \int_0^{w_1} \int_{\xi_2}^1 \left(\tilde{k}_n \tilde{\Phi}\right) \left(\xi, \binom{\xi_1 - w_1 + \eta_2}{\eta_2}\right) d\xi_1 d\eta_2 dw_1 d\xi_2 \\
+ &\int_\varepsilon^1 \int_0^{w_1} \int_0^{w_1} \int_{w_1}^1 \left(\tilde{k}_n \tilde{\Phi}\right) \left(\binom{\xi_1}{w_1 - w_2}, \binom{\xi_1 - w_1 + \eta_2}{\eta_2}\right) d\xi_1 dw_2 d\eta_2 dw_1 \Big\}.
\end{aligned} \tag{3.35}$$

By using the three-dimensional triangular coordinates

$\xi_2 = v2, w_1 = v_2 u_1, \eta_2 = v_2 u_1 u_2$	for the first integral,
$w_1 = v_2, w_2 = v_2 u_2, \eta_2 = v_2 u_1$	for the second integral,

we obtain

$$I = \int_0^1 \int_0^1 \int_0^1 \int_{v_2}^1 v_2^2 \left\{ u_1 (\tilde{k}_n \tilde{\Phi})(\xi^{(1)}, \eta^{(1)}) + (\tilde{k}_n \tilde{\Phi})(\xi^{(2)}, \eta^{(2)}) \right\} d\xi_1 dv_2 du_1 du_2,$$

where the integrand is non-singular. Since the integration domain E_2 is of same type as E_1, if we exchange ξ and η, we finally obtain (3.3).

If the mappings $\chi^{(\cdot)}$ are linear, then the integrands of (3.3) are polynomials in ξ_1 and v_2 and the correspondent inner integrations can be performed analytically.

Theorem 3.4. *Let π_i^h and π_j^h be two elements with one adjacent vertex, such that*

$$\chi^{(i)}(0) = \chi^{(j)}(0),$$

and let all other suppositions of Theorem 3.1 be fulfilled. Then the singular integral (3.2) is equal to the following regular *integral (which has bounded integrands)*

$$I = \int_0^1 \int_0^1 \int_0^1 \int_0^1 u_1^3 u_3 \left\{ (\tilde{k}_n \tilde{\Phi})(\xi, \eta) + (\tilde{k}_n \tilde{\Phi})(\eta, \xi) \right\} du_1 du_2 du_3 du_4, \quad (3.36)$$

where

$$\xi = \begin{pmatrix} u_1 \\ u_1 u_2 \end{pmatrix}, \qquad \eta = \begin{pmatrix} u_1 u_3 \\ u_1 u_3 u_4 \end{pmatrix}, \qquad (3.37)$$

and $(\tilde{k}_n \tilde{\Phi})(\xi, \eta) := \tilde{k}_n(\xi, \eta - \xi)\tilde{\Phi}(\xi, \eta)$ *with* $\chi(\xi) = \chi^{(i)}(\xi)$, $\chi(\eta) = \chi^{(j)}(\eta)$ *(see (3.7),(3.8)).*

Proof. Following the lines of the previous proof we also obtain in this case a weak singular integral (see (3.32,3.33,3.34))

$$I = \lim_{\varepsilon \to 0} \iint_{(E_1 \cup E_2) \setminus M_\xi(\varepsilon)} (\tilde{k}_n \tilde{\Phi})(\xi, \eta) d\eta d\xi, M_\xi(\varepsilon) := \{(\xi, \eta) \in \hat{\pi} \times \hat{\pi} : \xi_1 < \varepsilon\}.$$

Now we consider E_1 and introduce the generalized triangular coordinates (3.37) to obtain the first summand of the integrand in (3.36). Since E_2 is of same type as E_1, if we exchange ξ and η, the second summand of the integrand follows in the same manner.

The Gauss-Jacobi quadrature rule is suitable for the integrals (3.3),(3.36), since the integrands are of the type $w^n f(w)$ [9]. The error estimates are obtained in the same way as in the proof of Theorem 3.2 because of the use of Duffy's triangular coordinates.

3.3 Integration of regular integrals

Let us briefly consider the regular case 4 (see Fig. 3.1). Using the notations (3.7), (3.8) we have to compute integrals of the type

$$I = \int_0^1 \int_0^{\xi_1} \int_0^1 \int_0^{\eta_1} (\tilde{k}_n \tilde{\Phi})(\xi, \eta) d\eta_2 d\eta_1 d\xi_2 d\xi_1.$$

By using the triangular coordinates $\xi_1 = u_1, \xi_2 = u_1 u_2, \eta_1 = v_1, \eta_2 = v_1 v_2$ we obtain

$$I = \int_0^1 \int_0^1 \int_0^1 \int_0^1 u_1 v_1 \left(\tilde{k}_n \tilde{\Phi}\right) \left(\begin{pmatrix} u_1 \\ u_1 u_2 \end{pmatrix}, \begin{pmatrix} v_1 \\ v_1 v_2 \end{pmatrix} \right) dv_2 dv_1 du_2 du_1.$$

We approximate I by product formulas and use Gauss-Legendre quadrature rules with respect to u_2, v_2 and Gauss-Jacobi quadrature rules with respect to u_1, v_1. It is easy to prove that if the one-dimensional rules have degree d, the product formula has also degree d [24]. The degree d, which is necessary to obtain a given error bound C_{int}, can be roughly evaluated by estimations R_{min} of the minimal distance of the elements and the size of the elements H [15]:

$$(d+2)\left(\frac{H}{2R_{\text{min}}}\right)^{d+1} \leq C_{\text{int}}.$$

In the near field, subdivisions of the elements are necessary to avoid quadrature rules of very high order [15, 2]. Furthermore, integration methods for nearly singular surface integrals arising in the collocation method (see [7]) can be adapted to the Galerkin method.

4 Numerical example

The LEON problem (unbounded elastic domain, $\nu = 0.3$, with a spherical cavity under 1D tension) is approximately analyzed using the domain Ω (see Fig. 2). In all of the numerical examples, piecewise linear shape functions

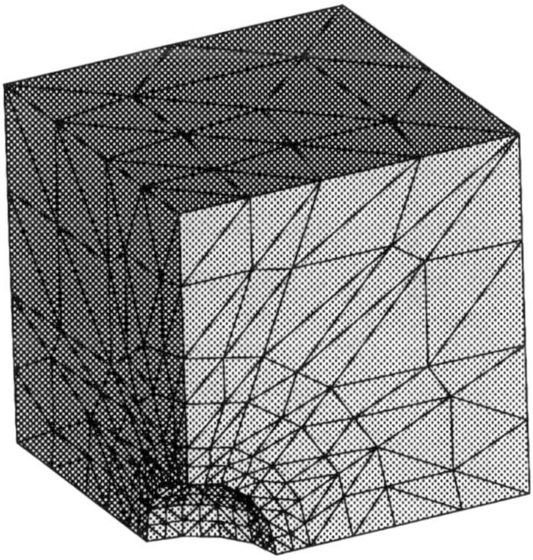

Fig. 4.1. Domain Ω

were used for the approximation of the displacements and piecewise constant shape functions for the tractions. In this case, some inner integrations can be performed analytically (see Table 4.1). Fig. 4 shows the approximated tractions $(t_h)_z$ in z-direction within the plane $z = 0$. The correspondent error is $e = \dfrac{\|t_z - (t_h)_z\|_{L_2}}{\|t_z\|_{L_2}} = 0.0385$. Table 4.2 shows the CPU time (SUN SPARCstation 10) of the matrix generation and the shares of the CPU time

Table 4.1. Number of analytical and numerical integrations

Case	analytical integration	numerical integration
coincident elements	3	1
edge-adjacent	2	2
vertex-adjacent	1	3
regular	0	4

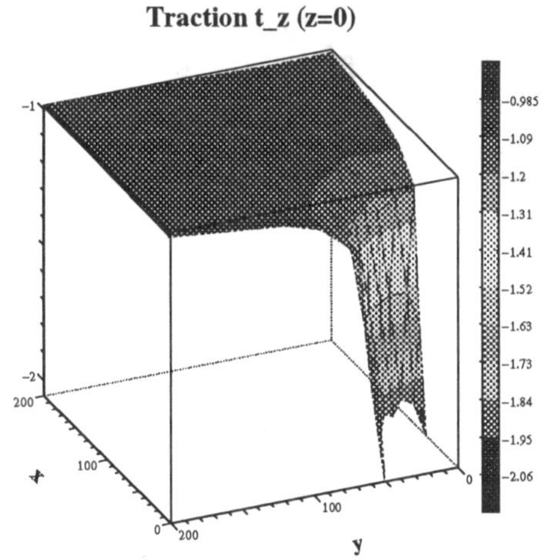

Fig. 4.2. Tractions $t_{h,z}$ within the plane $z = 0$

for the evaluation of the integrand function in each case. The percentage of computational labour consumed by the singular integration decreases rapidly in comparison with the regular integration.

Table 4.2. CPU time in sec. of the matrix generation

Nodes	Elements	DOF	$h_{\max}/h_{\min}$	coincid.	edge	vertex	regular
60	116	193	9.64	0.4	2	15	125
103	202	344	13.94	0.7	4	33	291
234	464	805	9.64	1.6	11	61	923
406	808	1419	13.94	2.8	19	117	2073

5 Conclusions

The mixed boundary value problem in three-dimensional linear elasticity is solved via a system of boundary integral equations. Differing from the well-known collocation approach, the symmetric GALERKIN BEM is used as approximation procedure.
The main result of this paper is a new general integration method for singular

double integrals, which can directly be applied to GALERKIN-type boundary element analysis of a wide range of BIEs on domains with curved (piecewise polynomial) boundary surfaces. Arbitrary polynomial shape functions can be considered. The singular double integrations for the symmetric GALERKIN BEM can be computed by a semi-analytical method efficiently in the case of polygonal bounded domains.

References

1. Bronstein, I.N., Semendjajew, K.A. (1989): Taschenbuch der Mathematik. Teubner, Leipzig
2. Doblare, M. (1987): Computational aspects of the boundary element method. In: Brebbia, C.A. (ed.) Topics in Boundary Element Research, volume 3. Springer, Berlin
3. Kupradze, V.D. (ed.) (1979): Three-dimensional Problems of the Mathematical Theory of Elasticity and Thermoelasticity. North-Holland Publ. Comp., Amsterdam New York Oxford
4. Haase, G., Heise, B., Kuhn, M., Langer, U. (1996): Adaptive domain decomposition methods for finite and boundary element equations. (In this book)
5. Hackbusch, W., Lage, C., Sauter, S.A. (1996): On the efficient realization of sparse matrix techniques for integral equations with focus on panel clustering, cubature and software design aspects. (In this book)
6. Hackbusch, W., Sauter, S.A. (1993): On the efficient use of the Galerkin method to solve Fredholm integral equations. Applications of Mathematics 38, 301–322
7. Hackbusch, W., Sauter, S.A. (1994): On numerical cubature of nearly singular surface integrals arising in bem collocation. Computing 52, 139–159
8. Hayami, K., Brebbia, C.A. (1987): A new coordinate transformation method for singular and nearly singular integrals over general curved boundary elements. In: Brebbia, C.A., Wendland, W.L., Kuhn, G. (eds.) Boundary Elements IX, vol. 1. Stuttgart, 9th Int. Conf, pp. 376–399
9. Hillion, P. (1977): Numerical integration on a triangle. Int. J. Num. Meth. Engng. 11, 797–815
10. Holzer, S.M. (1992): Das symmetrische Randelementverfahren: Numerische Realisierung und Kopplung mit der Finite-Element-Methode zur elastoplastischen Strukturanalyse. Berichte aus dem Konstruktiven Ingenieurbau, Technische Universität München
11. Kane, J.H., Balakrishna, C. (1993): Symmetric Galerkin boundary formulations employing curved elements. Int. J. Numer. Methods Eng. 36, 2157–2187
12. Kieser, R. (1991): Über einseitige Sprungrelationen und hypersinguläre Operatoren in der Methode der Randelemente. PhD thesis, Universität Stuttgart
13. Kieser, R., Schwab, C., Wendland, W.L. (1992): Numerical evaluation of singular and finite part integrals on curved surfaces using symbolic manipulation. Computing 49, 279–301
14. Krishnasamy, G., Schmerr, L.W., Rudolphi, T.J., Rizzo, F.J. (1990): Hypersingular boundary integral equations: Some applications in acoustic and elastic wave scattering. Trans. ASME 57, 404–414

15. Lachat, J.C., Watson, J.O. (1976): Effective numerical treatment of boundary integral equations: A formulation for three-dimensional elastostatics. Int. J. Numer. Meth. Eng. 10, 991–1005
16. Love, A.E.H. (1944): A Treatise on the Mathematical Theory of Elasticity. Dover
17. Maischak, M. (1995): The analytical computation of the Galerkin elements for the Laplace, Lamé and Helmholtz equation in 3d-bem. Preprint, Institut für angewandte Mathematik, Universität Hannover
18. Maischak, M., Stephan, E.P. (1996): The hp-Version of the BEM with Geometric Meshes in 3D. (In this book)
19. Polizzotto, C. (1988): An energy approach to the boundary element method, part i: elastic solids. Comp. Methods Appl. Mech. Eng. 69, 167–184
20. Rabinowitz, P., Richter, N. (1970): New error coefficients for estimating quadrature errors for analytic functions. Math. Comp. 24, 561–570
21. Sauter, S.A. (1992): Über die effiziente Verwendung von Galerkinverfahren zur Lösung Fredholmscher Integralgleichungen. Dissertation, Universität zu Kiel
22. Schwab, C., Wendland, W.L. (1992): Kernel properties and representation of boundary integral operators. Math. Nachr. 156, 187–218
23. Schwab, C., Wendland, W.L. (1992): On numerical cubatures of singular surface integrals in boundary element methods. Num. Math. 62, 343–369
24. Stroud, A.H. (1971): Approximate Calculation of Multiple Integrals. Prentice-Hall, Englewood Cliffs N.J.
25. von Petersdorff, T. (1989): Randwertprobleme der Elastizitätstheorie für Polyeder — Singularitäten und Approximation mit Randelementen. PhD thesis, Fachbereich Mathematik der TH Darmstadt
26. Wendland, W.L. (1985): On some mathematical aspects of boundary element methods for elliptic problems. In: The Mathematics of Finite Elements and Applications V. Academic Press, London, pp. 193–227

Parallel Coupling of FEM and BEM for 3D Elasticity Problems

Sz. Szikrai and E. Schnack

1 Introduction

In this paper we make use of the advantages of mixed variational FE/BE discretization and propose a parallel algorithm for solving the linear system of algebraic equations arise from discretizations of three dimensional linear elastostatic problems. The applied coupling technique was published in 1985 by Schnack [9]. The discretization leads to a linear system of algebraic equations, that can be reformulated in symmetric and indefinite form. This is very advantageous, because several efficient equation solvers can be applied. We can use Bramble/Pasciak's conjugate gradient method [1], or we can derive a positive definite and symmetric system (with building the Schur complement) and solve it with standard direct or iterative methods. Here, we solve the Schur complement system parallel, based on recursive substructuring and LU factorization. To decompose the mesh into subdomains, we use the simple and efficient method of Farhat [4]. To obtain the BE Schur complement matrices, one has to solve a Neumann boundary value problem in each BE subdomain locally. The BE discretization results in a linear system, where the solution is increasingly sensitive to perturbations (e.g. roundoff errors) and one has to resort to special techniques. We use the singular value decomposition (SVD), that enables us to intelligently handle the problem and to compute the solution to the linear system.

The parallelized algorithm presented is well suited for computations on MIMD computers with local memory and message passing principle (e.g. transputers). The communication architecture is a tree in the finite element domains and a hypercube in each of the boundary element domains.

2 The coupled 3D FE/BE variational formulation

2.1 Basic equations of linear elasticity

Let us consider a continuum in a bounded, three dimensional domain $\Omega \subset \mathbb{R}^3$ with a Lipschitz–continuous, piecewise smooth boundary $\Gamma = \partial\Omega$. We consider linear elastic materials, where the material properties are independent of position, i.e. the material is homogeneous, and the behavior of the material is the same in all directions, i.e. the material is isotropic. The boundary value problem is formulated by Navier's equation for the unknown displacement field u_i:

$$\Delta^* u = u_{j,kk} + \frac{1}{1-2\nu} u_{k,kj} = 0, \tag{2.1}$$

where ν is the Poisson's ratio. Equation (2.1) is a second order partial differential equation of elliptic type.

The solution of (2.1) has to fulfill the boundary conditions:

– Dirichlet (or first type) boundary condition

$$u_i = g_i, \quad \forall x \in \Gamma_D = \Gamma \backslash \bar{\Gamma}_N, \tag{2.2}$$

– Neumann (or second type) boundary condition

$$\sigma_{ij}(u) n_j = t_i, \quad \forall x \in \Gamma_N, \tag{2.3}$$

with the stress field σ_{ij} and the outward normal n_j. The Dirichlet and Neumann boundaries are denoted by Γ_D and Γ_N, respectively. With (2.1), (2.2) and (2.3) the mixed boundary value problem is completely defined. In the next section we reformulate (2.1) as a mixed domain decomposition variational problem with domain and boundary integrals.

2.2 DD variational formulation with non–overlapping substructures

There exist many techniques and ideas to reformulate the boundary value problem (2.1) as a mixed domain decomposition coupled domain and boundary integral variational problem, cf. [3], [13], [8] and others. Here we follow the method described in [9], [12], [5]:

Let us divide Ω into p non–overlapping, simply connected subdomains Ω_i $(i = 1, 2, ..., p)$ called also substructures such that

$$\bar{\Omega} = \bar{\Omega}_F \cup \bar{\Omega}_B \tag{2.4}$$

with

$$\bar{\Omega}_F := \bigcup_{i=1}^{q} \bar{\Omega}_i \quad \text{and} \quad \bar{\Omega}_B := \bigcup_{i=q+1}^{p} \bar{\Omega}_i, \tag{2.5}$$

where

$$\Omega_i \cup \Omega_j = 0, \quad \forall i \neq j. \tag{2.6}$$

The FE–substructures are denoted with the first q indices, and the BE–substructures (the so–called macro–elements) with the $p - q$ indices, respectively. Let us define some coupling boundaries (interfaces):

– Global coupling boundary:

$$\Gamma_C := \bigcup_{i=1}^{p} \Gamma_i \backslash \Gamma, \quad \text{with} \quad \Gamma_i := \partial\Omega_i, \quad i = 1, ..., p. \tag{2.7}$$

– FE/BE–coupling boundary:

$$\Gamma_0 := \partial\Omega_F \bigcap \partial\Omega_B. \tag{2.8}$$

– BE–boundary (skeleton of the macro–elements):

$$\Gamma_B := \bigcup_{i=q+1}^{p} \Gamma_i. \tag{2.9}$$

– FE–boundary:

$$\Gamma_F := \Gamma_C \backslash \Gamma_B. \tag{2.10}$$

Further on, we suppose that

$$\Gamma_B \bigcap \Gamma_N = 0 \tag{2.11}$$

for the simplicity. In the case that some macro–element boundaries contain parts of the Neumann boundary Γ_N, the proposed algorithm have to be slightly modified, see [5].

The energy test space corresponding to Ω_F will be denoted by

$$H^1_{\Gamma_D}(\Omega_F) = \{v \in H^1(\Omega_F) : v|_{\partial\Omega_F \bigcap \Gamma_D} = 0\}. \tag{2.12}$$

On the skeleton we introduce the mortar functions

$$\begin{aligned} \tilde{u} \in \left(H^{1/2}(\Gamma_B)\right)^3 :=& \{\tilde{w} = w|_{\Gamma_B} : w \in \left(H^1(\Omega)\right)^3, \\ & \|\tilde{w}\|_{\left(H^{1/2}(\Gamma_B)\right)^3} := \inf \|w\|_{(H^1(\Omega))^3}\}, \end{aligned} \tag{2.13}$$

and the product space

$$\mathbb{U} = \{(u, \tilde{u}) \in \left(H^1(\Omega_F)\right)^3 \times \left(H^{1/2}(\Gamma_B)\right)^3, \ u|_{\Gamma_0} = \tilde{u}|_{\Gamma_0}\} \tag{2.14}$$

equipped with the norm

$$\|(v, \tilde{v})\|_{\mathbb{U}} := \inf\{\|w\|_{(H^1(\Omega))^3} : w \in \left(H^1(\Omega)\right)^3, \ w|_{\Omega_F} = v, \ w|_{\Gamma_B} = \tilde{v}\}. \tag{2.15}$$

The test function space associated with $\mathbb{U}$ is denoted by

$$\begin{aligned} \mathbb{U}_0 = \{(u, \tilde{u}) \in \mathbb{U} : u \in \left(H^1_{\Gamma_D}(\Omega_F)\right)^3, \ \tilde{v} = w|_{\Gamma_B} \\ \text{with } w \in \left(H^1(\Omega)\right)^3 \text{ and } w|_{\Gamma_D} = 0\}. \end{aligned} \tag{2.16}$$

Let us introduce the trace spaces $\mathcal{V}$ and Λ on the skeleton:

$$\mathcal{V} = \{\nu = (\nu_{q+1}, ..., \nu_p) : \nu_i \in \left(H^{1/2}(\Gamma_i)\right)^3, \ i = q+1, ..., p\},$$

$$\begin{aligned} \Lambda &= \{\lambda = (\lambda_{q+1}, ..., \lambda_p) : \lambda_i \in \left(H_0^{-1/2}(\Gamma_i)\right)^3, \ i = q+1, ..., p\} = \\ &= \Lambda_{q+1} \times \Lambda_{q+2} \times \ldots \times \Lambda_p, \end{aligned} \tag{2.17}$$

with $\Lambda_i = \left(H_0^{-1/2}(\Gamma_i)\right)^3$, $i = q+1, \ldots, p$. Here, the subspace $\left(H_0^{-1/2}(\Gamma_i)\right)^3$ is defined as follows:

$$\left(H_0^{-1/2}(\Gamma_i)\right)^3 := \left\{ \chi \in \left(H^{-1/2}(\Gamma_i)\right)^3 : \int_{\Gamma_i} \chi \, d\Gamma = 0, \int_{\Gamma_i} r \times \chi \, d\Gamma = 0 \right\},$$

where $r \in \mathbb{R}^3$ denotes the distance vector to the origin.

The coupled domain and boundary integral variational problem now reads:

Find

$$((u, \tilde{u}), \lambda, \nu) \in \mathbb{V} = \mathbb{U} \times \Lambda \times \mathcal{V} \tag{2.18}$$

with

$$u|_{\Gamma_D \bigcap \partial\Omega_F} = g|_{\Gamma_D \bigcap \partial\Omega_F} \quad \text{and} \quad \tilde{u}|_{\Gamma_D \bigcap \Gamma_B} = g|_{\Gamma_D \bigcap \Gamma_B} \tag{2.19}$$

such that the global equations

$$a_F(u, v) + a_B(\lambda, \tilde{v}) = f(v), \quad \forall (v, \tilde{v}) \in \mathbb{U}_0, \tag{2.20}$$

the weak coupling conditions

$$< \tilde{u} - \nu_i, \ \eta_i >_{\Gamma_i} = 0, \qquad \forall \eta_i \in \left(H^{-1/2}(\Gamma_i)\right)^3, \quad q+1 \le i \le p \tag{2.21}$$

and the boundary integral equations

$$S_i \nu_i = \lambda_i, \quad q+1 \le i \le p \tag{2.22}$$

are satisfied. In (2.20) we used the following notations:

$$a_F(u, v) = \sum_{i=1}^{q} \int_{\Omega_i} \sigma_{kl}(u) \, \varepsilon_{kl}(v) \, d\Omega, \tag{2.23}$$

$$a_B(\lambda, \tilde{v}) = \sum_{i=q+1}^{p} \int_{\Gamma_i} \lambda_i \, \tilde{v} \, d\Gamma, \tag{2.24}$$

$$f(v) = \int_{\Gamma_N} t \, v \, d\Gamma. \tag{2.25}$$

In equation (2.21) the duality product $< \, . \, , \, . >_{\Gamma_i}$: $\left(H^{1/2}(\Gamma_i)\right)^3 \times \left(H^{-1/2}(\Gamma_i)\right)^3 \longmapsto \mathbb{R}$ was used as extension of the L_2 scalar product:

$$< \nu \, , \lambda >_{\Gamma_i} = \int\limits_{\Gamma_i} \nu \, \lambda \, d\Gamma, \quad \nu \in \left(H^{1/2}(\Gamma_i)\right)^3, \, \lambda \in \left(H^{-1/2}(\Gamma_i)\right)^3. \quad (2.26)$$

Furthermore, in (2.22), $S_i : \left(H^{1/2}(\Gamma_i)\right)^3 \longmapsto \left(H^{-1/2}(\Gamma_i)\right)^3$ denotes the Steklov–Poincaré operator, a Dirichlet–Neumann map, cf. [6]. It gives the relationship between the displacements ν_i and the tractions λ_i on the boundary Γ_i of the i–th BE–substructure, which are related by the two boundary integral equations, cf. [6]:

$$\nu_i = V_i\lambda_i + \left(\frac{1}{2}I_i - K_i\right)\nu_i, \quad (2.27)$$

$$\lambda_i = \left(\frac{1}{2}I_i + {K'}_i\right)\lambda_i + D_i\nu_i, \quad (2.28)$$

or in matrix form:

$$\begin{pmatrix} \nu_i \\ \lambda_i \end{pmatrix} = \begin{pmatrix} \frac{1}{2}I_i - K_i & V_i \\ D_i & \frac{1}{2}I_i + {K'}_i \end{pmatrix} \begin{pmatrix} \nu_i \\ \lambda_i \end{pmatrix} = C_i \begin{pmatrix} \nu_i \\ \lambda_i \end{pmatrix}, \quad (2.29)$$

where C_i denotes the Calderon projector and V_i, K_i, ${K'}_i$ and D_i are the four basic boundary integral operators:

- Weak–singular operator: $V_i\lambda_i(x) = \int\limits_{\Gamma_i} U(x,\xi)\, \lambda_i(\xi)\, d\Gamma_\xi$,
- Strong–singular operator: $K_i\nu_i(x) = \int\limits_{\Gamma_i} T(x,\xi)\, \nu_i(\xi)\, d\Gamma_\xi$,
- Transpose of the strong–singular operator: ${K'}_i\lambda_i(x) = \int\limits_{\Gamma_i} T'(\xi,x)\nu_i(\xi)d\Gamma_\xi$,
- Hyper–singular operator: $D_i\nu_i(x) = \int\limits_{\Gamma_i} S(x,\xi)\, \nu_i(\xi)\, d\Gamma_\xi$.

with the fundamental solution of Kelvin's problem $U(x,\xi)$ and the from U derived integral kernels T and S. The mapping properties on Sobolev spaces for these operators are well known (see e.g. [7]).

From (2.27)

$$\lambda_i = V_i^{-1}\left(\frac{1}{2}I_i + K_i\right)\nu_i = S_i\nu_i, \quad (2.30)$$

we get a Dirichlet–Neumann map, that defines the Steklov–Poincaré operator on the trace space $\left(H^{1/2}(\Gamma_i)\right)^3$, see [6]. Alternatively, from (2.27) and (2.28), one may also define the (symmetric) Steklov–Poincaré operator as

$$\lambda_i = D_i\nu_i + \left(\frac{1}{2}I_i + {K'}_i\right) V_i^{-1} \left(\frac{1}{2}I_i + K_i\right)\nu_i = S_i\nu_i. \quad (2.31)$$

The operator S_i is symmetric and $\left(H_0^{1/2}(\Gamma_i)\right)^3$–elliptic (i.e. positive definite) where

$$\left(H_0^{1/2}(\Gamma_i)\right)^3 := \left\{ \nu_i \in \left(H^{1/2}(\Gamma_i)\right)^3 \; : \; \int_{\Gamma_i} \nu_i \, d\Gamma = 0, \; \int_{\Gamma_i} r \times \nu_i \, d\Gamma = 0 \right\}. \tag{2.32}$$

Hence, S_i admits a right–inverse U_i on $\left(H_0^{-1/2}(\Gamma_i)\right)^3$, i.e.

$$S_i U_i \lambda = \lambda, \quad \forall \lambda \in \left(H_0^{-1/2}(\Gamma_i)\right)^3 \tag{2.33}$$

and

$$\exists \; a, b \in \mathbb{R}^3 : \quad U_i S_i \nu - \nu = \nu_R = a + b \times r, \quad \forall \nu \in \left(H^{1/2}(\Gamma_i)\right)^3. \tag{2.34}$$

The operator U_i is the Neumann–Dirichlet mapping named Poincaré–Steklov operator. It can also be expressed explicitly:

$$\nu_i = \left(\frac{1}{2} I_i - K_i\right)^{-1} V_i \lambda_i = U_i \lambda_i, \tag{2.35}$$

or

$$\nu_i = V_i \lambda_i + \left(\frac{1}{2} I_i - K_i\right) D_i^{-1} \left(\frac{1}{2} I_i - K'_i\right) \lambda_i = U_i \lambda_i. \tag{2.36}$$

The existence and uniqueness of the solution of the variational problem (2.20)–(2.22) is proved in [5]. In order to solve the variational problem numerically, we discretize the subdomains and approximate the solution in a finite dimensional subspace $\mathbb{V}_h \subset \mathbb{V}$. In the next section we define the coupled FE/BE–discretization in the nodal basis.

2.3 FE/BE–discretization

First, we divide the subdomains $\overline{\Omega}_i$ $(i = 1, ..., p)$ into finite elements, such that this discretization process results in a conform triangulation of $\overline{\Omega} = \overline{\Omega}_F \cup \overline{\Omega}_B$. Let us delete the FE–nodes and elements in the interior of the BE–subdomains Ω_i $(i = q + 1, ..., p)$, and let us form the BE–discretization as a refinement of the FE–triangulization on the skeleton Γ_B. Figure 2.1 shows an example, where tetrahedron finite elements (with quadratic basis functions to approximate the displacement field) and triangular boundary elements (with constant basis functions for displacements and for tractions as well) are used for the discretization. The tetrahedra generate the so–called coupling triangles (coarse grid with meshwidth H) on the boundaries of the BE–subdomains. The coupling triangles will then be decomposed into boundary elements (fine grid with meshwidth h).

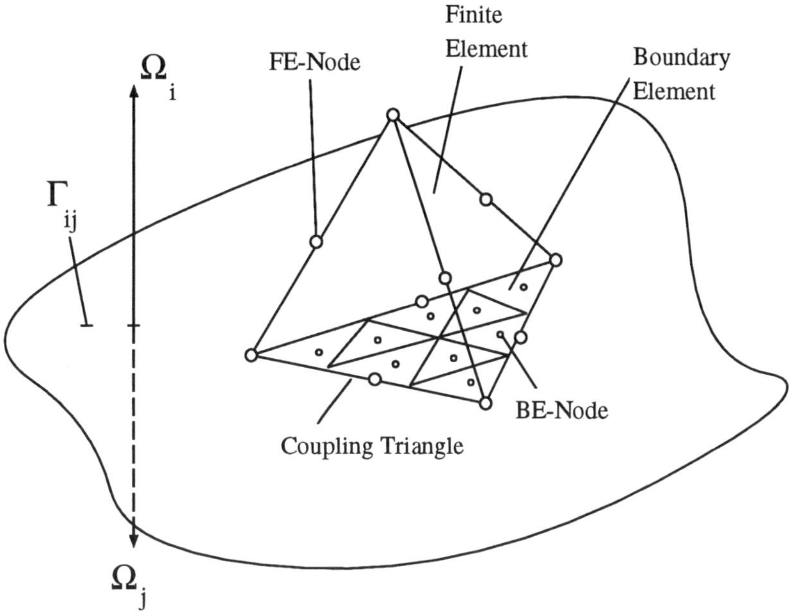

Fig. 2.1. FE– and BE–discretization on the interface Γ_{ij}, $1 \leq i \leq q < j \leq p$

The displacement field u (finite element substructures) is approximated with piecewise polynomial (degree $d \geq 2$) basis functions Φ_U:

$$u^H = \Phi_U \underline{u}_F \tag{2.37}$$

with

$$\Phi_U = [\Phi_I, \Phi_{C_F}, \Phi_{C_B}], \quad \underline{u}_F = \left[\underline{u}_I^T, \underline{u}_{C_F}^T, \underline{u}_{C_B}^T\right]^T. \tag{2.38}$$

The vector $\underline{u}_F$ is the usual finite element nodal displacement vector. Here and in the following the indices I, C_F and C_B mark quantities corresponding to the interior nodes of the FE–subdomains and to the coupling boundaries Γ_{FC}, and Γ_{BC} respectively.

The discretization of the traction field regarding the boundary elements is of the form:

$$\lambda_i^h = \Phi_{\Lambda_i} \underline{u}_{\Lambda_i}, \quad \text{span}\Phi_{\Lambda_i} \subset \left(H_0^{-1/2}(\Lambda_i)\right)^3, \quad q+1 \leq i \leq p. \tag{2.39}$$

The displacement field ν_i on Γ_i is approximated (like the tractions) with piecewise polynomial (degree $d' \geq 0$) basis functions:

$$\nu_i^h = \Phi_{\mathcal{V}_i} \underline{u}_{\mathcal{V}_i}, \quad \text{span}\Phi_{\mathcal{V}_i} \subset \left(H_0^{1/2}(\Gamma_i)\right)^3, \quad i = q+1, ..., p. \tag{2.40}$$

Here $\underline{u}_{\mathcal{V}_i}$ denotes the nodal displacement vector.

The nodal FE/BE basis

$$\Phi = [\Phi_U,\ \Phi_\Lambda,\ \Phi_\mathcal{V}] \tag{2.41}$$

and the FE/BE subspace

$$\mathbb{V} \equiv \mathbb{V}_h = \mathbb{U}_H \times \Lambda_h \times \mathcal{V}_h \tag{2.42}$$

is now defined by the finite dimensional subspaces

$$\mathbb{U}_H = \text{span}\Phi_U \subset \mathbb{U}, \tag{2.43}$$
$$\Lambda_h = \text{span}\Phi_\Lambda \subset \Lambda, \tag{2.44}$$
$$\mathcal{V}_h = \text{span}\Phi_\mathcal{V} \subset \mathcal{V}. \tag{2.45}$$

Once the basis Φ for $\mathbb{V}$ is chosen, we apply the Galerkin–method in the finite element subdomains and the

- Galerkin method or
- collocation method or
- classical Neumann series

in the boundary element subdomains to formulate the discretized version of the variational problem (2.20)–(2.22). This results in the system

$$K\underline{u} = \underline{f} \tag{2.46}$$

of coupled linear FE/BE–equations.

Taking into account the arrangement of the unknowns given in (2.41), we can rewrite system (2.46) in the block form

$$\begin{pmatrix} K_I & K_{IC_F} & K_{IC_B} & \mathbb{O} & \mathbb{O} \\ K_{C_FI} & K_{C_F} & K_{C_FC_B} & \mathbb{O} & \mathbb{O} \\ K_{C_BI} & K_{C_BC_F} & K_{C_B} & T^T & \mathbb{O} \\ \mathbb{O} & \mathbb{O} & -T & \mathbb{O} & S \\ \mathbb{O} & \mathbb{O} & \mathbb{O} & \tilde{B} & -A \end{pmatrix} \begin{pmatrix} \underline{u}_I \\ \underline{u}_{C_F} \\ \underline{u}_{C_B} \\ \underline{u}_\Lambda \\ \underline{u}_\mathcal{V} \end{pmatrix} = \begin{pmatrix} \underline{f}_I \\ \underline{f}_{C_F} \\ \underline{f}_{C_B} \\ \mathbb{O} \\ \mathbb{O} \end{pmatrix}. \tag{2.47}$$

By the elimination of $\underline{u}_{\mathcal{V}_i}$ in the BE subdomains we derive the following positive definite, but non-symmetric system:

$$\begin{pmatrix} K_I & K_{IC_F} & K_{IC_B} & \mathbb{O} \\ K_{C_FI} & K_{C_F} & K_{C_FC_B} & \mathbb{O} \\ K_{C_BI} & K_{C_BC_F} & K_{C_B} & T^T \\ \mathbb{O} & \mathbb{O} & -T & H \end{pmatrix} \begin{pmatrix} \underline{u}_I \\ \underline{u}_{C_F} \\ \underline{u}_{C_B} \\ \underline{u}_\Lambda \end{pmatrix} = \begin{pmatrix} \underline{f}_I \\ \underline{f}_{C_F} \\ \underline{f}_{C_B} \\ \mathbb{O} \end{pmatrix} \tag{2.48}$$

with

$$H := SA^{-1}\tilde{B} = SL. \tag{2.49}$$

The quadratic form of the matrix H corresponds on the continuous level to the energy bilinear form, cf. [9]:

$$\begin{aligned} a(u_i, u_i) &= \int\limits_{\Omega_i} \sigma_{kl}(u_i)\, \varepsilon_{kl}(u_i)\, d\Omega \\ &= \int\limits_{\Gamma_i} \lambda_i\, \nu_i\, d\Gamma \\ &= \underline{u}_{\Lambda_i}^T \int\limits_{\Gamma_i} R_{B_i}^T\, N_{B_i}\, d\Gamma\, \underline{u}_{\mathcal{V}_i} \\ &= \underline{u}_{\Lambda_i}^T\, S_i\, \underline{u}_{\mathcal{V}_i} \\ &= \underline{u}_{\Lambda_i}^T\, S_i\, L_i\, \underline{u}_{\Lambda_i} \\ &= (H_i\, \underline{u}_{\Lambda_i}, \underline{u}_{\Lambda_i}), \qquad i = q+1, ..., p, \end{aligned} \tag{2.50}$$

$$\begin{aligned} \forall u_i &\in \{u : \overline{\Omega}_i \longmapsto R^3 \,:\, u \in \left(H^1(\Omega_i)\right)^3, u|_{\overline{\Gamma}_i} \in \left(H^{1/2}(\overline{\Gamma}_i)\right)^3, u|_{\Gamma_D} = g\}, \\ \nu_i &\in \{\nu : \overline{\Gamma}_i \longmapsto R^3 \,:\, \nu \in \left(H^{1/2}(\Gamma_i)\right)^3, \nu = u_i|_{\Gamma_i}\}, \\ \lambda_i &\in \{\lambda : \overline{\Gamma}_i \longmapsto R^3 \,:\, \lambda \in \left(H^{-1/2}(\Gamma_i)\right)^3, \lambda = V^{-1}\left(\frac{1}{2}I + K\right)\nu_i\}, \end{aligned}$$

i.e. the matrix H is symmetric and positive definite. The strict symmetry and definiteness can be achieved although only approximately on the discrete level. Approximation errors occur in the numerical computation of the displacement field ν_i from the traction field λ_i by means of the boundary integral equation:

$$\underline{u}_{\mathcal{V}_i} = L_i\, \underline{u}_{\Lambda_i}, \tag{2.51}$$

where L_i is the approximation of the Poincaré–Steklov mapping U_i defined via the solution of (2.35) and/or (2.36) assuming that λ_i is given.

There are several possibilities to compute L_i:

- **A) Galerkin approximation:**
  In this case, we approximate the Poincaré–Steklov operator U_i by U_{ih}^G via the Galerkin approximation of equation (2.35):
  Find $\nu_i^h \in \mathcal{V}_{ih} = \text{span}\Phi_{\mathcal{V}_i} \subset \left(H_0^{1/2}(\Gamma_i)\right)^3$ such that

$$< V_i \lambda_i^h, \chi_i^h >_{\Gamma_i} = < (\frac{1}{2} I_i + K_i)\nu_i^h, \chi_i^h >_{\Gamma_i}, \quad \forall \chi_i^h \in \mathcal{V}_{ih}. \tag{2.52}$$

- **B) Collocation method:**
  Here, the test functions χ_i^h are the Dirac functions concentrated at suitable collocation points $p_{ik} \in \Gamma_i$, $k = 1, ..., M_i = \dim \mathcal{V}_{ih}$. We choose the BE–nodes as collocation nodes. The numerical problem to solve is as follows:

$$(V_i\lambda_i^h)(p_{ik}) = \frac{1}{2}\nu_i^h(p_{ik}) + (K_i\nu_i^h)(p_{ik}), \quad k = 1, ..., M_i, \tag{2.53}$$

whose solution provides us with the mapping $\lambda_i^h \mapsto \nu_i^h$. This again defines an approximate Poincaré–Steklov operator:

$$\lambda_i^h \mapsto \nu_i^h = U_{ih}^C \lambda_i^h. \tag{2.54}$$

– **C) Neumann series:**
Since the spectral radius of $(\frac{1}{2}I_i - K_i)$ in the space orthogonal to rigid body motions,

$$\left(H_0^0(\Gamma_i)\right)^3 := \{v_i \in (L_2(\Gamma_i))^3 \; : \; \int_{\Gamma_i} v_i \, d\Gamma = 0, \int_{\Gamma_i} r \times v_i \, d\Gamma = 0, \; \forall r \in R_i\}. \tag{2.55}$$

is less than one, the classical Neumann series can be used to compute the inverse

$$(\frac{1}{2}I_i + K_i)^{-1} = \sum_{l=0}^{\infty} (\frac{1}{2}I_i - K_i)^l \tag{2.56}$$

which converges in $\left(H_0^0(\Gamma_i)\right)^3$ and provides an explicit tool to solve the local Neumann problems in the macro–elements, i.e. to compute the Poincaré–Steklov operator

$$U_i = (\frac{1}{2}I_i + K_i)^{-1} V_i \tag{2.57}$$

as accurately as required, even on $\left(H_0^{-1/2}(\Gamma_i)\right)^3$. We denote this approximation by U_{ih}^N.
This method was developed and efficiently implemented by Türke in [12]. The symmetry and definiteness of the matrix H can be controlled adaptively. The algorithm was parallelized and implemented for 3D linear elastic problems, see [2].

For the definitions and the properties of the block matrices in (2.47) see [11]. The stability and convergence analysis of the discretized problem is described in [5].

2.4 The numerical solution of the local Neumann problems

In the parallel algorithm we use the collocation method to obtain the discretized Poincaré–Steklov operator, because the generation of the BE matrices A_i and B_i is 'cheaper' with the collocation technique, than with the Galerkin–scheme. In order to solve the local Neumann problems numerically, after generating the BE–matrices A_i and B_i, we have to compute the matrix L_i (see equation (2.51)). This involves the invertation of matrix A_i. The solution is increasingly sensitive to perturbations (e.g. roundoff errors). We apply the singular value decomposition (SVD), that is a useful tool to handle sensitive linear systems. The flop counts associated with the singular value decomposition algorithm is $4mn^2+8n^3$, which is quite large. The advantage of the method is however, that it is more accurate than other techniques, which produce comparably inaccurate results when applied to large, ill-conditioned problems.

Now we are interested in highly parallelizable and efficient methods for solving the system (2.48) of the coupled FE/BE equations.

2.5 Solution techniques

After eliminating $\underline{u}_V$ one can apply different parallel solution techniques to solve the block system (2.48) of linear equations:

1. *Bramble/Pasciak's CG:*
   Multiplying in (2.48) the first three block equations with (-1), we get the following symmetric, but indefinite system:

$$\begin{pmatrix} K_\Lambda & -K_{\Lambda C} & \mathbb{O} \\ -K_{C\Lambda} & -K_C & -K_{CI} \\ \mathbb{O} & -K_{IC} & -K_I \end{pmatrix} \begin{pmatrix} \underline{u}_\Lambda \\ \underline{u}_C \\ \underline{u}_I \end{pmatrix} = \begin{pmatrix} \underline{f}_\Lambda \\ -\underline{f}_C \\ -\underline{f}_I \end{pmatrix} \tag{2.58}$$

   with

$$K_\Lambda = H,\ K_{\Lambda C} = K_{C\Lambda}^T = (\mathbb{O} \quad T),\ \underline{f}_\Lambda = \mathbb{O},\ \underline{u}_C = \left(\underline{u}_{C_F}^T,\ \underline{u}_{C_B}^T\right)^T,$$

$$K_C = \begin{pmatrix} K_{C_F} & K_{C_F C_B} \\ K_{C_B C_F} & K_{C_B} \end{pmatrix},\ K_{CI} = K_{IC}^T = \begin{pmatrix} K_{C_F I} \\ K_{C_B I} \end{pmatrix}, \underline{f}_C = \begin{pmatrix} \underline{f}_{C_F} \\ \underline{f}_{C_B} \end{pmatrix}$$

   or in simplified form:

$$\begin{pmatrix} K_1 & K_{12} \\ K_{21} & -K_2 \end{pmatrix} \begin{pmatrix} \underline{u}_1 \\ \underline{u}_2 \end{pmatrix} = \begin{pmatrix} \underline{f}_1 \\ \underline{f}_2 \end{pmatrix} \tag{2.59}$$

   with the following definitions:

$$K_1 = K_\Lambda = H,\ K_{12} = K_{21}^T = (-K_{\Lambda C}\ \mathbb{O}),\ \underline{f}_1 = \mathbb{O},$$

$$K_2 = \begin{pmatrix} K_C & K_{CI} \\ K_{IC} & K_I \end{pmatrix}\ \underline{f}_2 = \begin{pmatrix} -\underline{f}_C \\ -\underline{f}_I \end{pmatrix}.$$

   The (symmetric) coupled FE/BE–algorithm by Langer leads to a block system of the same form, cf. [8]. One can apply Bramble and Pasciak's method to transform (2.59) into an equivalent system of linear equations with a positive definite and self–adjoint[1] system matrix, cf. [1]. In the transformation one needs an easily invertible, positive definite and symmetric matrix C_1, that is spectrally equivalent to the matrix K_1:

[1] with respect to a specially chosen inner product

$$\underline{\gamma} C_1 \le K_1 \le \overline{\gamma} C_1, \tag{2.60}$$

with positive spectral equivalence constants $\underline{\gamma}$ and $\overline{\gamma}$. Provided that besides C_1 a preconditioner C_2 for the matrix $K_2 + K_{21}K_1^{-1}K_{12}$ is known (e.g. by means of the Additive Schwarz Method, ASM), the Bramble/Pasciak's conjugate gradient method can efficiently be applied to solve the transformed system.

2. Iterative substructuring
Another approach obtaining a positive definite and symmetric system from (2.48) consists in the elimination of the unknowns $\underline{u}_I$ and $\underline{u}_\Lambda$. We obtain the full Schur complement system for the unknowns $\underline{u}_C$:

$$\begin{pmatrix} S_{C_F} & S_{C_F C_B} \\ S_{C_B C_F} & S_{C_B} \end{pmatrix} \begin{pmatrix} \underline{u}_{C_F} \\ \underline{u}_{C_B} \end{pmatrix} = \begin{pmatrix} \underline{f}^S_{C_F} \\ \underline{f}^S_{C_B} \end{pmatrix} \tag{2.61}$$

with

$$\begin{aligned} S_{C_F} &= K_{C_F} - K_{C_F I} K_I^{-1} K_{I C_F}, \\ S_{C_F C_B} &= K_{C_F C_B} - K_{C_F I} K_I^{-1} K_{I C_B}, \\ S_{C_B C_F} &= S^T_{C_F C_B}, \\ S_{C_B} &= K_{C_B} - K_{C_B I} K_I^{-1} K_{I C_B} + K_{C\Lambda} K_\Lambda^{-1} K_{\Lambda C}, \\ \underline{f}^S_{C_F} &= \underline{f}_{C_F} - K_{C_F I} K_I^{-1} \underline{f}_I, \\ \underline{f}^S_{C_B} &= \underline{f}_{C_B} - K_{C_B I} K_I^{-1} \underline{f}_I\,. \end{aligned} \tag{2.62}$$

The matrix in (2.61) is symmetric and positive definite, thus the preconditioned conjugate gradient (PCG) method can be applied to solve the system. As preconditioning one can use the wirebasket–based algorithms, that are constructed for three dimensional FE–DD problems, cf. [10].
In (2.62) there are inverse block entries, namely K_I^{-1} and K_Λ^{-1}. Therefore it is a good strategy to apply Bramble/Pasciak's CG for the system (2.59) in the case of large K_I and K_Λ matrices (e.g. there are only few processors with large memory), whereas for many substructures with few unknowns the Schur complement system (2.61) can be solved by means of the PCG.
In the conjugate gradient method the system matrix doesn't have to be available explicitly, but only in form of matrix–vector product. It is therefore sufficient to factorize the matrices K_I and K_Λ, and to apply the factorized matrices to the right hand side in each CG–step.
These algorithms are well suited for computations on MIMD computers with local memory and message passing communication principle. The hypercube seems to be the most suitable processor topology for both iterative solution techniques, cf. [8].

3. Recursive substructuring with static condensation
If we use a direct method to solve the positive definite and symmetric system (2.61) of linear equations, we can't avoid the expensive explicit

computation of the inverse matrices K_I^{-1} and K_Λ^{-1}. The advantage of the method however is in the recursion: With recursive organization of the computations can be achieved, that one has to solve systems with few unknowns at a time. Because of the recursion, the tree is the most suitable architecture for the communication of the processors.

We apply the third solution strategy, to solve the coupled FE/BE–system (2.47) of linear equations. In the next section, the decomposition of Ω_F is presented. Then, in Section 4 we define the recursive algorithm and in Section 5 we show some numerical results.

3 Decomposition of the FE mesh in substructures

In the recursive algorithm we need an automatic mesh decomposition algorithm, to achieve the FE subdomains from the original FE mesh. In our algorithm we implemented the automatic domain decomposer technique developed by *Farhat*, see [4]. The benefits of this method are, among others, as follows:

- The algorithm is suitable for two as well as three dimensional meshes and for arbitrary discretizations. It is able to handle eventually irregularities in the geometry as well.
- Each subdomain has equal number (n_{elem}) of finite elements. In the case, that the number of finite elements N_e cannot be divided by the number of subdomains q without a rest, one subdomain gets less elements: ($\tilde{n}_{elem} = N_e - (q-1)n_{elem}$), while the other $q-1$ domains consist of $n_{elem} = [\frac{N_e}{q}]+1$ elements. Therefore it yields a set of balanced subdomains.
- The number of interface nodes is kept low in order to reduce the cost of synchronization and/or message passing between the processors.
- It is easy to implement the algorithm into standard commercial FE preprocessors. The technique is simple and the storage requirement is low.

3.1 Examples

We have tested the above algorithm on some three dimensional finite element meshes. For the discretization we have used tetrahedron elements with 10 finite element nodes and each FE mesh has been decomposed into 8 subdomains. On the figures the thick lines denote the coupling boundaries. The number of elements (n_{elem}) and the number of FE nodes (interior and coupling: $n_{node} = n_I + n_C$) in the subdomains are summerized in a table.

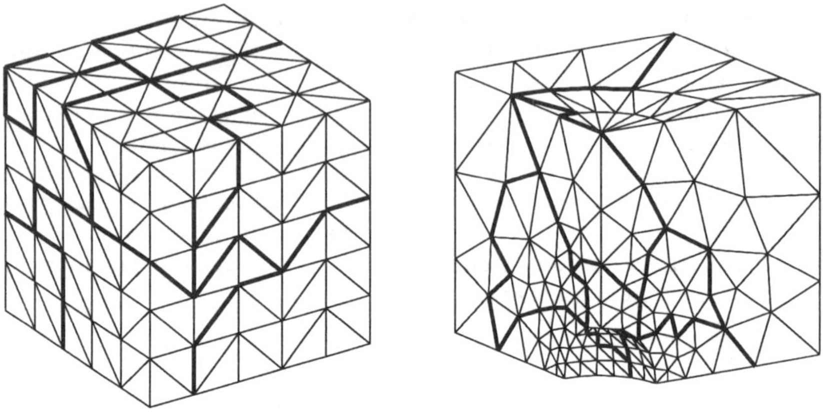

Fig. 3.1. Decomposition of a cube and Leon's cube into 8 FE subdomains

1. Example: Cube (750 Elements, 1331 Nodes)

Table 3.1. Distribution of finite elements and nodes of the subdomains (Ex. 1)

No.	n_{elem}	n_{node}	n_I	n_C	No.	n_{elem}	n_{node}	n_I	n_C
1	94	223	135	88	5	94	238	101	137
2	94	237	110	127	6	94	233	96	137
3	94	244	111	133	7	94	238	110	128
4	94	236	68	168	8	92	272	115	157

To outline the facts about the load balancing, we summerize some statistics (the minimum and maximum values, the averages $\overline{x}$ and the scatterings s) in another table. Here the following notations are used:

$$\overline{x} = \frac{1}{n} \sum_{i=1}^{n} x_i,$$

$$s = \frac{1}{n-1} \sum_{i=1}^{n} (x_i - \overline{x})^2, \quad (n = 8).$$

Table 3.2. Statistics about the load balancing among the subdomains (Ex. 1)

	min	max	$\overline{x}$	s
n_{elem}	92	94	93.7	0.7
n_{node}	223	272	240.1	14.2
n_I	68	135	105.7	19.1
n_C	88	168	134.4	23.6

Each subdomain (excepting the last one) has the same number of tetrahedron elements, thus the scattering of the element number is of course low. There are greater scattering values in the distribution of the interior nodes (I) and the coupling nodes (C).

2. Example: LEON's problem (1356 Elements, 2179 Nodes)

Table 3.3. Distribution of finite elements and nodes of the subdomains (Ex. 2)

No.	n_{elem}	n_{node}	n_I	n_C	No.	n_{elem}	n_{node}	n_I	n_C
1	170	360	252	108	5	170	366	191	175
2	170	378	161	217	6	170	358	169	189
3	170	356	188	168	7	170	384	151	233
4	170	360	222	138	8	166	372	228	144

The statistics are similar to the first example. The scattering of the elements is again less than that of the nodes with one order of magnitude. A good load–balancing between the processors requires low scattering values of the relevant data. For example, in the generation of the stiffness matrices, the number of finite elements (n_{elem}) is relevant. When computing the FE Schur complement matrices S_{C_F} (see Equation (2.62)), the number of interior and coupling nodes are relevant, since the dimensions of the matrices K_I and K_C are $3\,n_I$ and $3\,n_C$, respectively.

Table 3.4. Statistics about the load balancing among the subdomains (Ex. 2)

	min	max	$\overline{x}$	s
n_{elem}	166	170	169.5	1.4
n_{node}	356	384	366.7	10.2
n_I	151	252	195.2	35.6
n_C	108	233	171.5	41.6

4 The recursive substructuring algorithm

4.1 Basic idea

In this section we introduce an algorithm to solve the full Schur complement system (2.61) parallel, recursive, based on direct factorization. The basic idea of the recursion is as follows: After the unknowns corresponding to the interior nodes in the FE–subdomain Ω_i have been eliminated (with building the inverse of the matrix K_{I_i}), we get a "stiffness" matrix S_{C_i} (the FE Schur complement matrix) corresponding to the coupling nodes of the subdomain:

$$S_{C_i}\,\underline{u}_{C_i} = \underline{f}^S_{C_i}, \tag{4.1}$$

with

$$S_{C_i} = K_{C_i} - K_{CI_i} K_{I_i}^{-1} K_{IC_i}, \quad \underline{f}^S_{C_i} = \underline{f}_{C_i} - K_{CI_i} K_{I_i}^{-1} \underline{f}_{I_i}, \quad K_{CI_i} = K^T_{IC_i}.$$

We can treat the substructure as one large finite element with the stiffness matrix S_{C_i}, and these large elements can again be summarized into subdomains. The unknowns corresponding to the interior nodes of these new subdomains are eliminated with building the FE Schur complement matrix related to the coupling nodes of the larger subdomains. This technique can be further applied recursively.

Parallel to the FE–substructuring one local Neumann boundary value problem has to be solved in each BE subdomain (elimination of $\underline{u}_V$, see Equation (2.49)). After that, one has to compute the BE Schur complement matrices S_{B_i} (elimination of $\underline{u}_\Lambda$):

$$S_{B_i} = K_{C\Lambda_i} K_{\Lambda_i}^{-1} K_{\Lambda C_i}. \tag{4.2}$$

In the implementation we use the following communication structure:

The FE Schur complement matrices are computed in tree topology (with degree D and height H), whereas the BE substructure Ω_i in a hypercube of dimension N_i, see Figure 4.1:

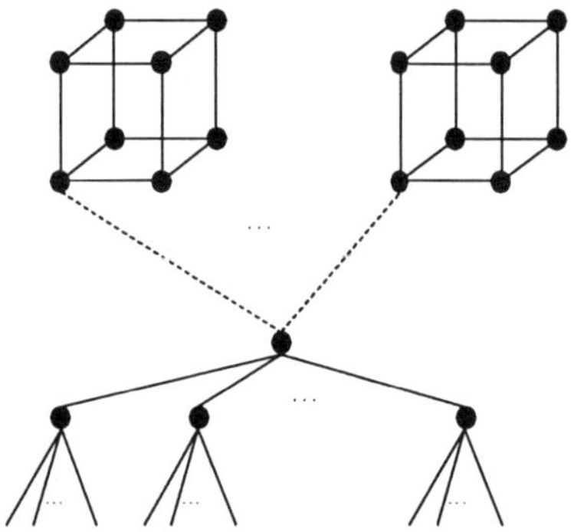

Fig. 4.1. Communication structure of the recursive algorithm

4.2 The parallel algorithm

In the next algorithm we use the following notations:

Notation:		Meaning:
ROUTINE $[l, p]$	$\longrightarrow$	The ROUTINE runs on processor (l, p) of the tree topology.
(for $i = k1 : s : k2$)	$\longrightarrow$	Loop with the variable i, that runs from $k1$ to $k2$ with step s.
(par $j = k1 : s : k2$): ROUTINE $[m, j]$	$\longrightarrow$	The ROUTINE runs parallel on the processors $[m, j]$ $(j = k1 : s : k2)$.

It is supposed in the following algorithm, that in the preprocessing the continuum under consideration is discretized into conform finite elements, and the finite elements and nodes within the BE substructures are deleted. This finest FE discretization is denoted with the indices [0,0]. The boundary conditions are also prescribed and stored together with the geometry data and material constants on the hard disk of the host computer.

Now we present the recursive algorithm:

- INPUT $[0,0]$: Reading the mesh data and boundary conditions for the whole problem from the host computer.
- SENDB $[0,0]$: The root of the tree topology sends the mesh data (the geometry data of the appropriate coupling triangles) and the boundary conditions to the processors with the id 0 of the hypercubes, see the broken lines on Figure 4.1.
- *FE substructures:*
  1. (for $i = 0:1:H-1$):
     - (par $k = 0:1:D^i - 1$) DECOMP $[i,k]$: Decomposition of the FE mesh $[i,k]$ into D substructures:
       FE mesh $[i+1,j]$, $(j = kD:1:(k+1)D-1)$. Here we implemented the mesh decomposition algorithm of Farhat [4], see Section 3
     - (par $k = 0:1:D^i - 1$) SENDF $[i,k]$: After the decomposition we send the mesh data of the j-th subdomain (FE mesh $[i+1, kD+j]$) to the j-th son (processor $(i+1, kD+j)$) $(j = 0:1:D-1)$.
     - (par $k = 0:1:D^{i+1} - 1$) RECVF $[i+1,k]$: The sons receive the mesh data from their father.
  2. (par $i = 0:1:D^H - 1$) MATGENF $[H,i]$: Generating the finite element stiffness matrices on the level H and assembling them in the subdomain stiffness matrix K_i. Here, the local numbering of the nodes is taken into account, i.e. the interior nodes come first, then the coupling nodes:

     $$K_{(H,i)} = \begin{pmatrix} K_{I_{(H,i)}} & K_{IC_{(H,i)}} \\ K_{CI_{(H,i)}} & K_{C_{(H,i)}} \end{pmatrix}$$

     The nodal load vector is also generated:

     $$\underline{f}_{(H,i)} = \begin{pmatrix} \underline{f}_{I_{(H,i)}} \\ \underline{f}_{C_{(H,i)}} \end{pmatrix}$$

     After that, the Dirichlet and Neumann boundary conditions are built in the matrix and the right–hand side.
  3. (for $i = H:-1:1$):
     - (par $k = 0:1:D^i - 1$) SCHURF $[i,k]$: Elimination of the unknowns in the interior nodes, that were not eliminated yet, with building the Schur complement matrix explicitly:

$$\begin{pmatrix} K_{I_{(i,k)}} & K_{IC_{(i,k)}} \\ \mathbb{O} & S_{C_{(i,k)}} \end{pmatrix} \begin{pmatrix} \underline{u}_{I_{(i,k)}} \\ \underline{u}_{C_{(i,k)}} \end{pmatrix} = \begin{pmatrix} \underline{f}_{I_{(i,k)}} \\ \underline{f}^S_{C_{(i,k)}} \end{pmatrix},$$

with

$$\begin{aligned} S_{C_{(i,k)}} &= K_{C_{(i,k)}} - K_{CI_{(i,k)}} K^{-1}_{I_{(i,k)}} K_{IC_{(i,k)}}, \\ \underline{f}^S_{C_{(i,k)}} &= \underline{f}_{C_{(i,k)}} - K_{CI_{(i,k)}} K^{-1}_{I_{(i,k)}} \underline{f}_{I_{(i,k)}}. \end{aligned}$$

- (par $k = 0 : 1 : D^i - 1$) SENDFSCHUR $[i, k]$: Sending the Schur complement matrix $S_{C_{(i,k)}}$ and the right–hand side $\underline{f}^S_{C_{(i,k)}}$ to the father processor.
- (par $k = 0 : 1 : D^{i-1} - 1$) RECVFSCHUR $[i - 1, k]$: The nodes receive the Schur complement matrices from their sons and assemble their own stiffness matrix $K_{(i-1,k)}$. The local numbering is taken into account:

$$K_{(i-1,k)} = \begin{pmatrix} K_{I_{(i-1,k)}} & K_{IC_{(i-1,k)}} \\ K_{CI_{(i-1,k)}} & K_{C_{(i-1,k)}} \end{pmatrix}$$

Here I denotes the unknowns in the interior nodes of the FE subdomain $[i - 1, k]$, that were not eliminated yet.

4. RECVBSCHUR $[0,0]$: The root processor of the tree receives the Schur complement matrices $S^{BE}_{C_i}$ of the BE subdomains:

$$S^{BE}_{C_i} = K_{CA_i} K^{-1}_{A_i} K_{AC_i}, \quad i = 0, ..., p - q - 1,$$

where $p - q$ is the number of the BE subdomains. These matrices will then be assembled in the FE stiffness matrix $K_{(0,0)}$.
5. SCHURF $[0,0]$: As the last step of the algorithm, we solve the Schur complement system concerning the matrix $K_{(0,0)}$.

The eliminated unknowns $\underline{u}_{I_{(i,j)}}$ can now be computed recursively with backward substituting the already computed vectors $\underline{u}_{C_{(i,j)}}$.

- *BE substructures:* (par $i = 0 : 1 : p - q - 1$)

The BE Schur complement matrix $S^{BE}_{C_i}$ is computed parallel, in a hypercube of the dimension N_i.

1. RECVB $[i]$: The processors with the id 0 of the k–th hypercube receive the geometry data of the coupling triangles of the k–th BE subdomain and the boundary conditions.
2. MATGENB $[i]$ and SOLVEB $[i]$: Applying the collocation method, the BE matrices A_i and B_i are generated. Then the nodal displacement vector $\underline{u}_{\mathcal{V}_i}$ is eliminated, i.e. we solve a Neumann problem on each BE subdomain, see the discretized boundary integral equation (2.27) and the Poincaré–Steklov operator (2.30). As a result, we achieve the matrix L_i (the discretized form of the Poincaré–Steklov operator) explicitly:

$$L_i = A_i^{-1} \tilde{B}_i = A_i^{-1} B_i R_{B_i}.$$

To compute the matrix L_i we use the singular value decompisiton, as mentioned in Section 2

3. SCHURB [i]: In this routine we first compute the matrix H_i, see Equation (2.49):
$$H_i = S_i L_i \quad (= S_i A_i^{-1} \tilde{B}_i).$$
Then we eliminate the nodal traction vector $\underline{u}_{\Lambda_i}$, i.e. compute the BE Schur complement matrix $S_{C_i}^{BE}$, see Equations (2.58) and (2.62):
$$S_{C_i}^{BE} = K_{C\Lambda_i} K_{\Lambda_i}^{-1} K_{\Lambda C_i}.$$
4. SENDBSCHUR [i]: The processor with the id 0 of the hypercube i sends the matrix $S_{C_i}^{BE}$ to the root processor of the tree topology.

5 Implementation and numerical results

5.1 Implementation on a distributed memory MIMD computer

The algorithm was implemented on a Multicluster 3DE Transputersystem of the firm Parsytec with 32 INMOS T805 Transputers (clock frequency: 30 MHz, 4 MB memory on each node, altogether 128 MB). The host computer with all the peripheral devices is a SUN SparcStation 10/30 with a connection board for the parallel computer. The developments of the parallel algorithms were made with an ANSI compatible C cross compiler for the runtime system PARIX using the versions 1.0 to 1.2 for T805 transputers. In Figure 5.1 two three dimensional hypercubes (for 2 BE subdomains) and one binary tree ($D = 2$) with the height $H = 3$ (for 8 FE domains) are shown. This virtual topology is used in the example in the same figure.

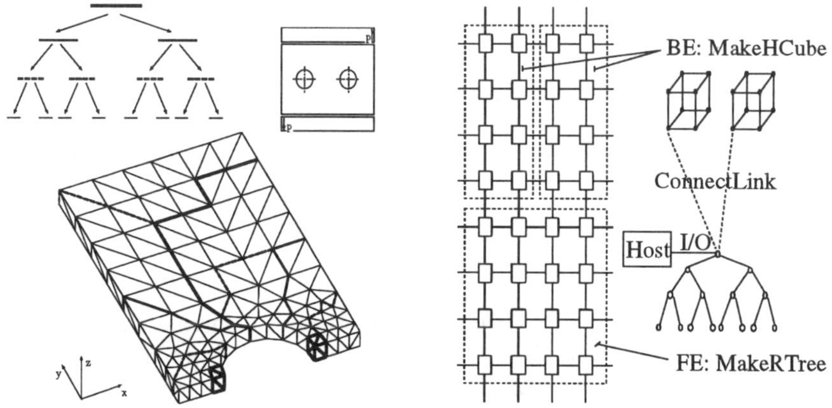

Fig. 5.1. Decomposition of a plate with two holes into 8 FE and 2 BE subdomains

The FE mesh consists of 480 elements (quadratical tetrahedra) and 1111 nodes, and there are 108 collocation points (constant triangles) per BE substructure. The minimal (Min), maximal (Max), average ($\overline{T}$) and strewing (S) values of the CPU–times needed for the computation of the FE and BE Schur complements are listed below:

Table 5.1. CPU–times of computing the FE Schur complements

Routine	Level	n_{proc}	Min [s]	Max [s]	$\overline{T}$ [s]	S [s]
MatGenF	3	8	10.79	14.67	12.06	1.33
SchurF	3	8	126.17	702.88	395.17	182.00
SchurF	2	4	10.46	38.01	21.52	11.77
SchurF	1	2	9.56	10.22	9.89	0.46
SchurF	0	1	15.82	15.82	15.82	0.00

Table 5.2. CPU–times of computing the BE Schur complements

Routine	HCube	n_{proc}	Min [s]	Max [s]	$\overline{T}$ [s]	S [s]
MatGenB	0	8	18.05	25.20	22.03	2.58
MatGenB	1	8	18.14	23.98	22.07	2.43
SolveB	0	1	197.82	197.82	197.82	0.00
SolveB	1	1	197.82	197.82	197.82	0.00
SchurB	0	1	38.12	38.12	38.12	0.00
SchurB	1	1	38.10	38.10	38.10	0.00

As expected, the explicit computation of the FE Schur complements on the lowest tree level requires the greatest computational cost.

In the routine MatGenB the BE matrices are generated in hypercube topology with row wise distribution. The Neumann problem is solved in the routine SolveB. The low scattering values in the parallel generation of the BE matrices mean a good load balancing between the hypercube processors. The most expensive part of the computations in the BE subdomains is (as expected) the SVD algorithm in the routine Solve. The great benefit of SVD is however, that the matrices H_i are positive definite, which was not the case when applying other solution algorithms.

6 Conclusions

In this paper a recursive substructuring algorithm for solving the linear system of equations arising from a coupled FE/BE discretization of three dimensional linear elastostatic problems on distributed memory parallel computers is presented. The elastic domain under consideration is decomposed into non–overlapping FE and BE subdomains. The proposed algorithm is well suited for stress concentration problems: The critical regions with high stress concentrations are computed with the boundary element method, while the rest part is approximated with finite elements.

In the FE region, the mesh decomposition algorithm of Farhat is used to create the FE subdomains. The decomposition is organized recursively, in tree processor topology. The root processor of the tree decomposes the mesh into two subdomains and sends the meshes to his sons. The sons further decompose their meshes and so on. One benefit of this recursive implementation of Farhat's mesh decomposition is that it results in compact, connected subdomains. (In the original algorithm of Farhat it is not guaranteed.) Another advantage of the recursion is, that only small systems of linear equations have to be solved on one level of the tree. Further, the mesh decomposition can be applied to quite arbitrary three dimensional domains. A drawback of the recursive substructuring algorithm is however, that because a direct equation solver is used, the Schur complement matrices have to be computed explicitly. Iterative solvers require the matrix usually only in form of a matrix–vector multiplication, so after factorization, the backward substitution doesn't have to be performed.

In the BE subdomains one has to solve local Neumann boundary value problems in order to compute the BE Schur complement matrices. Since the matrix of the linear system is close to singular and the solution is very sensitive to small perturbations, special technique has to be applied. The numerical tests show, that SVD is able to perform the theoretically guaranteed symmetry and positive definiteness of the matrix H. The drawback of the method however is the large number of floating point operations.

The algorithm was implemented on a parallel computer with message passing principle. With a suitable choice of the dimensions of the topologies the proposed domain decomposition algorithm can be applied to solve the mixed domain and boundary integral variational problem efficiently.

References

1. Bramble, J.H., Pasciak, J.E. (1988): A preconditioning technique for indefinite systems resulting from mixed approximations of elliptic problems. Mathematics of Computation 50(181), 1–17
2. Bundschuh, A. (1995): Parallele, iterative Lösung von Randintegralgleichungen für dreidimensionale Elastizitätsprobleme. Master's thesis, Universität Karlsruhe
3. Costabel, M. (1987): Symmetric methods for the coupling of the finite elements and boundary elements. In: Brebbia, C.A., Wendland, W.L., Kuhn, G. (eds.) Boundary Elements IX. Springer, pp.411–420
4. Farhat, C. (1988): A simple and efficient automatic fem domain decomposer. Computers & Structures 28(5), 579–602

5. Hsiao, G.C., Schnack, E., Wendland, W.L. (1995): A hybrid coupled finite–boundary element method. Technical Report 95–11, Mathematisches Institut A, Universität Stuttgart
6. Hsiao, G.C., Wendland, W.L. (1991): Domain decomposition in boundary element methods. In: Glowinski, R., et al. (eds.) Fourth International Symposium on Domain Decomposition Methods for Partial Differential Equations, Philadelphia, PA. SIAM. pp. 41–49
7. Hsiao, G.C., Wendland, W.L. (1977): A finite element method for some integral equations of the first kind. J. Math. Anal. Appl. 58, 449–481
8. Langer, U. (1992): Parallel iterative solution of symmetric coupled FE/BE-equations via domain decomposition. Technical Report 217, Technische Universität Chemnitz
9. Schnack, E. (1985): Stress analysis with a combination of HSM and BEM. In: Whiteman, J.R. (ed.) MAFELAP 1984 Conference on "The Mathematics of Finite Elements and Applications", Uxbridge. Academic Press, pp. 273–281
10. Smith, B.F. (1990): Domain Decomposition Algorithms for the Partial Differential Equations of Linear Elasticity. PhD thesis, Courant Institut, New York, NY
11. Szikrai, Sz. (1996): Parallel Coupling of FEM and BEM via Domain Decomposition for Three Dimensional Linear Elastostatic Problems. PhD thesis, Universität Karlsruhe
12. Türke, K. (1995): Eine Zweigitter–Methode zur Kopplung von FEM und BEM. PhD thesis, Universität Karlsruhe
13. Wendland, W.L. (1990): On the coupling of finite elements and boundary elements. In: Kuhn, G., Mang, H. (eds.) IUTUM/IACM Symposium in Vienna, 1989: "Discretization methods in structural mechanics". Springer, Berlin, pp. 405–414

Coupling of BEM and FEM for Elastic Structures

E. Schnack and K. Türke

1 Introduction

In this paper a numerical method for the solution of elliptic boundary value problems is presented. As an application we will analyze problems in linear elasticity. Especially we want to resolve zones with high stress gradients, for example the surrounding of notches in a loaded machine part. The stress peak may occur the failure of the whole structure.

In the field of numerical simulation the Finite Element Method (FEM) is the standard tool and already well established for industrial applications. Another approach is the Boundary Element Method (BEM) which offers the possibility to reduce the effort of discretization in comparison with the FEM by one dimension. In addition the BEM is able to approximate very accurately high stress concentrations in compact domains with a relative low number of degrees of freedoms. This leads us to the suggestion to combine the two methods FEM and BEM. Of special interest is the so called macro–element technique where the BEM substructures are interpreted as special finite elements. This strategy allows the implementation of the coupling method as an extension modul in existing FEM software.

The algorithm of this paper follows this stategy. A special characteristic of our coupling algorithm which has been developed by the research group *Schnack* (see literatur in Section 2) is the non–conformity of the discretizations. The introduction of a hybrid ansatz for the displacement and the traction fields on the surfaces of the macro–elements and the formulation of variational transmission conditions gives the possibility to use an independent discretization for each subdomain. Besides the larger flexibility concerning the mesh generation, this strategy is important for the parallelization of the algorithm. Here the stiffness matrices of the macro–elements and the FEM subdomains can be generated completely independent on different processors. A communication between the processors is only necessary during the solution of the global system.

The properties of the coupling matrix like symmetry and positive definiteness are controlled numerically within an adaptive scheme.

2 Basic system and continuous formulation

2.1 Model problem

We are analyzing interior domain problems in an isotropic domain Ω with a *Lipschitz* boundary Γ in $\mathbb{R}^m$, $m = 2, 3$, as shown in Fig. 2.1. In the next step

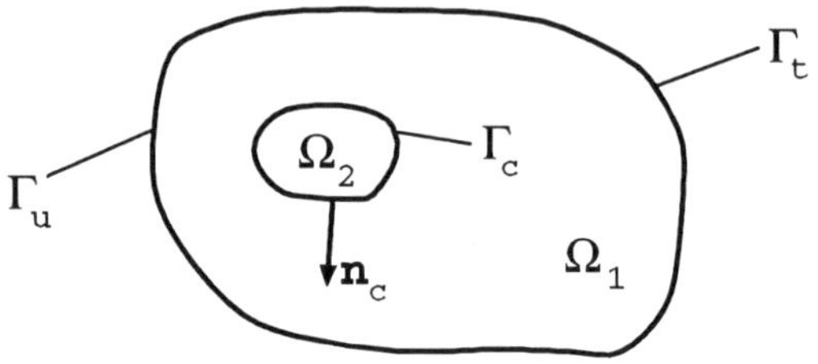

Fig. 2.1. Decomposed domain Ω

we define a non overlapping domain decomposition of Ω in a basic domain Ω_1 and several substructures Ω_i with $i = 2, \dots, p$ with

$$\Omega \subset \cup_{i=1}^{p}(\Omega_i \cup \Gamma_i), \quad \Omega_i \cap \Omega_j = \emptyset \quad \text{for } i \neq j, \quad \Gamma = \partial\Omega = \Gamma_t \cup \Gamma_u \tag{2.1}$$

and

$$\Gamma_i \cap \Gamma = \emptyset \qquad \text{and} \qquad \Gamma_1 = \Gamma \cup \left(\cup_{i=2}^{p} \Gamma_i \right) \quad \forall \;\; i \geq 2. \tag{2.2}$$

For the sake of simplicity we restrict ourself now to the case $p = 2$ and define

$$\Gamma_c = \partial\Omega_1 \cap \partial\Omega_2 \tag{2.3}$$

as the coupling interface (see Fig. 2.1). We introduce the following notation: Values on Γ_c coming from Ω_2 will be denoted with a "+" symbol and values coming from Ω_1 with a "-" symbol.

As a boundary value problem in the given domain Ω we have the *Lamé* system of linear elasticity with mixed boundary conditions. This leads us to the *Navier* equation

$$\Delta^* \mathbf{u} = 0 \qquad \text{for } \boldsymbol{\xi} \in \Omega \tag{2.4}$$

as the describing differential equation for the displacement field $\boldsymbol{u} \in [H^1(\Omega)]^m$. Δ^* denotes an elliptic partial differential operator. As boundary conditions we have given displacements on the *Dirichlet* boundary Γ_u and given tractions on the *Neumann* boundary Γ_t:

$$\boldsymbol{u}(\boldsymbol{\xi}) = \bar{\boldsymbol{u}}(\boldsymbol{\xi}) \qquad \text{for } \boldsymbol{\xi} \in \Gamma_u = \Gamma \setminus \Gamma_t \tag{2.5}$$

$$\boldsymbol{t}(\xi) = \Pi\, \boldsymbol{u}(\xi) = \bar{\boldsymbol{t}}(\xi) \qquad \text{for } \boldsymbol{\xi} \in \Gamma_t. \tag{2.6}$$

Π denotes the traction operator.

As a result of the domain decomposition we have to fulfill additionally the kinematic and static transmission conditions

$$\boldsymbol{u}_+(\xi) = \boldsymbol{u}_-(\xi) \tag{2.7}$$

$$\xi \in \Gamma_c$$

$$\boldsymbol{t}_+(\xi) = \boldsymbol{t}_-(\xi). \tag{2.8}$$

2.2 Set of coupling equations

The base of our coupling method is given by a hybrid stress method, formulated 1973 by *Schnack* [25] and *Tong, Pian* and *Lasry* [33], which realizes a mixed FE method by the transformation of domain integrals in boundary integrals. Starting with a modified *Reissner* functional and by formulating independent trial functions for the kinematic and the static group one results in compatibility and equilibrium conditions in a variational form. The stiffness matrices of the hybrid elements can be generated analytically, if the problem is restricted to special element geometries and to a small number of degrees of freedom in each element.

1983 *Schnack* [26] realized a generalization of this method by incorporating the boundary integral equation in the procedure. This extension allows the definition of arbitrary shaped "macro–elements" with an unrestricted number of degrees of freedom. In the following years further developments and applications on two and three dimensional problems in elasticity were published by *Schnack* [27], [28], *Carmine* [1] and *Karaosmanoglu* [14]. A stability and convergence analysis for the coupling method is done by *Hsiao, Schnack* and *Wendland* [13]. Comparisons with other coupling strategies can be found in the papers of *Hsiao* [11], [12] and *Polizzotto* and *Zito* [22]. A comparable strategy to our coupling method was applied for example by *Grannell* [7], who also uses hybrid elements for coupling FEM and BEM.

In the following we will define the four fundamental equations for the coupling algorithm on the continuous level.

1. The first basic equation is the variational formulation of the given boundary value problem (2.4),(2.5),(2.6), split up in two parts for the subdomains Ω_1 and Ω_2:

   $$a_1(\boldsymbol{u}_1, \boldsymbol{v}_1) + a_2(\boldsymbol{u}_2, \boldsymbol{v}_1) = f(\boldsymbol{v}_1) \qquad \forall \quad \boldsymbol{v}_1 \in \left[H_0^1(\Omega)\right]^m, \tag{2.9}$$

   where a denotes the positive definite and symmetric energy bilinear form

   $$a_i(\boldsymbol{u}, \boldsymbol{u}) = \int\limits_{\Omega_i} \sigma_{ij}(\boldsymbol{u})\, e_{ij}(\boldsymbol{u})\, d\Omega \tag{2.10}$$

   with the stress tensor $\boldsymbol{\sigma}$ and the strain tensor $\boldsymbol{e}$. The test function $\boldsymbol{v}_1$ is defined on the whole domain Ω. The right hand side is a boundary integral on the *Neumann* boundary Γ_t

$$f(\boldsymbol{v}_1) = \int_{\Gamma_t} (\overline{\boldsymbol{t}})^T \, \boldsymbol{v}_1 \, d\Gamma. \tag{2.11}$$

Eq. (2.9) is the well known energy functional also used in the classical FEM formulation and represents a weak equilibrium condition.

2. Additionally we need a second bilinear form to realize the nonconformity of the grids on the interface Γ_c. We define

$$a_2(\boldsymbol{v}_2, \boldsymbol{u}_2) = g(\boldsymbol{v}_2) \qquad \forall \quad \boldsymbol{v}_2 \in \left[H_0^1(\Omega_2)\right]^m \tag{2.12}$$

with the test functions $\boldsymbol{v}_2$ only defined on the subdomain Ω_2. The right hand side is once again a boundary integral now defined on the interface Γ_c. With the notation

$$\int_{\Gamma_c} \boldsymbol{x}^T \, \boldsymbol{y} \, d\Gamma = < \boldsymbol{x} \, , \boldsymbol{y} > \tag{2.13}$$

we have

$$g(\boldsymbol{v}_2) = < \tilde{\boldsymbol{t}_+} \, , \boldsymbol{u}_- > \qquad \text{with } \tilde{\boldsymbol{t}_+} = \Pi[\boldsymbol{v}_2]. \tag{2.14}$$

3. As a side condition we require that the static equilibrium

$$\sigma_{ij,i} = 0 \qquad \forall \quad \xi \in \Omega_2 \tag{2.15}$$

holds in each point of the subdomain Ω_2.

4. The last basic equation results from the *Calderon* projector, introduced by *Costabel* and *Wendland* [4]

$$\begin{pmatrix} \boldsymbol{u}_+(\boldsymbol{\xi}) \\ \boldsymbol{t}_+(\boldsymbol{\xi}) \end{pmatrix} = \begin{pmatrix} \frac{1}{2}\boldsymbol{I} - \boldsymbol{K} & \boldsymbol{V} \\ \boldsymbol{D} & \frac{1}{2}\boldsymbol{I} + \boldsymbol{K}^* \end{pmatrix} \begin{pmatrix} \boldsymbol{u}_+(\boldsymbol{\xi}) \\ \boldsymbol{t}_+(\boldsymbol{\xi}) \end{pmatrix} \qquad \boldsymbol{\xi} \in \Gamma_c. \tag{2.16}$$

In Eq. (2.16) $\boldsymbol{V}$ denotes the weak singular integral operator resulting from *Kelvins* fundamental solution, $\boldsymbol{K}$ and $\boldsymbol{K}^*$ are singular integral operators and $\boldsymbol{D}$ is a hypersingular integral operator. For our coupling method we can use an arbitrary projection of the *Neumann* data $\boldsymbol{t}_+$ in the *Dirichlet* data $\boldsymbol{u}_+$ on Γ_c. So we choose the singular formulation of the *Poincaré–Steklov* operator resulting from the the *Calderon* projector (2.16)

$$\boldsymbol{u}_+ = \Pi^{-1}\boldsymbol{t}_+ = \left[\left(\frac{1}{2}\boldsymbol{I} + \boldsymbol{K} \right)^{-1} \boldsymbol{V} \right] \boldsymbol{t}_+. \tag{2.17}$$

With this set of equations we can formulate the basic relationships of the coupling method. By applying the 1. *Greens* formula on the bilinear form a_2 we get

$$a_2(\boldsymbol{u}_2, \boldsymbol{v}_1) = \int_{\Omega_2} \sigma_{ij}(\boldsymbol{u}_2) \, e_{ij}(\boldsymbol{v}_1) \, d\Omega = < \boldsymbol{t}_+ \, , \boldsymbol{v}_- > - \int_{\Omega_2} \sigma_{ij,i}(\boldsymbol{u}_2) \, v_{1_j} \, d\Omega. \tag{2.18}$$

Under consideration of the side condition (2.15) the last domain integral in Eq. (2.18) disappears and (2.9) can be transformed to

$$a_1(\boldsymbol{u}_1, \boldsymbol{v}_1) + < \boldsymbol{t}_+ , \boldsymbol{v}_- > = f(\boldsymbol{v}_1). \tag{2.19}$$

In the same way the second coupling equation (2.12) is equal to the interface integral

$$< \tilde{\boldsymbol{t}_+} , (\boldsymbol{u}_+ - \boldsymbol{u}_-) > = 0. \tag{2.20}$$

The displacement field $\boldsymbol{u}_+$ on Γ_c can be replaced by the *Poincaré–Steklov* operator (2.17), so that we finally receive the coupling equation

$$< \tilde{\boldsymbol{t}_+} , \left[\left(\frac{1}{2}\boldsymbol{I} + \boldsymbol{K} \right)^{-1} \boldsymbol{V} \right] \boldsymbol{t}_+ > - < \tilde{\boldsymbol{t}_+} , \boldsymbol{u}_- > = \boldsymbol{0}, \tag{2.21}$$

that defines a variational compatibility condition on Γ_c consisting of positive definite and symmetric energy terms. From (2.21) we can calculate the traction field $\boldsymbol{t}_+$ in dependance of $\boldsymbol{u}_-$ and insert the result in (2.19), so that finally there remains a variational form depending only on test and trial functions for the diplacements in the basic domain Ω_1 and on the interface Γ_c.

3 Discrete formulation

In this section we will derive an equivalent discrete formulation of the continuous coupling equations (2.19) and (2.21) for the implementation in a numerical algorithm. In Fig. 3.1 you can see in the upper part an example of a decomposed structure Ω in 2D with a coarse grid H which consists of triangular finite elements in Ω_1 and of line elements on the boundary of Ω_2. On the right hand side the subdomain is depicted again with the coarse grid H and additionally with an independent fine grid h.

The configuration on the interface Γ_c for the 3D case is depicted in the lower part of Fig. 3.1. Here the domain discretization H in Ω_1 is realized with tetrahedra which generate coupling triangles on Γ_c. The fine grid consists of triangular boundary elements, in Fig. 3.1 we have nine boundary elements for one coupling triangle.

Now we realize the spaces for the trial and test functions. We start with the space for the displacement field $\boldsymbol{u}_H$ on the interface Γ_c

$$U := \{H^{1/2}(\Gamma_c)\}. \tag{3.1}$$

The displacement field is approximated by the spline formulation

$$U_H := \{\boldsymbol{u}_H \in U \mid \boldsymbol{u}_H \in S_d^0(\Delta H)\}. \tag{3.2}$$

S_d^0 denotes the space of continuous spline functions of the order d. In an explicit form this leads to the piecewise polynomial expansion

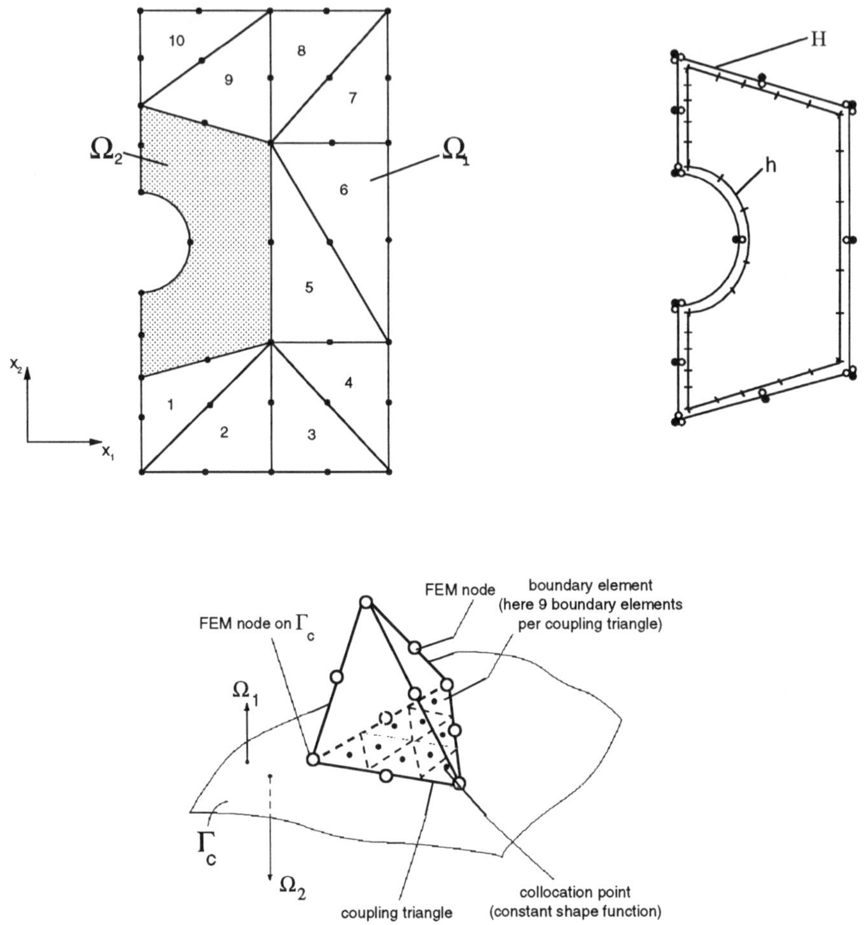

Fig. 3.1. Discretization in 2D and in 3D

$$\boldsymbol{u}_H = \boldsymbol{N}_c \boldsymbol{\alpha}_c \qquad \text{on } \Gamma_c. \tag{3.3}$$

Because of the hybrid formulation we have to define an independent space for the traction field on Γ_c by

$$W := \{H^{-1/2}(\Gamma_c), \quad < \boldsymbol{t} \ , \ \boldsymbol{I} \ >= 0, \quad < \boldsymbol{r} \times \boldsymbol{t} \ , \ \boldsymbol{I} \ >= \boldsymbol{0}\} \tag{3.4}$$

with the unit operator $\boldsymbol{I}$. The two side conditions guarantee that the traction formulation leads to a stress field in Ω_2 that fulfills the condition (2.15) for the static equilibrium. An explicit formulation of these side conditions will be derived in Section 4.

On the approximation level we have

$$W_H := \{\boldsymbol{t}_H \in W \mid \boldsymbol{t}_H \in S^{-1}_{d-1}(\Delta H)\}. \tag{3.5}$$

S_{d-1}^{-1} denotes the spline space with a reduced order by one because the tractions are computed with the derivatives of the displacements. Additionally we use so called double nodes for the formulation of the traction ansatz. This leads to a discontinuous spline space. The double nodes are depicted in the right part of Fig. 3.1 by the white dots at the corner points. The explicit expansion is given by

$$\boldsymbol{t}_H = \boldsymbol{R}_H \, \boldsymbol{\beta} \qquad \text{on } \Gamma_c, \tag{3.6}$$

where the parameters $\boldsymbol{\beta}$ are independent of the parameters $\boldsymbol{\alpha}_c$. With Eq. (3.3) and (3.6) we have realized the hybrid ansatz on the coarse grid H. In order to work on the fine grid h, we have to realize a projection of the coarse grid trial function $\boldsymbol{t}_H$ to $\boldsymbol{t}_h$. If we use the same spline space W_H on the fine grid level, we obtain a transformation

$$\boldsymbol{t}_h = \boldsymbol{C} \, \boldsymbol{t}_H, \tag{3.7}$$

with a constant projection operator $\boldsymbol{C}$ containing the coordinates of the fine grid elements h on the coarse grid elements H. As a result we obtain on the fine grid level the trial functions

$$\boldsymbol{t}_h = \boldsymbol{R}_h \, \boldsymbol{\beta} \tag{3.8}$$

with

$$\boldsymbol{R}_h = \boldsymbol{C} \, \boldsymbol{R}_H. \tag{3.9}$$

We need the corresponding displacements $\boldsymbol{u}_h$ depending on the *Neumann* data $\boldsymbol{t}_h$. This is realized by the initial formulation

$$\boldsymbol{u}_h = \boldsymbol{N}_h \, \boldsymbol{\gamma} \qquad \text{on } \Gamma_c. \tag{3.10}$$

$\boldsymbol{u}_h$ belongs to the same spline space U_H (see Eq. (3.2)) as the displacements $\boldsymbol{u}_H$ on the coarse grid. the parameter groups $\boldsymbol{\beta}$ and $\boldsymbol{\gamma}$ are connected by the *Poincaré–Steklov* operator (2.17)

$$\boldsymbol{u}_h = \left[\left(\frac{1}{2} \boldsymbol{I} + \boldsymbol{K} \right)^{-1} \boldsymbol{V} \right] \boldsymbol{t}_h. \tag{3.11}$$

On the discrete level this leads to

$$\boldsymbol{\gamma} = \boldsymbol{A}^{-1} \, \boldsymbol{B} \, \boldsymbol{\beta}. \tag{3.12}$$

This set of trial and test functions can now be used to fulfill the coupling equations (2.19) and (2.20) on the discrete level. This leads to the equivalent formulation

$$a_1(\boldsymbol{u}_H, \boldsymbol{v}_H) + < \boldsymbol{t}_H \, , \, \boldsymbol{v}_H > = f(\boldsymbol{v}_H) \tag{3.13}$$

and

$$< \tilde{\boldsymbol{t}}_H \, , \, \boldsymbol{u}_H > - < \tilde{\boldsymbol{t}}_h \, , \, \boldsymbol{u}_h > = 0. \tag{3.14}$$

Inserting (3.3),(3.6),(3.7) and (3.10) in (3.13) and (3.14) under consideration of (3.12) we finally get the following block structure system:

$$\begin{bmatrix} \boldsymbol{K}_{ii} & \boldsymbol{K}_{ic} & 0 \\ \boldsymbol{K}_{ci} & \boldsymbol{K}_{cc} & \boldsymbol{T}^T \\ 0 & \boldsymbol{T} & -\boldsymbol{H} \end{bmatrix} \begin{bmatrix} \boldsymbol{\alpha}_i \\ \boldsymbol{\alpha}_c \\ \boldsymbol{\beta} \end{bmatrix} = \begin{bmatrix} \boldsymbol{F}_i \\ \boldsymbol{F}_c \\ \boldsymbol{0} \end{bmatrix} . \tag{3.15}$$

$\boldsymbol{\alpha}_i$ are the parameters of the displacement ansatz similar to (3.3) in the interior of the basic domain Ω_1. The matrices $\boldsymbol{K}$ are the usual global stiffness matrices of the finite elements in Ω_1 and on Γ_c resulting from the discretization of the bilinear form $a_1(\boldsymbol{u}_1, \boldsymbol{v}_1)$. The right hand side of (3.15) contains the load vector $\boldsymbol{F}$, given by the discretization of the boundary integral $f(\boldsymbol{v}_1)$ on the *Neumann* boundary Γ_t.
The coupling equation (3.14) generates the integral operators

$$\boldsymbol{S} = < \boldsymbol{R}_h \, , \mathrm{N}_h > \tag{3.16}$$

and

$$\boldsymbol{T} = < \boldsymbol{R}_H \, , \mathrm{N}_c > . \tag{3.17}$$

Under consideration of the solution of the *Neumann* problem (3.12) the operator $\boldsymbol{H}$ is defined by

$$\boldsymbol{H} = \boldsymbol{S}\, \boldsymbol{A}^{-1}\, \boldsymbol{B}. \tag{3.18}$$

$\boldsymbol{H}$ is the interface coupling matrix which corresponds to the symmetric energy bilinear form on the continuous level

$$a_2(\boldsymbol{u}_2, \boldsymbol{u}_2) = \int\limits_{\Omega_2} \sigma_{ij}(\boldsymbol{u}_2)\, e_{ij}(\boldsymbol{u}_2)\, d\Omega = < \boldsymbol{t}_+ \, , \boldsymbol{u}_+ > . \tag{3.19}$$

On the discrete level we have the equivalence

$$< \boldsymbol{t}_h \, , \boldsymbol{u}_h > = \boldsymbol{\beta}^T\, \boldsymbol{H}\, \boldsymbol{\beta}. \tag{3.20}$$

The symmetry and definiteness of $\boldsymbol{H}$ depends on the numerical accuracy for the computation of the *Poincaré–Steklov* operator (3.11). In Section 6 we will develop an iterative procedure to control this accuracy. With a sufficient small numerical error we obtain the symmetric, but indefinite system (3.15). This structure can be transformed into a positive definite system by eliminating the parameter $\boldsymbol{\beta}$ with the *Schur* complement, so that we get with

$$\overline{\boldsymbol{K}}_{cc} = \boldsymbol{K}_{cc} + \boldsymbol{T}^T\, \boldsymbol{H}^{-1}\, \boldsymbol{T} \tag{3.21}$$

a symmetric and positive definite global stiffness relation

$$\begin{bmatrix} \boldsymbol{K}_{ii} & \boldsymbol{K}_{ic} \\ \boldsymbol{K}_{ci} & \overline{\boldsymbol{K}}_{cc} \end{bmatrix} \begin{bmatrix} \boldsymbol{\alpha}_i \\ \boldsymbol{\alpha}_c \end{bmatrix} = \begin{bmatrix} \boldsymbol{F}_i \\ \boldsymbol{F}_c \end{bmatrix} . \tag{3.22}$$

System (3.22) can be solved by standard direct or iterative solvers for FE systems.

4 Trial functions for the traction field

In this section we will derive an explicit form of the side conditions for the space W of the traction field on Γ_c (see Eq. (3.4)) for the 2D case.
Let us consider a piecewise smooth boundary Γ_c of the subdomain Ω_2 with ne_c smooth subboundaries generated by the coarse grid elements H. In each element l $(1 \leq l \leq ne_c)$ we define

$$\boldsymbol{t}_H^{(l)} = \boldsymbol{M}^{(l)} \overline{\boldsymbol{\beta}}^{(l)} \tag{4.1}$$

with a polynomial interpolation matrix $\boldsymbol{M}^{(l)}$ and the free parameters $\overline{\boldsymbol{\beta}}^{(l)}$. If we use polynoms of second degree for the interpolation we receive

$$\overline{\boldsymbol{\beta}}^{(l)} = [\beta_1^{(l)} \quad \beta_2^{(l)} \quad \beta_3^{(l)} \quad \beta_4^{(l)} \quad \beta_5^{(l)} \quad \beta_6^{(l)}]^T \tag{4.2}$$

and

$$\boldsymbol{M}^{(l)} = \begin{bmatrix} \boldsymbol{g}_\beta^T & \boldsymbol{0} \\ \boldsymbol{0} & \boldsymbol{g}_\beta^T \end{bmatrix} \tag{4.3}$$

with

$$\boldsymbol{g}_\beta^T = \left[\frac{1}{2}\zeta\,(\zeta-1) \quad 1-\zeta^2 \quad \frac{1}{2}\zeta\,(\zeta+1)\right] = \left[m_1(\zeta) \quad m_2(\zeta) \quad m_3(\zeta)\right]. \tag{4.4}$$

The interpolation functions are defined in a reference element $\zeta \in [-1,1]$. The superposition

$$\boldsymbol{t}_H = \sum_{l=1}^{ne_c} \boldsymbol{t}_H^{(l)} = \sum_{l=1}^{ne_c} \boldsymbol{M}^{(l)} \overline{\boldsymbol{\beta}}^{(l)} \tag{4.5}$$

leads to double nodes at the corners between two elements as shown in the right part of Fig. 3.1.
To ensure that the static equilibrium (2.15) and *Cauchys* symmetry condition

$$\sigma_{ij} = \sigma_{ji} \tag{4.6}$$

are exactly fulfilled in Ω_2 we have to modify the polynomial ansatz (4.5). In the first step we formulate global side conditions, which are sufficient, if we have a completely smooth boundary Γ_c without double nodes. Condition (2.15) leads for the trace of the stress tensor to an equilibrium of forces

$$< \boldsymbol{t}\,, \boldsymbol{I} >= 0 \qquad \text{or} \qquad \sum_{l=1}^{ne_c} \int_{\Gamma_l} t_i^{(l)}\, L_l \, d\Gamma = 0 \quad , \quad i = 1,2, \tag{4.7}$$

where L_l is the length of the element l. From the symmetry condition (4.6) we get on the boundary an equilibrium condition of momentums

$$< \boldsymbol{t} \times \boldsymbol{r}\,, \boldsymbol{I} >= 0 \qquad \text{or} \qquad \sum_{l=1}^{ne_c} \int_{\Gamma_l} \left(x_1^{(l)}\, t_2^{(l)} - x_2^{(l)}\, t_1^{(l)}\right) L_l \, d\Gamma = 0. \tag{4.8}$$

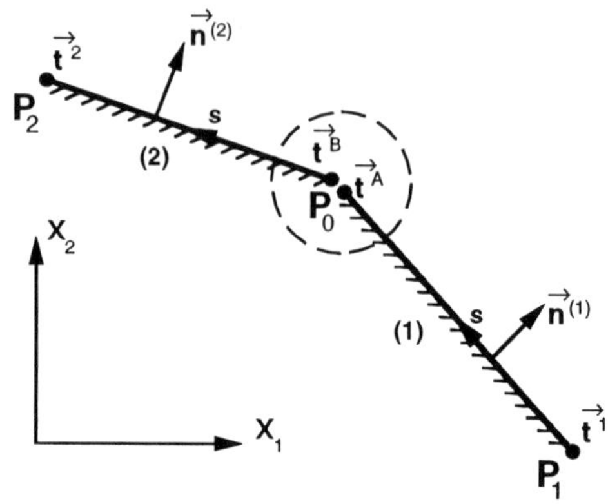

Fig. 4.1. Boundary Γ_c at the corner point P_0

Additionally we have to take under consideration local side conditions if we have a boundary Γ_c with corners that cause the generation of double nodes (see point P_0 in Fig. 4.1). The corner is characterized by a jump of the normal unit vector $\boldsymbol{n}$. With the definition of the traction vector

$$t_k = \sigma_{km}\, n_m \tag{4.9}$$

we can formulate the components of the stress tensor in P_0 depending on the trace vectors $\boldsymbol{t}^A$ and $\boldsymbol{t}^B$

$$\sigma_{1k} = \frac{1}{d}\left(n_2^{(2)}\, t_k^A - n_2^{(1)}\, t_k^B\right), \tag{4.10}$$

$$\sigma_{2k} = \frac{1}{d}\left(n_1^{(1)}\, t_k^B - n_1^{(2)}\, t_k^A\right); \quad k = 1, 2. \tag{4.11}$$

In (4.10) and (4.11) the abbreviation

$$d = n_1^{(1)}\, n_2^{(2)} - n_2^{(1)}\, n_1^{(2)} \tag{4.12}$$

is used. By inserting (4.10) in the divergence theorem (2.15) we receive two equilibrium conditions

$$n_2^{(2)}\, \frac{\partial t_i^A}{\partial x_1} - n_2^{(1)}\, \frac{\partial t_i^B}{\partial x_1} + n_1^{(1)}\, \frac{\partial t_i^B}{\partial x_2} - n_1^{(2)}\, \frac{\partial t_i^A}{\partial x_2} = 0 \quad , \quad i = 1, 2. \tag{4.13}$$

The symmetry condition (4.6) delivers one more local side condition

$$n_2^{(2)}\, t_2^A - n_2^{(1)}\, t_2^B = n_1^{(1)}\, t_1^B - n_1^{(2)}\, t_1^A. \tag{4.14}$$

Chaudonneret [2] has also introduced compatibility equations for the traction formulation at corner points. In the formulation of *Chaudonneret* the stress tensor is required to be symmetric and additionally the volume dilatation has to be constant as a kinematic side condition. In comparison with *Chaudonneret* the equations (4.13) and (4.14) define compatibility conditions only depending on the traction field.

In summary we have to modify the polynomial traction ansatz (4.5) by fulfilling the global equilibrium conditions (4.7),(4.8) and additionally in each corner point the local conditions (4.13),(4.14). This leads to a decrease of the number of free parameters $\overline{\boldsymbol{\beta}} \to \boldsymbol{\beta}$ according to the number of side conditions. We end up with the expansion

$$\boldsymbol{t}_H = \boldsymbol{R}_H \, \boldsymbol{\beta}. \tag{4.15}$$

For details concerning the elimination of free parameters we refer to *Carmine* [1].

5 Calculation of boundary integrals

For the numerical realization of the *Poincaré–Steklov* operator (3.11) we have to calculate integrals of the type

$$I^{(k)}(\boldsymbol{\xi}_i) = \int_{\Gamma_k} K(\boldsymbol{\xi}_i, \overline{\boldsymbol{x}}) \, \varphi(\overline{\boldsymbol{x}}) \, d\Gamma(\overline{\boldsymbol{x}}) \tag{5.1}$$

with the i.g. regular shape function $\varphi(\overline{\boldsymbol{x}})$ and the strong or weak singular kernel function $K(\boldsymbol{\xi}_i, \overline{\boldsymbol{x}})$, depending if we have to integrate the operator $\boldsymbol{K}$ or $\boldsymbol{V}$. Fig. 5.1 shows the different configurations in the two or three dimensional case.

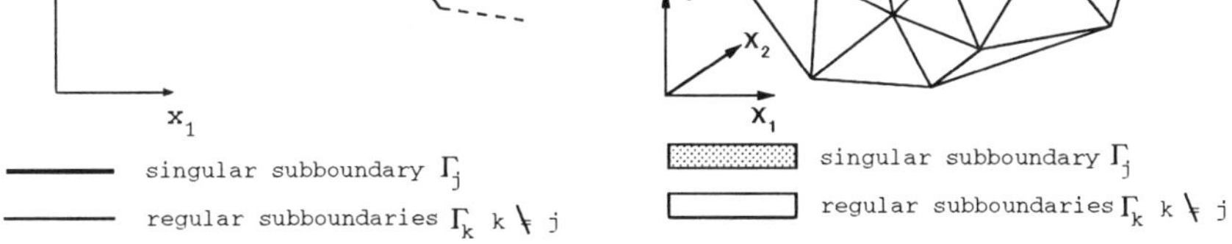

Fig. 5.1. Geometrical approximation in $\mathbb{R}^2$ and $\mathbb{R}^3$

We restrict ourself to the case that the evaluation point $\boldsymbol{\xi}_i$ is always located in the interior of a boundary element, so that we have no projection of a singularity to neighbour elements. This can be easily realized by using the iterative solution scheme for the *Poincaré–Steklov* operator descriped in Section 6. Another simplification is the use of only plaine elements, so that a curved surface has to be discretized by a sufficient number of plain elements.

5.1 Singular integrals

We give an overview on the methods used to solve integrals in the case $\boldsymbol{\xi}_i \in \Gamma_j$, that means that the evaluation point is located inside the integration area. If the shape function $\varphi(\overline{\boldsymbol{x}})$ is a constant or a linear polynom the integrals can be solved analytically. In $\mathbb{R}^2$ we have to solve line integrals. In the case of weak singular integrals ($\sim \ln r$) we calculate the limit of an improper integral, singular integrals ($\sim r^{-1}$) can be easily handled by using the *Cauchy* principal value. In $\mathbb{R}^3$ analytical solutions of the weak and strong singular boundary integrals over triangular elements were derived by *Cruse* [5] and *Li, Han, Mang, Torzick* [19].
In the case of using arbitrary regular shape functions $\varphi(\overline{\boldsymbol{x}})$ (for example if we deal with so called augmented elements) we also implemented a numerical version for the solution of singular integrals. In $\mathbb{R}^2$ the weak singular integrals can be solved by using a nonlinear coordinate transformation introduced by *Telles* [32]. The strong singular integrals are treated with the Clenshaw–Curtis algorithm (*Clenshaw, Curtis* [3], *Davis, Rabinowitz* [6]). In $\mathbb{R}^3$ the weak singular integrals over a triangle can be regularized by the transformation to polar coordinates. The remaining regular integral can be solved by a standard *Gauss* quadrature. For the strong singular integrals we also use polar coordinates and calculate the finite part of the integral in radial direction with the method of *Kutt* [17]. More general and efficient algorithms for this problem are published by *Guiggiani, Gigante* [8] and *Kieser, Schwab, Wendland* [15].

For details of the formulation and implementation we refer to *Türke* [34].

5.2 Regular integrals

For the calculation of regular boundary integrals ($\boldsymbol{\xi}_i \notin \Gamma_j$) we use a general numerical algorithm with an adaptive mesh refinement due to an estimate of the integration error. This strategy was introduced by *Lachat* and *Watson* [18]. Here we will explain the basic features for the integration over triangular boundary elements Γ_k.
Applying the coordinate transformations depicted in Fig. 5.2 on Γ_k we receive

$$I^{(k)}(\xi_i) = \int_{\Gamma_k} K(\boldsymbol{\xi}_i, \overline{\boldsymbol{x}})\, \varphi(\overline{\boldsymbol{x}})\, d\Gamma(\overline{\boldsymbol{x}}) = \int_{\Gamma_k} \hat{K}(\boldsymbol{\xi}_i, \overline{\boldsymbol{x}})\, d\Gamma(\overline{\boldsymbol{x}}) = \tag{5.2}$$

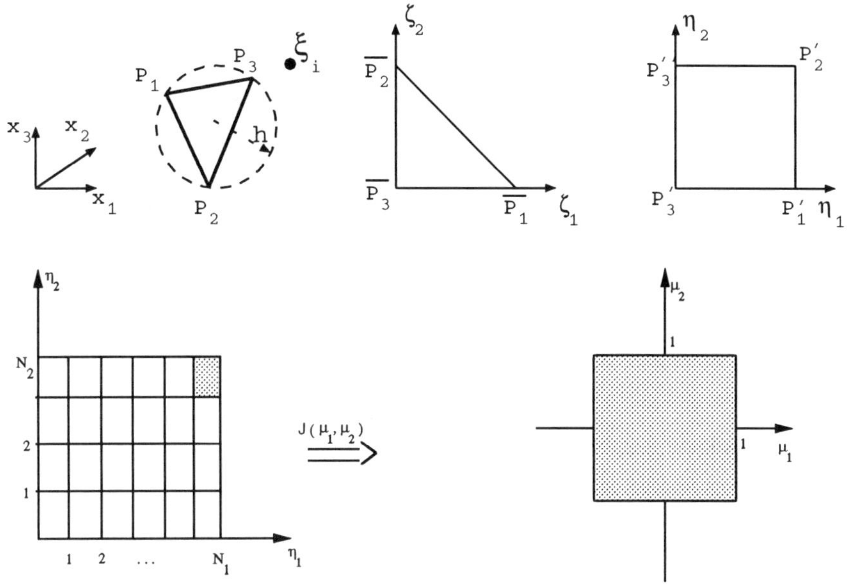

Fig. 5.2. Transformation of the boundary element Γ_k

$$= \sum_{m_1=1}^{N_1} \sum_{m_2=1}^{N_2} \int_{-1}^{+1} \int_{-1}^{+1} \hat{K}(\boldsymbol{\xi}_i, \overline{\boldsymbol{x}}(\mu_1, \mu_2)) \, J_1(\zeta_1, \zeta_2) \, J_2(\eta_1, \eta_2) \, J_3(\mu_1, \mu_2) \, d\mu_1 \, d\mu_2 .$$

The two line integrals over the unit square in the (μ_1, μ_2) system can be approximated by a two dimensional standard *Gauss* quadrature formula. If $A(\mu_1, \mu_2)$ is the abbreviation for the whole kernel function in the last part of (5.2) we finally obtain the approximation

$$I^{(k)}(\xi_i) \approx Q^{(k)}(\xi_i) = \sum_{m_1=1}^{N_1} \sum_{m_2=1}^{N_2} \sum_{\lambda=1}^{n_2} \sum_{\kappa=1}^{n_1} A(\mu_1(\kappa), \mu_2(\lambda)) \, w_\kappa \, w_\lambda . \tag{5.3}$$

To find appropriate values for n_1, n_2, N_1 and N_2 we use the error estimate derived by *Stroud* and *Secrest* [30]

$$\left| \int_{-1}^{+1} \int_{-1}^{+1} A(\mu_1, \mu_2) \, d\mu_1 \, d\mu_2 - \sum_{\lambda=1}^{n_2} \sum_{\kappa=1}^{n_1} A(\mu_1(\kappa), \mu_2(\lambda)) \, w_\kappa \, w_\lambda \right| \leq 2 \sum_{p=1}^{2} e_p \, H_p \tag{5.4}$$

with

$$\left| \frac{\partial^{2n_p} A(\mu_1, \mu_2)}{\partial {\mu_p}^{2n_p}} \right| \leq H_p \quad ; \quad e_p \approx \frac{G}{2^{2n_p} (2\, n_p)!} . \tag{5.5}$$

G is an appropriate constant. Under consideration of some simplifications (cf [34]) we obtain for kernel functions with the characetristic (r^{-2}) the estimate

$$(2n_p + 1)\left(\frac{L_p}{2\,d_{min}}\right)^{2n_p} \leq C_{int} \qquad ; \qquad p = 1, 2. \tag{5.6}$$

L_p is the characteristic length of the actual subelement in μ_p direction and d_{min} is the minimal Euklidian distance between the evaluation point $\boldsymbol{\xi}_i$ and the actual subelement. For a free chosen upper bound of the admissible numerical integration error C_{int} formula (62) can be used to determine the required number of subelements N_1 and N_2 and the number of *Gauss* points n_1 and n_2 in each subelement.

Numerical tests (cf [34]) have shown that the estimate (5.6) is too pessimistic. *Schwab* and *Wendland* have found in [29] that the error estimate (5.4) is only correct in η_1 or μ_1 direction but has to be replaced by a derivative free error estimate in η_2 or μ_2 direction. This is due to the *Duffy* coordinate transformation $J_2(\eta_1, \eta_2)$ in η_2 direction. Appropriate derivative free error estimates were introduced by *Rabinowitz* and *Richter* [23]. For the given problem we obtain a precise numerical integration scheme by the determination of n_1 with formula (5.6) for a given bound C_{int}. The number n_2 of *Gauss* points in η_2 direction has to be chosen by

$$n_2 \geq \frac{|\ln h|}{\ln \delta}\, n_1 \,. \tag{5.7}$$

Here h is one half of the element length (cf Fig. 5.2) of the integration element Γ_k in the real space (x_1, x_2, x_3) and δ is a geometrical value depending on the tangential vectors $\chi_{|1}$ and $\chi_{|2}$ at the corner point P_3 of Γ_k which has the smallest distance d_{min} to the evaluation point $\boldsymbol{\xi}_i$. For details we refer to [29], [34].

This numerical tool is the base for a precise and efficient numerical treatment of the *Poincaré–Steklov* operator.

6 A Neumann series for elasticity problems

In this chapter we will derive an iterative scheme for the numerical solution of the singular boundary integral equation (2.17). The base for this method goes back to the classical *Neumann* series for the *Laplace* equation by *C. Neumann* [21] in 1877. In 1965 and 1969 *W.L. Wendland* [35] and *W. Haack, W.L. Wendland* [9] used the expansion of the *Laplace* equation in a *Neumann* series with singular potentials for the construction of a numerical algorithm.

We start with the interior domain problem in the subdomain $\Omega_2 \subset \mathbb{R}^3$ (cf Figure 2.1) with given *Neumann* data $\boldsymbol{t}_+$ on the boundary Γ_c. For a consistent notation we change the index $\Omega_2 \to \Omega_+$.
As a result we want to obtain the solution of the differential equation

$$\Delta^* \boldsymbol{u}(\boldsymbol{\xi}) = 0 \quad \forall \quad \boldsymbol{\xi} \in \Omega_+ . \tag{6.1}$$

Remark 1: In the following deriviations we use the work of *Kupradze* et.al. [16] who requires for the solution $\boldsymbol{u}(\boldsymbol{\xi})$ of Equation (6.1) the smoothness

$$\boldsymbol{u} \in C^1(\overline{\Omega}_+) \cap C^2(\Omega_+) \tag{6.2}$$

with $\overline{\Omega}_+ = \Omega_+ \cup \Gamma_c$. For the boundary Γ_c *Kupradze* requires

$$\Gamma_c \in \Pi_1(\alpha) \quad , \quad \alpha \geq 0, \tag{6.3}$$

where $\Pi_1(0)$ is the class of smooth surfaces and $\Pi_1(\alpha)$, $\alpha > 0$ is the class of the *Liapunov* surfaces. An extension of these requirements to *Lipschitz* surfaces was derived by *Costabel* [4]. *Costabel* uses the regularity results of *Nečas* [20] for the *Neumann* and the *Dirichlet* problem. If we choose for the *Neumann* problem the tractions on Γ_c from the *Sobolev* space $\boldsymbol{t}_+ \in H^{-1/2+\sigma}(\Gamma_c)$ with $\sigma \in [0, \frac{1}{2}]$, we obtain the displacement field on the boundary in $\boldsymbol{u} \in H^{+1/2+\sigma}(\Gamma_c)$. Because *Bettis* theorem remains valid for *Lipschitz* boundaries, we can apply the energy relations of *Kupradze* also on general *Lipschitz* boundaries.

The solution of (6.1) can be obtained as follows: The *Calderon* projector (2.16) in the singular formulation gives a mapping of the boundary values on Γ_c:

$$\boldsymbol{u}(\boldsymbol{\xi}) = \big(\frac{1}{2}\boldsymbol{I} - \boldsymbol{K}\big)\, \boldsymbol{u}(\boldsymbol{\xi}) + \boldsymbol{V}\, \boldsymbol{t}_+(\boldsymbol{\xi}) \quad \forall \quad \boldsymbol{\xi} \in \Gamma_c = \partial\Omega_+. \tag{6.4}$$

With (6.4) we can calculate the boundary displacements on Γ_c so that all boundary values are known. The displacement field in the interior domain Ω_+ can be calculated with *Somiglianas* formula.
The solution $\boldsymbol{u}(\boldsymbol{\xi})$ in the interior domain Ω_+ can be extended to the exterior domain $\boldsymbol{\xi} \in \Omega_-$ with $\Omega_- = \mathbb{R}^3 \backslash \overline{\Omega_+}$ by *Somiglianas* formula because all integrals for radius $R \to \infty$ disappear. This is due to the assymptotic behavior of the singular fundamental solutions. While the displacements are continuous at the interface Γ_c we obtain a discontinuity of the tractions $\boldsymbol{t}_+ \neq \boldsymbol{t}_-$ known as jump relation of the double layer potential.

As a result we obtain a continuous displacement field $\boldsymbol{u}(\boldsymbol{\xi})$ in the full space $\boldsymbol{\xi} \in \mathbb{R}^3 = \overline{\Omega_+ \cup \Omega_-}$. By applying the stress operator on the displacement field $\boldsymbol{u}(\boldsymbol{\xi})$ the stress field $\boldsymbol{\sigma}(\boldsymbol{\xi})$ in the interior and the exterior domain is achieved as first derivatives of the displacement field.

Definition 1: The linear function space X is defined as the space of elastic states that includes all stress and deformation states in the interior domain Ω_+ and in the exterior domain Ω_- caused by a loading of Ω_+ on Γ_c with an admissible traction field $\boldsymbol{t}_+$. Admissible traction vectors are charcterized by the requirement that $\boldsymbol{t}_+$ leads to an equilibrium system in Ω_+ (cf Equation (2.15)).

For the explicit formulation of $\boldsymbol{t}_+$ we refer to Chapter 4.

Definition 2: We call the static field $\boldsymbol{\sigma}(\boldsymbol{\xi})$ and the kinematic field $\boldsymbol{u}(\boldsymbol{\xi})$ in the exterior and the interior domain the representatives of the elastic states and denote the elements of the space X in general with $\boldsymbol{g} \in X$.

If the elastic state is represented by the displacement field $\boldsymbol{u}(\boldsymbol{\xi})$, the solution is not unique due to the constant vector $\boldsymbol{c}$, that includes the rigid body motions. In $\mathbb{R}^3$ $\boldsymbol{c}$ consists of three translatoric and three rotatoric components.

Definition 3: We define X' as a subspace of X, that contains the vector $\boldsymbol{c}$ of the free constants: $\boldsymbol{c} \in X'$. Each element $\boldsymbol{g} \in X$ is associated with a class of elements

$$[\boldsymbol{g}] = \{\boldsymbol{g} + \boldsymbol{c} \, : \, \boldsymbol{c} \in X'\} \tag{6.5}$$

and is denoted as a linear manifold. As a result two linear manifolds are equal if

$$\boldsymbol{g}_1 - \boldsymbol{g}_2 \in X'.$$

The set of linear manifolds $[\boldsymbol{g}]$ also defines a linear solution space and is called the factor space

$$\overset{\circ}{X} = X \,/\, X'. \tag{6.6}$$

The solution $\boldsymbol{u}(\boldsymbol{\xi})$ is now defined in the factor space $\overset{\circ}{X}$.

With a scalar product $(\boldsymbol{g}, \boldsymbol{g})$ and a norm

$$\|\boldsymbol{g}\| = \sqrt{(\boldsymbol{g}, \boldsymbol{g})} \tag{6.7}$$

of the elements $\boldsymbol{g}$ of the space $\overset{\circ}{X}$ the elastic states define a real *Hilbert* space, since it can be shown (cf *Rieder* [24]) that the space $\overset{\circ}{X}$ is a real, linear, closed and complete function space.

Definition 4: For two displacement fields $\overline{\boldsymbol{u}}(\boldsymbol{\xi})$ and $\overline{\overline{\boldsymbol{u}}}(\boldsymbol{\xi})$ with $\boldsymbol{\xi} \in \overline{\Omega_+ \cup \Omega_-}$ the scalar product of the two associated elastic states $\overline{\boldsymbol{g}}$ and $\overline{\overline{\boldsymbol{g}}}$ is defined by the positive semidefinite elastic energy in the interior and the exterior domain (cf Equation (2.10))

$$(\overline{\boldsymbol{g}}, \overline{\overline{\boldsymbol{g}}}) = a_+(\overline{\boldsymbol{u}}, \overline{\overline{\boldsymbol{u}}}) + a_-(\overline{\boldsymbol{u}}, \overline{\overline{\boldsymbol{u}}}) = \int\limits_{\Omega_+} \sigma_{ij}(\overline{\boldsymbol{u}})\, e_{ij}(\overline{\overline{\boldsymbol{u}}})\, d\Omega + \int\limits_{\Omega_-} \sigma_{ij}(\overline{\boldsymbol{u}})\, e_{ij}(\overline{\overline{\boldsymbol{u}}})\, d\Omega. \tag{6.8}$$

By applying the first *Greens* formula on (6.8) the domain integrals can be transformed to integrals on the interface Γ_c. We obtain

$$(\overline{\boldsymbol{g}}, \overline{\overline{\boldsymbol{g}}}) = \int\limits_{\Gamma_c} t_{+\,j}(\overline{\boldsymbol{u}})\, \overline{\overline{u}}_j \, d\Gamma - \int\limits_{\Gamma_c} t_{-\,j}(\overline{\boldsymbol{u}})\, \overline{\overline{u}}_j \, d\Gamma. \tag{6.9}$$

In the function space $\overset{\circ}{X}$ we investigate the properties of the projection (6.4) with a given traction vector $\boldsymbol{t}_+(\boldsymbol{\xi})$ and an unknown boundary displacement vector $\boldsymbol{u}(\boldsymbol{\xi})$.

The following spectral analysis of (6.4) refers to *Kupradze* et.al. [16] and *Taylor* [31]. For a consistent notation we define

$$\mathcal{T} = \left(\frac{1}{2}\boldsymbol{I} - \boldsymbol{K}\right) = -\int_S [T(\partial_y, n)\,\Gamma(y-x)]' \ldots d_y S. \tag{6.10}$$

The formulation $-\int_S \ldots d_y S$ is used by *Kupradze* (cf p. 351, Equation (2.1), (2.2)). The displacement vector $\boldsymbol{u}(\boldsymbol{\xi})$ may be the representative of the elastic state. From (6.4) we receive the formulation

$$\boldsymbol{g} - \lambda \mathcal{T} \boldsymbol{g} = \boldsymbol{V}\,\boldsymbol{t}_+ \tag{6.11}$$

with

$$\lambda = +1 \tag{6.12}$$

which is consistent to *Kupradze* (p. 351, Equation (I)). The associated eigenvalue problem (cf *Kupradze*, p. 362, Equation. (4.4)) is defined by

$$\boldsymbol{g} - \lambda \mathcal{T} \boldsymbol{g} = \boldsymbol{0}. \tag{6.13}$$

With

$$\mu = \frac{1}{\lambda} \tag{6.14}$$

and

$$\boldsymbol{g}^* = \mu \boldsymbol{V}\,\boldsymbol{t}_+ \tag{6.15}$$

we obtain from (6.11)

$$\boldsymbol{g}^* = (\mu - \mathcal{T})\,\boldsymbol{g} \tag{6.16}$$

which is consistent to *Taylor* (p. 253). Since the function space $\overset{\circ}{X}$ is a complete and closed space, the domain $D(\mathcal{T})$ and the range $W(\mathcal{T})$ of the operator $\mathcal{T}$ lies in $\overset{\circ}{X}$. We denote according to *Taylor* $\mathcal{T} \in [\overset{\circ}{X}]$. The solution $\boldsymbol{g}$ can be calculated with (6.16) by

$$\boldsymbol{g} = (\mu - \mathcal{T})^{-1}\,\boldsymbol{g}^* \quad \text{with} \quad \mu = 1. \tag{6.17}$$

The existence and uniqueness of the solution of this *Neumann* problem in $\mathbb{R}^3$ is shown by *Kupradze* for the continuity conditions in Remark 1. Now we investigate, if (6.17) can also be obtained by the expansion in a *Neumann* series. This is possible, if a resolvent R_μ with

$$(\mu - \mathcal{T})^{-1} = R_\mu = \sum_{m=1}^{\infty} \mu^{-m}\,\mathcal{T}^{m-1} \tag{6.18}$$

can be represented by a convergent series in the space $\overset{\circ}{X}$.

Definition 5: If the scalar μ in (6.16) is of the kind that the range of $(\mu - \mathcal{T})$ is dense in $\overset{\circ}{X}$ and if $(\mu - \mathcal{T})$ has a continuous inverse then μ is in the set of resolvents $\rho(\mathcal{T})$ of $\mathcal{T}$. All values of μ, that are not in $\rho(\mathcal{T})$ are called the spectrum $\sigma(\mathcal{T})$ of $\mathcal{T}$.

Definition 6: If $\sigma(\mathcal{T})$ is bounded and not empty, we define

$$r_\sigma(\mathcal{T}) = \sup_{\mu \in \sigma(\mathcal{T})} |\mu| \tag{6.19}$$

and call $r_\sigma(\mathcal{T})$ the spectral radius of $\mathcal{T}$.

In the following we use the property that $\overset{\circ}{X}$ is a real *Banach* space (cf *Rieder* [24]). With this property we can formulate a sufficient convergence criteria for the expansion (6.18) in Theorem 1:

Theorem 1: If $\mathcal{T} \in [\overset{\circ}{X}]$ and $\overset{\circ}{X}$ is a (complex) *Banach* space, the resolvent R_μ is given by

$$R_\mu = \sum_{m=1}^{\infty} \mu^{-m} \mathcal{T}^{m-1}, \tag{6.20}$$

if $|\mu| > r_\sigma(\mathcal{T})$ holds. This series also represents R_μ, if the series is convergent and $|\mu| = r_\sigma(\mathcal{T})$. The series is divergent for $|\mu| < r_\sigma(\mathcal{T})$.

The proof of Theorem 1 can be found in *Taylor*, p. 262.
To proof the convergence of the series (6.20) and the existence of a solution for problem (6.16) with a *Neumann* series, we have to show according to Theorem 1 with $\mu = +1$ that the spectral radius holds

$$r_\sigma < 1. \tag{6.21}$$

Since $\mu = 1/\lambda$ (cf (6.14)) the convergence criteria (6.21) is equivalent with the condition that all eigenvalues of the eigenvalue problem (6.13) hold

$$\lambda_i > 1 \quad \forall \quad i = 1, 2, \dots . \tag{6.22}$$

The validity of inequality (6.22) for the integral operator $\mathcal{T}$ can be shown by considering three theorems of *Kupradze* [16] (pp. 362–364):

Theorem 2: All eigenvalues λ of the kernel of the integral operator $\mathcal{T}$ are real.

Theorem 3: The eigenvalues λ are not smaller than unity in absolute value.

Proof: From Theorem 2 we have $\lambda \in \mathbb{R}$. As a consequence the eigenvalues can be determined by the formula

$$|\lambda| = \frac{a_+(\boldsymbol{u},\boldsymbol{u}) + a_-(\boldsymbol{u},\boldsymbol{u})}{|a_-(\boldsymbol{u},\boldsymbol{u}) - a_+(\boldsymbol{u},\boldsymbol{u})|} =$$

$$= \frac{\int\limits_{\Omega_+} \sigma_{ij}(\boldsymbol{u})\, e_{ij}(\boldsymbol{u})\, d\Omega + \int\limits_{\Omega_-} \sigma_{ij}(\boldsymbol{u})\, e_{ij}(\boldsymbol{u})\, d\Omega}{|\int\limits_{\Omega_-} \sigma_{ij}(\boldsymbol{u})\, e_{ij}(\boldsymbol{u})\, d\Omega - \int\limits_{\Omega_+} \sigma_{ij}(\boldsymbol{u})\, e_{ij}(\boldsymbol{u})\, d\Omega|} = \tag{6.23}$$

$$= \frac{\int\limits_{\Gamma_c} t_{+\,j}(\boldsymbol{u})\, u_j\, d\Gamma - \int\limits_{\Gamma_c} t_{-\,j}(\boldsymbol{u})\, u_j\, d\Gamma}{|-\int\limits_{\Gamma_c} t_{-\,j}(\boldsymbol{u})\, u_j\, d\Gamma - \int\limits_{\Gamma_c} t_{+\,j}(\boldsymbol{u})\, u_j\, d\Gamma|}.$$

Because the energy bilinear forms a_+ and a_- are positive semidefinite we find

$$|\lambda| \geq 1, \tag{6.24}$$

and the proof of Theorem 3 is complete.

We investigate the cases $\lambda = +1$ and $\lambda = -1$:

a) $\lambda = +1$: According to formula (6.23) for the eigenvalues in terms of an energy quotient, for $\lambda = +1$ we must have either $a_+(\boldsymbol{u},\boldsymbol{u}) = 0$ or $a_-(\boldsymbol{u},\boldsymbol{u}) = 0$. From the physical point of view this is equivalent to a vanishing state of deformation in the interior or the exterior domain, that means a rigid body motion of the interior or the exterior domain. Because the solutions are defined in the factor space $\overset{\circ}{X}$ (cf (6.6)) the rigid body motion is equivalent to a zero element in $\overset{\circ}{X}$. It follows that $\lambda = +1$ is a regular value of the problem (6.11).
b) $\lambda = -1$: Since the energy bilinear forms $a(\boldsymbol{u},\boldsymbol{u})$ are positive semidefinite, negative eigenvalues are not possible according to the formula (6.23).

Theorem 4: The eigenvalues λ are simple poles of the resolvent.

As a consequence of Theorem 4 all eigenvalues are isolated and $\lambda = +1$ is not an accumulation point of the eigenvalues.

In summary we can conclude from Theorem 2,3 and 4:

$$\lambda_i \geq \lambda_{i-1} \geq \ldots \geq \lambda_1 > 1. \tag{6.25}$$

As a consequence the convergence criteria (6.22) holds.

Additionally with $\mu = 1/\lambda$ for $\lambda = +1$ the inequality (6.21) is valid and according to Theorem 1 we can compute the solution $\boldsymbol{g}$ by the convergent *Neumann* series

$$\boldsymbol{g} = (1 - \mathcal{T})^{-1}\, \boldsymbol{g}^* = \sum_{m=1}^{\infty} \mathcal{T}^{m-1}\, \boldsymbol{g}^*. \tag{6.26}$$

With the initial notation of the *Calderon* projector (6.4) we obtain the *Neumann* series

$$\boldsymbol{u}(\boldsymbol{\xi}) = \lim_{k\to\infty} \boldsymbol{u}^{(k)}(\boldsymbol{\xi}) = \sum_{m=1}^{\infty} \left(\frac{1}{2}\boldsymbol{I} - \boldsymbol{K}\right)^{m-1} \boldsymbol{V}\,\boldsymbol{t}_+(\boldsymbol{\xi}) \quad \forall \boldsymbol{\xi} \in \Gamma_c. \tag{6.27}$$

From that a recursive scheme for approximations $\boldsymbol{u}^{(k)}(\boldsymbol{\xi})$ can be formulated by

$$\boldsymbol{u}^{(k)}(\boldsymbol{\xi}) = \left(\frac{1}{2}\boldsymbol{I} - \boldsymbol{K}\right) \boldsymbol{u}^{(k-1)}(\boldsymbol{\xi}) + \boldsymbol{V}\,\boldsymbol{t}_+(\boldsymbol{\xi}) \tag{6.28}$$

with the initial value

$$\boldsymbol{u}^{(0)} = \boldsymbol{0}. \tag{6.29}$$

With the formulation (6.28), (6.29) we can implement a numerical scheme as shown in the following chapter.

7 Numerical results

First we want to study the convergence behaviour of the iteration rule (6.28) derived in Section 6. Therefore we analyze a *Neumann* problem, for example a thick plate under constant tension σ_0 with a circular cutout (cf Figure 7.1).

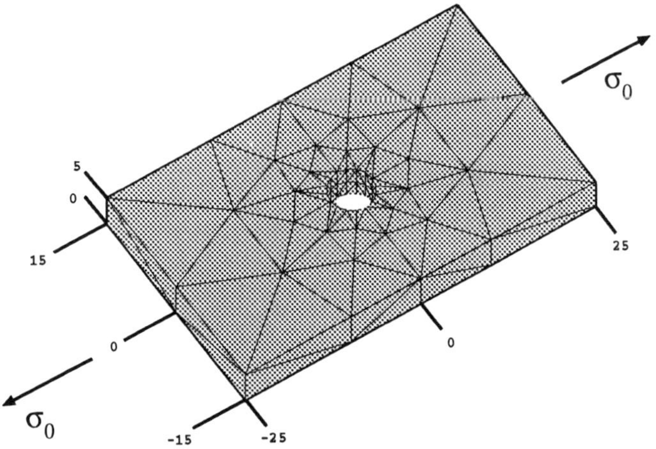

Fig. 7.1. Discretized plate under tension

The surface of the plate is discretized by 224 triangular boundary elements. With a trivial start solution we calculate the surface displacement field iteratively by using the expansion (6.28).

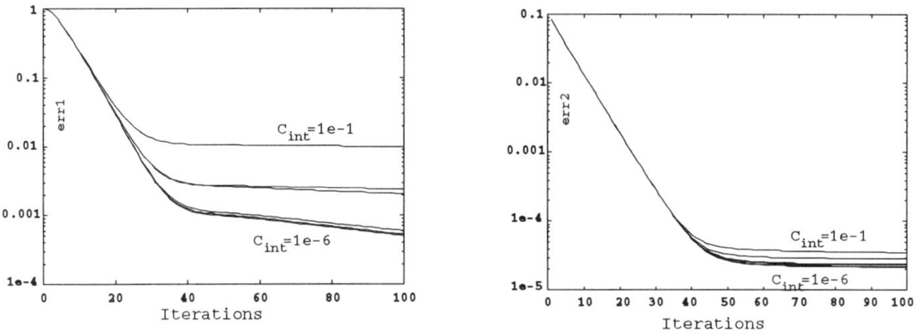

Fig. 7.2. Convergence of the *Dirichlet* data

Fig. 7.2 shows two different error norms of the displacement field while increasing the number of iterations (k) with the definitions

$$err1 = \frac{|\boldsymbol{u}^{FEM}(\boldsymbol{P}) - \boldsymbol{u}_h(\boldsymbol{P})|}{\boldsymbol{u}^{FEM}(\boldsymbol{P})} \quad ; \quad err2 = \frac{\|\boldsymbol{u}_h^{(k+1)} - \boldsymbol{u}_h^{(k)}\|_0}{\|\boldsymbol{u}_h^{(k)}\|_0}. \tag{7.1}$$

In the left part of Fig. 7.2 we compute a characteristic displacement vector $\boldsymbol{u}_h$ on the notch surface in $\boldsymbol{P}$ with the *Neumann* series and compare it with a reverence value $\boldsymbol{u}^{FEM}$ generated by a FEM analysis of the same problem with a very fine grid (30237 degrees of freedom). In the right part we compute the L_2 norm of the numerical displacement field $\boldsymbol{u}_h$ on the whole surface of the plate and observe the variation from iteration (k) to $(k+1)$. The parameter C_{int} denotes different upper bounds for the integration error (cf Section 5).

Since we observe a good convergence behaviour of the *Neumann* series in the first example we can incorporate this iterative algorithm in the coupling method. In the next numerical test we analyze a two dimensional notch problem with domain decomposition under consideration of the compatibility conditions for the traction field derived in Section 4. Fig. 7.3 shows the macro–

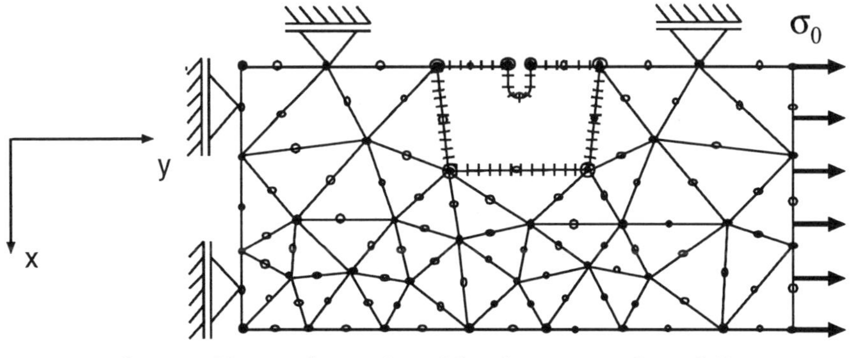

Fig. 7.3. Plate under tension with substructures Ω_1 and Ω_2

element in the near field of the notch and the discrization of the skeleton with triangular finite elements. For the numerical experiments we use 187 fine grid elements for the discretization of the coupling interface Γ_c. As an error indicator we define a relative symmetry defect

$$MSD = \frac{\sum_{i=1}^{q} \sum_{j=1}^{i-1} \mid (h_{ij} - h_{ji}) \mid}{\sum_{i=1}^{q} \sum_{j=1}^{i-1} \mid (h_{ij} + h_{ji}) \mid} 100\,(\%) \tag{7.2}$$

to control the symmetry of the coupling operator $\boldsymbol{H}$ (cf Eq. (3.20)). q denotes the dimension of the quadratic matrix $\boldsymbol{H}$ with the elements (h_{ij}). A second error indicator to control the definiteness of $\boldsymbol{H}$ is the smallest eigenvalue λ_{min}. For the notch problem with an infinite long plate in two directions we can solve the problem with the *Kolosov Mushelishwili* representation. This leads us to an analytical solution for the stress peak value

$$\alpha_{exact} = 4.7632, \tag{7.3}$$

which is used as comparison with the numerical results, see *Hahn* [10]. To compare the numerical and the exact values of the stress concentration we define the relative error

$$\varepsilon_\alpha = \frac{|\alpha_{num} - \alpha_{exact}|}{\alpha_{exact}}. \tag{7.4}$$

As shown in Table 7.1, we have a high accuracy for the stress peak after 38 iteration with a deviation of only 0.26%.

Table 7.1. Stress peak, symmetry defect, lowest eigenvalue for notch problem

iter	1	10	20	25	30	36	37	38
α_{num}	0.54	3.53	4.32	4.49	4.61	4.73	4.74	4.7505
ε_α [%]	88.7	25.9	9.30	5.74	3.21	0.70	0.49	0.26
MSD $[10^{-4}]$	5.05	5.45	6.17	6.36	6.50	6.61	6.63	6.64
$\lambda_{\min}$ $[10^{-2}]$	0.99	2.83	3.19	3.26	3.30	3.335	3.338	3.341

We can control automatically $\lambda_{\min}$ and MSD, so that we can control local stability of the method providing high accuracy and reliability. In Table 1 one can see the dependence of the minimum eigenvalue and the symmetry defect of the problem on the number of iterations; $\lambda_{\min}$ increases with the number of iterations in the first phase rapidly while the symmetry defect remains on a low level between $4 \cdot 10^{-4}$ and $7 \cdot 10^{-4}$. This allows us to define a priori an upper bound for MSD and a lower bound for $\lambda_{\min}$. Then the algorithm automatically increases the fine grid density and the number of iteration steps for the *Neumann* series until the given requirements of symmetry and stability are fulfilled.

For further numerical investigations and an extension of the coupling method to three dimensional problems we refer to [14], [34] and [13].

8 Summary and conclusions

The numerical experiments have shown that the coupling method gives the possibility to obtain a high resolution of stress concentrations with only a coarse global FE discretization by using singular integral operators. With the *Neumann* series approach we are able to control the numerical characters of the coupling operator as symmetry and stability within an adaptive algorithm.

For future developments the method can be used to analyze nonlinear problems in continuum mechanics iteratively and to solve coupled problems in general field theory.

References

1. Carmine, R. (1989): Ein Kopplungsverfahren von FEM und BEM zur Berechnung ebener Spannungskonzentrationsprobleme. Doctoral Thesis, Karlsruhe University
2. Chaudonneret, M. (1977): Sur la discontinité du vecteur contrainte dans les calculs de structure par la méthode des équations intégrales. CRAS, Paris, Série A 284, 463–466
3. Clenshaw, C.W., Curtis, A.R. (1960): A method for numerical integration on an automatic computer. Numer. Math. 2, 197–205
4. Costabel, M., Wendland, W.L. (1986): Strong ellipticity of boundary integral operators. Krelles Journal für die Reine und Angewandte Mathematik 372, 34–63
5. Cruse, T.A. (1969): Numerical Solutions in Three Dimensional Elastostatics. Int. J. Solids Struct. 5, 1259–1273
6. Davis, P.J., Rabinowitz, P. (1984): Methods of numerical integration. Second edition, Academic Press, New York
7. Grannell, J.J. (1987): On simplified hybrid methods for coupling of finite elements and boundary elements. In: Brebbia, C.A., Wendland, W.L., Kuhn, G. (eds.) Boundary elements IX, Vol. 1. Springer, Berlin, pp. 447–460
8. Guiggiani, M., Gigante, A. (1990): A general algorithm for multidimensional Cauchy principal value integrals in the boundary element method. ASME J. Applied Mechanics 57, 907–915
9. Haack, W., Wendland, W.L. (1969): Vorlesungen über Partielle und Pfaffsche Differentialgleichungen. Birkhäuser Verlag, Basel Stuttgart
10. Hahn, H.G. (1963): Über den Einflußdes Flankenwinkels auf die Spannungskonzentration an Kerben. Doctoral Thesis, University of Munich
11. Hsiao, G.C. (1988): The Coupling of BEM and FEM – A Brief Review. In: Brebbia, C.A. et al. (eds.) Boundary Elements X, Vol. 1. Springer, Berlin, pp. 431-445
12. Hsiao, G.C. (1990): The Coupling of Boundary Element and Finite Element Methods. ZAMM, Z. angew. Math. Mech. 70, 493–503
13. Hsiao, G.C., Schnack, E., Wendland, W.L. (1995): A Hybrid Coupled Finite–Boundary Element Method. Preprint 95–11, Universit"at Stuttgart, Mathematisches Institut A

14. Karaosmanoglu, N. (1989): Kopplung von Randelement– und Finite–Element–Verfahren für dreidimensionale elastische Strukturen. Doctoral Thesis, Karlsruhe University
15. Kieser, R., Schwab, C., Wendland, W.L. (1992): Numerical evaluation of singular and finite part integrals on curved surfaces using symbolic manipulation. Computing 49, 279–301
16. Kupradze, V.D., Gegelia, T.G., Basheleishvili, M.O., Burchuladze, T.V. (1979): Three dimensional problems of the mathematical theory of elasticity and thermoelasticity. North–Holland Series in Applied Mathematics and Mechanics, Vol. 25, North–Holland Publishing Company, Amsterdam New York Oxford
17. Kutt, H.R. (1975): The numerical evaluation of principal value integrals by finite–part integration. Numer. Math. 24, 205–210
18. Lachat, J.C., Watson, J.O. (1976): Effective Numerical Treatment of Boundary Integral Equations: A Formulation for Three- Dimensional Elastostatics. Int. J. Num. Meth. Engng. 10, 991–1005
19. Li, H.B., Han, G.H., Mang, A.H., Torzick, P. (1986): A New Method for Coupling of Finite Elements and Boundary Elements Discretized Subdomains of Elastic Bodies. Computer Methods in Applied Mechanics Engng. 54, 161–185
20. Nečas, J. (1967): Les méthodes directes en théorie des équations elliptiques. Masson, Paris
21. Neumann, C. (1877): Untersuchungen über das logarithmische und Newtonsche Potential. Teubner, Leipzig
22. Polizzotto, C., Zito, M. (1994): Variational Formulations for Coupled BE/FE Methods in Elastostatics. ZAMM, Z. angew. Math. Mech. 74, 533–543
23. Rabinowitz, P., Richter, N. (1970): New error coefficients for estimating quadrature errors for analytic functions. Math. Comp. 24, 561–570
24. Rieder, G. (1962): Iterationsverfahren und Operatorgleichungen in der Elastizitätstheorie. Abhandlungen der Braunschweigischen Wissenschaftlichen Gesellschaft, Band XIV, Fried. Vieweg & Sohn Verlag, Braunschweig
25. Schnack, E. (1973): Beitrag zur Berechnung rotationssymmetrischer Spannungskonzentrationsprobleme mit der Methode der Finiten Elemente. Doctoral Thesis, Technical University of Munich
26. Schnack, E. (1985): Stress Analysis with a Combination of HSM and BEM. Proceedings of the MAFELAP 1984 Conference on "The Mathematics of Finite Elements and Applications", Academic Press, pp. 273–281
27. Schnack, E. (1987): A Hybrid BEM-Model. Int. J. Num. Meth. Engng. 24, No. 5, 1015–1025
28. Schnack, E. (1990): Macro-Elements for 2D– and 3D–Elasticity with BEM. In: Tanaka, M., Brebbia, C.A., Honma, T. (eds.) Boundary Elements XII, Vol. I. Springer, Berlin Heidelberg, pp. 21–31
29. Schwab, C., Wendland, W.L. (1992): On numerical cubatures of singular surface integrals in boundary element methods. Numer. Math. 62, 343–369
30. Stroud, A.H., Secrest, D. (1966): Gaussian quadrature formulas. Prentice–Hall Inc., Englewood Cliffs, New York
31. Taylor, A.E. (1958): Introduction to functional analysis. John Wiley & Sons, New York London Sydney
32. Telles, J.C.F. (1987): A Self- Adaptive Co-ordinate Transformation for Efficient Numerical Evaluation of General Boundary Element Integrals. Int. J. Num. Meth. Engng. 24, 959-973
33. Tong, P., Pian, T.H.H., Lasry, S.J. (1973):: A hybrid element approach to crack problems in plane elasticity. Int. J. Num. Meth. Engng. 7, 297–308

34. Türke, K. (1995): Eine Zweigitter–Methode zur Kopplung von FEM and BEM. Doctoral Thesis, University of Karlsruhe
35. Wendland, W.L. (1965): Lösung der ersten und zweiten Randwertaufgaben des Innen– und Außengebietes für die Potentialgleichung im $\mathbb{R}^3$ durch Randbelegungen. Doctoral Thesis, Technical University of Berlin

Fast Algorithms for the Solution of Pseudodifferential Equations

D. Berthold and B. Silbermann

1 Introduction

The paper deals with the fast solution of pseudodifferential equations $Au = f$ of order zero given on a certain curve Γ. Such equations often arise from boundary value problems. Let, for example, $\Omega \subset \mathbb{R}^2$ be a finite domain having a smooth closed boundary Γ and let us given a partial differential equation (or a system of such equations) in Ω together with boundary conditions on Γ. The *Boundary Element Method* (BEM) being based on an integral formulation of the original problem is a powerful tool for its solution. Since we have to discretize only the boundary Γ, the BEM is even more efficient than the Finite Difference Method or the Finite Element Method for some classes of problems.

In many cases the BEM produces an integral equation having a kernel singularity of the type x^{-1}, a so–called Cauchy type singular integral equation given on the smooth closed curve Γ

$$a(t)u(t) + \frac{b(t)}{\pi i} \int_\Gamma \frac{u(\tau)d\tau}{\tau - t} + \int_\Gamma k(t,\tau)u(\tau)d\tau = f(t), \quad t \in \Gamma, \tag{1.1}$$

where $k(t,\tau)$ is smoother than $(\tau - t)^{-1}$. The present paper is mainly devoted to the solution of such kind of equations.

As a second example, we consider the case when Γ is an interval of the real axis, say $\Gamma = [-1, 1]$. Then we shall call equation (1.1) the *airfoil equation*, because it plays an important role in aerodynamics, but also in fracture and crack mechanics and elasticity theory.

There is an enormous number of papers dealing with approximate methods for solving (1.1) and with methods for the solution of the algebraic systems of equations arising from the approximation methods. The most familiar approximation methods are Galerkin, collocation, and quadrature methods. However, from these methods there always arise algebraic systems having unstructurized and non–sparse coefficient matrices. Thus, for the first look we need $O(n^3)$ operations in order to get n unknowns of the approximate solution. But in this situation, the multigrid methods developed in the last 15 years prove to be an advantageous alternative to the direct methods, see, for example, [11, 12, 22, 2]. Using multigrid strategies we can reduce the computational complexity to $O(n^2)$ operations.

Moreover, one can develope strategies whose amount of computational work is only $O(n \log n)$. Such a strategy being based on the *Fast Fourier Transformation* was described in [1] for the first time. In this paper, equation (1.1) with Γ being a smooth closed curve has been solved by a special approximation method with trigonometric ansatz functions and its convergence rate has been shown to be optimal in the scale of the usual Sobolev spaces $\mathbf{H}^s(\Gamma)$, $s \in \mathbb{R}$. Here and in the following the term *optimal* means that if u_n denotes the approximate solution of $Au = f$ the error $\|u - u_n\|_{\mathbf{H}^t(\Gamma)}$ behaves like $O(n^{t-s})$ for all t : $t_0 \leq t \leq s$ with a certain $t_0 \in \mathbb{R}$, supposed that $f \in \mathbf{H}^s(\Gamma)$. Unfortunately, we can get error estimates only for $t_0 > \frac{1}{2}$ from the strategy in [1]. On the other hand, just in the case of the BEM, one is very interested in a good convergence behaviour of the approximate solution u_n in Sobolev norms with $t < 0$, because the error of the approximate solution of the original boundary value problem can be estimated (even in uniform norms in the interior of Ω) by the $\mathbf{H}^t(\Gamma)$–norm on the boundary Γ.

In the author's paper [4], there is proposed the so–called *corrected collocation method* as an alternative to the qualocation method (see [7, 23, 24]) for equation (1.1). Both methods aim at the improvement of the approximation error in "negative" Sobolev norms. However, fast algorithms for the solution are not available. In the present paper we give an algorithm called *Parametrix–Galerkin method* being based on the corrected collocation method and on the strategy of [1]. Its computational complexity is only $O(n \log n)$ and it has a better convergence behaviour in "negative" Sobolev norms than the collocation method. Further, we describe some other fast algorithms for equation (1.1) given on smooth closed curves Γ or on the interval $[-1, 1]$ and prove their convergence in several scales of function spaces, which we define in Section 2. In Section 3 we deal with special properties of equation (1.1). The "classical" approximation methods are described in Section 4, whereas in Section 5 the fast algortihms are proposed. All of the algorithms consist in two steps. In the first step, a parametrix for the pseudodifferential operator A is used in order to get an approximate solution on a fine grid. In a second step this solution is corrected in a certain manner by the solution on a coarse grid. Mostly, we only give an overview of the algorithms since the results are partly published yet, see [2, 3, 4, 5]. The error estimates in Hölder–Zygmund norms (Subsection 5.3) are new. With some remarks on further investigations (Section 6) and numerical results (Section 7) we close our paper.

2 Scales of Spaces

Let us introduce some function spaces. All what follows is well–known and can be found, for example, in [16, 26, 27]. For Γ being a smooth closed curve let $\mathbf{L}^2(\Gamma)$ be the Lebesgue space on Γ provided with the norm

$$\|u_\Gamma\|_{\mathbf{L}^2(\Gamma)} := \left(\int_\Gamma |u_\Gamma(x)|^2 |dx|\right)^{\frac{1}{2}},$$

where $|dx|$ refers to the arc length differential. With the help of the (regular) 1-periodic parametric representation $\gamma : \mathbb{R} \to \Gamma$ of the curve Γ we have a one-to-one correspondence between functions u_Γ defined on Γ and 1-periodic functions u defined on $\mathbb{R}$. We put $u(x) = u_\Gamma(\gamma(x)),\ x \in \mathbb{R}$. We denote by $\mathbf{L}^2$ the Hilbert space of all square integrable 1-periodic functions endowed with the scalar product and the norm

$$\|u\|_{\mathbf{L}^2} := \sqrt{(u,u)}, \qquad (u,v) := \int_0^1 u(x)\overline{v(x)}dx,$$

respectively. Note that $\|u_\Gamma\|_{\mathbf{L}^2(\Gamma)} = \|u\sqrt{\gamma'}\|_{\mathbf{L}^2}$. Further, let $\mathbf{C}$ be the Banach space of all 1-periodic continuous functions on $\mathbb{R}$ and $\mathbf{C}^\infty$ the set of all 1-periodic infinitely differentiable functions, where

$$\|u\|_{\mathbf{C}} := \max\{|u(x)| :\ x \in \mathbb{R}\} = \max\{|u(x)| :\ x \in [0,1)\}$$

defines the (maximum) norm in $\mathbf{C}$. We denote by $\mathbf{H}^s, s \in \mathbb{R}$, the periodic Sobolev space of order s, i.e. the completion of $\mathbf{C}^\infty$ with respect to the norm

$$\|u\|_{\mathbf{H}^s} := \left(|(u,e_0)|^2 + \sum_{k\in\mathbb{Z}^*} |k|^{2s}|(u,e_k)|^2\right)^{\frac{1}{2}}, \tag{2.1}$$

where the expressions (u, e_k) are the usual Fourier coefficients of the function u with respect to the orthonormal system $\{e_k\}_{k\in\mathbb{Z}}$ of the trigonometric polynomials $e_k(x) = e^{2\pi ikx}$ and $\mathbb{Z}^* := \mathbb{Z}\backslash\{0\}$. Here are some basic properties of the periodic Sobolev spaces.

- The $\mathbf{H}^s$ are Hilbert spaces with the scalar product $(u,v)_{\mathbf{H}^s} = (u,e_0)\overline{(v,e_0)} + \sum_{k\in\mathbb{Z}^*} |k|^{2s}(u,e_k)\overline{(v,e_k)}$.
- The space $\mathbf{H}^{-s}$, $s \in \mathbb{R}$, is the dual space to $\mathbf{H}^s$ and the relation $\|u\|_{\mathbf{H}^{-s}} = \sup\{|(u,w)| :\ w \in \mathbf{H}^s, \|w\|_{\mathbf{H}^s} = 1\}$ holds.
- If $t < s$ then $\mathbf{H}^s$ is a dense subset of $\mathbf{H}^t$ and $\|u\|_{\mathbf{H}^t} \le \|u\|_{\mathbf{H}^s}$ for each $u \in \mathbf{H}^s$. Moreover, the imbedding $\mathbf{H}^s \subset \mathbf{H}^t$ is compact. For this reason, the family $\{\mathbf{H}^s\}_{s\in\mathbb{R}}$ proves to be a Hilbert scale.
- If $s > \frac{1}{2}$ then $\mathbf{H}^s$ is continuously imbedded in $\mathbf{C}$. Furthermore, $\mathbf{H}^s$ forms an algebra on which the operator aI of multiplication is continuous.

Let us introduce another scale of functions spaces. We denote by $\mathcal{D}$ the operator of (distributional) differentation and put $\Delta_h^1 u(x) := u(x+h) - u(x)$ and $\Delta_h^k u = \Delta_h^1(\Delta_h^{k-1}u),\ k = 2,3,\ldots$ With the help of Δ_h^k we define the Hölder–Zygmund seminorm by

$$[u]_\gamma := \begin{cases} \sup_{h>0} \dfrac{\|\Delta_h^1 u\|_{\mathbf{C}}}{h^\gamma}, & 0 < \gamma < 1, \\[2ex] \sup_{h>0} \dfrac{\|\Delta_h^2 u\|_{\mathbf{C}}}{h}, & \gamma = 1, \end{cases}$$

and the Hölder–Zygmund space $\mathcal{H}^s$, $s = m + \gamma$, $m \in \mathbb{N} \cup \{0\}$, $0 < \gamma \leq 1$, by

$$\mathcal{H}^s := \{u \in \mathbf{C} : \ \mathcal{D}^k u \in \mathbf{C}; \ k = 0, ..., m; \ [\mathcal{D}^m u]_\gamma < \infty\}$$

endowed with the norm

$$\|u\|_{\mathcal{H}^s} := \sum_{k=0}^{m} \|\mathcal{D}^k u\|_{\mathbf{C}} + [\mathcal{D}^m u]_\gamma.$$

Here we list some properties of the Hölder–Zygmund spaces.

- The $\mathcal{H}^s$ are Banach spaces.
- If $0 < t < s$ then $\mathcal{H}^s$ is a dense subset of $\mathcal{H}^t$ and $\|u\|_{\mathcal{H}^t} \leq \|u\|_{\mathcal{H}^s}$ for each $u \in \mathcal{H}^s$. Moreover, the imbedding $\mathcal{H}^s \subset \mathcal{H}^t$ is compact.
- For each $s > 0$, the space $\mathcal{H}^s$ forms an algebra on which the operator aI of multiplication is continuous.

Finally, we are going to introduce a scale of (nonperiodic) spaces defined on the interval $(-1, 1)$. Let $\rho(x) = (1-x)^\alpha (1+x)^\beta$, $\alpha, \beta > -1$, be a Jacobi weight function, $x \in (-1, 1)$, and $\mathbf{L}^2_\rho = \mathbf{L}^2_\rho(-1, 1)$ be the Hilbert space of all square integrable functions on $(-1, 1)$ with respect to the Jacobi weight ρ endowed with the scalar product and the norm

$$\|u\|_{\mathbf{L}^2_\rho} := \sqrt{(u, u)_\rho}, \qquad (u, v)_\rho := \int_{-1}^{1} u(x)\overline{v(x)}\rho(x)dx,$$

respectively. We denote by $\mathbf{C}[a, b]$ the Banach space of all continuous functions on the interval $[a, b]$ endowed with the maximum norm

$$\|u\|_{\mathbf{C}[a,b]} := \max\{|u(x)| : \ x \in [a, b]\}.$$

Further, let $\{p^\rho_n\}_{n=0}^\infty$ be the sequence of the orthonormal Jacobi polynomials (with respect to the Jacobi weight ρ) with positive leading coefficients and $\deg p_n = n$. For any real number $s \geq 0$ we define the Sobolev type subspace $\mathbf{L}^2_{\rho,s}$ of $\mathbf{L}^2_\rho$ as follows

$$\mathbf{L}^2_{\rho,s} := \{u \in \mathbf{L}^2_\rho : \|u\|_{\mathbf{L}^2_{\rho,s}} < \infty\},$$

where

$$\|u\|_{\mathbf{L}^2_{\rho,s}} := \left(|(u, p^\rho_0)_\rho|^2 + \sum_{k=1}^{\infty} k^{2s} |(u, p^\rho_k)_\rho|^2 \right)^{\frac{1}{2}}. \tag{2.2}$$

The expressions $(u, p^\rho_k)_\rho$ can be interpreted as the Fourier coefficients of u with respect to the orthonormal system $\{p^\rho_k\}_{k \geq 0}$ of the Jacobi polynomials . Note that $\mathbf{L}^2_{\rho,0} = \mathbf{L}^2_\rho$. Here we list some properties of the spaces $\mathbf{L}^2_{\rho,s}$ (see [3]):

- The $\mathbf{L}^2_{\rho,s}$ are Hilbert spaces with the scalar product $(u, v)_{\mathbf{L}^2_{\rho,s}} = (u, p^\rho_0)_\rho \overline{(v, p^\rho_0)_\rho} + \sum_{k=1}^\infty k^{2s} (u, p^\rho_k)_\rho \overline{(v, p^\rho_k)_\rho}$.

- If $0 \le t < s$ then $\mathbf{L}^2_{\rho,s}$ is a dense subset of $\mathbf{L}^2_{\rho,t}$ and $\|u\|_{\mathbf{L}^2_{\rho,t}} \le \|u\|_{\mathbf{L}^2_{\rho,s}}$ for each $u \in \mathbf{L}^2_{\rho,s}$. Moreover, the imbedding $\mathbf{L}^2_{\rho,s} \subset \mathbf{L}^2_{\rho,t}$ is compact. For this reason, the family $\{\mathbf{L}^2_{\rho,s}\}_{s\ge 0}$ proves to be a Hilbert scale.
- If $s > \frac{1}{2}$ then the functions $u \in \mathbf{L}^2_{\rho,s}$ are in $\mathbf{C}[a,b]$ for all $[a,b] \subset (-1,1)$. Moreover, if $\alpha, \beta \le \frac{1}{2}$ then $\mathbf{L}^2_{\rho,s}$ is continuously imbedded in $\mathbf{C}[-1,1]$. (More precisely, there is a continuous representative in the equivalent class containing u.)

If $s \in \mathbb{N}$ then the spaces $\mathbf{L}^2_{\rho,s}$ can be described in another and very simple way ([3]). Indeed, the space $\mathbf{L}^2_{\rho,s}$ is the same as the subspace of all functions $u \in \mathbf{L}^2_{\rho}$ having the property that $\|\mathcal{D}^k u \cdot \varphi^k\|_{\mathbf{L}^2_\rho}$ is finite for all $k = 0, ..., s$, where φ is defined by $\varphi(x) := \sqrt{1-x^2}$. Moreover, the norm $\|u\|_{\mathbf{L}^2_{\rho,s}}$ is equivalent to the norm $\sum_{k=0}^{s} \|\mathcal{D}^k u \cdot \varphi^k\|_{\mathbf{L}^2_\rho}$.

3 Pseudodifferential equations

3.1 Pseudodifferential equations on smooth closed curves

Let Γ be a smooth closed curve in the complex plane and let be given on Γ a pseudodifferential equation of order $\beta = 0$

$$A_\Gamma u_\Gamma = f_\Gamma, \quad f_\Gamma \in \mathbf{L}^2(\Gamma). \tag{3.1}$$

The Cauchy singular integral operator on Γ

$$S_\Gamma u_\Gamma(t) = \frac{1}{\pi i} \int_\Gamma \frac{u_\Gamma(\tau) d\tau}{\tau - t}, \quad t \in \Gamma, \tag{3.2}$$

is an example of such an operator. Let $\mathbb{T} := \{z \in \mathbb{C} : |z| = 1\}$ be the unit circle. The action of $S_\mathbb{T}$ on $\mathbf{L}^2(\mathbb{T})$ can be described (in terms of Fourier coefficients) by

$$S_\mathbb{T}(\textstyle\sum_{k\in\mathbb{Z}} c_k t^k) \mapsto \sum_{k\ge 0} c_k t^k - \sum_{k<0} c_k t^k.$$

Obviously, $S_\mathbb{T}$ has the property $S_\mathbb{T}^2 = I$. For this reason the operators $P_\mathbb{T} = (I + S_\mathbb{T})/2$ and $Q_\mathbb{T} = (I - S_\mathbb{T})/2$ prove to be projections on $\mathbf{L}^2(\mathbb{T})$. Note that, for given functions $a_\mathbb{T}$ and $b_\mathbb{T}$ on $\mathbb{T}$, the singular integral operator $a_\mathbb{T} I + b_\mathbb{T} S_\mathbb{T}$ equals $c_\mathbb{T} P_\mathbb{T} + d_\mathbb{T} Q_\mathbb{T}$, where $c_\mathbb{T} = a_\mathbb{T} + b_\mathbb{T}$, $d_\mathbb{T} = a_\mathbb{T} - b_\mathbb{T}$. Using the well–known Agranovič representation of any periodic pseudodifferential operator of order 0

$$A_\Gamma = c_\mathbb{T} P_\mathbb{T} + d_\mathbb{T} Q_\mathbb{T} + K, \tag{3.3}$$

with K being a certain compact operator, we can restrict our investigations to the case when $\Gamma = \mathbb{T}$. Thus, we only have to deal with singular integral operators with compact perturbation $A_\mathbb{T} = a_\mathbb{T} I + b_\mathbb{T} S_\mathbb{T} + K$. Finally, using the parametrization $\gamma : \mathbb{R} \to \mathbb{T}$, $y \mapsto e^{2\pi i y}$ we get

$$\begin{aligned} S_{\mathbb{T}} u_{\mathbb{T}}(t) &= 2\int_0^1 \frac{u_{\mathbb{T}}(e^{2\pi iy})\, e^{2\pi iy}}{e^{2\pi iy} - e^{2\pi ix}} dy \\ &= \int_0^1 u(y)\, [1 + i\, \cot \pi(x-y)]\, dy =: Su(x), \quad x \in [0,1), \quad (3.4) \end{aligned}$$

where $u(y) = u_{\mathbb{T}}(e^{2\pi iy})$. Thus, in the sequel we shall investigate the singular integral equation

$$Au = au + bSu + Ku = f, \quad f \in \mathbf{L}^2, \tag{3.5}$$

being equivalent to (3.1). Note that S acts on trigonometric polynomials according to the rule $Se_k = e_k,\ k \geq 0$ and $Se_k = -e_k,\ k < 0$.

For two Banach spaces $\mathbf{X}$ and $\mathbf{Y}$ let $\mathcal{L}(\mathbf{X},\mathbf{Y})$ be the Banach space of all bounded linear operators $A : \mathbf{X} \to \mathbf{Y}$ and $\mathcal{K}(\mathbf{X},\mathbf{Y})$ the set of all compact operators $K : \mathbf{X} \to \mathbf{Y}$. Further, the symbol $\mathrm{G}\mathcal{L}(\mathbf{X},\mathbf{Y})$ denotes the open subset of $\mathcal{L}(\mathbf{X},\mathbf{Y})$ of all operators $A \in \mathcal{L}(\mathbf{X},\mathbf{Y})$ with im $A = \mathbf{Y}$ having an inverse $A^{-1} \in \mathcal{L}(\mathbf{Y},\mathbf{X})$. If $\mathbf{X} = \mathbf{Y}$ we simply write $\mathcal{L}(\mathbf{X}), \mathrm{G}\mathcal{L}(\mathbf{X})$, and $\mathcal{K}(\mathbf{X})$, respectively.

We are going to study some properties of A in Sobolev spaces and Hölder–Zygmund spaces. Put $c := a + b$ and $d := a - b$ and denote by $[a, S]$ the commutator $aS - Sa$.

Lemma 3.1 ([4, 19, 20, 21]). *Let $\delta > 0$ be an arbitrarily small real number.*

(i) If $a \in \mathbf{H}^r,\ r \geq 0$, then $[a,S] \in \mathcal{K}(\mathbf{H}^s, \mathbf{H}^r)$ for all $s > \frac{1}{2}$.
(ii) If $a \in \mathcal{H}^r,\ r > 0$, then $[a,S] \in \mathcal{K}(\mathcal{H}^s, \mathcal{H}^{r-\delta})$ for all $s > 0$. Moreover, if $r < 1$ then $[a,S] \in \mathcal{K}(\mathcal{H}^s, \mathcal{H}^r)$ for all $s > 0$.

Denote by P and Q the analogue of $P_{\mathbb{T}}$ and $Q_{\mathbb{T}}$ in the space $\mathbf{L}^2$. They can be defined by $Pu = \sum_{k \geq 0} (u, e_k)\, e_k$ and $Q = I - P$. Obviously, P and Q act on trigonometric polynomials as

$$Pe_l = \begin{cases} e_l, & l \geq 0, \\ 0, & l < 0, \end{cases} \qquad Qe_l = \begin{cases} 0, & l \geq 0, \\ e_l, & l < 0. \end{cases} \tag{3.6}$$

It is easy to see that $aS - Sa = 2QaP - 2PaQ$. Thus $[a,S]$ is compact if and only if both QaP and PaQ are so.

Lemma 3.2 ([21]).

(i) If $c, d \in \mathbf{H}^r,\ r > \frac{1}{2}$, are invertible then $B := c^{-1}P + d^{-1}Q \in \mathcal{L}(\mathbf{H}^s)$ for all $s \leq r$.
(ii) If $c, d \in \mathcal{H}^r,\ r > 0$, are invertible then $B \in \mathcal{L}(\mathcal{H}^s)$ for all $s :\ 0 < s \leq r$.

The operator B is a so-called parametrix for the operator A. Define $T := I - BA$. Since $T = c^{-1}(QcP - PdQ) + d^{-1}(PdQ - QcP) - BK$, one has that

(i) $T \in \mathcal{K}(\mathbf{H}^s, \mathbf{H}^r)$ for all $s > \frac{1}{2}$ if $K \in \mathcal{K}(\mathbf{H}^s, \mathbf{H}^r)$ and
(ii) $T \in \mathcal{K}(\mathcal{H}^s, \mathcal{H}^{r-\delta})$ for all $s > 0$ if $K \in \mathcal{K}(\mathcal{H}^s, \mathcal{H}^{r-\delta})$

under the assumptions of Lemma 3.2.

3.2 The airfoil equation on the interval

Let us consider the Cauchy-type singular integral equation with constant real coefficients a and $b \neq 0$ and an additional integral term with logarithmic singularity on $\mathbb{J} = [-1, 1]$, the so-called airfoil equation,

$$av(x) + \frac{b}{\pi} \int_{-1}^{1} \frac{v(y)dy}{y-x} - \frac{\nu}{\pi} \int_{-1}^{1} \log|x-y|v(y)dy + \int_{-1}^{1} k(x,y)v(y)dy = f(x), \tag{3.7}$$

$x \in (-1, 1)$. For simplicity assume that $a^2 + b^2 = 1$. Further, ν is a complex constant, and k and f are given continuous complex-valued functions on $\mathbb{J} \times \mathbb{J}$ and $\mathbb{J}$, respectively.

Following [8, 15, 21], we form a Jacobi weight function $\sigma(x) = (1-x)^\alpha(1+x)^\beta$ in dependence on the coefficients a and b, where $\alpha = \lambda^+ + g$, $\beta = \lambda^- - g$, $g = \frac{1}{\pi}\arg(a - ib) \in (-1, 1)$, and λ^+, λ^- are two integers being chosen such that $-1 < \alpha, \beta < 1$. Doing this, the operator $A_0 := a\sigma I + bS_{\mathbb{J}}\sigma I$ possesses mapping properties which are very advantageous for applying approximation methods. Here I denotes the identity and $S_{\mathbb{J}}$ is the Cauchy singular integral operator defined by

$$S_{\mathbb{J}}u(x) = \frac{1}{\pi} \int_{-1}^{1} \frac{u(y)dy}{y-x}, \quad x \in (-1, 1). \tag{3.8}$$

Lemma 3.3 ([8, 15, 18]). *The operator A_0 turns out to be a Fredholm operator in the pair of weighted Lebesgue spaces $(\mathbf{L}^2_\sigma, \mathbf{L}^2_\mu)$, $\mu = \sigma^{-1}$, with $\kappa :=$ ind $A_0 = -(\lambda^+ + \lambda^-) \in \{-1, 0, 1\}$. Furthermore, A_0 maps polynomials into polynomials; more precisely, we have $A_0 p_n^\sigma(x) = p_{n-\kappa}^\mu(x)$, $x \in (-1, 1)$. The operator $\hat{A}_0 := a\mu I - bS_{\mathbb{J}}\mu I$ fulfils the inverse relation $\hat{A}_0 p_n^\mu(x) = p_{n+\kappa}^\sigma(x)$, $x \in (-1, 1)$.*

Having this in mind, we shall consider in the sequel the equation

$$Au = A_0 u + \nu W u + K u = f \tag{3.9}$$

being equivalent to (3.7), where $Wu(x) = -\frac{1}{\pi}\int_{-1}^{1} \log|x - y|u(y)\,\sigma(y)\,dy$, $v = u\sigma$ and $Ku(x) = \int_{-1}^{1} k(x,y)u(y)\sigma(y)dy$. Note that A is a Fredholm operator in the same pair of spaces as A_0 and that ind $A =$ ind A_0.

One particalar case of equation (3.7) is closely connected with the so-called Chebyshev polynomials of the first, second, third, and fourth kind. Indeed, if $a = 0$ then one gets that $|\alpha| = |\beta| = \frac{1}{2}$. Let Π_n denote the set of all algebraic polynomials of degree less than or equal to $n - 1$.

Lemma 3.4 ([2, 9]). *If $|\alpha| = |\beta| = \frac{1}{2}$ then the operator W is a bounded linear one from $\mathbf{L}^2_{\sigma,s}$ into $\mathbf{L}^2_{\mu,s+1}$ for all $s \geq 0$ and it is compact from $\mathbf{L}^2_\sigma$ into $\mathbf{C}[-1,1]$. Moreover, W sends polynomials from Π_n into $\Pi_{n+\alpha+\beta+1}$.*

If $a \neq 0$ then the investigation of the properties of W is more involved. For this reason, we restrict ourselves to the two cases when $a = 0$ or $\nu = 0$, respectively. Furthermore, in the sequel we only deal with $\lambda^+ = \lambda^- = \kappa = 0$. For $\kappa \neq 0$ see [3].

4 Approximation methods

4.1 Approximation methods for periodic pseudodifferential equations

For a natural number n, the operator P_n is defined for any $u \in \mathbf{L}^2$ by $P_n u = \sum_{k\in\mathbb{Z}_n}(u, e_k)\, e_k$, where $\mathbb{Z}_n := \{-n, -n+1, ..., n\}$. Note that P_n is the orthogonal projection onto the set of the trigonometric polynomials $\mathcal{T}_n := \text{lin}\{e_k,\ k \in \mathbb{Z}_n\}$. An equivalent definition for P_n is that $(u - P_n u, \pi_n) = 0\ \forall u \in \mathbf{L}^2\ \forall \pi_n \in \mathcal{T}_n$. Furthermore, we define the interpolation operators $R_n : \mathcal{R} \to \mathcal{T}_n$ by $R_n u(x_{n,j}) = u(x_{n,j}),\ j \in \mathbb{Z}_n$, where $\mathcal{R}$ denotes the class of all 1-periodic bounded and Riemann integrable functions on $\mathbb{R}$ and $x_{n,j} = j/(2n+1)$. In the following we shall denote by C several positive constants.

Lemma 4.1 (cf. [17, 20]).

(i) There exists a positive constant C depending only on t and s such that for any $u \in \mathbf{H}^s$ the following estimates hold

$$\begin{aligned} \|u - P_n u\|_{\mathbf{H}^t} &\leq C\, n^{t-s}\, \|u\|_{\mathbf{H}^s}, && \forall t: \quad t \leq s, \\ \|u - R_n u\|_{\mathbf{H}^t} &\leq C\, n^{t-s}\, \|u\|_{\mathbf{H}^s}, && \forall t: \quad 0 \leq t \leq s; \quad s > \tfrac{1}{2}. \end{aligned}$$

If $t = s$ then both norms on the left–hand side tend to zero (i.e. they are not only bounded).

(ii) There exists a positive constant C depending only on t and s such that for any $u \in \mathcal{H}^s$ the following estimates hold ($n \geq 2$)

$$\begin{aligned} \|u - P_n u\|_{\mathcal{H}^t} &\leq C\, n^{t-s} \log n\, \|u\|_{\mathcal{H}^s}, && \forall t: \quad 0 < t \leq s, \\ \|u - R_n u\|_{\mathcal{H}^t} &\leq C\, n^{t-s} \log n\, \|u\|_{\mathcal{H}^s}, && \forall t: \quad 0 < t \leq s. \end{aligned}$$

Now, let us describe some approximation methods for the pseudodifferential equation (3.5). We restrict our further investigations to equations with continuous coefficients c and d and compact operators $Ku(x) = \int_0^1 k(x,y)u(y)dy$, $x \in [0,1)$, having certain smoothing properties, because the fast algorithms in Section 5 below only work in this case. More precisely, we make the following assumptions on equation (3.5), which are assumed to be true throughout this paper:

(i) $c, d \in \mathbf{H}^r$, $r > \frac{1}{2}$, and $K \in \mathcal{K}(\mathbf{L}^2, \mathbf{H}^r)$ in the "Sobolev case" and
(ii)$c, d \in \mathcal{H}^r$, $r > 0$, and $K \in \mathcal{K}(\mathbf{C}, \mathcal{H}^r)$ in the "Hölder–Zygmund case".

Further, we suppose that $c(x) \neq 0$ and $d(x) \neq 0$ for all $x \in \mathbb{R}$ and that the winding number of c/d around the origin $[\arg c(x)/d(x)]_{x=0}^{x=1}$ equals zero. Then $cP + dQ$ is in $\mathrm{G}\mathcal{L}(\mathbf{H}^t)$, $t \leq r$ under condition (i) and in $\mathrm{G}\mathcal{L}(\mathcal{H}^t)$, $0 < t \leq r$ under condition (ii), [20].

The *Galerkin method* defines an approximate solution $y_n \in \mathcal{T}_n$ by

$$(Ay_n, \pi_n) = (f, \pi_n) \ \ \forall \pi_n \in \mathcal{T}_n, \text{ or, equivalently, by } P_n A y_n = P_n f, \tag{4.1}$$

whereas the *collocation method* determines $z_n \in \mathcal{T}_n$ by

$$Az_n(x_{n,j}) = f(x_{n,j}), \ \ j \in \mathbb{Z}_n, \text{ or, equivalently, by } R_n A y_n = R_n f. \tag{4.2}$$

The *quadrature method* consists in finding a $z_n \in \mathcal{T}_n$ by

$$c(x_{n,j})Pz_n(x_{n,j}) + d(x_{n,j})Qz_n(x_{n,j}) + K_n z_n(x_{n,j}) = f(x_{n,j}), \ \ j \in \mathbb{Z}_n, \tag{4.3}$$

where

$$\begin{aligned} K_n z_n(x) &= \tfrac{1}{2n+1} \textstyle\sum_{k\in\mathbb{Z}_n} k(x, x_{n,k}) z_n(x_{n,k}) \\ &= \int_0^1 k_n(x,y) z_n(y) dy, \\ k_n(x,\cdot) &:= R_n k(x,\cdot) \text{ for all } x \in [0,1). \end{aligned}$$

The last equality is a consequence of the quadrature rule

$$\int_0^1 u(x)dx \approx \frac{1}{2n+1} \sum_{k\in\mathbb{Z}_n} u(x_{n,k})$$

being exact for $u \in \mathcal{T}_{2n}$. Occasionally, we shall refer to both the collocation and the quadrature method by "approximation methods of collocation type".

Theorem 4.1 (cf. [17, 20, 21]). *If n is sufficiently large and the corresponding assumptions are fulfilled then each of the approximation methods defined above possesses a uniquely determined solution y_n and z_n, respectively. Moreover, we have the error estimates given in the table below, where u denotes the exact solution of equation (3.5). Remember that r denotes the degree of smoothness of c, d, and the image of K.*

	Sobolev case	Hölder–Zygmund case
Assumptions	$A \in \mathrm{G}\mathcal{L}(\mathbf{H}^s)$, $f \in \mathbf{H}^s$, $s \leq r$	$A \in \mathrm{G}\mathcal{L}(\mathcal{H}^s)$, $f \in \mathcal{H}^s$, $s < r$ If $s < 1$ then $s = r$ is allowed.
Galerkin method	$\Vert u - y_n\Vert_{\mathbf{H}^t} \leq C\, n^{t-s}\, \Vert f\Vert_{\mathbf{H}^s}$ $t \leq s$	$\Vert u - y_n\Vert_{\mathcal{H}^t} \leq C\, n^{t-s} \log n\, \Vert f\Vert_{\mathcal{H}^s}$ $0 < t < s$
Collocation type methods	$\Vert u - z_n\Vert_{\mathbf{H}^t} \leq C\, n^{t-s}\, \Vert f\Vert_{\mathbf{H}^s}$ $0 \leq t \leq s$; $s > \frac{1}{2}$	$\Vert u - z_n\Vert_{\mathcal{H}^t} \leq C\, n^{t-s} \log n\, \Vert f\Vert_{\mathcal{H}^s}$ $0 < t < s$

Proof. Let us sketch the proof for the Hölder–Zygmund case. For details see [21]. At first, we consider the Galerkin method. The assumptions ensure that we can factorize the coefficients $c = c_+c_-$ and $d = d_-d_+$, such that $(c_+, e_k) = (d_+, e_k) = 0$ for all $k < 0$ and $(c_-, e_k) = (d_-, e_k) = 0$ for all $k > 0$. We define the operator $M = M_+M_-$, where $M_+ = Pc_+P + Qd_-Q$ and $M_- = Pc_-P + Qd_+Q$ having the inverses $M_+^{-1} = Pc_+^{-1}P + Qd_-^{-1}Q$ and $M_-^{-1} = Pc_-^{-1}P + Qd_+^{-1}Q$. Note that $M_+^{\pm 1}, M_-^{\pm 1} \in G\mathcal{L}(\mathcal{H}^r)$. One can show that $A = M_+M_- + T$ with $T \in \mathcal{K}(\mathcal{H}^t, \mathcal{H}^s)$. Obviously, the equations $Au = f$ and $M_+^{-1}Au = (M_- + M_+^{-1}T)u = M_+^{-1}f$ are equivalent. The property $P_nM_+^{\pm 1} = P_nM_+^{\pm 1}P_n$ yields that the approximate equations $P_nAu_n = P_nf$ and $P_n(M_- + M_+^{-1}T)u_n = P_nM_+^{-1}f$ are also equivalent. Moreover, using additionally the property $P_nM_-^{\pm 1}P_n = M_-^{\pm 1}P_n$, the last equation equals $(M_- + P_nM_+^{-1}T)u_n = P_nM_+^{-1}f$.

Let us proceed with the collocation method. By the property of the interpolation operator $R_n(g \cdot h) = R_n(g \cdot R_nh)$ for any functions $g, h \in \mathcal{R}$, we can assume without loss of generality that $d = 1$. Define $M_+ = c_+I$ and $M_- = Pc_-P + Qc_+^{-1}Q$. In an analogous manner as above we again obtain $A = M_+M_- + T$ with (another) $T \in \mathcal{K}(\mathcal{H}^t, \mathcal{H}^s)$, the equivalence of $Au = f$ and $(M_- + M_+^{-1}T)u = M_+^{-1}f$, and the equivalence of $R_nAu_n = R_nf$ and $(M_- + R_nM_+^{-1}T)u_n = R_nM_+^{-1}f$.

Let L_n be one of the projections P_n and R_n.
Since $\|L_nM_+^{-1}T - M_+^{-1}T\|_{\mathcal{H}^t \to \mathcal{H}^t} \to 0$ as $n \to \infty$, it is easy to show that $M_- + L_nM_+^{-1}T$ is invertible for all $n \geq n_0$ and that the sequence $\{(M_- + L_nM_+^{-1}T)^{-1}\}_{n \geq n_0}$ is uniformly bounded in $\mathcal{H}^t$. Furthermore, we have the identity

$$\begin{aligned}(M_- + L_nM_+^{-1}T)(u - u_n) &= M_-u + L_nM_+^{-1}Tu - L_nM_+^{-1}f \\ &= (I - L_n)M_+^{-1}f + (L_n - I)M_+^{-1}Tu.\end{aligned}$$

Putting these things together and using Lemma 4.1 we get the assertion.

4.2 Approximation methods for the airfoil equation

Let n be a natural number and let ρ refer to one of the Jacobi weights σ and μ. We denote by P_n^ρ the orthogonal (with respect to the scalar product $(\cdot,\cdot)_\rho$) projection $P_n^\rho : u \mapsto \sum_{k=0}^{n-1}(u, p_n^\rho)_\rho \, p_n^\rho$ onto Π_n having the property $(u - P_n^\rho u, \pi_n) = 0 \; \forall u \in \mathbf{L}_\rho^2 \; \forall \pi_n \in \Pi_n$. Further, let $R_n^\rho : \mathcal{R}^\rho \to \Pi_n$ be the interpolation operator $R_n^\rho u(x_{n,j}^\rho) = u(x_{n,j}^\rho)$, $j = 0, ..., n-1$, where $\mathcal{R}^\rho$ denotes the class of all bounded and Riemann integrable (with respect to the weight ρ) functions on $\mathbb{J}$ and the $x_{n,j}^\rho$ are the zeros of the orthogonal polynomial p_n^ρ.

Lemma 4.2. *There exists a positive constant C depending only on t and s such that for any $u \in \mathbf{L}_{\rho,s}^2$ the following estimates hold*

$$\|u - P_n^\rho u\|_{\mathbf{L}^2_{\rho,t}} \leq C\, n^{t-s}\, \|u\|_{\mathbf{L}^2_{\rho,s}}, \qquad \forall t: \ 0 \leq t \leq s.$$

If $|\alpha| = |\beta| = \frac{1}{2}$ *then*

$$\|u - R_n^\rho u\|_{\mathbf{L}^2_{\rho,t}} \leq C\, n^{t-s}\, \|u\|_{\mathbf{L}^2_{\rho,s}}, \qquad \forall t: \ 0 \leq t \leq s; \quad s > \tfrac{1}{2},$$

whereas in the remaining cases we have

$$\|u - R_n^\rho u\|_{\mathbf{L}^2_{\rho,t}} \leq C\, n^{t-s}\, \|u\|_{\mathbf{L}^2_{\rho,s}}, \qquad \forall t: \ 0 \leq t \leq s; \quad s \geq 1.$$

If $t = s$ *then all norms on the left–hand side tend to zero (i.e. they are not only bounded). Furthermore, for all* $u_n \in \Pi_n$ *the Bernstein type inequality hold*

$$\|u_n\|_{\mathbf{L}^2_{\rho,t}} \leq n^{t-q}\, \|u_n\|_{\mathbf{L}^2_{\rho,q}} \qquad 0 \leq q \leq t.$$

Proof. The first and the very last relation obviously follow from the definition of P_n^ρ and the $\mathbf{L}^2_{\rho,t}$–norm. For Jacobi weights ρ with $|\alpha| = |\beta| = \frac{1}{2}$, the second estimate is due to the authors [3]. The third one has been recently proved by Capobianco and Mastroianni [6].

Now, we are going to describe some approximation methods for the airfoil equation (3.9). For the same reason as in the previous subsection, we make the smoothness assumption $K \in \mathcal{L}(\mathbf{L}^2_\sigma, \mathbf{L}^2_{\mu,r})$, $r > \frac{1}{2}$. Sufficient conditions for this property are, for example,

$$\|k(\cdot, y)\|_{\mathbf{L}^2_{\mu,r}} \leq C_1 \text{ for all } y \in \mathbb{J} \text{ and } \|k(x, \cdot)\|_{\mathbf{L}^2_{\sigma,r}} \leq C_2 \text{ for all } x \in \mathbb{J}. \quad (4.4)$$

Because we restrict our investigations to operators A_0 having index zero, A_0 is invertible in each pair of spaces $(\mathbf{L}^2_{\sigma,t}, \mathbf{L}^2_{\mu,t})$, $t \geq 0$, having the inverse $A_0^{-1} := \hat{A}_0$, see [3]. At first we consider the case when $\nu = 0$.

The *Galerkin method* defines an approximate solution $y_n \in \Pi_n$ by

$$P_n^\mu A y_n = P_n^\mu f. \quad (4.5)$$

Further, the *collocation method* determines $z_n \in \Pi_n$ by

$$R_n^\mu A z_n = R_n^\mu f, \quad (4.6)$$

and the *quadrature method* by

$$R_n^\mu (A_0 + K_n) z_n = R_n^\mu f, \quad (4.7)$$

where $K_n z_n(x) = \sum_{k=0}^{n-1} k(x, x_{n,k}^\sigma) z_n(x_{n,k}^\sigma) \sigma_{n,k} = \int_{\mathbb{J}} k_n(x,y) z_n(y) \sigma(y) dy$, $k_n(x, \cdot) := R_n^\sigma k(x, \cdot)$ for all $x \in \mathbb{J}$, and the $\sigma_{n,k}$ are the Christoffel numbers of the orthogonal polynomial p_n^σ, i.e. the weights in the Gaussian quadrature rule $\int_{\mathbb{J}} u(x)\sigma(x)dx \approx \sum_{k=0}^{n-1} u(x_{n,k}^\sigma)\sigma_{n,k}$ being exact for $u \in \Pi_{2n}$.

Theorem 4.2. *Let $\nu = 0$ in (3.9). If $A \in G\mathcal{L}(\mathbf{L}^2_{\sigma,t}, \mathbf{L}^2_{\mu,t})$, $f \in \mathbf{L}^2_{\mu,s}$, $0 \le t \le s < r$, then for all sufficiently large n each of the approximation methods defined above possesses a uniquely determined solution y_n and z_n, respectively. Moreover, we have the error estimates*

$$\|u - y_n\|_{\mathbf{L}^2_{\sigma,t}} \le C\, n^{t-s}\, \|f\|_{\mathbf{L}^2_{\mu,s}}, \qquad \forall t:\ 0 \le t \le s. \tag{4.8}$$

If $|\alpha| = |\beta| = \frac{1}{2}$ then

$$\|u - z_n\|_{\mathbf{L}^2_{\sigma,t}} \le C\, n^{t-s}\, \|f\|_{\mathbf{L}^2_{\mu,s}}, \qquad \forall t:\ 0 \le t \le s;\ s > \frac{1}{2}, \tag{4.9}$$

whereas in the remaining cases we have

$$\|u - z_n\|_{\mathbf{L}^2_{\sigma,t}} \le C\, n^{t-s}\, \|f\|_{\mathbf{L}^2_{\mu,s}}, \qquad \forall t:\ 0 \le t \le s;\ s \ge 1.$$

where u denotes the exact solution of equation (3.9).

Proof. The proofs can be found in [3]. For arbitrary Jacobi weights, however, only a weaker estimate is given. Using the results of [6], we also get the optimal estimate in this case. Furthermore, approximation methods for the airfoil equation are also considered in [13].

Let us, for example, sketch the proof for the quadrature method. The other proofs are analogous. At first, one verifies the stability of the sequence $\{R^\mu_n(A_0 + K_n)P^\sigma_n\}$ or, what is the same, the invertibility of $A_0 + R^\mu_n K_n$ for all $n \ge n_0$. This can be done simply by showing that $\|(A_0 + K) - (A_0 + R^\mu_n K_n)\|_{\mathbf{L}^2_{\sigma,t}\to\mathbf{L}^2_{\mu,t}} = \|K - R^\mu_n K_n\|_{\mathbf{L}^2_{\sigma,t}\to\mathbf{L}^2_{\mu,t}}$ tends to zero as $n \to \infty$. Now the error estimate immediately follows from the relation $(A_0 + R^\mu_n K_n)(u - z_n) = A_0 u + R^\mu_n K_n u - R^\mu_n f = (f - R^\mu_n f) + (R^\mu_n K_n u - Ku)$.

Since the estimate for $\|(K - R^\mu_n K_n)u\|_{\mathbf{L}^2_{\mu,t}}$ was shown under the stronger condition $r > 1$ and only for $|\alpha| = |\beta| = \frac{1}{2}$ in [3], we are going to give the proof in the general case in the following. Let $u \in \mathbf{L}^2_\sigma$. From Lemma 4.2 we get that

$$\|(K - R^\mu_n K_n)u\|_{\mathbf{L}^2_{\mu,t}} \le \|(I - R^\mu_n)Ku\|_{\mathbf{L}^2_{\mu,t}} + n^t\, \|R^\mu_n(K - K_n)u\|_{\mathbf{L}^2_\mu}.$$

Applying the Gaussian quadrature rule $\int_{\mathbb{J}} f(x)\mu(x)dx \approx \sum_{j=0}^{n-1} f(x^\mu_{n,j})\mu_{n,j}$ being exact for $f \in \Pi_{2n}$, we obtain

$$\begin{aligned}\|R^\mu_n(K - K_n)u\|^2_{\mathbf{L}^2_\mu} &\le C \sup_{x\in\mathbb{J}} \left| \int_{\mathbb{J}} [k(x,y) - k_n(x,y)]u(y)\sigma(y)dy \right|^2 \\ &\le C \sup_{x\in\mathbb{J}} \|(I - R^\sigma_n)k(x,\cdot)\|^2_{\mathbf{L}^2_\sigma}\, \|u\|^2_{\mathbf{L}^2_\sigma}.\end{aligned}$$

Thus,

$$\|(K - R^\mu_n K_n)u\|_{\mathbf{L}^2_{\mu,t}} \le \|(I - R^\mu_n)Ku\|_{\mathbf{L}^2_{\mu,t}} + C\, n^t \sup_{x\in\mathbb{J}} \|(I - R^\sigma_n)k(x,\cdot)\|_{\mathbf{L}^2_\sigma}\, \|u\|_{\mathbf{L}^2_\sigma},$$

from which the estimate

$$\|(K - R_n^\mu K_n)u\|_{\mathbf{L}^2_{\mu,t}} \le C\, n^{t-r}\, \|u\|_{\mathbf{L}^2_\sigma}, \quad \forall t:\ 0 \le t \le r,$$

follows, where $r > \frac{1}{2}$, if $|\alpha| = |\beta| = \frac{1}{2}$, and $r \ge 1$, otherwise, has to be assumed.

Now, let us consider the case when $\nu \ne 0,\ a = 0$. For details see [2, 3, 9, 10, 14]. Here we always have $|\alpha| = |\beta| = \frac{1}{2}$. Note that the quadrature method in this case is defined by

$$R_n^\mu (A_0 + \nu W + K_n) z_n = R_n^\mu f.$$

In particular, we do not need to apply an additional quadrature rule because of the mapping properties of W with respect to polynomials.

Theorem 4.3. *Let $a = 0$ in (3.9) and let the assumptions of Theorem 4.2 be fulfilled. Then for all sufficiently large n each of the approximation methods defined above possesses a uniquely determined solution y_n and z_n, respectively, and we have the error estimates (4.8) and (4.9).*

5 Fast algorithms

Seeking an approximate solution $u_n \in \mathcal{T}_n$ of the equation (3.5) consists actually in finding n Fourier coefficients (u_n, e_k), $k \in \mathbb{Z}_n$, which are related in some sense to the corresponding Fourier coefficients (u, e_k) of the exact solution u.

Studying the classical approximation method from Section 4 more closely, one find that the best approximation can be obtained using the Galerkin method. Moreover, the approximation error is optimal in each Sobolev norm, that means $\|u - u_n\|_{\mathbf{H}^t} \le C\, n^{t-s}\, \|f\|_{\mathbf{H}^s}$, $t \le s$. Consequently, each summand $|k|^t |(u - u_n, e_k)|$ (which is the error of the k–th Fourier coefficient, apart from the factor $|k|^t$) of the series (2.1) again fulfils such an optimal estimate.

In the cases when collocation type methods are used we do not have this optimal convergence on the whole Sobolev scale. It is true that the error rate is optimal if we measure it in norms with $t \ge 0$. However, for negative t we have no more increase in the convergence rate. Nevertheless, by an observation of the authors in [4], the higher Fourier coefficients of the approximate solution also converge optimally in cases when $t < 0$, whereas only the lower Fourier coefficients cause the worse convergence behaviour in "negative" Sobolev norms. Indeed, if $t < 0$ and m denotes a fixed positive integer less than n, we have

$$\begin{aligned} \|u - z_n\|_{\mathbf{H}^t} &\le \|P_m(u - z_n)\|_{\mathbf{H}^t} + \|Q_m(u - z_n)\|_{\mathbf{H}^t} \\ &\le \|P_m(u - z_n)\|_{\mathbf{H}^t} + C\ m^t\, \|u - z_n\|_{\mathbf{L}^2} \\ &\le \|P_m(u - z_n)\|_{\mathbf{H}^t} + C\ n^{t-s}\, \|f\|_{\mathbf{H}^s}. \end{aligned} \tag{5.1}$$

Since $\|u - z_n\|_{\mathbf{H}^t}$ is only of the order $O(n^{-s})$, we conclude that $\|P_m(u - z_n)\|_{\mathbf{H}^t}$ can not be better than $O(n^{-s})$.

Another observation concerning the higher Fourier coefficients of the collocation solution $z_n = (R_n A R_n)^{-1} R_n f$ (in the case when $t \geq 0$) is due to Amosov [1]. In his paper he states that if we replace $(R_n A R_n)^{-1}$ by $R_n B R_n$, where B is the parametrix to A (see Lemma 3.2) then the higher Fourier coefficients still converges optimally. We only have to assume that the coefficients c and d and the image of K in (3.5) are smooth enough.

Further, using an appropriate correction of z_n one can also achieve optimal convergence for the lower Fourier coefficients and, consequently, for the approximate solution u_n itself. Amosov's correction is based upon a small quadrature scheme with a modified right–hand side. Of course, we do not have optimal convergence in all Sobolev norms $\|\cdot\|_{\mathbf{H}^t}$, but at least for t varying in $[t_0, s]$ with a certain $t_0 > \frac{1}{2}$.

Amosov has been developed a *fast algorithm* for solving equations of the type (3.5). First of all, the application of $R_n B R_n$ instead of $(R_n A R_n)^{-1}$ to $R_n f$ reduces the amount from, in general, $O(n^3)$ to $O(n^2)$. But, in the present situation we can use the *Fast Fourier Transformation*, which reduces the amount once more to $O(n \log n)$.

Originally, the authors have been used the above observations for another goal. In their paper [4] they have been proposed *corrected collocation type methods*. The purpose of these methods is to improve the convergence rate in "negative" Sobolev norms. Having (5.1) in mind, the lower Fourier coefficients of the collocation type solution z_n are replaced by the Fourier coefficients of the solution of a small Galerkin scheme. Thus, we have somewhat more expense in computing, but it results that the convergence rate can be improved, say, by $O(n^{t/3})$ (in dependence on the size of the Galerkin scheme). For the corrected collocation type methods we also need some smoothness assumptions on c, d, and the image of K.

In a further paper [5], the authors have been brought these two things together, i.e. they have been established a method, having both a good convergence behaviour in "negative" Sobolev norms and a low computational complexity. Moreover, the case of spline approximation methods has been considered (see also [4]).

5.1 Amosov's fast algorithm

At first, we briefly describe Amosov's method [1]. Let us denote the complementary projection of P_n by Q_n $(= I - P_n)$ and let m be an integer less than n. The main idea consists in the calculation of the higher and the lower Fourier coefficients of an approximation $u_n \in \mathcal{T}_n$ of the exact solution u by two different methods.

Amosov's Method or Parametrix–Quadrature Method

$$u_n := P_m w_m + Q_m z_n \; \in \mathcal{T}_n, \tag{5.2}$$

where $z_n \in \mathcal{T}_n$ is defined by

$$z_n := R_n B R_n f \tag{5.3}$$

and $w_m \in \mathcal{T}_m$ is the solution of the quadrature scheme

$$R_m(cP + dQ + K_m)w_m = R_m g, \quad g := f - (cP + dQ)Q_m z_n. \tag{5.4}$$

The following algorithm calculating u_n with an amount of only $O(n \log n)$ operations is due to Amosov. For details see [1, 5].
Let $F_n = [e^{2\pi ijk/(2n+1)}]_{j,k\in\mathbb{Z}_n}$ be the Fourier matrix of order $2n+1$, the inverse of which can be written as $F_n^{-1} = [\frac{1}{2n+1}e^{-2\pi ijk/(2n+1)}]_{k,j\in\mathbb{Z}_n}$. It is well–known that the multiplication of F_n or F_n^{-1} by a vector (the so–called *Fourier transformation* and *inverse Fourier transformation*, respectively) can be realized with $O(n \log n)$ operations. For a function g we denote by $\tilde{g}_n$ the vector, whose entries are the values of g at the points $x_{n,j}$, $j \in \mathbb{Z}_n$, and by $\hat{g}_n$ the vector containing the Fourier coefficients (g, e_k), $k \in \mathbb{Z}_n$. If $g \in \mathcal{T}_n$, then $\tilde{g}_n = F_n \hat{g}_n$.

The computation of u_n can be split up into three steps. Firstly, the computation of z_n, secondly, the computation of the right–hand side $R_m g$, and thirdly, the solution of the collocation scheme (5.4). The first one can be done as follows

$$\hat{z}_n = F_n^{-1}(C_n^{-1}F_n I_n^+ + D_n^{-1}F_n I_n^-)F_n^{-1}\tilde{f}_n, \tag{5.5}$$

where $I_n^+ = \begin{pmatrix} 0 & 0 \\ 0 & I_{n+1} \end{pmatrix}$, $I_n^- = I_{2n+1} - I_n^+$, $C_n = \operatorname{diag}_{j\in\mathbb{Z}_n}\tilde{c}_{n,j}$, $D_n = \operatorname{diag}_{j\in\mathbb{Z}_n}\tilde{d}_{n,j}$, and I_n is the unit matrix of order n. The second step consists in

$$\tilde{g}_m = \tilde{f}_m - (C_m I_{m,n} F_n I_n^+ + D_m I_{m,n} F_n I_n^-) I_n^m \hat{z}_n, \tag{5.6}$$

where I_n^m is the diagonal matrix with the entries $[I_n^m]_{j,k} = 1$, if $j = k \in \mathbb{Z}_n \setminus \mathbb{Z}_m$, and $[I_n^m]_{j,k} = 0$, otherwise, $j, k \in \mathbb{Z}_n$, and $I_{m,n}$ is the (non–quadratic) matrix $[I_{m,n}]_{i,j} = 1$, if $x_{m,i} = x_{n,j}$, and $[I_{m,n}]_{i,j} = 0$, otherwise, $i \in \mathbb{Z}_m; j \in \mathbb{Z}_n$. (Here we suppose that (2n+1)/(2m+1) is an integer.) Finally, the third step is the solution of the $m \times m$ algebraic system of equation. In general, we need $O(m^3)$ operations since the matrix is neither sparse nor well–structurized. However, both in the first and in the second step there occur only diagonal and Fourier matrices. Thus both can be realized with $O(n \log n)$. Choosing $m \sim \sqrt[3]{n}$ then the calculation of u_n needs really only $O(n \log n)$ operations. Here and in the following the notation $a_n \sim b_n$ means that there are positive constants C_1 and C_2 such that for all n the relation $C_1 \leq a_n/b_n \leq C_2$ holds.

5.2 Error estimates in Sobolev spaces

In this subsection we suppose that the right–hand side of equation (3.5) is an element of $\mathbf{H}^s$, $s > \frac{1}{2}$ and that the operator A is in $G\mathcal{L}(\mathbf{H}^s)$. Then the uniquely determined solution (being always referred to by u in the following) is again in $\mathbf{H}^s$. Since $\|u-u_n\|_{\mathbf{H}^t} \leq \|P_m(u-u_n)\|_{\mathbf{H}^t} + \|Q_m(u-u_n)\|_{\mathbf{H}^t}$ we are able to investigate the convergence behaviour of the higher and of the lower Fourier coefficients separately.

Lemma 5.1. *Let $r \geq s > \frac{1}{2}$. Then for sufficiently large n and m we have the estimate*

$$\|Q_m(u-z_n)\|_{\mathbf{H}^t} \leq \begin{cases} C\ (n^{t-s} + m^{t-r})\ \|f\|_{\mathbf{H}^s}, & 0 \leq t \leq s, \\ C\ (m^t\ n^{-s} + m^{t-r})\ \|f\|_{\mathbf{H}^s}, & t < 0. \end{cases}$$

Proof. For $0 \leq t \leq s$ the proof is due to Amosov ([1], Lemma 5.1.). The case $t < 0$ has been proved by the authors in [5].

Lemma 5.2 ([1]). *Let $r \geq s > \frac{1}{2}$. If the collocation method (5.4) is stable then for sufficiently large n and m we have the estimate (note that w_m actually depends on n via (5.4))*

$$\|P_m(u-w_m)\|_{\mathbf{H}^t} \leq C\ (n^{t-s} + m^{t-r})\ \|f\|_{\mathbf{H}^s} \text{ for all } t : \tfrac{1}{2} < t \leq s. \qquad (5.7)$$

Choosing $m \sim \sqrt[3]{n}$, a straightforward calculation shows that the error estimate is optimal, i.e.

$$\|u-u_n\|_{\mathbf{H}^t} \leq C\ n^{t-s}\|f\|_{\mathbf{H}^s} \text{ for all } t > \tfrac{1}{2} :\ s - \tfrac{r-s}{2} \leq t \leq s. \qquad (5.8)$$

Thus, this method is very suitable if we have sufficiently smooth coefficients c and d (being in $\mathbf{H}^r$) and if the image of K is also smooth enough ($\subset \mathbf{H}^r$). Note that the right–hand side f must not belong to $\mathbf{H}^r$ but only to $\mathbf{H}^s$.

Unfortunately, if we want to measure the approximation error $u - u_n$ in "negative" Sobolev norms, this method does not give any improvement of the convergence rate. The reason is that (5.7) can not be extended to $t < 0$, since we use a collocation type method for the correction of z_n. On the other hand, Lemma 5.1 has been extended to the case of negative t by the authors. Furthermore, in [4] the authors succeeded in correcting the collocation type solution by the solution of a certain small Galerkin scheme with regard to better convergence rates for $t < 0$. More precisely, the corrected collocation method consists in the evaluation of the higher Fourier coefficients of the approximate solution by the usual collocation method, whereas the lower Fourier coefficients are evaluated by the usual Galerkin method. Both methods are applied to the whole equation (3.5).

Corrected Collocation Method

$$u_n := P_m y_M + Q_m z_n \in \mathcal{T}_n, \tag{5.9}$$

where $z_n \in \mathcal{T}_n$ is the solution of the collocation scheme

$$R_n A z_n = R_n f \tag{5.10}$$

and $y_M \in \mathcal{T}_M$ is the solution of the Galerkin scheme

$$P_M A y_M = P_M f. \tag{5.11}$$

Here M is an integer lying between m and n (in practice one can choose $M := 2m$, for example). For details see [4]. If $M \ll n$ then the costs for the Galerkin scheme (5.11) are defensible. Since $\|Q_m(u - z_n)\|_{\mathbf{H}^t} \leq C\ m^t\ \|u - z_n\|_{\mathbf{L}^2} \leq C\ m^t\ n^{-s}\ \|f\|_{\mathbf{H}^s}\ \forall t < 0$, the higher Fourier coefficients of u_n converge better than the collocation solution z_n itself. The following lemma is devoted to the convergence of the lower Fourier coefficients. Let K^* denote the $\mathbf{L}^2$–adjoint to K.

Lemma 5.3 ([4]). *Let $r \geq s > \frac{1}{2}$, $K^* \in \mathcal{L}(\mathbf{H}^{-t}, \mathbf{H}^r)$, and the Galerkin method be stable. Then for sufficiently large n and m we have the estimate*

$$\|P_m(u - y_M)\|_{\mathbf{H}^t} \leq C\ (M - m)^{-r}\ \|f\|_{\mathbf{H}^s} \text{ for all } t : -r \leq t < -\tfrac{1}{2}. \tag{5.12}$$

Proof. The estimate (5.12) proves to be one of the crucial points of the corrected collocation methods. Therefore, we sketch main ideas of the proof, although it is given already in [4]. By the famous Kozak formula and $P_m P_M = P_m$ we find that

$$\begin{aligned} P_m(u - y_M) &= P_m A^{-1} f - P_m (P_M A P_M)^{-1} f \\ &= P_m A^{-1} Q_M (I + (Q_M A^{-1} Q_M)^{-1} Q_M A^{-1} P_M) f \end{aligned}$$

It can be shown that the operator sequence $I + (Q_M A^{-1} Q_M)^{-1} Q_M A^{-1} P_M$ is uniformly bounded in $\mathbf{H}^s$. Furthermore, we have

$$P_m A^{-1} Q_M = P_m B Q_M + P_m T A^{-1} Q_M = P_m (P c^{-1} P + Q d^{-1} Q) Q_M + P_m T' Q_M$$

with $T' \in \mathcal{K}(\mathbf{H}^q, \mathbf{H}^r)$, $\forall q > \frac{1}{2}$. Let $v \in \mathbf{H}^s$. A little thought shows that the identity

$$\|P_m P c^{-1} P Q_M v\|^2_{\mathbf{H}^t} = \sum_{0 \leq k \leq m/2} (1 + |k|)^{2t} \left| \sum_{l > M/2} (v, e_l)\ (c^{-1}, e_{k-l}) \right|^2$$

holds. Using the Cauchy–Schwarz inequality and the relation $k - l < \frac{1}{2}(M - m) < 0$ we obtain

$$\begin{aligned} \|P_m P c^{-1} P Q_M v\|_{\mathbf{H}^t} &\leq C\ \|Q_{M-m} c^{-1}\|_{\mathbf{L}^2} \|Q_M v\|_{\mathbf{H}^s} \\ &\leq C\ (M - m)^{-r}\ \|c^{-1}\|_{\mathbf{H}^r} \|v\|_{\mathbf{H}^s}. \end{aligned}$$

The same applies to $\|P_m Q d^{-1} Q Q_M v\|_{\mathbf{H}^t}$.
Finally, we have $\|P_m T' Q_M\|_{\mathbf{H}^s \to \mathbf{H}^t} \le C\ M^{-r-s}$. Thus, the estimate

$$\|P_m A^{-1} Q_M\|_{\mathbf{H}^s \to \mathbf{H}^t} \le C\ (M-m)^{-r} \tag{5.13}$$

is proved and we are done. The last relation can be interpreted as some kind of *localness of A^{-1} with respect to the family of the projections* $\{P_m\}_{m\in\mathbb{N}}$. Note that (5.13) is an essential ingredient of the corrected collocation methods, since we can not succeed without such a property.

Choosing again $m \sim \sqrt[3]{n}$ and $M = 2m$, we arrive at

$$\begin{aligned}\|u-u_n\|_{\mathbf{H}^t} &\le C\ (m^t n^{-s} + (M-m)^{-r})\ \|f\|_{\mathbf{H}^s} \\ &\le C\ (\ n^{\frac{t}{3}-s} + n^{-\frac{r}{3}}\ )\ \|f\|_{\mathbf{H}^s} \text{ for all } t:\ -r \le t < -\tfrac{1}{2}.\end{aligned} \tag{5.14}$$

One can show that if $r > \max\{3s, s+|t|\}$ then the last expression for the convergence rate improve both the rate of the pure collocation type method (4.2) and (4.3), respectively, and the rate of the pure Galerkin method (5.11) in each negative Sobolev norm $\|\cdot\|_{\mathbf{H}^t}$.

Nevertheless, for solving the collocation scheme (5.10) we have still a complexity of $O(n^3)$ operations. The following algorithm unites the advantages of Amosov's $O(n \log n)$–algorithm and author's corrected collocation method.

Parametrix–Galerkin Method

$$u_n := P_m y_M + Q_m z_n\ \in \mathcal{T}_n, \tag{5.15}$$

where $z_n \in \mathcal{T}_n$ is defined by

$$z_n := R_n B R_n f \tag{5.16}$$

and $y_M \in \mathcal{T}_M$ is the solution of the Galerkin scheme

$$P_M A y_M = P_M f. \tag{5.17}$$

Choosing again $m \sim \sqrt[3]{n}$ and $M = 2m$, the Lemmata 5.1 and 5.3 lead to

$$\begin{aligned}\|u-u_n\|_{\mathbf{H}^t} &\le C\ (m^t n^{-s} + m^{t-r} + (M-m)^{-r})\ \|f\|_{\mathbf{H}^s} \\ &\le C\ (\ n^{\frac{t}{3}-s} + n^{-\frac{r}{3}}\ )\ \|f\|_{\mathbf{H}^s} \text{ for all } t:\ -r \le t < -\tfrac{1}{2},\end{aligned} \tag{5.18}$$

i.e. the same rate as for the corrected collocation method. Thus, all conclusions from above apply again. Moreover, it is clear that the complexity for the computation of u_n by the parametrix–Galerkin method is $O(n \log n)$, except for the calculation of the entries of the matrix A_M and the right–hand side f_M of (5.17). In order to reduce the complexity somewhat, one should make use of the fact that the main part of A_M has a diagonal–circulant structure.

5.3 Error estimates for Amosov's method in Hölder-Zygmund spaces

Now let the right–hand side of equation (3.5) be an element of $\mathcal{H}^s$, $s > 0$, and the operator A be in $\mathrm{G}\mathcal{L}(\mathcal{H}^s)$. The uniquely determined solution will be denoted by $u \in \mathcal{H}^s$. Consider the method (5.2) – (5.4).

Lemma 5.4. *Let $r > s > 0$. Then for sufficiently large n and m we have the estimate*

$$\|Q_m(u - z_n)\|_{\mathcal{H}^t} \leq C\ (n^{t-s}(\log n)^2 + m^{t-r+\delta}) \log m\ \|f\|_{\mathcal{H}^s}, \quad 0 < t < s.$$

Proof. (cf. [1]) We have

$$Q_m u - Q_m z_n = Q_m u - Q_m R_n B R_n f = Q_m T u + Q_m h_n,$$

where T is the compact operator from Subsection 3.1 and $h_n := (I - R_n)Bf + B(I - R_n)f + (R_n - I)B(I - R_n)f$.
Thus, $\|Q_m(u - z_n)\|_{\mathcal{H}^t} \leq C\ m^{t-r+\delta} \log m\ \|Tu\|_{\mathcal{H}^{r-s}} + C\ \log m\ \|h_n\|_{\mathcal{H}^t}$. From the Lemmata (3.2)(ii) and (4.1)(ii) we get that $\|h_n\|_{\mathcal{H}^t} \leq C\, n^{t-s}(\log n)^2 \|f\|_{\mathcal{H}^s}$. Finally, we use the smoothing property of T to get $\|Tu\|_{\mathcal{H}^{r-s}} \leq C\ \|f\|_{\mathcal{H}^s}$.

Lemma 5.5. *Let $r > s > 0$. Then for sufficiently large n and m we have the estimate (note that w_m actually depends on n via (5.4))*

$$\|P_m(u - w_m)\|_{\mathcal{H}^t} \leq C\ (n^{t-s}(\log n)^2 + m^{t-r+\delta})\ (\log m)^2\ \|f\|_{\mathcal{H}^s}, \quad 0 < t < s. \tag{5.19}$$

Proof. We use the notations from that part of the proof of Theorem 4.1 concerning the collocation method. Put $A_0 := cP + dQ$. First of all, the equations $Au = f$ and $R_m(A_0 + K_m)w_m = R_m g$ are equivalent to $(M_- + M_+^{-1}T)u = M_+^{-1}f$ and $(M_- + R_m M_+^{-1} T_m)w_m = R_m M_+^{-1} g$, respectively. The operators T_m are defined by $T_m := T - K + K_m$ and prove to be in $\mathcal{K}(\mathcal{H}^t, \mathcal{H}^{r-\delta})$ for all m. We have the identity

$$\begin{aligned}
&(M_- + R_m M_+^{-1} T_m) P_m (u - w_m)\\
&= (M_- + R_m M_+^{-1} T_m) P_m u - R_m M_+^{-1} f + R_m M_+^{-1} A_0 Q_m z_n\\
&= (M_- + R_m M_+^{-1} T_m) P_m u - R_m (M_- + M_+^{-1} T) u\\
&\quad + R_m M_+^{-1} (M + T - K) Q_m z_n\\
&= (M_- P_m u - R_m M_- P_m u) - R_m M_- (Q_m u - Q_m z_n)\\
&\quad + R_m M_+^{-1} (T_m P_m u - T u) + R_m M_+^{-1} (T_m - K_m) Q_m z_n\\
&= - R_m M_- (Q_m u - Q_m z_n) - R_m M_+^{-1} T_m (Q_m u - Q_m z_n)\\
&\quad + R_m M_+^{-1} (K_m - K) u - R_m M_+^{-1} K_m Q_m z_n,
\end{aligned}$$

where we have used that $M_- P_m u \in \mathcal{T}_m$, $R_m M_+^{-1} = R_m M_+^{-1} R_m$, and $T_m - T = K_m - K$. Since $K_m Q_m z_n(x) = \int_0^1 k_m(x,y) Q_m z_n(y) dy$ and $k_m(x,\cdot)$ is orthogonal to $Q_m z_n$ for all $x \in [0,1)$, the last summand vanishes. From the proof of Theorem 4.1 we know that $M_- + R_m M_+^{-1} T_m$ is invertible in $\mathcal{H}^t$ for all $m \geq m_0$ and that the sequence of the inverses is uniformly bounded. Therefore, we can estimate

$$\|P_m(u - w_m)\|_{\mathcal{H}^t} \leq C \ \|R_m(M_- + M_+^{-1} T_m)\|_{\mathcal{H}^t \to \mathcal{H}^t} \|Q_m u - Q_m z_n\|_{\mathcal{H}^t}$$
$$+ C \ \|R_m M_+^{-1} R_m (K_m - K) u\|_{\mathcal{H}^t}.$$

Proposition 8.13 in [21] yield that the estimate

$$\|R_m(K_m - K)u\|_{\mathcal{H}^t} = \|R_m(T_m - T)u\|_{\mathcal{H}^t} \leq C \ m^{t-r} \log m \ \|u\|_{\mathcal{H}^s}$$

is valid and, thus, $\|R_m(M_- + M_+^{-1} T_m)\|_{\mathcal{H}^t \to \mathcal{H}^t} \leq C \ \log m$. Now from Lemma 5.4 the assertion follows.

Choosing $m \sim \sqrt[3]{n}$, a straightforward calculation shows that the error estimate is almost optimal, namely

$$\|u - u_n\|_{\mathcal{H}^t} \leq C \ n^{t-s} (\log n)^4 \ \|f\|_{\mathcal{H}^s} \text{ for all } t > 0 : \ s - \tfrac{r-\delta-s}{2} \leq t < s. \quad (5.20)$$

5.4 A fast algorithm for the airfoil equation

Now we are going to describe a fast algorithm (being based on the discrete cosine transformation) for the airfoil equation (3.9). Let $Q_m^\sigma := I - P_m^\sigma$, where m is an integer less than n. One essential assumption for the algorithm is that the set of the collocation points has the property $\{x_{m,i}^\mu, i = 0, ..., m-1\} \subset \{x_{n,j}^\mu, j = 0, ..., n-1\}$. This can be guaranteed without any difficulties in the cases when $|\alpha| = |\beta| = \frac{1}{2}$. For example, if $\alpha = -\frac{1}{2}, \beta = \frac{1}{2}$ then $(2n+1)/(2m+1) \in \mathbb{Z}$ ensures the above inclusion. Thus, we shall only consider equation (3.9) with $a = 0$.

Obviously, the operator $B := (A_0 + \nu W)^{-1}$ proves to be a parametrix to A, since $I - BA := -BK$ is compact. An easily verifiable criterion for the invertibility of $A_0 + \nu W$ can be found in [2], Thm. 3.7. In particular, if $\nu \in \mathbb{R}$ then $A_0 + \nu W$ is always invertible.

Parametrix–Quadrature Method

$$u_n := P_m^\sigma w_m + Q_m^\sigma z_n \ \in \Pi_n, \quad (5.21)$$

where $z_n \in \Pi_n$ is defined by

$$z_n := (R_n^\mu (A_0 + \nu W) P_n^\sigma)^{-1} R_n^\mu f \quad (5.22)$$

and $w_m \in \Pi_m$ is the solution of the quadrature scheme

$$R_m^\mu (A_0 + \nu W + K_m) w_m = R_m^\mu g, \quad g := f - R_n^\mu (A_0 + \nu W) Q_m^\sigma z_n. \quad (5.23)$$

Let us compare (5.3) and (5.22). Both equations realize simplifications of the quadrature method directed to the construction of fast algorithms. However, in the present case we can get a "fast" method already by omitting the operator K (in fact, we get the collocation method for $A_0+\nu W$) because of the good mapping properties of $A_0+\nu W$ on polynomials, whereas the same procedure in the case of non–constant coefficients as in (5.3) still does not work.

Now, let us describe the realization of the method on the computer. For details see [3]. For $\rho=\sigma$ or $\rho=\mu$ let $G_n^\rho=[p_k^\rho(x_{n,j}^\rho)]_{j,k=0,\dots,n-1}$ be the analogue of the Fourier matrix of order n with respect to the system $\{p_k^\rho\}_{k=0}^\infty$, a so–called generalized Vandermonde matrix. Using the Gauss type quadrature rule $\int_{\mathbb{J}} v(x)\rho(x)dx \approx \sum_{k=0}^{n-1} v(x_{n,k}^\rho)\rho_{n,k}$, one can show that the relation $(G_n^\rho)^T D_n^\rho G_n^\rho = I_n$ is valid, where $D_n^\rho=\mathrm{diag}_{k=0,\dots,n-1}\rho_{n,k}$. Thus, the inverse of G_n^ρ can be written as $(G_n^\rho)^T D_n^\rho$. For a function g we denote by $\tilde{g}_n^\rho$ the vector, whose entries are the values of g at the points $x_{n,j}^\rho$, $j=0,\dots,n-1$, and by $\hat{g}_n^\rho$ the vector containing the Fourier coefficients $(g,p_n^\rho)_\rho$, $k=0,\dots,n-1$. If $g\in\Pi_n$, then $\tilde{g}_n^\rho=G_n^\rho\hat{g}_n^\rho$.

The mapping properties of A_0 and W on polynomials lead to a representation H_n of $R_n^\mu(A_0+\nu W)P_n^\sigma$ in the pair of the bases $(\{p_k^\sigma\}_{k=0}^{n-1}\,,\,\{p_k^\mu\}_{k=0}^{n-1})$ being a tridiagonal matrix, see [2]. Note that solving an algebraic system of equations having a tridiagonal matrix can be done with $O(n)$ operations. We compute z_n as follows.

$$\hat{z}_n^\sigma = H_n^{-1}(G_n^\mu)^{-1}\tilde{f}_n^\mu = H_n^{-1}(G_n^\mu)^T D_n^\mu \tilde{f}_n^\mu. \tag{5.24}$$

For the right–hand side of (5.23) we have

$$\tilde{g}_m^\mu = \tilde{f}_m^\mu - I_{m,n}G_n^\mu H_n I_n^m \hat{z}_n^\sigma, \tag{5.25}$$

where I_n^m is the diagonal matrix with the entries $[I_n^m]_{j,k}=1$, if $m\le j=k\le n-1$, and $[I_n^m]_{j,k}=0$, otherwise, $j,k=0,\dots,n-1$, and $I_{m,n}$ is the (non–quadratic) matrix $[I_{m,n}]_{i,j}=1$, if $x_{m,i}^\mu=x_{n,j}^\mu$, and $[I_{m,n}]_{i,j}=0$, otherwise, $i=0,\dots,m-1; j=0,\dots,n-1$.

Finally, we have to solve the quadrature scheme (5.23), which needs $O(m^3)$ operations. Using the *discrete cosine transformation*, see [25], we can do the computations (5.24) and (5.25) with an amount of $O(n\log n)$ operations. Choosing $m\sim\sqrt[3]{n}$ then the calculation of u_n needs really only $O(n\log n)$ operations.

5.5 Error estimates

Here we suppose that the right–hand side f of equation (3.9) is in $\mathbf{L}^2_{\mu,s}$, $s>\frac{1}{2}$, and that the operator A is in $G\mathcal{L}(\mathbf{L}^2_{\sigma,s},\mathbf{L}^2_{\mu,s})$. Further, let (4.4) be fulfilled. The uniquely determined solution is denoted by $u\in\mathbf{L}^2_{\sigma,s}$. For detailed proofs of the following two lemmata see [3].

Lemma 5.6. *Let $r \geq s > \frac{1}{2}$. Then for sufficiently large n and m we have the estimate*

$$\|Q_m^\sigma(u - z_n)\|_{\mathbf{L}^2_{\sigma,t}} \leq C\,(n^{t-s} + m^{t-r})\,\|f\|_{\mathbf{L}^2_{\mu,s}}, \quad 0 \leq t \leq s.$$

Proof. Use the decomposition $Q_m^\sigma(u - z_n) = Q_m^\sigma(I - BA)u + Q_m^\sigma h_n$, where $h_n := B(f - R_n^\mu f) + \nu B(R_n^\mu - I)Wz_n$. Since the collocation method for $(A_0 + \nu W)$ is stable, the assertion follows immediately.

Lemma 5.7. *Let $r \geq s > \frac{1}{2}$ and let n and m be such that $(2n+1)/(2m+1)$ is an integer. Then for sufficiently large n and m we have the estimate (note that w_m actually depends on n via (5.23))*

$$\|P_m^\sigma(u - w_m)\|_{\mathbf{L}^2_{\sigma,t}} \leq C\,(n^{t-s} + m^{t-r})\,\|f\|_{\mathbf{L}^2_{\mu,s}} \quad \textit{for all } t: \ \tfrac{1}{2} < t \leq s.$$

Proof. We have the identity

$$\begin{aligned} &R_m^\mu(A_0 + \nu W + K_m)P_m^\sigma(u - w_m) \\ &= R_m^\mu(A_0 + \nu W + K_m)P_m^\sigma u - R_m^\mu f + R_m^\mu(A_0 + \nu W)Q_m^\sigma z_n \\ &= R_m^\mu(K_m - K)u - R_m^\mu(A_0 + \nu W)Q_m^\sigma(u - z_n) - R_m^\mu K_m Q_m^\sigma u. \end{aligned}$$

Due to an orthogonality argument, the last summand vanishes. Using the stability of the quadrature method, after a straightforward calculation we arrive at the assertion.

Choosing $m \sim \sqrt[3]{n}$, we arrive at the optimal error estimate

$$\|u - u_n\|_{\mathbf{L}^2_{\sigma,t}} \leq C\,n^{t-s}\,\|f\|_{\mathbf{L}^2_{\mu,s}} \text{ for all } t > \tfrac{1}{2}: \ s - \tfrac{r-s}{2} \leq t \leq s, \tag{5.26}$$

which is completely analogous to (5.8).

6 Concluding remarks

1° *Spline approximation methods:* It is also possible to construct fast algorithms for solving the equation (3.5) if we seek the approximate solution in form of a spline function. In [5] there is given the analogue of the parametrix–Galerkin method (see Subsection 5.2 of the present paper) in the spline case. This algorithm can also be realized with an amount of only $O(n \log n)$ operations and it possesses a better convergence behaviour than the spline collocation method in "negative" Sobolev norms.

2° *Multigrid methods:* In the previous section we always have supposed that the small schemes of the fast algorithms are solved by a direct method with an amount of $O(m^3)$ operation. However, there are a couple of papers dealing with multigrid methods for equation (3.5), see for example [11, 12, 22], and for equation (3.9), see the paper of the authors [2]. In both cases we can reduce the computational complexity to $O(m^2)$ if we use multigrid methods.

Therefore, we have still an amount of $O(m^2 + n \log n)$ operations for the realization of the fast algorithms. If we choose n and m such that $m \sim \sqrt[3]{n}$ then the total amount of computational work for calculating the approximate solution u_n is again $O(n \log n)$. However, the error estimates (5.8) and (5.26) become optimal even for all $t > \frac{1}{2} : \; s - (r - s) \leq t \leq s$.

3° *Systems of equations:* The results of the present paper can be extended to systems of pseudodifferential equations without any difficulty.

4° *Multidimensional case:* In the present paper we have only dealt with one–dimensional pseudodifferential equations. However, the ideas leading to the corrected collocation method and to the parametrix–Galerkin method are not restricted to the one–dimensional case. It is possible to establish these methods also in the multidimensional case, if the underlying "classical" approximation methods are proved to be convergent and if the operator A^{-1} possesses a property like (5.13).

7 Numerical results

We have applied the fast algorithms from the Subsections 5.1 and 5.2 to a Cauchy singular integral equation on the unit circle $\mathbb{T}$. Using the parametric representation $\gamma : \mathbb{R} \to \mathbb{T}, \; y \mapsto e^{2\pi i y}$, such an equation has the form

$$Au := (aI + bS)u = f, \tag{7.1}$$

where S is the Cauchy singular integral operator (3.4) and a, b, u, and f are 1-periodic functions.

In the following table we give the results for equation (7.1) with $a(x) = 2 + |\sin 2\pi x|^5$ and $b(x) = 1$, such that the functions $c = a + b$ and $d = a - b$ are (invertible) elements of $\mathbf{H}^r$, $\forall r < 11/2$. The right–hand side f is chosen in such a way that (7.1) possesses the solution $u(x) = \cos 2\pi x + i * |\sin 2\pi x|$ being an element of $\mathbf{H}^s$, $\forall s < 3/2$.

We have done our computations on a Pentium PC (90 MHz) using a TURBO–Pascal 7.0 program code. In Table 7.1, there is written down the approximation error in the $\mathbf{H}^t$-norm ($t = 1, 0, -1, -2, -3$) of the parametrix–quadrature method (i.e. Amosov's method) and of the parametrix–Galerkin method. Furthermore, we give the results for the pure Galerkin and the pure collocation method in order to compare with (only for small n). Note that for equation (7.1) the quadrature method coincide with the collocation method. All algebraic systems of equations has been solved by the Gaussian algorithm.

The column "*cpu*" of the table contains the cpu–time needed (in seconds) for calculating the approximate solution, including the time needed for establishing the matrices but without the time needed for computing the right–hand sides. Note that we have defined the collocation points here by $x_{n,j} = (j + \frac{1}{2})/n, \; j = 0, ..., n-1$, instead of $x_{n,j} = j/(2n+1), \; j = -n, ..., n$, as in Section 4. For each of the methods, the numbers C and α characterize the convergence rate $\|u - u_n\|_{\mathbf{H}^t} \approx C\, n^{-\alpha}$. These two numbers were determined by the least square method from the sequence of the computed errors $\|u - u_n\|_{\mathbf{H}^t}$ and the corresponding integers n. According to (5.8) we can expect a rate of $O(n^{t-1.5})$ for $0.5 < t \leq 1.5$ in case of Amosov's method. For the parametrix– Galerkin method the theoretical convergence rate in the present case is $O(n^{\frac{t}{3}-1.5})$ for $-1 \leq t < -0.5$ and $O(n^{-1.833})$ for $-5.5 \leq t < -1$, see (5.18). The numbers α_{th} in Table 7.1 refer to these theoretical convergence rates.

Table 7.1 : **Errors and cpu – time**

n	M	m	cpu	$t=1$	$t=0$	$t=-1$	$t=-2$	$t=-3$
Amosov's Method								
80		16	0.38	0.13417316	0.00337379	0.00029899	0.00014776	0.00012408
216		24	1.93	0.08027192	0.00072422	0.00003509	0.00001974	0.00001696
480		32	5.88	0.05356986	0.00021505	0.00000675	0.00000399	0.00000343
1000		40	49.92	0.03703535	0.00007145	0.00000154	0.00000092	0.00000079
1584		48	83.92	0.02940520	0.00003583	0.00000061	0.00000037	0.00000032
	C			1.2	2.6	2.5	1.0	0.8
	α			−0.51	−1.52	−2.07	−2.01	−2.00
	α_{th}			−0.50	−1.00	−1.00	−1.00	−1.00
Parametrix – Galerkin Method								
80	16	8	0.22	0.14510365	0.01061950	0.00191730	0.00037883	0.00007957
216	24	12	1.37	0.08074067	0.00125040	0.00012795	0.00001675	0.00000234
480	32	16	4.12	0.05413013	0.00085010	0.00008826	0.00000959	0.00000106
1000	40	20	34.00	0.03704252	0.00009239	0.00000484	0.00000040	0.00000003
1584	48	24	57.67	0.02940949	0.00005166	0.00000279	0.00000021	0.00000002
	C			1.4	23.4	25.2	19.6	17.8
	α			−0.53	−1.76	−2.18	−2.50	−2.84
	α_{th}					−1.83	−1.83	−1.83
Collocation Method								
16			0.05	0.33612030	0.04038787	0.00814864	0.00394639	0.00320271
24			0.28	0.26115328	0.02094460	0.00333428	0.00169198	0.00141092
32			0.66	0.22086943	0.01329585	0.00179492	0.00093631	0.00078912
40			1.49	0.19482968	0.00938736	0.00111737	0.00059414	0.00050354
48			3.13	0.17624279	0.00707894	0.00076116	0.00041045	0.00034903
64			7.69	0.15092927	0.00454864	0.00041748	0.00022951	0.00019591
80			17.75	0.13410252	0.00323407	0.00026295	0.00014640	0.00012523
96			41.58	0.12188266	0.00244991	0.00018059	0.00010146	0.00008690
112			99.20	0.11249070	0.00193835	0.00013159	0.00007444	0.00006381
	C			1.5	3.0	2.8	1.1	0.8
	α			−0.56	−1.56	−2.11	−2.04	−2.01
	α_{th}			−0.50	−1.50	−1.50	−1.50	−1.50
Galerkin Method								
16			0.06	0.22762698	0.01693653	0.00171368	0.00019165	0.00002466
24			0.27	0.18473136	0.00901615	0.00059466	0.00004309	0.00000328
32			0.61	0.15964009	0.00580911	0.00028484	0.00001530	0.00000086
40			1.38	0.14264252	0.00414076	0.00016173	0.00000691	0.00000031
48			2.97	0.13014137	0.00314329	0.00010206	0.00000362	0.00000013
64			7.36	0.11264102	0.00203718	0.00004949	0.00000131	0.00000004
80			17.19	0.10072114	0.00145616	0.00002827	0.00000060	0.00000001
96			40.54	0.09193086	0.00110708	0.00001790	0.00000032	$5.8e-9$
112			97.06	0.08510296	0.00087821	0.00001216	0.00000018	$2.9e-9$
	C			0.92	1.12	1.90	3.58	8.03
	α			−0.50	−1.52	−2.54	−3.56	−4.62
	α_{th}			−0.50	−1.50	−2.50	−3.50	−4.50

References

1. Amosov, B.A. (1990): On the approximate solution of elliptic pseudodifferential equations on a smooth closed curve (in russian). ZfAA 9, 545–563
2. Berthold, D., Hoppe, W., Silbermann, B. (1992): The numerical solution of the generalized airfoil equation. J. Int. Eq. Appl. 4:3, 309–335
3. Berthold, D., Hoppe, W., Silbermann, B. (1992): A fast algorithm for solving the generalized airfoil equation. J. Comp. Appl. Math. 43, 185–219
4. Berthold, D., Silbermann, B. (1995): Corrected collocation methods for periodic pseudodifferential equations. Numer. Math. 70, 397–425
5. Berthold, D., Silbermann, B. (1996): The fast solution of periodic pseudodifferential equations. To appear in Math. Meth. Appl. Sc.
6. Capobianco, M.R., Mastroianni, G. (1995): Uniform boundedness of Lagrange operators in some weighted Sobolev type spaces. To appear in Math. Nachr.
7. Chandler, G.A., Sloan, I.H. (1990): Spline qualocation methods for boundary integral equations. Numer. Math. 58, 537–567
8. Elliott, D. (1982): Orthogonal polynomials associated with singular integral equations having a Cauchy kernel. SIAM J. Math. Anal. 13, 1041–1052
9. Fromme, J.A., Golberg, M.A. (1978): Numerical solution of a class of integral equations arising in two–dimensional aerodynamics. J. Opt. Th. Appl. 24:1, 169–206
10. Golberg, M.A. (ed.) (1990): Numerical solution of integral equations. Plenum Press, New York
11. Hackbusch, W. (1981): Die schnelle Auflösung der Fredholmschen Integralgleichung zweiter Art. Beitr. Num. Math. 9, 47–62
12. Hackbusch, W. (1985): Multigrid methods and applications. Springer, Berlin
13. Hartmann, T., Stephan, E.P. (1994): Rates of convergence for collocation with Jacobi polynomials for the airfoil equation. J. Comp. Appl. Math. 51, 179–191
14. Junghanns, P. (1994): Product integration for the generalized airfoil equations. In: Schock, E. (ed.) Beiträge zur Angew. Analysis und Informatik. Shaker, Aachen, pp. 171–188
15. Junghanns, P., Silbermann, B. (1981): Theorie der Näherungsverfahren für singuläre Integralgleichungen auf Intervallen. Math. Nachr. 103, 199–244
16. Kufner, A., John, O., Fučik, S. (1977): Function spaces. Prague
17. McLean, W., Wendland, W.L. (1989): Trigonometric approximation of solutions of periodic pseudodifferential equations. Oper. Th.: Advances and Applications 41, 359–383
18. Muskhelishwili, N.I. (1953): Singular integral equations. Noordhoff, Groningen
19. Prößdorf, S. (1978): Some classes of singular equations. North–Holland Publ. Comp., Amsterdam New York Oxford
20. Prößdorf, S., Silbermann, B. (1977): Projection methods and the approximate solution of singular equations (in german). BSB B.G. Teubner Verlagsges., Leipzig
21. Prößdorf, S., Silbermann, B. (1991): Numerical analysis for integral and related operator equations. Akademie–Verlag, Berlin. Birkhäuser–Verlag, Basel
22. Schippers, H. (1983): Multiple grid methods for equation of the second kind with application in fluid mechanics. Math. Centre Tracts 163, Amsterdam
23. Sloan, I.H. (1988): A quadrature–based approach to improving the collocation method. Numer. Math. 54, 41–56
24. Sloan, I.H., Wendland, W.L. (1989): A quadrature–based approach to improving the collocation method for splines of even degree. ZfAA 8, 361–376

25. Tasche, M. (1993): Fast algorithms for discrete Chebyshev-Vandermonde transforms and applications. Numer. Algorithms 5, 453–464
26. Triebel, H. (1978): Interpolation theory, function spaces, differential operators. North–Holland Publ. Comp., Amsterdam New York Oxford
27. Zygmund, A. (1959): Trigonometric series, Vol. I–II. Cambridge Univ. Press, London

Different Methodologies for Coupled BEM and FEM with Implementation on Parallel Computers

U. Brink, M. Kreienmeyer, and E. Stein

1 Introduction

The boundary element method (BEM) and the finite element method (FEM) may be computationally expensive if complex problems are to be solved, thus there is the need of implementing them on fast computer architectures, especially parallel computers. Because BEM and FEM are complementary to each other, the coupling of these methods is widely used. In this paper, a parallel FE–BE-code and its implementation on a distributed memory system (Parsytec MultiCluster2) is described. As model problems, we assume linear elasticity for the boundary element method and elastoplasticity for the finite element method. The efficiency of our implementation is shown by various test examples. Two coupling methods are proposed. First, we show by numerical examples that a multiplicative SCHWARZ method for coupling BEM with FEM is very well suited for parallel implementation. Furthermore, we treat a symmetric coupling of GALERKIN-BEM and mixed finite elements of RAVIART–THOMAS type leading to improved stress approximation.

This paper treats different methodologies of the coupling of FEM and BEM and the parallel implementation on a distributed memory system, the Parsytec MultiCluster2. Most parallel computers can be divided into two categories: shared memory and distributed memory computers. Shared memory systems typically have few very powerful processors, whereas distributed memory computers (like Intel Hypercube, nCube, PowerXplorer) normally consist of a large number of moderate performance processors with local memory which requires communication between the processors, achieved by message passing. The efficient implementation of numerical algorithms on these two architectures differs substancially. On shared memory computers, it is analogous to the implementation on vector computers, whilst distributed memory architectures imply a finer grain parallelism with a larger number of communication steps.

In particular, the parallelization is performed by global data decomposition which normally leads to finer grain algorithms compared with the well-known domain decomposition method. The same algorithm as on a sequential computer is implemented, and data have to be distributed in such a way that the communication and synchronization costs are minimized.

The outline of this paper is as follows: In Section 2, we briefly describe the model problem (linear elasticity and elastoplasticity for BEM and FEM, respectively) and its discretization by FEM and BEM. In Section 3, the parallel implementation of the two methods is treated, including some results from various test examples. Solving the linear systems of equations causes the highest communication costs and is described in detail. Section 4 contains the BEM–FEM coupling by a multiplicative Schwarz method and a symmetric coupling of GALERKIN BEM with mixed finite elements of the RAVIART-THOMAS type.

2 The model problem

2.1 Linear elasticity and rate-independent elastoplasticity

The linearized strain $\boldsymbol{\epsilon} = \frac{1}{2}(\nabla \boldsymbol{u} + (\nabla \boldsymbol{u})^T)$ is decomposed into an elastic part $\boldsymbol{\epsilon}^e$ and a plastic part $\boldsymbol{\epsilon}^p$,

$$\boldsymbol{\epsilon}^e = \boldsymbol{\epsilon} - \boldsymbol{\epsilon}^p. \tag{2.1}$$

In the frame of linear elasticity, a free energy function is postulated in the form

$$\Psi = \frac{1}{2}\boldsymbol{\epsilon}^e : \mathcal{C} : \boldsymbol{\epsilon}^e \ , \tag{2.2}$$

with the elastic strains $\boldsymbol{\epsilon}^e$ and the symmetric and positive definite elastic tensor $\mathcal{C}$. Then the stresses follow as

$$\boldsymbol{\sigma} = \partial_{\boldsymbol{\epsilon}^e}\Psi(\boldsymbol{\epsilon}^e) \ . \tag{2.3}$$

As a model problem, we treat the classical elastoplastic VON MISES model (see e. g. [16]) with the yield function

$$\Phi(\boldsymbol{\sigma}, \alpha) = \| \operatorname{dev}\boldsymbol{\sigma} \| - \sqrt{\frac{2}{3}}K(\alpha) \ , \tag{2.4}$$

where $K(\alpha)$ denotes the isotropic hardening function. For linear hardening we have

$$K(\alpha) = y_0 + H\alpha \ , \tag{2.5}$$

with the hardening modulus $H \geq 0$ and the initial yield stress $y_0 \geq 0$. The associated flow rule is described by

$$\dot{\boldsymbol{\epsilon}^p} = \lambda \partial_{\boldsymbol{\sigma}}\Phi(\boldsymbol{\sigma}, \alpha) \tag{2.6}$$

and the evolution of the internal variable α by

$$\dot{\alpha} = \lambda\sqrt{\frac{2}{3}}. \tag{2.7}$$

The loading–unloading conditions can be expressed in the Kuhn–Tucker form as

$$\lambda \geq 0, \ \Phi(\boldsymbol{\sigma}, \alpha) \leq 0, \ \lambda\Phi(\boldsymbol{\sigma}, \alpha) = 0 \ . \tag{2.8}$$

The solution of the coupled elastoplastic problem is obtained by a standard return mapping algorithm [6, 22].

2.2 Discretization with boundary elements

In the boundary element domain Ω_B linear elasticity is assumed. We start from KELVIN's fundamental solution which for plane strain reads

$$G(x, y) = \frac{-1}{8\pi(1-\nu)G} \left\{ (3-4\nu)\ln(r)\delta_{ij} - r_{,i}r_{,j} \right\} \tag{2.9}$$

with

$$r_i = y_i - x_i, \quad r = \sqrt{\langle r_i, r_i \rangle}, \quad r_{,i} = \frac{r_i}{r}.$$

G is the shear modulus. The displacements at any interior point of Ω_B are given by the BETTI formula

$$u(x) = -\int_{\Gamma_B} G(x, y)t(y)ds(y) + \int_{\Gamma_B} (T_y G(x, y))^T u(y)ds(y) \tag{2.10}$$

where T_y denotes the traction operator, and $t(y) = T_y u(y)$ are the tractions corresponding to the displacements u at a point y. Taking the limit $x \to \Gamma_B$ we obtain the first boundary integral equation

$$\frac{1}{2}u = -Vt + Ku \; . \tag{2.11}$$

Here,

$$(Vt)(x) = \int_{\Gamma_B} G(x, y)t(y)ds(y), \tag{2.12}$$

is a weakly singular operator and

$$(Ku)(x) = \int_{\Gamma_B} (T_y G(x, y))^T u(y)ds(y), \tag{2.13}$$

is Cauchy-singular. Applying the traction operator T_x yields the second boundary integral equation

$$\frac{1}{2}t = -K't - Wu \tag{2.14}$$

where K' is the adjoint of K and

$$(Wu)(x) = -T_x \int_{\Gamma_B} (T_y G(x, y))^T u(y)ds(y), \tag{2.15}$$

is a hypersingular operator. There are different numerical methods, namely collocation, BUBNOV-GALERKIN, NYSTRÖM and least-square method. With the collocation method, using the first integral equation 2.11 one obtains the following linear system of equations

$$H\,u = G\,t \;\; , \tag{2.16}$$

where $\boldsymbol{u}$ and $\boldsymbol{t}$ contain the unknown and the known displacements and tractions in the collocation points. Reordering yields

$$\boldsymbol{A}\,\boldsymbol{x} = \boldsymbol{b} \tag{2.17}$$

with a nonsymmetric and dense matrix $\boldsymbol{A}$. Using both integral equations (2.11) and (2.14) the GALERKIN method leads to a symmetric but in general indefinite system matrix $\boldsymbol{A}$. For this reason, standard iterative solvers for symmetric positive definite matrices may not work. In [12, 14, 18], some iterative solvers are tested. It turns out that for the collocation method Bi-CGSTAB [25] and GMRES [19] are favourable.

In the following, we compare collocation and BUBNOV-GALERKIN method with respect to accuracy. The regular integrals in the collocation code are computed with a GAUSSian rule, the singular integrals analytically. For the GALERKIN BEM, we use an implementation [17] where the inner integration is performed analytically whereas for the outer integrals a GAUSSian rule is used. The non-regular integrals are evaluated analytically.

As a test example, we consider a circular cavity with radius $1\,cm$ in an infinite elastic medium under horizontal tension of magnitude $1\frac{kN}{cm^2}$. The problem is analyzed under plane strain conditions, the material parameters are Young's modulus $E = 1\frac{kN}{cm^2}$ and Poisson's ratio $\nu = 0.2$. Linear trial functions are used for the unknown displacements. In Table 2.1, the computed horizontal stresses $\sigma_{x,h}$ from the collocation and the GALERKIN method at the point (0, 1.02) are given for different numbers of regulary distributed elements. In Figure 2.1, the geometry of the problem is given.

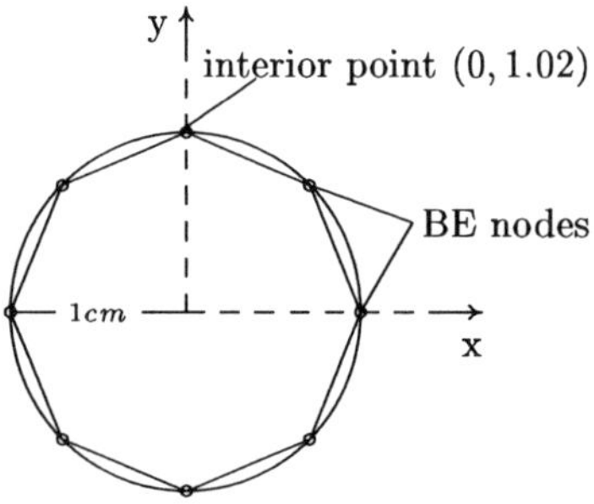

Fig. 2.1. Geometry and coarse discretization of circular cavity problem

n, n' denote the number of elements on two different discretization levels, and $\kappa_n = \frac{log(E_{n'}/E_n)}{log(n/n')}$ is the order of convergence with $n' < n$ and $E_n = |\sigma_x - \sigma_{x,h}|$. The GALERKIN method yields higher accuracy and higher order of convergence (exact solution $\sigma_x = 2.8663525\frac{kN}{cm^2}$).

Table 2.1. $\sigma_{x,h}$ at the point (0, 1.02)

	GALERKIN		Collocation	
n	$\sigma_{x,h}$	κ_n	$\sigma_{x,h}$	κ_n
8	2.7877299		3.3673909	
16	2.8632967	4.69	3.0131097	1.77
32	2.8655840	1.99	2.904475	1.94
64	2.8659372	0.89	2.8781045	1.70
128	2.8662519	2.05	2.8687889	2.10
256	2.8663456	3.84	2.8669831	2.03
512	2.8663524	4.88	2.8665075	2.02

2.3 Discretization with finite elements

The displacement-based finite element method is a GALERKIN method or mechanically the principle of virtual work which for the discretized domain without body forces reads

$$\sum_{i=1}^{n} \int_{\Omega_i} \sigma_h(\boldsymbol{u}_h) : \epsilon_h(\boldsymbol{v}_h)\, d\Omega = \sum_{i=1}^{n} \int_{\partial\Omega_i \cap \Gamma_t} \bar{\boldsymbol{t}}_h \boldsymbol{v}_h\, dA \tag{2.18}$$

with

$$\boldsymbol{u}_h, \boldsymbol{v}_h \in H^1(\Omega),\ \boldsymbol{v}_h = 0 \text{ on } \Gamma_u,\ \boldsymbol{u}_h = \bar{\boldsymbol{u}} \text{ on } \Gamma_u, \tag{2.19}$$

$$\bar{\boldsymbol{t}}_h \in L_2(\Gamma_t)^2, \tag{2.20}$$

and Ω_i is the element domain, n denotes the total number of elements, The trial displacements $\boldsymbol{u}_h$ within the element i are approximated by linear shape functions $\boldsymbol{N}$, written in condensed matrix form as

$$\boldsymbol{u}_h = \boldsymbol{N}\boldsymbol{u}^i \text{ in } \Omega_i\ , \tag{2.21}$$

where $\boldsymbol{u}^i$ are the nodal displacements. In the same way, the test functions $\boldsymbol{v}_h$ are represented. This leads to element stiffness matrices $\boldsymbol{k}^i$ and element force vectors $\boldsymbol{f}^i$, which are assembled to the global algebraic system

$$\mathbf{K}\boldsymbol{u} = \boldsymbol{f}\ , \tag{2.22}$$

where $\mathbf{K}$ is symmetric, positive definite and sparse. This system can be solved by various direct and iterative solvers.

In the case of elastoplastic materials a linearization is necessary if NEWTON-type methods are used. Then, in each time step i (or load step i) the linear system

$$\boldsymbol{K}_T(\boldsymbol{u}_{j+1}^i)\Delta\boldsymbol{u}_{j+1}^i = \boldsymbol{R}(\boldsymbol{u}_{j+1}^i) \tag{2.23}$$

has to be solved. $\boldsymbol{K}_T(\boldsymbol{u}_{j+1}^i)$ and $\boldsymbol{R}(\boldsymbol{u}_{j+1}^i)$ are the tangent stiffness matrix and the residual force vector at the configuration defined by $\boldsymbol{u}_{j+1}^i$. Here, subscripts refer to the load step and superscripts to iteration within the load step.

3 Parallel implementation

On distributed memory computers, each processor has access to its own local memory only. Data exchange between different processors is done by message passing, and the time needed for this interprocessor communication is to be considered. For parallelization, data are distributed onto the processors in such a way that communication costs should be as low as possible. One approach is the domain decomposition method which yields high efficiency if it is possible to divide the domain into subdomains of nearly equal amount of work. In the case of non-linearities or adaptive mesh refinement, this can be extremely complicated. In this paper, we discuss the parallelization by data decomposition. It yields an algorithm with very good load-balancing and high efficiency as shown by the examples in this section.

3.1 Parallel implementation of the BEM-code

The parallel program starts by reading the output file of a preprocessor, and then data are distributed amongst the processors. The parallel BEM-code consists of three parts:

- assembling of the system matrix and the right-rand side
- solving the linear system of equations
- computing displacements and stresses at interior points

The parallel implementation of the first and the last part can be done very simply in different ways, but solving the system of equations causes high communication costs. Due to this fact, the data distribution remains unchanged during all steps of computation and depends on the implementation of the solver.

3.1.1 Parallel assembling of the system matrix

For the boundary element method, only a discretization of the boundary is needed. This is a small amount of data in comparison to domain discretizing algorithms like the finite element method. For 2D-problems it is possible to store all data needed for the description of the problem (geometry, boundary conditions, elements, nodes etc.) in each local memory. As a result, each processor can now compute any element of the system matrix or the right–hand side independently of the other processors, i.e. without communication. In the case of linear elements in 2D linear elasticity we have to compute 2x4 element matrices. Some of the computations can be done for all 8 entries of the element matrix. If the columns of **A** are distributed these computations have to be done four times, in case of a row–wise distribution only two times. Another advantage of the row–oriented version is that it is possible to compute the strongly singular integrals, which occur in the **H**–matrix of the collocation method by a row summation without communication. In

comparison to the row–oriented assembling, the column–oriented version is about 1.3 times slower. So, each processor has to perform n/p rows of the system matrix and n/p elements of the right–hand side vector which gives a well-balanced workload.
If a direct solver is applied, a wrap–around–distribution is used. In the case of an iterative solution, a blockwise storage of the matrix **A** is performed, see Figure 3.1.

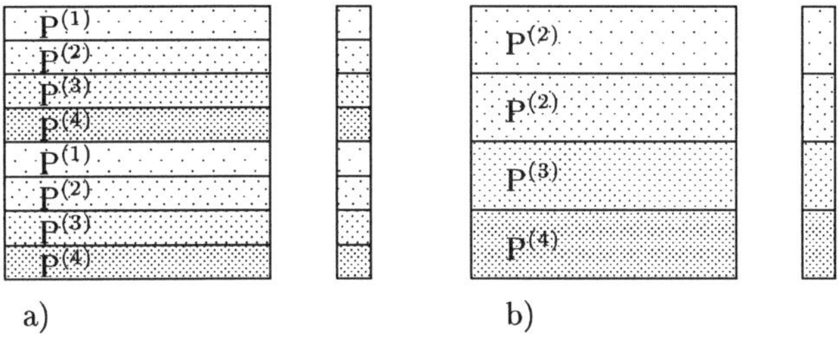

Fig. 3.1. Storage scheme a) wrap–around-distribution b) blockwise distribution

3.1.2 Computing displacements and stresses at interior points

After solving the system, the computed data, i.e. tractions and displacements on the boundary, are distributed on the local memories of the processors. Then, each processor is able to compute displacements and stresses at any given interior point. If we want to compute results at m points with p processors, each has to perform m/p points, and a good load–balancing is achieved. The computed results on the boundary and at interior points are collected on one processor which sends these data to the host computer where the post-processing is done. The collection and broadcasting of data is implemented in a tree of processors.

3.1.3 Parallel implementation of the solver

We have to solve a linear system with a full and nonsymmetric or symmetric but indefinite system matrix. This can be done by a direct method, i.e. LU-decomposition. If we employ p processors for the solution of a linear system with n equations, each processor receives n/p rows and n/p elements of each vector in its local memory. These rows of **A** are distributed in a cyclic manner via the so–called wrap–around–map so that an equally balanced workload can be achieved. This distribution is not done via communication but the rows are computed locally as needed for the LU–decomposition, i. e., processor $\mathrm{P}^{(i)}$ computes rows with index j with $j \bmod p = i$. For details, see [1, 9].
For the parallel implementation of an iterative solver on p processors, each processor receives a block of n/p rows of the matrix **A** in its local memory, i. e., processor $\mathrm{P}^{(i)}$ computes rows with index j with $(n/p) \cdot i \leq j < (n/p) \cdot (i+1)$. The most time consuming part of an iterative solver is the matrix vector

product. This is performed in a ring of processors. In a C-like pseudo code, the product $\mathbf{A}\ \mathbf{q} = \mathbf{v}$ reads for each processor $\mathrm{P}^{(i)}$:

```
v_local = A_local*q_local;
tmp_local = q_local;
for(k=0; k<Number_of_Processors; k++){
   Send tmp_local to the right neighbor;
   Receive tmp1_local from the left neighbor;
   Compute v_local+=A_local*tmp1_local;
   tmp_local = tmp1_local;
}
```

The communicational amount is of order $\mathcal{O}(n)$, the arithmetical costs are of order $\mathcal{O}(n^2)$.

3.1.4 Performance results

In this section, some results on the parallel efficiency of the BE-code (collocation method) are given. To compute the parallel efficiency E_p

$$E_p = \frac{Speedup}{Number\ of\ processors}$$

we use the definition for the parallel speedup S_p

$$S_p = \frac{Time\ on\ one\ processor}{Time\ on\ p\ processors} \quad .$$

As a test example, the problem described in Figure 3.2 is discretized with a varying number of elements. The material parameters are $E = 21000 \frac{kN}{cm^2}$, $\nu = 0.1$, and the internal pressure is $p_r = 750 \frac{kN}{cm^2}$.

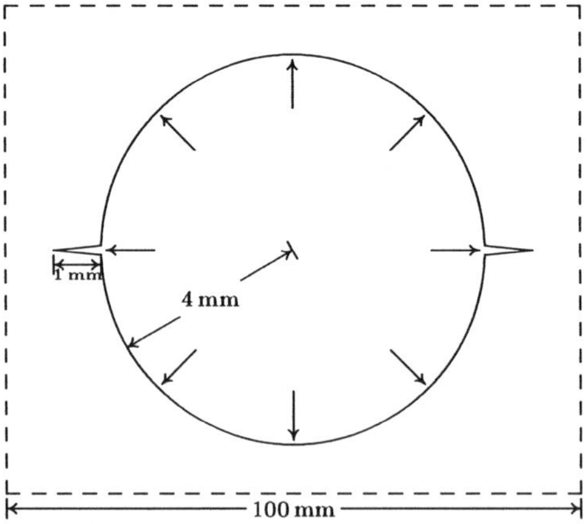

Fig. 3.2. Bore hole with two radial cracks (plane strain)

In Table 3.1 we show some results of the performance of different iterative solvers and the LU-decomposition.

Table 3.1. Performance results for direct (LU) and iterative (Bi–CGSTAB, GMRES) solvers, varying granularity, 32 processors

p=32	n=400	n=800	n=1600	n=2000
E_p(LU)	0.40	0.59	0.80	0.90
E_p(Bi-CGSTAB)	0.54	0.72	0.91	0.93
E_p(GMRES)	0.51	0.71	0.92	0.95

These results show that a considerable efficiency in parallel computing can be obtained also for systems of moderate size, i.e. fine granularity. For larger systems, the iterative solver becomes superior. In Table 3.2, we give results for different numbers of processors for assembling the system matrix and the right–hand side.

Table 3.2. Speedup and efficiency for assembling the system

p	1	3	4	5	8	12	16	32
Speedup	1	2.94	3.95	4.95	7.8	11.65	15.6	31.2
Efficiency	1.00	0.98	0.99	0.99	0.98	0.97	0.98	0.98

The parallel efficiency and the speedup for the calculation of stresses and displacements at interior points are given in Table 3.3.

Table 3.3. Speedup and efficiency for calculation at interior points

p	1	3	4	5	8	12	16	32
Speedup	1	2.9	3.8	4.75	7.7	11.3	15.05	30.1
Efficiency	1.00	0.97	0.95	0.95	0.96	0.94	0.94	0.94

The performance of a parallel algorithm depends on the ratio of arithmetic work and the amount of communication, i.e. the granularity of the problem. Assembling the system matrix, computing the right–hand side and the computation at internal points imply very low communication costs. But solving the system by LU–decomposition or iterative solvers requires a finer granularity for this part of the code. The parallel efficiency of assembling the system and of the evaluation of stresses and displacements at internal points does not depend as strongly on the granularity as it is the case for solving the system. For the latter, the efficiency slows down if more and more processors are used for a problem of constant size.

3.2 Parallelization of FEM

A displacement–based finite element program consists of the following parts:

- generation of element stiffness matrices and force vectors,
- assembling and solution of the global algebraic system,
- calculations of strains and stresses within each element.

In nonlinear analysis, these modules interact in an iterative manner. The parallelization of FEM is done similarly to the BEM implementation, i.e. by data distribution [10, 11]. Since FEM is a domain-discretizing method, the amount of data for describing the problem is much larger than for BEM. Each processor can receive only a part of the whole data in its local memory. Starting with a coarse mesh, macro elements (i.e. problem data concerning element patches) are distributed on p processors and then refined regularly so that a hierarchical mesh is obtained. Then each processor computes a local stiffness matrix for these elements. This can be done in parallel without any communication, and each processor assembles a local matrix $\mathbf{K}_i$. The global stiffness matrix can be written as

$$\mathbf{K} = \sum_{i=1}^{p} \boldsymbol{B}_i^T \mathbf{K}_i \boldsymbol{B}_i \ , \tag{3.1}$$

where $\boldsymbol{B}_i$ is a Boolean matrix but $\mathbf{K}$ is not assembled. In comparison with the domain decomposition method [8], each processor receives more than one macro element (see Figure 3.3) which results in a finer grain algorithm with higher communicational amount. But in case of local nonlinearities which occur in elastoplastic FE analysis or adaptive mesh refinement, a better load–balancing is achieved. The system is solved with a parallel preconditioned

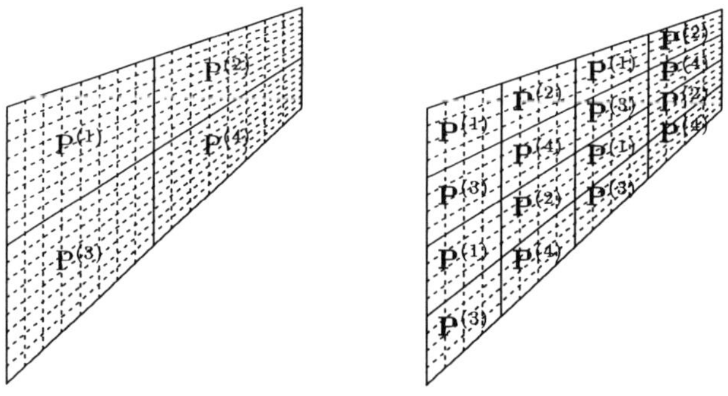

(a) (b)

Fig. 3.3. Distribution of 256 elements on 4 processors, a) 1 macro element b) 4 macro elements on each processor

conjugate gradient method. As a preconditioner we use the concept of hierarchical bases [26, 27]. Therefore no additional communication (in comparison with the unpreconditioned cg) is needed for this method.

3.2.1 Parallel implementation of the pcg–solver

In what follows, the parallel implementation of the conjugate gradient method is described, for details see [15]. Communication is only needed for nodes at the vertices of macro elements and for computing global scalar products. Applying hierarchical bases as preconditioning method, a hierarchical mesh as in the multigrid method is required. The refinement is done locally on each processor. Then, if $\mathcal{V}_j$ is the standard FE nodal basis on the refinement level j, the hierarchical basis $\mathcal{S}_j$ is constructed by

$$\mathcal{S}_j = \mathcal{S}_{j-1} \oplus \mathcal{V}_j \ , \tag{3.2}$$

where the functions in $\mathcal{V}_j$ vanish at the nodes associated with $\mathcal{S}_{j-1}$, and the coarse grid space satisfies

$$\mathcal{S}_0 = \mathcal{V}_0 \ . \tag{3.3}$$

This leads to a transformed system matrix

$$\boldsymbol{K}_{hb} = \boldsymbol{S}^T \boldsymbol{K} \boldsymbol{S} \ , \tag{3.4}$$

where $\boldsymbol{S}$ and $\boldsymbol{S}^T$ are basis transformation matrices. Due to the low condition number of $\boldsymbol{K}_{hb}$, namely

$$\kappa(\boldsymbol{K}_{hb}) = \mathcal{O}(|\, ln\ h\,|^2) \tag{3.5}$$

and

$$\kappa(\boldsymbol{K}_{hb}) = \kappa(\boldsymbol{S}^T \boldsymbol{K} \boldsymbol{S}) = \kappa(\boldsymbol{S}^T \boldsymbol{S} \boldsymbol{K}) \ , \tag{3.6}$$

it follows that $\boldsymbol{C}^{-1} = \boldsymbol{S}^T \boldsymbol{S}$ is a very efficient preconditioning matrix.

During mesh refinement, each processor stores the information concerning the generated nodes and elements. With this information at hand, the required matrix multiplications $\boldsymbol{S}\boldsymbol{y}, \boldsymbol{S}^T \boldsymbol{r}$ can be performed very efficiently. In parallel, one computes locally on each processor $P^{(i)}$

$$\boldsymbol{y}_i = \boldsymbol{S}_i^T \boldsymbol{r}_i \ . \tag{3.7}$$

For the transformations, one proceeds from the finest to the coarsest mesh. Because all refinement steps have been performed locally on each processor $P^{(i)}$, all data are available to compute this step without communication. Only on the coarsest grid, there are nodes belonging to more than one macro element, i.e. to more than one processor. Now, the additive vector $\boldsymbol{r}_i$ has to be transferred to an overlapping one by

$$\boldsymbol{y} = \sum_{i=1}^{p} \boldsymbol{B}_i^T \boldsymbol{y}_i \ . \tag{3.8}$$

After this, the back transformation from the coarsest to the finest mesh can be computed with

$$\boldsymbol{w}_i = \boldsymbol{S}_i \tilde{\boldsymbol{y}}_i \text{ with } \tilde{\boldsymbol{y}} = \boldsymbol{B}_i \boldsymbol{y} \ . \tag{3.9}$$

Of course summations only have to be performed between processors which have common macro element vertices. For regular meshes it is easy to map neighboring elements on neighboring processors, so that no global communication has to be performed. The summation of the global scalar product can be implemented in a very efficient way in a tree of processors. For more complex meshes (e.g. due to adaptive mesh refinement), the mapping can be performed by recursive bisection algorithms (see [23]). For the parallel implementation of an adaptive finite element code with hierarchical preconditioning, see [11].

3.3 Performance results

For the kind of data distribution described in 3.2, it is important to analyze how the granularity of the distribution affects the efficiency of assembling the local stiffness matrices and the solver. Working with more and more macro elements on each processor leads to a better load-balancing but to less efficiency for the pcg because the amount of communication increases. Furthermore, there will be a loss of numerical efficiency, since for finer starting meshes, the pcg needs more iteration steps.

As a test example, we choose Cook's membrane problem (see Figure 3.4) with different numbers of macro elements on each processor.

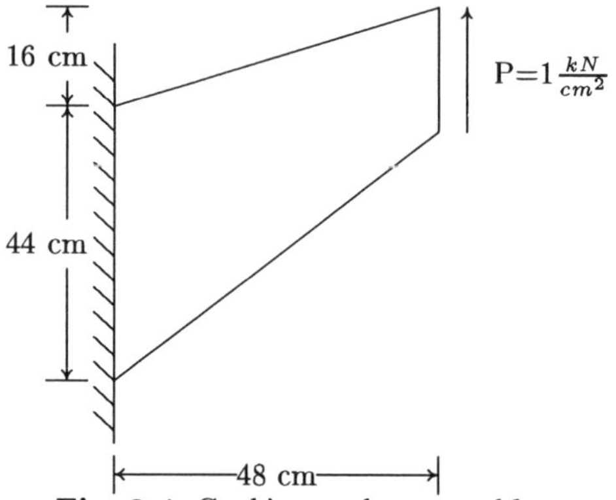

Fig. 3.4. Cook's membrane problem

The coarsest mesh is given in Figure 3.3. For this plane strain problem with von Mises plasticity and linear isotropic hardening. The material parameters are

- Young's Modulus $E = 70\frac{kN}{cm^2}$, Poisson's ratio $\nu = 0.33333$
- $H = 120.0\frac{kN}{cm^2}$, $y_0 = 0.45\frac{kN}{cm^2}$.

On the finest mesh, we have 65536 elements, i.e. 131584 unknowns, and 32 processors are used. In the case of elastoplasticity, the best distribution of macro elements depends very strongly on the element type. If the evaluation of the element stiffness matrix is cheap, a worse load–balancing during the assembly has a smaller effect on the overall efficiency of the algorithm than the loss in efficiency in the pcg–solver. But if higher–order elements with highly non–linear material laws are used, the load–balancing becomes more and more important.

Table 3.4. Efficiency for assembling the stiffness matrix (32 processors)

time step	14	15	16	17	18	19	20
1 macro element	0.98	0.93	0.89	0.87	0.85	0.84	0.80
4 macro elements	0.99	0.96	0.93	0.92	0.92	0.94	0.95
16 macro elements	0.99	0.97	0.96	0.97	0.98	0.96	0.96

Table 3.5. Efficiency and Speedup for solving the system (32 processors)

macro elements	Efficiency	Speedup
1	0.90	28.8
4	0.85	27.4
16	0.80	25.6

4 Coupling of BEM and FEM

4.1 Multiplicative SCHWARZ method in two overlapping domains

As a coupling algorithm, we use the multiplicative SCHWARZ method in two overlapping domains [21]. See Figure 4.1 for the notations. Ω_1 denotes the FE-domain, Ω_2 denotes the BE-domain. Then, the iteration scheme reads

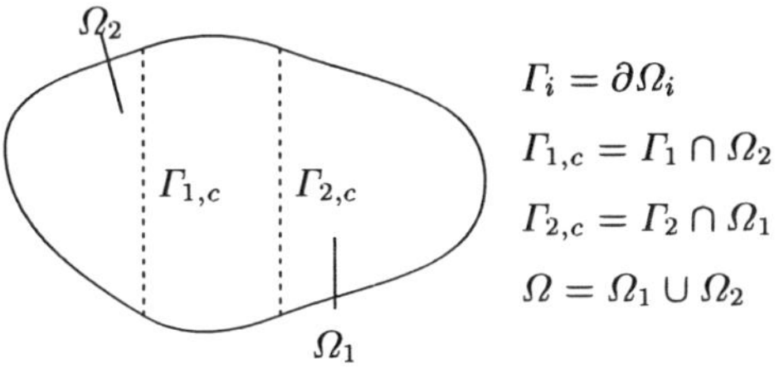

Fig. 4.1. Overlapping domain decomposition

Start with $\boldsymbol{u}_{1,c}^0 = 0$

$\boldsymbol{u}^m \to \boldsymbol{u}^{m+1}$

Solve $\quad \boldsymbol{K}\,\boldsymbol{u}_1^{m+\frac{1}{2}} = \boldsymbol{f}_1$

with $\boldsymbol{u}_{1,c}^{m+1} = \boldsymbol{u}_{2,c}^m$ on $\Gamma_{1,c}$

Solve $\quad \boldsymbol{A}\,\boldsymbol{x}^{m+1} = \boldsymbol{b}$

from $\boldsymbol{H}\,\boldsymbol{u}_2^{m+1} = \boldsymbol{G}\,\boldsymbol{t}_2^{m+1}$

$$\text{with } \boldsymbol{u}_{2,c}^{m+1} = \boldsymbol{u}_{1,c}^{m+\frac{1}{2}} \text{ on } \Gamma_{2,c}$$

$$\text{Compute } \boldsymbol{u}_{2,c}^{m+1} = \widehat{\boldsymbol{G}}\, \boldsymbol{t}_2^{m+1} - \widehat{\boldsymbol{H}}\, \boldsymbol{u}_2^{m+1} \text{ on } \Gamma_{1,c}$$

If the displacements have to be computed at k interior points, $\widehat{\boldsymbol{G}}$ and $\widehat{\boldsymbol{H}}$ are $2k \times 2n$ matrices resulting from the discretization of BETTI's formula (2.10). The advantage of this iteration scheme compared with other coupling methods is that for each subdomain optimal solvers can be implemented, and the parallel BEM- and FEM-codes can be used without changes. The system matrix $\boldsymbol{A}$ remains the same during the nonlinear computation. Note that the evaluation of the matrices $\widehat{\boldsymbol{G}}$ and $\widehat{\boldsymbol{H}}$ is performed only in the first iteration step.

In the following, we assume linear elastic behavior in Ω_2 and an elastoplastic material in Ω_1. The nonlinear problem is solved by NEWTON's method. In each step of the NEWTON algorithm, one iteration step of the multiplicative SCHWARZ method is performed. The iteration scheme then reads:

$$\left.\begin{array}{l} \text{Solve } \boldsymbol{K}_T\, \Delta \boldsymbol{u}_1^i - \boldsymbol{f} = 0 \\ \qquad \text{with } \boldsymbol{u}\Big|_{\Gamma_{1,c}} = \boldsymbol{u}_2^{i-1}\Big|_{\Gamma_{1,c}} \quad \left(\boldsymbol{u}_2^{i-1}\Big|_{\Gamma_{1,c}} = 0 \text{ if } i = 1 \right) \\ \text{Solve } \boldsymbol{A}\, \boldsymbol{x}_2 = \boldsymbol{b} \text{ with } \boldsymbol{u}\Big|_{\Gamma_{2,c}} = \boldsymbol{u}_1^i\Big|_{\Gamma_{2,c}} \end{array}\right\} \tag{4.1}$$

As a test example the elastoplastic behavior of a square block compressed by two opposite perfectly rigid punches is studied. The problem is analyzed under plane strain conditions and the material is assumed to be perfectly plastic according to Section 2.1. The problem data are

- Young's modulus $E = 6890 \frac{kN}{cm^2}$, Poisson's ratio $\nu = 0.33$,
- yield stress $y_0 = 0.89 \frac{kN}{cm^2}$,
- hardening parameter $H = 0$,
- pressure $p_y = 25 \frac{kN}{cm^2}$,
- 10 load steps.

In Figure 4.2, a coarse discretization is shown. For the parallel computation, 5184 finite elements, 316 boundary elements and 215 interior points are used. In Figure 4.3, the plastic zone after 10 load steps is shown.
For the computing times on different numbers of processors, see Table 4.1.

Table 4.1. Computing time in seconds, varying processor numbers p

p	8	16	32
Time	197.7	116.3	82.5
Efficiency	0.90	0.77	0.54

In comparison to the pure BE- and FE-computations, the parallel efficiency is now lower due to the fact that data exchanges are needed and the

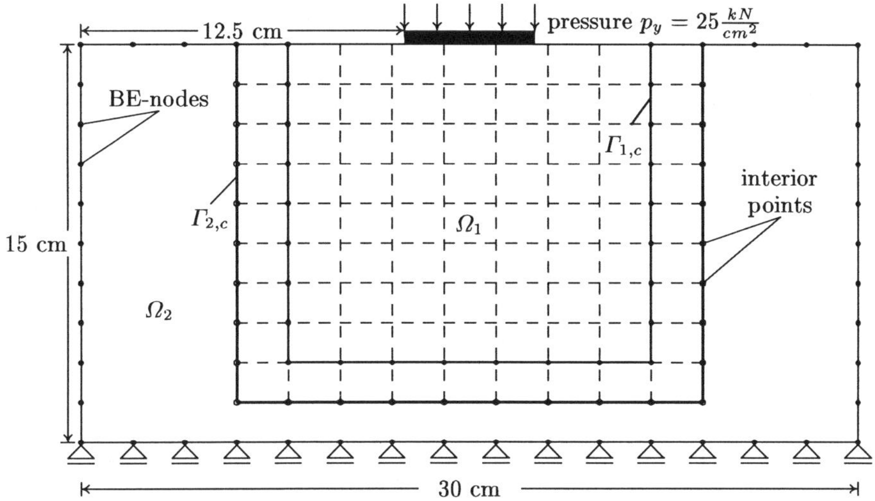

Fig. 4.2. Geometry and coarse discretization for rough punch problem, FE discretization in Ω_1, BE discretization on $\Gamma_2 = \partial\Omega_2$

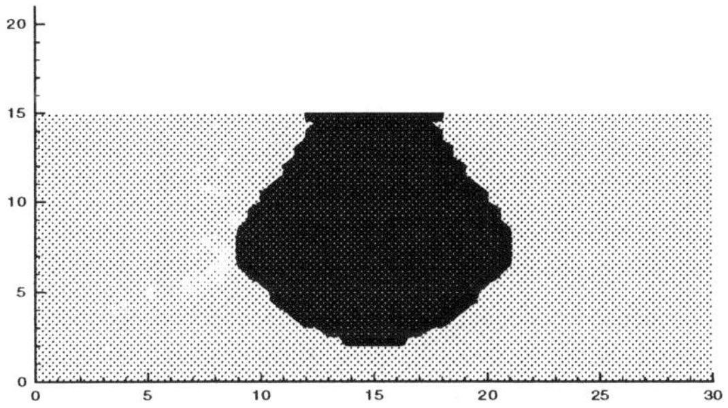

Fig. 4.3. Total spread of plastic zone

BE-system is solved only in the first iteration step. For further load steps, only a back- and forward substitution is performed which requires less CPU-time but is of lower parallel efficiency.

In [12], numerical experiments concerning different sizes of overlaps are reported. Larger overlapping implies faster convergence of the multiplicative SCHWARZ method but higher computational amount in each iteration step, too.

4.2 Symmetric coupling of GALERKIN-BEM with mixed finite elements

In comparison to standard FEM, one of the main benefits of boundary elements is the high accuracy in stress computations. An alternative approach is to use mixed finite elements with stresses rather than displacements as primary unknowns. By these means one may also overcome locking when Poisson's ratio ν approaches 0.5.

For simplicity, we consider the linear elasticity problem

$$\left.\begin{aligned} \operatorname{div}\boldsymbol{\sigma} &= -\boldsymbol{f} && \text{in } \Omega \\ \boldsymbol{\sigma} - \mathcal{C}\boldsymbol{\varepsilon}(\boldsymbol{u}) &= \mathbf{0} && \text{in } \Omega \\ \boldsymbol{u} &= \mathbf{0} && \text{on } \Gamma_u \\ \boldsymbol{\sigma}\boldsymbol{n} &= \mathbf{0} && \text{on } \Gamma_t \end{aligned}\right\} \tag{4.2}$$

on the domain $\Omega = \Omega_F \cup \Gamma \cup \Omega_B$ in $\mathbb{R}^d$, $d = 2$ or $d = 3$ (see Figure 4.4).

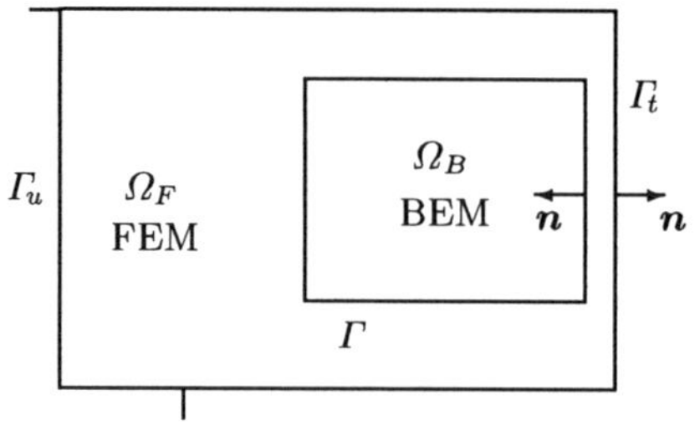

Fig. 4.4. Notation

We assume that the body load $\boldsymbol{f} \in L_2(\Omega)^d$ vanishes on Ω_B for simplicity and that the Lamé coefficients are constant on Ω_B.

The mixed FEM is based on an extended complementary Hellinger–Reissner principle. Thus the displacement $\boldsymbol{u}$ and the stress $\boldsymbol{\sigma}$ are independent unknowns. Moreover, the stress tensor $\boldsymbol{\sigma}$ is not a priori assumed to be symmetric. Symmetry will be enforced in a weak sense by a Lagrange multiplier. The second equation in (4.2) is weighted by test stresses $\boldsymbol{\tau}$ with $\boldsymbol{\tau}\boldsymbol{n} = \mathbf{0}$ on Γ_t:

$$\int_{\Omega_F} \boldsymbol{\tau} : \mathcal{C}^{-1}\boldsymbol{\sigma}\, d\Omega - \int_{\Omega_F} \boldsymbol{\tau} : \boldsymbol{\varepsilon}(\boldsymbol{u})\, d\Omega = 0\,. \tag{4.3}$$

Integration by parts yields

$$\int_{\Omega_F} \boldsymbol{\tau} : \mathcal{C}^{-1}\boldsymbol{\sigma}\, d\Omega + \int_{\Omega_F} \operatorname{div}\boldsymbol{\tau} \cdot \boldsymbol{u}\, d\Omega + \int_{\Omega_F} \boldsymbol{\tau} : \boldsymbol{\gamma}\, d\Omega = \int_{\Gamma} \boldsymbol{\tau}\boldsymbol{n} \cdot \boldsymbol{u}\, ds \tag{4.4}$$

with a third unknown

$$\boldsymbol{\gamma} = \tfrac{1}{2}(\boldsymbol{\nabla}\boldsymbol{u} - (\boldsymbol{\nabla}\boldsymbol{u})^T),$$

representing rotations. The variational formulation in the FEM domain Ω_F is given by

$$\left.\begin{array}{rcll}\int_{\Omega_F}\boldsymbol{\tau}:\mathcal{C}^{-1}\boldsymbol{\sigma}\,d\Omega+\int_{\Omega_F}\operatorname{div}\boldsymbol{\tau}\cdot\boldsymbol{u}\,d\Omega & + & \int_{\Omega_F}\boldsymbol{\tau}:\boldsymbol{\gamma}\,d\Omega & \\ & = & \int_{\Gamma}\boldsymbol{\tau}\boldsymbol{n}\cdot\boldsymbol{\phi}\,ds & \forall\boldsymbol{\tau}\in\mathcal{H}_{\mathrm{div}} \\ \int_{\Omega_F}\operatorname{div}\boldsymbol{\sigma}\cdot\boldsymbol{v}\,d\Omega & = & -\int_{\Omega_F}\boldsymbol{f}\cdot\boldsymbol{v}\,d\Omega & \forall\boldsymbol{v}\in\mathcal{L} \\ \int_{\Omega_F}\boldsymbol{\sigma}:\boldsymbol{\eta}\,d\Omega & = & 0 & \forall\boldsymbol{\eta}\in\mathcal{W}\end{array}\right\}\tag{4.5}$$

with the unknowns

$$\begin{aligned}\boldsymbol{\sigma}\in\mathcal{H}_{\mathrm{div}} &:= \{\boldsymbol{\tau}\in L_2(\Omega_F)^{d\times d}:\operatorname{div}\boldsymbol{\tau}\in L_2(\Omega_F)^d,\quad \boldsymbol{\tau}\boldsymbol{n}=\boldsymbol{0}\text{ on }\Gamma_t\}\\ \boldsymbol{u}\in\mathcal{L} &:= L_2(\Omega_F)^d\\ \boldsymbol{\gamma}\in\mathcal{W} &:= \{\boldsymbol{\eta}\in L_2(\Omega_F)^{d\times d}:\boldsymbol{\eta}+\boldsymbol{\eta}^T=0\},\end{aligned}$$

and $\boldsymbol{\phi}$ denoting displacements on Γ. $\mathcal{H}_{\mathrm{div}}$ is a Hilbert space with the norm $(\|\boldsymbol{\tau}\|^2_{L_2(\Omega_F)}+\|\operatorname{div}\boldsymbol{\tau}\|^2_{L_2(\Omega_F)})^{1/2}$.

Our objective is to establish a canonical coupling with Galerkin BEM. Since the displacements $\boldsymbol{u}$ are not in $H^1(\Omega_F)$, we cannot insert the traces into the boundary integral equations as it is done in the classical symmetric coupling, see Costabel [5]. The condition $\boldsymbol{\sigma}\in\mathcal{H}_{\mathrm{div}}$ implies that the tractions $\boldsymbol{\sigma}\boldsymbol{n}$ on Γ lie in $H^{-1/2}(\Gamma)$ and can be inserted into the boundary integral equations. In detail, we proceed as follows: For the displacements $\boldsymbol{\phi}$ on the interface Γ we insert a boundary integral expression, namely

$$\boldsymbol{\phi}=-\boldsymbol{V}(\boldsymbol{\sigma}\boldsymbol{n})+(\tfrac{1}{2}\boldsymbol{I}+\boldsymbol{K})\boldsymbol{\phi}$$

which is obtained from the first integral equation (2.11). To obtain a symmetric coupling, also the second integral equation (2.14) is used on the interface. The coupled formulation is a typical saddle point problem: Find $(\boldsymbol{\sigma},\boldsymbol{\phi},\boldsymbol{u},\boldsymbol{\gamma})\in\mathcal{H}_{\mathrm{div}}\times H^{1/2}(\Gamma)\times\mathcal{L}\times\mathcal{W}$ such that for all $(\boldsymbol{\tau},\boldsymbol{\psi},\boldsymbol{v},\boldsymbol{\eta})\in\mathcal{H}_{\mathrm{div}}\times H^{1/2}(\Gamma)\times\mathcal{L}\times\mathcal{W}$

$$\left.\begin{array}{rcl}a(\boldsymbol{\sigma},\boldsymbol{\phi};\boldsymbol{\tau},\boldsymbol{\psi})+b(\boldsymbol{\tau};\boldsymbol{u},\boldsymbol{\gamma}) & = & 0\\ b(\boldsymbol{\sigma};\boldsymbol{v},\boldsymbol{\eta}) & = & -\int_{\Omega_F}\boldsymbol{f}\cdot\boldsymbol{v}\,d\Omega\end{array}\right\}\tag{4.6}$$

with the bilinear forms

$$\begin{aligned}a(\boldsymbol{\sigma},\boldsymbol{\phi};\boldsymbol{\tau},\boldsymbol{\psi}) := &\int_{\Omega_F}\boldsymbol{\tau}:\mathcal{C}^{-1}\boldsymbol{\sigma}\,d\Omega\\ &+\langle\boldsymbol{\tau}\boldsymbol{n},\boldsymbol{V}(\boldsymbol{\sigma}\boldsymbol{n})\rangle-\langle\boldsymbol{\tau}\boldsymbol{n},(\tfrac{1}{2}\boldsymbol{I}+\boldsymbol{K})\boldsymbol{\phi}\rangle\\ &-\langle\boldsymbol{\psi},\boldsymbol{W}\boldsymbol{\phi}\rangle-\langle\boldsymbol{\psi},(\tfrac{1}{2}\boldsymbol{I}+\boldsymbol{K}')(\boldsymbol{\sigma}\boldsymbol{n})\rangle\end{aligned}$$

and

$$b(\boldsymbol{\sigma};\boldsymbol{v},\boldsymbol{\eta}) := \int_{\Omega_F}\operatorname{div}\boldsymbol{\sigma}\cdot\boldsymbol{v}\,d\Omega+\int_{\Omega_F}\boldsymbol{\sigma}:\boldsymbol{\eta}\,d\Omega\,.$$

The bilinear form a is symmetric, i.e.

$$a(\boldsymbol{\sigma},\boldsymbol{\phi};\boldsymbol{\tau},\boldsymbol{\psi})=a(\boldsymbol{\tau},\boldsymbol{\psi};\boldsymbol{\sigma},\boldsymbol{\phi})\,.$$

Exploiting properties of the boundary integral operators and applying Brezzi's theory of saddle point problems we obtain the following theorem.

Theorem[3]: For every $\boldsymbol{f} \in L_2(\Omega_F)^d$ the saddle point problem (4.6) has a unique solution satisfying

$$\|\boldsymbol{\sigma}\|_{L_2(\Omega_F)} + \|\operatorname{div}\boldsymbol{\sigma}\|_{L_2(\Omega_F)} + \|\boldsymbol{\phi}\|_{H^{1/2}(\Gamma)} + \|\boldsymbol{u}\|_{L_2(\Omega_F)} + \|\gamma\|_{L_2(\Omega_F)} \\ \leq C\,\|\boldsymbol{f}\|_{L_2(\Omega_F)}$$

with a constant C independent of $\boldsymbol{f}$.

For the discretization of (4.6) there are several possibilities, including a family of mixed finite elements proposed and analyzed by Stenberg [24]. In [3] we prove optimal order convergence of the coupled method.

In the FEM domain Ω_F the tractions across element boundaries must be continuous. Our coupled discrete scheme also yields continuous tractions across the FEM–BEM interface. In contrast, continuity of displacements across the interelement sides and the interface is satisfied in a weak sense only. On the surface of the BEM domain, the displacements are continuous. Continuity of tractions is achieved by Lagrange multipliers. Then, in the FEM domain there are no continuity requirements except for the Lagrange multipliers. Hence all other unknowns can be eliminated on each element before assembling the global linear system, which, in its final form, can be written as

$$\left[\begin{array}{cc|cc} \mathbf{K}_F & \mathbf{K}_{FC} & & \\ \mathbf{K}_{FC}^T & \mathbf{K}_C & \mathbf{C} & \\ \hline & \mathbf{C}^T & -\mathbf{A}_V & \mathbf{A}_K \\ & & \mathbf{A}_K^T & \mathbf{A}_W \end{array}\right] \left[\begin{array}{c} \boldsymbol{\Lambda}_F \\ \boldsymbol{\Lambda}_C \\ \hline \mathbf{T} \\ \boldsymbol{\Phi} \end{array}\right] = \left[\begin{array}{c} \tilde{\mathbf{f}}_1 \\ \tilde{\mathbf{f}}_2 \\ \hline \mathbf{0} \\ \mathbf{0} \end{array}\right]. \tag{4.7}$$

Here, the degrees of freedom in the Lagrange multiplier are partitioned into $\boldsymbol{\Lambda}_F$ and $\boldsymbol{\Lambda}_C$, the latter denoting the unknowns on the interface Γ. The discrete tractions and displacements in the BEM are denoted by $\mathbf{T}$ and $\boldsymbol{\Phi}$, respectively. The submatrices $\mathbf{A}_V$, $\mathbf{A}_K$ and $\mathbf{A}_W$ arise from the boundary integral operators. The matrix $\mathbf{C}$ is a simple mass matrix. For more details on the implementation we refer to [4].

In the following example, we investigate the stress distribution around two neighboring cavities in an infinite elastic continuum under uniform vertical pressure of magnitude 1. The two cavities have circular cross-section and radius 1. The distance between the cavities is 1 (cf. Figure 4.5). For this plane strain problem the material parameters are Young's modulus $E = 1$ and Poisson's ratio $\nu = 0.2$. Body forces are neglected. The discretization involves lowest order mixed finite elements and linear boundary elements (cf. [4]).

Figure 4.6 shows the normal stress σ_y across the line AB (cf. Figure 4.5). The mesh refinement is performed uniformly. The highest pressure occurs at the points A and B and is in good agreement with the analytic solution -3.264 [20].

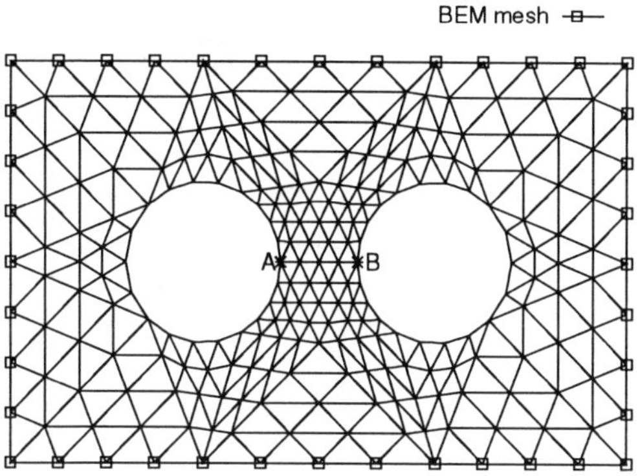

Fig. 4.5. Two neighboring cavities in an infinite elastic continuum, mesh with 416 mixed finite elements and 40 boundary elements

In [3] we investigated examples for which solutions are available in closed form and present experimental orders of convergence.

A posteriori error estimators for mixed elements including Stenberg's lowest order method [24] are studied in [2]. As our results in [4] confirm, the coupling with adaptively refined mixed finite elements is very efficient, especially for the treatment of corner singularities.

5 Conclusions

We presented parallel algorithms for boundary and finite element methods and their implementation on an MIMD parallel computer. The results show that a parallelization by data decomposition yields high parallel efficiency for BE and FE computations.

For the coupling of the two methods, two different schemes are discussed. The multiplicative Schwarz method in two overlapping domains is very well suited for parallel implementation because the efficient codes from pure BE and FE computations can be used with low additional data exchange. For this method, there is still the need of theoretical investigations especially in the non-linear case. The convergence depends on the overlap of the two domains. A larger overlap leads to faster convergence but higher computational costs. If the overlap is small many SCHWARZ iteration steps have to be performed, but because optimal solvers for each subdomain can be used, the method remains very efficient.

To solve linear systems, iterative methods often are superior to GAUSSian elimination, in particular on parallel computers [14]. When using iterative solvers GALERKIN BEM becomes attractive since it yields symmetric matrices. The authors established a symmetric coupling of GALERKIN BEM and

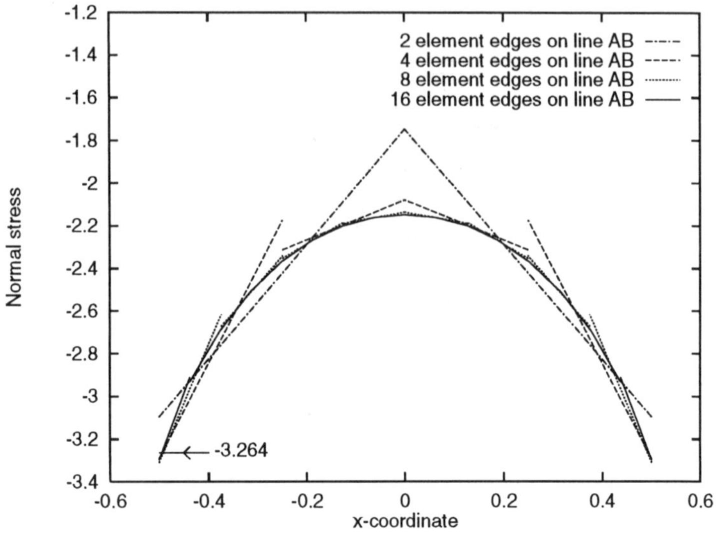

Fig. 4.6. Normal stress σ_y across the line AB

mixed finite elements of the RAVIART-THOMAS type. We presented theoretical and computational results for problems of linear elasticity. A corresponding discretization for nonlinear problems is subject of further research.

References

1. Bader, G., Gehrke, E. (1991): On the performance of transputer networks for solving linear systems of equations. Parallel Computing 17, 1397–1407
2. Braess, D., Klaas, O., Niekamp, R., Stein, E., Wobschal, F. (1995): Error indicators for mixed finite elements in 2-dimensional linear elasticity. Comput. Methods Appl. Mech. Engrg. 127, 345–356
3. Brink, U., Carstensen, C., Stein, E. (1996): Symmetric coupling of boundary elements and Raviart–Thomas-type mixed finite elements in elastostatics. Numer. Math., in print
4. Brink, U., Klaas, O., Niekamp, R., Stein, E. (1995): Coupling of adaptively refined dual mixed finite elements and boundary elements in linear elasticity. Advances in Engineering Software 24, 13–26
5. Costabel, M. (1988): A symmetric method for the coupling of finite elements and boundary elements. In: Whiteman, J.R. (ed.) The Mathematics of Finite Elements and Applications IV, MAFELAP 1987. Academic Press, London
6. Gruttmann, F., Stein, E. (1988): Tangentiale Steifigkeitsmatrizen bei der Anwendung von Projektionsverfahren in der Elastoplastizitätstheorie. Ing.-Arch. 58, 15–24
7. Haase, G., Langer, U., Meyer, A. (1990): A new approach to the Dirichlet domain decomposition method. Proc. '5th Multigrid Seminar', R-MATH-09/90, Karl–Weierstrass–Institut für Mathematik, Berlin

8. Haase, G., Langer, U., Meyer, A. (1992): Parallelisierung und Vorkonditionierung des CG–Verfahrens durch Gebietszerlegung. In: Bader, G., Rannacher, R., Wittum, G. (eds.) Numerische Algorithmen auf Transputer–Systemen. Teubner Skripten zur Numerik. Teubner, Stuttgart
9. Klaas, O., Kreienmeyer, M., Stein, E. (1993): Numerische Lösung ebener Probleme der linearen Elastizitätstheorie mit der direkten Randelementmethode auf einem MIMD–Parallelrechner. In: Baumann, M., Grube, D. (eds.) Parallele Datenverarbeitung mit dem Transputer. Springer, Berlin Heidelberg New York, pp. 265–274
10. Klaas, O., Kreienmeyer, M., Stein, E. (1994): Elasto–plastic finite element analysis on a MIMD parallel–computer. Eng. Analysis — Int. J. Comp. Aided Eng. Software 11, (4), 347–355
11. Klaas, O., Niekamp, R., Stein, E. (1995): Parallel Adaptive Finite Element Computations with Hierarchical Preconditioning. Int. J. Comp. Mech. 16, 45–52
12. Kreienmeyer, M. (1995): Kopplung von Rand- und Finite-Element-Methodsn für ebene Elastoplastizität mit Implementierung auf Parallelrechnern. Doctoral Thesis, Universität Hannover. Submitted
13. Kreienmeyer, M., Stein, E. (1993): Parallel implementation of BEM, FEM and coupled FEM-BEM on a transputer network. IBNM-Bericht 93/9
14. Kreienmeyer, M., Stein, E. (1995): Parallel implementation of the boundary element method for linear elastic problems on a MIMD parallel computer. Int. J. Comp. Mech. 15, (4), 342–349
15. Law, K.H. (1986): A parallel finite element solution method. Computers & Structures 23, (6), 845–858
16. Lubliner, J. (1990): Plasticity Theory. Macmillan Publishing Company, New York
17. Maischak, M. (1992): Eine Randelement-Diskretisierung und projiziertes SOR-Verfahren für ein Kontaktproblem der Elastostatik. Diplomarbeit, Institut für Angewandte Mathematik, Universität Hannover
18. Romate, J.E. (1993): On the use of conjugate gradient-type methods for boundary integral equations. Int. J. Comp. Mech. 12, 214–233
19. Saad, Y., Schultz, M.H. (1986): GMRES: A generalized minimal residual method for solving nonsymmetric linear systems. SIAM J. Sci. Statist. Comput. 7, 856–869
20. Sawin, G.N. (1956): Spannungserhöhung am Rande von Löchern. VEB Verlag Technik, Berlin
21. Schwarz, H.A. (1870): Vierteljahresschrift der Naturforschenden Gesellschaft in Zürich, Band 15
22. Simo, J., Taylor, R.L. (1986): A return mapping algorithm for plane stress elastoplasticity. Int. J. Num. Meth. Eng. 22, 649–670
23. Simon, H.D. (1991): Partitioning of unstructured problems for parallel processing. Comp. Systems Eng. 2, 135–148
24. Stenberg, R. (1988): A family of mixed finite elements for the elasticity problem. Numer. Math. 53, 513–538
25. van der Vorst, H.A. (1992): BI–CGSTAB: A Fast and Smoothly Converging Variant of BI–CG for the solution of Nonsymmetric Linear Systems. SIAM J. Sci. Statist. Comput. 13, 631–640
26. Yserentant, H. (1991): Hierarchical bases. In: O'Malley, R.E. (ed.) ICIAM 91. SIAM, Philadelphia.
27. Yserentant, H. (1986): On the Multi–Level Splitting of Finite Element Spaces. Numer. Math. 49, 379–412

Error Analysis for the BEM on Nonsmooth Curves

T. Hartmann and E.P. Stephan

1 Introduction

In the following several methods for approximating the solution of integral equations on polygons are considered. These methods and the techniques of proving stability and error estimates have to take care of the fact that the solutions of the equations considered contain singularities. In Section 2 we derive stability conditions for a quadrature method using splines for a weakly singular integral equation. The same equation is considered in the next two sections. In Section 3 collocation methods using Jacobi weights and polynomials were extended from an interval to a polygon and in Section 4, local error estimates for Galerkin's method are given. In Section 5 a discrete collocation method using a nonlinear transform is considered for solving a hypersingular integral equation.

2 Operator algebra methods

In the following we derive stability conditions for the quadrature method of the single layer potential on a polygon Γ

$$V_\Gamma u(x) := -\frac{1}{\pi}\int_\Gamma \log|x-y|u(y)\,ds_y = f(x) \qquad x \in \Gamma \tag{2.1}$$

with piecewise constant trial functions using a uniform mesh. If not explicitly mentioned, the proofs and further details are given in [10]. Let the Cauchy-singular integral operator S_Γ and the double layer potential K_Γ be defined as

$$S_\Gamma u(t) := \frac{1}{\pi i}\int_\Gamma \frac{1}{\tau - t}u(\tau)ds_\tau$$

$$K_\Gamma u(t) := -\frac{1}{\pi}\int_\Gamma \frac{\partial}{\partial n_\tau}\log|\tau - t|u(\tau)ds_\tau .$$

Then there holds for $t \in \Gamma$:

$$S_\Gamma u(t) + K_\Gamma u(t) = -iV_\Gamma u'(t).$$

Due to this relation, we examine a quadrature method with piecewise linear trial functions for the equation $S_\Gamma u + K_\Gamma u = f$. This quadrature method corresponds to a quadrature method with piecewise constant trial functions for $V_\Gamma u = f$. The equation $S_\Gamma u + K_\Gamma u = f$ is replaced by a discrete operator equation $R_n u_n = f_n$ where R_n is an approximation operator of $R := S_\Gamma + K_\Gamma$ acting in the space X_n of piecewise linear functions on a uniform mesh and $f_n \in X_n$ interpolates f. Such a method is called stable if R_n is invertible for n sufficient large and $\sup \|R_n\| < \infty$. If the method is stable, f Riemann integrable, and R_n converges strongly to R, the approximate solutions u_n converge to u. Thus, the crucial point is the proof of stability of the scheme. This is done by showing stability of a corresponding model problem on an angle, making use of a localization principle by Gohberg and Krupnik, e.g. [8]. Therefore we consider only the case of Γ being an infinite angle $\Gamma_\omega = \mathbb{R}^+ \bigcup e^{i\omega}\overline{\mathbb{R}^+}$ with opening $0 < \omega < 2\pi$. The general case of a polygon follows by localization arguments. We fix $n \in \mathbb{N}$, $0 < \varepsilon, \delta < 1$ and for $k \in \mathbb{Z}$ choose the quadrature points t_n^k and collocation points τ_n^k as follows. We introduce with $\Delta_k := 1$ if $k \geq 0$ and $\Delta_k := -e^{i\omega}$ if $k < 0$:

$$t_n^k = \frac{k+\delta}{n}\Delta_k \qquad\qquad \tau_n^k = \frac{k+\varepsilon}{n}\Delta_k.$$

By using a rectangular rule as the quadrature formula we obtain for a discretization of

$$S_{\Gamma_\omega} u + K_{\Gamma_\omega} u = f \tag{2.2}$$

the system

$$-\, i\cot(\pi(\varepsilon-\delta))u(t_k^n) - \frac{i}{\pi n} Re\left\{\sum_{j\in\mathbb{Z}} \frac{u(t_j^n)}{t_j^n - \tau_k^n}\right\}\Delta_j = f(\tau_k^n). \tag{2.3}$$

If there exists a solution $(u(t_k^n)), k \in \mathbb{Z}$ we consider a piecewise linear approximation for the solution $u \in L^2(\Gamma_\omega)$ of (2.2). This approximation is represented by the usual 'hat'-functions ψ_k^n

$$u_n(t) = \sum_{k\in\mathbb{Z}} \xi_k^n \psi_k^n(t).$$

We obtain for the values $u_n(t_k^n)$

$$u_n(t_k^n) = (1-\delta)\xi_k^n + \delta\xi_{k+1}^n \qquad k \in \mathbb{Z} \tag{2.4}$$

and in order to determine the approximate solution u_n we first have to solve (2.3) and to calculate the coefficients ξ_k^n with (2.4). Now we have to show that for each $f \in L^2(\Gamma_\omega)$ there exists a unique solution of this procedure. To prove this, some results on Toeplitz operators are needed which are due to Krupnik, see e.g. [9]. Let A denote the smallest algebra containing all Toeplitz operators $T(a)$ with $a \in PC(\Pi)$ ($PC(\Pi)$ denotes the class of piecewise continuous function on the unit circle Π). Then $A_{n\times n} \subset L(l^2)_{n\times n}$ is an algebra of continuous operators in $l^2_{n\times n}$. There exists a multiplicative linear mapping

$\mathcal{A}_{n\times n} \in B \to \mathcal{A}_B$ into the algebra of bounded $n\times n$-matrix functions over $\Pi \times [0,1]$. The symbol $\mathcal{A}_B$ of $B = (B_{k,j})_{k,j=1}^n$, $B_{k,j} \in A$, is equal to $(\mathcal{A}_{B_{k,j}})_{k,j=1}^n$ Furthermore, $B \in A_{n\times n}$ is a Fredholm operator if and only if $\det \mathcal{A}_B(\tau,\mu) \neq 0$ for all $\tau \in \Pi$ and $0 \leq \mu \leq 1$.

Let S_1, K_1 denote the parts of the matrix R_1 corresponding to the operators S_{Γ_ω}, K_{Γ_ω}. The symbol for S_1 was calculated by Rathsfeld and Prössdorf [15] and for K_1 by Costabel, Ervin and Stephan [2]. With the definitions

$$z_0 := \frac{1}{2} + \frac{i}{2\pi}\log\left(\frac{\mu}{1-\mu}\right) \qquad 0 \leq x \leq 1$$

$$f^{\varepsilon-\delta}(e^{2\pi i x}) := \frac{2e^{-i\pi(\varepsilon-\delta)(x-1)}\sin(\pi(\varepsilon-\delta)x)}{\sin(\pi(\varepsilon-\delta))} - 1 \qquad 0 \leq x \leq 1$$

the symbol of R_1 has the following form:

$$\mathcal{A}_{R_1}(\tau,\mu) = \begin{cases} \begin{pmatrix} f^{\varepsilon-\delta}(\tau) & 0 \\ 0 & -f^{\delta-\varepsilon}(\tau) \end{pmatrix} & \tau \neq 1 \quad 0 \leq \mu \leq 1 \\ \begin{pmatrix} 1-2\mu & \frac{i\cos(\pi-\omega)z_0}{\sin \pi z_0} \\ -\frac{-i\cos(\pi-\omega)z_0}{\sin \pi z_0} & 2\mu - 1 \end{pmatrix} & \tau = 1 \quad 0 \leq \mu \leq 1 \end{cases}$$

Further development of the symbol yields the following theorem:

Theorem 2.1. *1. The operator $R_1 \in L(\tilde{l}^2)$ is a Fredholm operator with index zero $\Leftrightarrow \varepsilon \neq \delta$ and $\delta \neq 1/2$.*

2. The operator R_1 is invertible in $L(\tilde{l}^2)$, if $-1 \notin \Phi$, where Φ is an at most countable set $\Phi \subset \mathbb{C} \setminus \Omega$ with accumulation points only in Ω.

3. If $\omega = \pi$, then $\Phi = \emptyset$.

Now a quadrature method on a polygon can be introduced. Using a quadrature formula due to (2.3) it turns out that this method will be stable if the corresponding model problem for each angle of the polygon is stable, i.e. Theorem 2.1 is fullfilled. However to check assertion 2. is still an open problem. For further details see [10].

3 Collocation methods using Jacobi polynomials

We consider the integral equation (2.1) in the case that $\Gamma = [-1,1]$ is an interval

$$V_I\psi(x) := -\frac{1}{\pi}\int_{-1}^{1} \log|x-y|\psi(y)\,dy = f(x) \qquad x \in \Gamma$$

and rewrite its solution $\psi = \varrho\phi$ using a function $\varrho(x) := (1-x^2)^{-1/2}$. For ψ we expect a singular behaviour near the endpoints of the form $r^{-1/2}$, where r denotes the distance to the corner. Since the function ϱ contains the singularity of the solution we expect ϕ to be more smooth than ψ. Now ψ is approximated by

$$\psi_N = \varrho\phi_N = \varrho \sum_{k=0}^{N-1} a_{kN} T_k. \tag{3.1}$$

Here, T_k denote the kth Chebychev polynomials of first kind and a_{kN} unkown real numbers. In order to obtain error estimates, we define non-standard Sobolev spaces. For $f = \sum_{k=0}^{\infty} \tilde{f}(k) T_k$ we define

$$\|f\|_{s,\varrho}^2 := \sum_{k=1}^{\infty} k^{2s} |\tilde{f}(k)|^2 + |\tilde{f}(0)|^2$$

and the spaces $\mathcal{H}_\varrho^s$ to be the closure of the set of all polynomials with respect to the norm $\|\cdot\|_{s,\varrho}$. Using these spaces and norms stability and error estimates for the Galerkin method (3.2) and the collocation method (3.3) can be obtained. These spaces and techniques are introduced by Sloan and Stephan in [18] for a weakly singular integral equation. They are further applied for a hypersingular integral equation [7] and the airfoil eqution [12].

$$\text{Find } \psi_N^G \text{ such that } \quad \langle V_I \psi_N^G, \varphi \rangle_\varrho = \langle f, \varphi \rangle_\varrho \quad \forall \varphi \in \mathbb{P}_{N-1} \tag{3.2}$$

$$\text{Find } \psi_N^C \text{ such that } \quad V_I \psi_N^C(x_{jN}) = f(x_{jN}) \quad j = 1, ..., N \tag{3.3}$$

Here $x_{jN} = \cos \frac{2j-1}{2N}\pi$ denote the zeroes of the Nth Chebychev polynomial. The scalar product $\langle \cdot, \cdot \rangle_\varrho$ is given by

$$\langle f, g \rangle_\varrho := \int_{-1}^{1} f(x) g(x) \varrho(x)\, dx.$$

Theorem 3.1. *[12] [18] Let $f \in \mathcal{H}_\varrho^{t+1}$. Then for $N \geq N_0$ there exists a unique solution of (3.3) and (3.2), respectively. Moreover, if $t > s$ there holds (C is a constant independent of N):*

$$\|\phi - \phi_N^G\|_{s,\varrho} \leq C N^{s-t} \|f\|_{t+1,\varrho} \qquad s \leq t$$

$$\|\phi - \phi_N^C\|_{s,\varrho} \leq C N^{s-t} \|f\|_{t+1,\varrho} \qquad 0 \leq s \leq t.$$

These assertions are confirmed by numerical results [18].

The integral equation

$$V_\Gamma \tilde{\psi}(t) = -\frac{1}{\pi} \int_\Gamma \log|\tau - t| \tilde{\psi}(\tau) ds_\tau \tag{3.4}$$

on a semi circle $\Gamma = \{t | t = ie^{-ix\pi/2}; -1 < x < 1\}$ with $\psi(x) := \tilde{\psi}(t)$, $t = ie^{-ix\pi/2}$ can be written as

$$V_\Gamma \tilde{\psi}(t) = V_I \psi(x) + K\psi(x) = g(t)$$

$$K\psi(x) := -\frac{1}{\pi} \int_{-1}^{1} \log \frac{2 \sin \frac{\pi}{4}(x-y)}{x-y} \psi(y) dy$$

i.e. the integral equation on the semi circle can be transfomed in an integral equation on an interval $[-1, 1]$ with an additional compact operator K.

Theorem 3.2. *Let $f \in \mathcal{H}_\varrho^{t+1}$. Then for $N \geq N_0$ there exists a unique solution $\psi_N = \varrho\phi_N$ of*

$$V_I \psi_N(x_{jN}) + K\psi_N(x_{jN}) = f(x_{jN}) \quad j = 1, ..., N$$

Moreover, if $t > s$ there holds (C is a constant independent of N):

$$\|\phi - \phi_N\|_{s,\varrho} \leq CN^{s-t} \|f\|_{t+1,\varrho} \qquad 0 \leq s \leq t.$$

Further we use a two level algorithm, which is proposed by Berthold, Hoppe and Silbermann for integral equations on intervals [1]. Therefore the coefficients a_{kN} of $\phi_N = \sum_{k=0}^{N-1} a_{kN} T_k$ are calculated in the following manner:

1. Choose $M > 0$ with $M << N$.
2. Solve the collocation equations $V_I \psi_N(x_{lN}) = f(x_{lN})$ for $l = 1, \ldots, N$ with $\psi_N = \varrho \sum_{k=0}^{N-1} b_{kN} T_k$. Due to the mapping property of the integral operator this yields a diagonal matrix. Take these coefficients b_{kN} for $k \geq M$ as coefficients a_{kN} of the approximate solution ϕ_N.
3. Solve the collocation equations for $l = 1, \ldots, M$

$$V_I \psi_M(x_{lM}) + K\psi_M(x_{lM}) = f(x_{lM}) - V_I \hat{\psi}_N(x_{lM})$$

with $\psi_M = \varrho \sum_{k=0}^{M-1} c_{kM} T_k$, where $\hat{\psi}_N = \varrho \sum_{k=M}^{N-1} b_{kN} T_k$ denotes the part of the solution which has been calculated in step 2. This yields a system of linear equations of dimension M which are solved using Gauß' procedure. Take the coefficients c_{kM} with $k < M$ as coefficients a_{kN} of the approximate solution ϕ_N.

The two level algorithm yields the same convergence rate with less costs. If M can be choosen as $M^3 \approx N$ the costs are of the order $O(N \log N)$.

Let Γ be a polygon with straight line segments Γ_j, $\Gamma = \sum_{j=1}^{J} \Gamma_j$, corners τ_j and interior angles ω_j. The solution of the integral equation $V_\Gamma \psi = f$ defined by (2.1) has a singular behaviour near the corners τ_j as r^{α_j}, where r denotes the distance to the corner and α_j is defined by

$$\alpha_j := \min \left\{ \frac{\pi}{\omega_j}, \frac{\pi}{2\pi - \omega_j} \right\} - 1. \tag{3.5}$$

The singular behaviour depends on the angle and can be of the form r^γ with $-1/2 < \gamma < 0$. According to the collocation methods for weakly singular integral equations on an interval or on a smooth open curve we approximate the solution ψ of $V_\Gamma\psi = f$ at each straight line segment Γ_j by a weight function ϱ^j multiplied by a polynomial. Thus we can write for the approximation ψ_N:

$$\psi_N = \sum_{j=1}^{J} \psi_N^j = \sum_{j=1}^{J} \varrho^j \sum_{k=0}^{N-1} a_{kN}^j P_k^j \tag{3.6}$$

where $\psi_N^j = \psi_N|_{\Gamma_j}$, a_{kN}^j are real numbers and P_k^j are certain polynomials. The collocation points on the segments Γ_i are denoted by (x_{lN}^i) with $l = 1, ..., N$. Thus, the collocation method can be described as:
Find ψ_N, i.e. real numbers a_{kN}^j, for j=1,...,J and k=1,...,N, such that:

$$\sum_{j=1}^{J} \sum_{k=0}^{N-1} V\left(\varrho^j a_{kN}^j P_k^j\right)(x_{lN}^i) = f(x_{lN}^i) \qquad i = 1, ..., J \quad l = 1, ..., N \tag{3.7}$$

Different choices of the collocation points x_{lN}, the weight functions and the polynomials P_k^j are considered in [10, 13]. Further fast algorithms due to [1] are tested in [10, 13], for brevity we refer to these papers.

4 Local error estimates for Galerkin's method

In the following we consider Galerkin's method for solving Symm's integral equation (2.1) on a polygon. It is well known that the behaviour of the approximation depends heavily on the characteristics of the settings of the problem. Non-smoothness of domains and singularities of given data produce singularities of the exact solution, and reduce the global rate of convergence of the approximations. It is then natural to ask whether the accuracy of the approximation is better in regions where the exact solution is smooth.

For strongly elliptic integral equations on a smooth and closed curve in $\mathbb{R}^2$, local error estimates for the Galerkin method were proved by Saranen [16]. With adaption of the techniques, the estimates were proved for the case of open surfaces in $\mathbb{R}^3$ by Stephan and Tran [19]. The proofs of the following theorems concerning the case of a polygon are based on these works and can be found in [14].

Let Γ^j be any fixed straight line segment of the polygon Γ with nested subsegments Γ_0 and Γ_*

$$\Gamma_0 \subset\subset \Gamma_* \subset\subset \Gamma^j \qquad \text{and let} \qquad d := dist(\partial\Gamma_*, \partial\Gamma_0)$$

where $X \subset\subset Y$ means that the closure of X is contained in the interior of Y.

Since local error estimates are only valuable when the exact solution of the equation is in some subsegments smoother than on the whole polygon Γ, it is worth considering local regularity of the solution. This is shown in the following theorem:

Theorem 4.1. *Let $f \in H^s(\Gamma_*)$ with $s > 1/2$. Then there holds for the solution ψ of $V_\Gamma \psi = f$*

$$\|\psi\|_{H^{s-1}(\Gamma_0)} \leq C\{\|f\|_{H^s(\Gamma_*)} + \|\psi\|_{H^{-1/2}(\Gamma)}\}.$$

We will consider the following spline spaces

$$S_h^{r,0}(\Gamma_*) = \{\phi \in S_h^r : \text{supp}\phi \subset \Gamma_j\}$$

$$S_h^r(\Gamma_*) = \{v \in H^{-1/2}(\Gamma) : v|_{\Gamma_*} = \phi|_{\Gamma_*} \text{ for some } \phi \in S_h^r\}$$

and we are concerned with the error $\|\psi - \psi_h\|_{t,\Gamma_0}$, where ψ and ψ_h satisfy

$$\begin{aligned} &\psi \in H^{-1/2}(\Gamma) \cap H^s(\Gamma_*) \qquad \text{for } -1/2 \leq s \leq r \\ &\psi_h \in S_h^r(\Gamma_*) \\ &\langle V(\psi - \psi_h), \phi \rangle = 0 \qquad \text{for all } \phi \in S_h^{r,0}(\Gamma_*). \end{aligned} \tag{4.1}$$

Note that we only assume in (4.1) that ψ_h is a spline on the subsegment Γ_* which can be arbitrarily extended onto Γ provided that the extension belongs to $H^{-1/2}$.

Theorem 4.2. *Let $\psi \in H^s(\Gamma_*)$ and $\beta < -1/2$ arbitrary but fixed. Then for $-r-1 \leq t \leq s \leq r$ and $t < r - 1/2$ we have*

$$\|\psi - \psi_h\|_{t,\Gamma_0} \leq C\{h^{s-t}\|\psi\|_{s,\Gamma_*} + h^\sigma \|\psi - \psi_h\|_{\beta,\Gamma}\} \tag{4.2}$$

where $\sigma = 0$ if $t \leq -1/2$ and $\sigma = -1/2 - t$ if $-1/2 < t$.

Example: If piecewise constant trial functions (r=1) are used we expect the global error in the L^2 norm to be of order $O(h^{s_0 - 1/2})$, where s_0 depends on the geometry and takes values between 1/2 and 1. In the $H^{-1/2-s_0}$ norm the highest order is achieved which is $O(h^{2s_0})$. Due to Theorem 4.2 we expect a local error in the L^2 norm of order $O(h^1)$ for the first term in (4.2) and of order $O(h^{2s_0-1/2})$ for the second term in (4.2).

A better approximation can be obtained by averaging the values of the Galerkin solution, using the K-operator. The action of the K-operator on the approximate solution ψ_h is denoted by $K_h(\psi_h)$ and is given by convolution with a carefully chosen spline $k_{h,l}^q$. This spline is defined by a combination of $2q-1$ splines each being a $(l-1)$ times convolution of a B-spline. For more details see [14, 21].

Theorem 4.3. *Let $l \geq r+1$ and $q > r + 1/2$. For the solution ψ of $V\psi = f$ there holds $\psi \in H^{2r+1}(\Gamma_*)$. Let β be arbitrary but fixed. Then there holds with C independent of h:*

$$\|K_h(\psi_h) - \psi\|_{0,\Gamma_0} \leq C\{h^{2r+1}\|\psi\|_{2r+1,\Gamma_*} + \|\psi_h - \psi\|_{\beta+r+1,\Gamma}\}. \tag{4.3}$$

Since in (4.1) it is assumed that ψ_h is a spline only on a subsegment Γ_* we can use a grid which is uniform on Γ_* and graded near to the corners. This yields an approximation for which the second term in (4.3) is of higher order than the first term. In this case we found that the error for piecewise constant trial functions is of order 3.

Further the local error estimates are used to show that the local residual is a local error indicator and can be used for constructing adaptive algorithms. In the case of smooth closed curves this is shown by Saranen and Wendland [19]. With an approach different from that used in [19] similar results can be obtained for polygons. The proof adapts techniques used for finite element methods by Johnson and Eriksson, see for example [6], and can be found in [10].

The strategy of the method considered is to adapt the mesh size to the size of the residual so as to equidistribute the element contributions to the global error. For a choosen error tolerance TOL the algorithm looks as follows:
1. Start with a coarse and quasiuniform mesh.
2. Compute the corresponding Galerkin solution ψ_h.
3. Calculate the residual $R := f - V_\Gamma \psi_h$.
4. If there holds for any element I (N number of elements)

$$\|R\|_{1,I} \le TOL/N \tag{4.4}$$

then accept ψ_h and stop. Otherwise, halve all elements I for which (4.4) does not hold and go back to step 2..

Let H and h denote the global and local meshsize, i.e.,

$$H := \max\left\{\tilde{h}|\tilde{h} \text{ is meshsize of an element on } \Gamma\right\}$$

$$h := \max\left\{\tilde{h}|\tilde{h} \text{ is meshsize of an element on } \Gamma_*\right\}$$

Of course, there holds $h \le H$.

For this algorithm local efficiency can be proved.

Theorem 4.4. *Assume that* $0 < s < 1$ *and for a positive constant* C_1

$$\|R\|_{1,\Gamma_0} \ge C_1\, TOL$$

then there exists for sufficient small h *a subsegment* Γ_* *with*

$$h^s \|\psi\|_{s,\Gamma_*} \ge C\; TOL.$$

Note: For brevity the local error estimates and the K-operator are here formulated for Symm's integral equation only. However, similar results can be obtained for more general operators which are pseudodifferential operators restricted on polygons, see [14].

5 Discrete collocation using nonlinear transform

In the following we consider the hypersingular integral equation $Wu = f$

$$Wu(x) := \frac{1}{\pi}\frac{\partial}{\partial n_x}\int_\Gamma \frac{\partial}{\partial n_\xi}\log|x-\xi|u(\xi)\,ds_\xi \tag{5.1}$$

on a teardrop shaped domain Γ with single corner at x_0. Following the development in [3, 4, 5], we now rewrite (5.1) using an appropriate nonlinear parametrization $\gamma : [0,1] \to \Gamma$ which varies more slowly than arc-length parametrization in the vicinity of x_0. Consider a parametrization $\gamma_0 : [0,1] \to \Gamma$ such that $\gamma_0(0) = \gamma_0(1) = x_0$ and $|\gamma_0'(s)| > 0$ for all $0 \le s \le 1$. Choosing a grading exponent $q \in \mathbb{N}$ and selecting a function ν such that

$$\nu \in C^\infty[0,1], \quad \nu(0) = 0, \quad \nu(1) = 1, \quad \nu'(s) > 0, \quad 0 \le s \le 1, \tag{5.2}$$

we define the mesh grading transformation

$$\gamma(s) = \gamma_0(\omega(s)), \quad \text{where} \quad \omega(s) = \frac{\nu^q(s)}{\nu^q(s) + \nu^q(1-s)}.$$

The chosen parametrization γ is graded with exponent q near the corner. The simplest choice of ν satisfying (5.2) is, of course, $\nu(s) = s$, for further details see [4].

Using this transform we rewrite (5.1) with $(x = \gamma(s),\ \xi = \gamma(\sigma))$, multiply with $|\gamma'(s)|$ and achieve for $s \in [0,1]$

$$Kw(s) := \frac{1}{\pi}\frac{\partial}{\partial n_x}\int_0^1 \frac{\partial}{\partial n_\xi}\log|x-\xi|\,|\gamma'(\sigma)||\gamma'(s)|w(\sigma)\,d\sigma = g(s), \tag{5.3}$$

where

$$w(\sigma) := u(\gamma(\sigma)), \qquad g(s) := |\gamma'(s)|f(\gamma(s)).$$

If the right hand side f is smooth, the solution w of (5.3) becomes arbitrarily smooth by choosing a grading exponent large enough. Therefore trigonometric polynomials are a good choice to approximate w. Since there holds $Wu = 0$ for any constant u, an additional constraint is neccessary to ensure a unique solution. We look for solutions w of $Kw = f$ satisfying $\int_0^1 w = 0$. In the following we define some operators and spaces which will be used below.

$$\overset{\circ}{H}{}^s(\Gamma) := \{f \in H^s(\Gamma) \mid \int_\Gamma f = 0\}$$

$$Aw(s) = -2\int_0^1 \log|2e^{-1/2}\sin(\pi(s-\sigma))|w(\sigma)\,d\sigma,$$

$$Bw(s) = Kw(s) - A^{-1}w(s) = \int_0^1 b(s,\sigma)w(\sigma)\,d\sigma, \tag{5.4}$$

Further let $Dv(s) = v'(s)$ denote the derivate of v. The following mapping properties can be obtained:

Theorem 5.1. *The operators* $A^{-1}K : \overset{\circ}{H}{}^1 \to \overset{\circ}{H}{}^{-1}$ *and* $K : \overset{\circ}{H}{}^1 \to \overset{\circ}{H}{}^0$ *are continuously invertible. Further there holds a Gårding's inequality:*

$$Re\langle (A^{-1}K + T)v, v\rangle \geq c\|Dv\|_0^2 \qquad v \in \overset{\circ}{H}{}^1,$$

with a compact operator T.

In order to solve (5.3) we choose as collocation points

$$s_j = jh, \quad j = 1, ..., 2n, \quad \text{where} \quad h := 1/(2n+1),$$

and denote by $\mathcal{T}_h^0$ the space of trigonometic polynomials ϕ of degree $\leq n$ satisfying $\int \phi = 0$. We look for an approximation w_h defined by

$$Kw_h(s_j) = g(s_j) \qquad j = 1, ..., 2n. \tag{5.5}$$

For any continuous 1-periodic function v with $\int v = 0$ we define an interpolation operator Q_h onto $\mathcal{T}_h^0$ by

$$(Q_h v)(s_j) = v(s_j), \quad j = 1, ..., 2n.$$

Since Q_h and A^{-1} commutate on $\mathcal{T}_h^0$ we write (5.5) as

$$(A^{-1} + Q_h B)w_h = Q_h g, \qquad w_h \in \mathcal{T}_h^0. \tag{5.6}$$

Further we can show for $w_h \in \mathcal{T}_h^0$

$$A^{-1}(A^{-1} + Q_h B)w_h = -D^2 w_h + R_h A^{-1} B w_h$$

using $R_h := A^{-1} Q_h A$. We derive for (5.6) an equivalent version:

$$(-D^2 + R_h A^{-1} B)w_h = R_h e \qquad w_h \in \mathcal{T}_h^0, \tag{5.7}$$

where $e = A^{-1}g$. We can only prove stability for a slightly modified equation. Therefore a cutoff operator T_r is introduced:

$$T_\tau(v)(s) = \begin{cases} v(s), & s \in (\tau, 1-\tau), \\ 0, & s \in (0,\tau) \cup (1-\tau, 1) \end{cases}.$$

Using the cutoff for (5.7) we consider

$$(-D^2 + R_h A^{-1} B T_{i^* h})w_h = R_h e \qquad w_h \in \mathcal{T}_h^0. \tag{5.8}$$

with fixed i^* independent on h. If $i^* = 0$ then (5.8) and (5.7) are equivalent. For a discrete collocation method we use the trapezoidal rule and a regularization to replace B in (5.6) by

$$B_h v(s) := h \sum_{j=0}^{2n} b(s, \sigma_j)(v(\sigma_j) - v(0)).$$

and prove stability and error estimates.

Theorem 5.2. *Let $q \geq 2$ and i^* large enough. Then there holds:*

1. *The discrete collocation equations are stable, i.e*

$$\|(-D^2 + R_h A^{-1} B_h T_{i^* h})v\|_{-1} \geq c\|Dv\|_0 \qquad v \in \mathcal{T}_h^0$$

for h small enough, with c independent of h and v.

2. *If f is smooth enough and there holds $w \in H^l$, we have the error estimate*

$$\|w - w_h\|_1 \leq c\, h^{l-1}$$

with c being a constant depending on i^*, but independent of h.

A similar theorem can be proved concerning the collocation method (5.8) without replacing B by B_h. For further details see [11].

References

1. Berthold, D., Hoppe, W., Silbermann, B. (1992): A fast algorithm for solving the generalized airfoil equation. J. Comp. Appl. Math 43, 185-219
2. Costabel, M., Ervin, V., Stephan, E.P. (1993): Quadrature and Collocation Methods for the Single Layer Potential on Polygons. Zeitschrift für Analysis und ihre Anwendungen 12, 699-707
3. Elschner, J., Graham, I.G. (1995): An optimal order collocation method for first kind boundary integral equations on polygons. Numer. Math. 70, 1-31
4. Elschner, J., Graham, I.G.: Quadrature methods for Symm's integral equation on polygons. Submitted to Math. Comp.
5. Elschner, J., Stephan, E.P. (1995): A discrete collocation method for Symm's integral equation on curves with corners. Preprint 143, WIAS Berlin
6. Eriksson, K.: An adaptive finite element method with efficient maximum error control for elliptic problems. Preprint, University of Goteborg, Sweden
7. Ervin, V.J., Stephan, E.P. (1992): Collocation with Chebyshev Polynomials for a hypersingular Integral Equation on an Interval. J. Comp. Appl. Math. 43, 221-229
8. Gohberg, I., Krupnik, N.Y.: Einführung in die Theorie eindimensionaler singulärer Integraloperatoren. Birkhäuser Verlag, Basel Stuttgart
9. Krupnik, N.Y. (1987): Banach algebras with symbols and singular integral operators. Birkhäuser Verlag, Basel Boston
10. Hartmann, T. (1995): Verbesserte Konvergenzordnungen für die näherungsweise Lösung von Integralgleichungen auf Polygonen. PhD thesis, Hannover
11. Hartmann, T., Stephan, E.P. (1996): A discrete collocation method for a hypersingular integral equation on curves with corners. Preprint, Institut für Angewandte Mathematik, Universität Hannover
12. Hartmann, T., Stephan, E.P. (1994): Rates of Convergence for Collocation with Jacobi polynomials for the Airfoil equation. J. Comp. Appl. Math. 51, 179-191
13. Hartmann, T.; Stephan, E.P.: Jacobi polynomial methods for solving singular integral equations on an interval or on a polygon. Preprint, Institut für Angewandte Mathematik, Universität Hannover, in preparation

14. Hartmann, T.; Stephan, E.P. (1996): Local error estimates and post processing for the Galerkin boundary element method on polygons. Preprint, Institut für Angewandte Mathematik, Universität Hannover
15. Prössdorf, S., Rathsfeld, A. (1988): Mellin techniques in the numerical analysis for one-dimensonal singular integral equation. Report 06/88 Akademie der Wissenschaften der DDR
16. Saranen, J. (1977): Local error estimates for some Petrov-Galerkin methods applied to strongly elliptic equations on curves. Math. Comp. 48, 485-502.
17. Saranen, J., Wendland, W.L. (1993): Local residual-type error estimates for adaptive boundary element methods on closed curves. Appl. Anal. 48, 37-50.
18. Sloan, I.H., Stephan, E.P. (1992): Collocation with Chebyshev Polynomials for Symm's Integral Equation on an Interval. J. Australian Math. Soc. Ser. B 34, 199-211
19. Stephan, E.P., Tran, T. (1994): Localization and post processing for the Galerkin boundary element method applied to three-dimensional screen problems. Preprint Uni NSW, to appear in J. Integral Eqns. Appl.
20. Tran, T. (1993): Local error estimates for the Galerkin method applied to strongly elliptic integral equations on open curves. J. Int. Eq. Appl. 5, 405-428
21. Tran, T. (1995): The K-Operator and the Galerkin method for strongly elliptic equations on smooth curves: Local estimates. Math. Comp. 64, 501-513

The hp-Version of the BEM with Geometric Meshes in 3D

M. Maischak and E. P. Stephan

1 Introduction

The solutions of three-dimensional elliptic boundary value problems in polyhedral domains or on screens have special singular forms at corners and edges. If those problems are converted via the direct method into boundary integral equations, the solutions of the latter inherit those edge and corner singularities. The singularities disturb the convergence of numerical schemes, e.g. finite element or boundary element methods. Numerical schemes converge as fast as the solution is approximated by the chosen trial functions due to the quasioptimality of the Galerkin method. Thus the rate of convergence is determined by the regularity of the solution. Even for smooth data, the regularity is reduced by edge and corner singularities, and therefore the convergence of these procedures is much slower than in case of smooth domains.

In 2D boundary value problems we have only corner singularities (see [2]). From [24, 25, 26] which are based on [5] it is known that the solutions of three-dimensional elliptic boundary value problems also contain edge and corner-edge singularities. While edge singularities can be analyzed using one dimensional approximation results (see [8]), the corner-edge singularities demand further investigation. Therefore we have defined countably normed spaces which contain not only the well-known corner singularities but also the edge-singularities and even more the corner-edge singularities. As we do not only have point singularities, all approximation results of [7, 8] have to be extended and generalized considerably. First results for the hp-version of the finite element method for 3D problems have been announced in [3]. Because in this case the trace theorem gives no result for screen problems, those results are not suitable for the hp-version of the boundary element method considered here. Also the mesh defined in [3] is more complicated than the simple tensor product mesh defined here (see Figure 3.1) and demands much more effort to implement. From Lemma 3.1 it can be seen that the singularities described in [24, 25, 26] are contained in our countably normed spaces. From [24, 25, 26] it is known that for opening edge angles $\omega < 2\pi$ of the polyhedral, i.e. we have no screen problem, the singularities are square integrable, which can be analyzed directly using our countably normed spaces. The case of screen problems for the Laplacian is analyzed in [9] and for the Helmholtz operator in [10]. The case of the Lamé operator is analyzed in [17].

In this paper, it will be shown that under proper assumptions usually satisfied in applications, the hp-version with a geometric mesh for the Dirichlet problem of the Laplacian has exponential rate of convergence with respect to the number of degrees of freedom. In Section 2 we consider the Dirichlet screen problem (for Dirichlet and Neumann problems in polyhedrons see [11]). The countably normed spaces introduced in Section 3 are motivated by the form of the corner-edge singularities described in [25].

In Section 3 we define the mesh and the appropriate countably normed spaces. For the proof of our main result (Theorem 3.3 and Theorem 3.4), the exponential convergence of the hp-version of the BEM for 3D problems, we define an equivalent countably normed space on a reference element using cartesian coordinates. In Section 4 we describe some numerical results.

In the following, C denotes a generic constant.

2 The screen problem and the Galerkin procedure

For brevity let us consider the Dirichlet screen problem with $\Gamma = (0,2)^2$: Given f on the screen Γ find u in $\Omega_\Gamma = \mathbb{R}^3 \backslash \bar{\Gamma}$ satisfying

$$\Delta u = 0 \quad \text{in} \quad \Omega_\Gamma, \tag{2.1}$$

$$u = f \quad \text{on} \quad \Gamma \tag{2.2}$$

and

$$u = \mathcal{O}\left(\frac{1}{|x|}\right) \quad \text{as} \quad |x| \to \infty. \tag{2.3}$$

In [1, 4, 21] this problem is converted into the integral equation

$$V\Psi(x) := \frac{1}{4\pi} \int_\Gamma \frac{1}{|x-y|} \Psi(y)\, ds_y = f(x), \quad x \in \Gamma, \tag{2.4}$$

for the Dirichlet problem. Here V is the single layer potential.

It is known from [4], [21] that the unique solution of (2.4) belongs to a certain Sobolev space of distributions concentrated on Γ. Namely, given $f \in H^{1/2}(\Gamma)$ there exists exactly one solution $\Psi \in \tilde{H}^{-1/2}(\Gamma)$ of (2.4). Furthermore, problem (2.1), (2.2), (2.3) and equation (2.4) are equivalent, i.e. $u \in H^1_{\text{loc}}(\mathbb{R}^3 \backslash \bar{\Gamma})$ solves (2.1), (2.2), (2.3) if and only if the jump $[\frac{\partial u}{\partial n}]|_\Gamma \in \tilde{H}^{-1/2}(\Gamma)$ is the solution Ψ of (2.4).

The regularity theory in [25, 26] shows that even for smooth right hand side f the solution Ψ of the integral equation (2.4) becomes unbounded near the edges of Γ, it has so-called edge and corner-edge singularities. From [25, Corollary 5] and [26, Theorem 2.7 and (2.38)] we present the regularity result for the screen $\Gamma = (0,2)^2$:

Theorem 2.1. *Let $\Gamma = (0,2)^2$ and $f \in H^2(\Gamma)$ in (2.4) be given. The unique solution $\Psi \in \tilde{H}^{-1/2}(\Gamma)$ of (2.4) has near the origin a decomposition into edge and corner-edge singularities of the form*

$$\Psi = \Psi_0 + \chi(r) a_1 r^{\gamma-1} \omega(\theta) + \tilde{\chi}(\theta) e_1(r) r^{-1} \sin^{-1/2}\theta + \tilde{\chi}\left(\frac{\pi}{2} - \theta\right) e_2(r) r^{-1} \cos^{-1/2}\theta \quad (2.5)$$

with $\Psi_0 \in H^{1-\varepsilon}(\Gamma)$, $a_1 \in \mathbb{R}$, $\omega(\theta) \in H^{1-\varepsilon}[0,\pi/2]$, $e_i(r) = b_i r^\gamma + c_i$, $r^{-1/2} c_i(r) \in H^1(\mathbb{R}_+)$, $r^{-3/2} c_i(r) \in L^2(\mathbb{R}_+)$, $b_i \in \mathbb{R}$, $i = 1,2$. Corresponding decompositions hold at the other vertices of Γ.

Here (r,θ) are plane polar coordinates concentrated at the origin. χ and $\tilde{\chi}$ are C^∞ cut-off-functions with $\chi \equiv 1$ for $r < 1/4$, $\tilde{\chi} \equiv 1$ for $\theta < \pi/4$. The exponent $\gamma \approx 0.2966$ is obtained in [20].

Remark 2.1. The analysis in [25] and [26] shows that for smoother data f in (2.4) expansions like (2.5) hold with smoother remainder terms Ψ_0 and smoother contributions to the edge and corner-edge singularities.

It is shown in [4] that the operator V in (2.4) is positive definite, i.e. there exists a constant $\eta > 0$ such that for all $\varphi \in \tilde{H}^{-1/2}(\Gamma)$

$$\langle V\varphi, \varphi\rangle_{L^2(\Gamma)} \geq \eta \|\varphi\|^2_{\tilde{H}^{-1/2}(\Gamma)}. \quad (2.6)$$

It is well known that this property of V guarantees the convergence of the boundary element Galerkin solution for (2.4) if conforming, approximating subspaces X_N of $\tilde{H}^{-1/2}(\Gamma)$ are used.

Then the Galerkin scheme for (2.4) reads:
Find $\Psi_N \in X_N$ such that for any $\varphi \in X_N$

$$\langle V\Psi_N, \varphi\rangle_{L^2(\Gamma)} = \langle f, \varphi\rangle_{L^2(\Gamma)}. \quad (2.7)$$

Application of the general results of the Galerkin procedure for strongly elliptic pseudodifferential operators (see [23]) yields the following result.

Proposition 2.1. *The Galerkin equation (2.7) is uniquely solvable for any N. Let Ψ be the solution of (2.4) and Ψ_N its Galerkin approximation defined by (2.7). Then there holds*

$$\|\Psi - \Psi_N\|_{\tilde{H}^{-1/2}(\Gamma)} \leq C \inf\left\{\|\Psi - \varphi\|_{\tilde{H}^{-1/2}(\Gamma)} ; \varphi \in X_N\right\}. \quad (2.8)$$

Here C is a positive constant independent of N.

3 The hp-version with geometric mesh

Let Γ be the boundary of a simply connected bounded polyhedron Ω in $\mathbb{R}^3$. We first introduce appropriate countably normed spaces $\mathcal{B}^l_\beta(\Gamma)$, $(0<\beta<1)$ (cf. [6]). Let F be a face of Ω having the corner points $e_1,\dots,e_m$. For each i, f_i denotes the edge of F connecting e_i with e_{i+1}, where the periodicity convention $e_{m+1}=e_1$, $f_{m+1}=f_1$ is adopted. Consider a covering of F by neighbourhoods U_i $(i=1,\dots,m)$ of e_i not containing e_j, $j\neq i$, and introduce polar coordinates (r_i,Θ_i) of origin e_i in U_i such that $r_i=\operatorname{dist}(x,e_i)$ and the edges f_{i-1} and f_i are given by $\Theta_i=\omega_i$ and $\Theta_i=0$, respectively. For $0<\beta<1$, let

$$\begin{aligned}\mathcal{B}^l_\beta(F)=\{u\,|\,u\in H^{l-1}(F):\exists C,d>0 \text{ independent of } k \text{ such that}\\ \|r_i^{\beta+\alpha_r-l}(\Theta_i(\omega_i-\Theta_i))^{(\beta+\alpha_\Theta-l)_+}(\partial/\partial r_i)^{\alpha_r}(\partial/\partial\Theta_i)^{\alpha_\Theta}u\|_{L^2(U_i)}\le\\ \le C\,d^{\alpha_r+\alpha_\Theta-l}(\alpha_r+\alpha_\Theta-l)!, \qquad (3.1)\\ \alpha_r+\alpha_\Theta=k=l,l+1,\dots,i=1,\dots,m\},\\ \mathcal{B}^l_\beta(\Gamma)=\{u\,|\,u\in H^{l-1}(\Gamma),u\in B^l_\beta(F)\text{ for all faces } F \text{ of } \Gamma\}\end{aligned}$$

with $(a)_+:=\max(a,0)$. If we would like to emphasize the dependence on the constants C,d we will write $\mathcal{B}^l_\beta(F)=\mathcal{B}^l_{\beta,C,d}(F)$, etc.

Now we define the geometric mesh on the faces of the polyhedron. We assume that the face F is a triangle. This is no loss of generality because every polygonal domain can be decomposed into triangles. We divide this triangle into three parallelograms and three triangles where each parallelogram lies in a corner of the face F and each triangle lies at an edge of F apart of the corners. By linear transformations φ_i we can map the parallelograms onto the reference square $Q=[0,1]^2$ such that the vertices of the face F are mapped to $(0,0)$. The triangles can be mapped by linear transformations $\tilde\varphi_i$ onto the reference triangle $\tilde Q=\{(x,y)\in Q\,|\,y\le x\}$ such that the corner point of the triangle in the interior of the face F is mapped to $(1,1)$ of the reference triangle. By Definition 3.1 the geometric mesh and appropriate spline spaces are defined on the reference element Q. Analogously the geometric mesh can be defined on the reference triangle $\tilde Q$ (see Figure 3.1). Via the transformations $\varphi_i^{-1},\tilde\varphi_i^{-1}$ the geometric mesh Γ^n_σ is also defined on the faces of the polyhedron. The approximation on the reference square is the more interesting case because it handles the corner-edge singularities. Therefore we deal in the following only with the approximation on the reference square.

Definition 3.1 (geometric mesh). *Let $Q=[0,1]\times[0,1]$. For $0<\sigma<1$ we use the partition Q^n_σ of Q into n^2 subsquares R_{kl}*

$$R_{kl}=[x_{k-1},x_k]\times[x_{l-1},x_l],\quad(k,l=1,\dots,n),\qquad Q=\bigcup_{k,l=1}^n R_{kl}\qquad(3.2)$$

where

$$x_0 = 0, \qquad x_k = \sigma^{n-k}, \quad k = 1, \ldots, n. \tag{3.3}$$

With Q_σ^n *we associate a degree vector* $p = (p_1, \ldots, p_n)$ *and define* $S^{p,r}(Q_\sigma^n)$ *as the vectorspace of all piecewise polynomials* $v(x,y)$ *on* Q *having degree* p_k *in* x *and* p_l *in* y *on* $[x_{k-1}, x_k] \times [x_{l-1}, x_l]$, $k, l = 1, \ldots, n$, *i.e.* $v|_{[x_{k-1},x_k]\times[x_{l-1},x_l]} \in P_{p_k,p_l}(R_{kl})$. *For the differences* $h_k = x_k - x_{k-1}$ *we have with* $\lambda = (1-\sigma)/\sigma$

$$h_k = x_k - x_{k-1} = x_{k-1}(\frac{1}{\sigma} - 1) \le x(\frac{1}{\sigma} - 1) = x\lambda, \quad \forall x \in [x_{k-1}, x_k] \; (2 \le k \le n) \tag{3.4}$$

The index $r \in \{0, 1\}$ *in* $S^{p,r}(Q_\sigma^n)$ *determines the regularity of the piecewise polynomials, i.e. discontinuity in case of* $r = 0$ *and continuity in case of* $r = 1$.

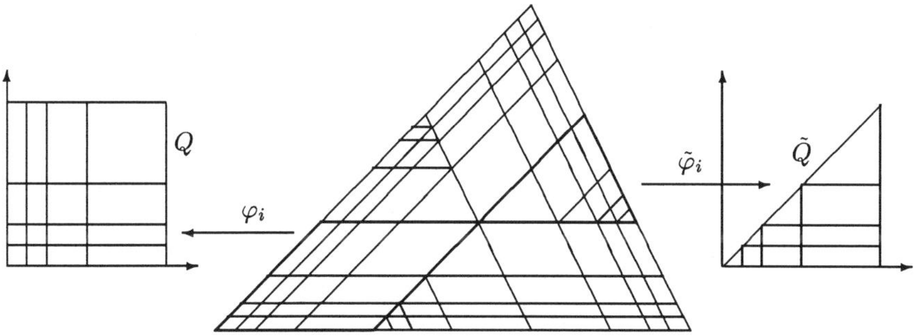

Fig. 3.1. Geometric mesh with $\sigma = 0.5$ on a polyhedral face

Now we define countably normed spaces on the reference element Q using cartesian coordinates

Definition 3.2 (countably normed spaces $B_\beta^l(Q)$).
Let β *be a real number with* $0 < \beta < 1$. *The weight function* $\Phi_{\beta,\alpha,l} = \Phi_{\beta,\alpha,l}(x,y)$ *is for* $\alpha = (\alpha_1, \alpha_2)$ *and an integer* $l \ge 1$ *defined by*

$$\Phi_{\beta,\alpha,l} = x^\beta \sum_{\gamma_1 = \max(\alpha_1 - l, 0)}^{\min(\alpha_1 - 1, \alpha_1 + \alpha_2 - l)} x^{\gamma_1} y^{\alpha_1 + \alpha_2 - l - \gamma_1} + y^\beta \sum_{\gamma_2 = \max(\alpha_2 - l, 0)}^{\min(\alpha_2 - 1, \alpha_1 + \alpha_2 - l)} x^{\alpha_1 + \alpha_2 - l - \gamma_2} y^{\gamma_2} \tag{3.5}$$

Let

$$D^\alpha = \frac{\partial^{|\alpha|}}{\partial x^{\alpha_1} \partial y^{\alpha_2}} = \partial_x^{\alpha_1} \partial_y^{\alpha_2}$$

The countably normed spaces for $l \ge 1$ *are defined by*

$$B^l_\beta(Q) = \Big\{u : u \in H^{l-1}(Q), \|\Phi_{\beta,\alpha,l} D^\alpha u\|_{L^2(Q)} \leq C\, d^{k-l}(k-l)! \\ \textit{for } |\alpha| = k = l, l+1, \ldots, \\ \textit{with } C \geq 1, d \geq 1 \textit{ independent of } k\Big\}. \tag{3.6}$$

If we would like to emphasize the dependence on the constants C, d we will write $B^l_\beta(Q) = B^l_{\beta,C,d}(Q)$, etc.

Theorem 3.1. *[14, 15] Let Q be the reference element and let φ be the linear transformation from a parallelogram, which lies in a corner of the face F, to the reference element Q. Then it follows for $l = 1, 2$ from*

$$u \in \mathcal{B}^l_{\beta,\tilde{C},\tilde{d}}(\varphi(Q)) \tag{3.7}$$

that we have

$$u \circ \varphi^{-1} \in B^l_{\beta,C,d}(Q) \tag{3.8}$$

where C, d (resp. $\tilde{C}, \tilde{d}$) are the constants in the definition of $B^l_\beta(Q)$ (resp. $\mathcal{B}^l_\beta(\varphi(Q))$).

Remark 3.1. In the case $l = 1$ we have also equivalence.

Now we show that singularities (which typically appear in solutions of boundary integral equations on polyhedral domains Ω [24, 25, 26]) are contained in the countably normed spaces considered here. The arising singularities have the form (3.11) near the corners of the faces of Ω. Therefore we need countably normed spaces which describe singularities in the angular variable Θ.

Definition 3.3 (countably normed spaces $B^l_\beta([0,\omega])$).
The countably normed spaces $B^l_\beta([0,\omega])$ for $l \geq 1$ are defined by

$$B^l_{\beta,C,d}([0,\omega]) = \Big\{u \, : \, u \in H^{l-1}([0,\omega]), \|(\Theta(\omega-\Theta))^{\beta+\alpha-l}\partial^\alpha_\Theta u\|_{L^2([0,\omega])} \\ \leq C\, d^{\alpha-l}(\alpha-l)! \textit{ for } \alpha = l, l+1, \ldots, \\ \textit{with } C \geq 1, d \geq 1 \textit{ independent of } \alpha\Big\}. \tag{3.9}$$

Lemma 3.1 (corner-edge singularities). *[18] Let $0 < \beta < 1$. Let $(a_k)_k$ be a sequence of real numbers. Let $(\lambda_k)_k$ be a sequence of real numbers with $\lambda_{k+1} - \lambda_k \geq 0$ for all $k \geq 1$ and $\lambda_1 > l - \beta - 1$. Let $w_k \in B^l_{\beta,C_k,d_k}([0,\omega])$ with $d_k \leq \bar{d} < \infty$ for all $k \geq 1$ and $\|w_k\|_{H^{l-1}([0,\omega])} \leq C_k$. Let*

$$\limsup_{k\to\infty} \frac{C_{k+1}|a_{k+1}|}{C_k|a_k|} 2^{|\lambda_{k+1}|-|\lambda_k|} R^{\lambda_{k+1}-\lambda_k} \leq q < 1 \tag{3.10}$$

Then we have

$$\sum_{k=1}^\infty a_k r^{\lambda_k} w_k(\Theta) \in \mathcal{B}^l_{\beta,C,d}(S^R_\omega) \tag{3.11}$$

with the sector $S^R_\omega = \{(r,\Theta)\,|\,0 < r < R, 0 < \Theta < \omega\}$.

Remark 3.2. $\mathcal{B}^l_{\beta,C,d}(S^R_\omega)$ is the restriction of $\mathcal{B}^l_\beta(F)$ (see (3.1)) to the sector S^R_ω on the face F.

Corollary 3.1. *[18] Let $0 < \beta < 1$ and $\nu > l - 1/2 - \beta$, $\lambda > l - 1 - \beta$, $\lambda > -1$, $\nu > -1/2$. Then we have for $l \geq 1$*

$$r^\lambda \theta^\nu \in \mathcal{B}^l_\beta(S^R_\omega) \tag{3.12}$$

The following result is used to estimate the approximation error for smooth remainder terms in the asymptotic expansions of the solutions of the integral equation (2.4).

Theorem 3.2. *[9, Lemma 5] Given $u \in H^t(Q)$ there exists $\phi \in S^{p,0}(Q^n_\sigma)$ such that*

$$\|u - \phi\|_{L^2(Q)} \leq C\, N^{(1-t)/4} \tag{3.13}$$

with C independent of $N = \dim S^{p,0}(Q^n_\sigma)$

The exponentially good approximation properties of splines on our geometric meshes for general functions $u \in B^l_\beta(Q)$ ($l = 1, 2$) are given by the following theorem.

Theorem 3.3. *[18, 19, 14]*

(i) Let $u(x,y) \in B^1_\beta(Q)$ with $0 < \beta < 1$. Then there is a spline function $u_N(x,y) \in S^{p,0}(Q^n_\sigma)$ and constants $C_1, b_1 > 0$ independent of N, but depending on σ, μ, β such that

$$\|u(x,y) - u_N(x,y)\|_{L^2(Q)} \leq C_1\, e^{-b_1 \sqrt[4]{N}} \tag{3.14}$$

with $p = (p_1, \ldots, p_n)$, $p_k = [\mu(k-1)]$ for $\mu > 0$ and $N = \dim S^{p,0}(Q^n_\sigma)$. Moreover $u_N(x,y)$ can be chosen to be the L^2-projection of u on $S^{p,0}(Q^n_\sigma)$.

(ii) Let $v(x,y) \in B^2_\beta(Q) \cap C^0(Q)$ with $0 < \beta < 1$. Then there is a spline function $v_N \in S^{p,1}(Q^n_\sigma)$ and constants $C_2, b_2 > 0$ independent of N, but dependent on σ, μ, β such that

$$\|v(x,y) - v_N(x,y)\|_{H^1(Q)} \leq C_2\, e^{-b_2 \sqrt[4]{N}} \tag{3.15}$$

with $p = (p_1, \ldots, p_n)$, $p_1 = 1$, $p_k = \max(2, [\mu(k-1)] + 1)$ $(k > 1)$ for $\mu > 0$ and $N = \dim S^{p,1}(Q^n_\sigma)$.

Next we give the energy estimate for the Galerkin solution of the integral equation (2.4) computed via the hp-version. In the following let $X_N = S^{p,0}(\Gamma^n_\sigma)$ denote the subspace of piecewise polynomials on the geometric mesh Γ^n_σ and let X_N be locally defined as in Definition 3.1.

Theorem 3.4. *[9, Theorem 2] Let the right hand side f in equation (2.4) be piecewise analytic and Ψ be defined by (2.4) and $\Psi_N \in X_N = S^{p,0}(\Gamma_\sigma^n)$ its Galerkin approximation be defined by (2.7). Then, with $N = \dim X_N$, there holds for any $\alpha > 0$*

$$\|\Psi - \Psi_N\|_{\tilde{H}^{-1/2}(\Gamma)} \leq C\, e^{-b\sqrt[4]{N}} + \mathcal{O}(N^{-\alpha}) \tag{3.16}$$

with constants $C > 0$ and $b > 0$, independent of N but depending on σ, μ and α.

Proof. Due to Theorem 2.1 we need only to consider the behavior near the corners and edges of $\Gamma = (0,2)^2$. With respect of the symmetry of Γ it is sufficient to look at the origin. Hence let us consider the most singular part of the exact solution near the origin, namely the corner-edge singularity in $(0,0)$. Since $\omega(\theta) \sim \sin^{1/2}\theta$ we have to approximate the function

$$\Psi_Q := r^{\gamma-1}\sin^{1/2}\theta + r^{\gamma-1}\sin^{-1/2}\theta \quad (\gamma \approx 0.2966) \tag{3.17}$$

by a spline-function $\Psi_g = \psi_1 + \psi_2 \in S^{p,0}(Q_\sigma^n)$. The term $r^{\gamma-1}\sin^{1/2}\theta \sim r^{\gamma-1}\theta^{1/2}$ can be approximated exponentially fast by splines $\psi_1 \in S^{p,0}(Q_\sigma^n)$ in the L^2-norm due to Corollary 3.1 and Theorem 3.3. In [9] it was shown that the term $r^{\gamma-1}\sin^{-1/2}\theta$ can be approximated exponentially fast by splines $\psi_2 \in S^{p,0}(Q_\sigma^n)$ in the $\tilde{H}^{-1/2}(Q)$-norm

$$\|r^{\gamma-1}\sin^{-1/2}\theta - \psi_2(x,y)\|_{\tilde{H}^{-1/2}(Q)} \leq C\, e^{-b\sqrt[4]{N}}. \tag{3.18}$$

Now the remaining part of Ψ, namely $\Psi - \Psi_Q$ is smoother than the most singular terms of Ψ. Since f is piecewise analytic, the difference $\Psi - \Psi_Q$ can be decomposed into a sum of higher edge and higher corner-edge singularities and an arbitrarily smooth remainder term. This can be seen by further expanding the functions v_k and $\tilde{b}^j_{m,p}$ in the decomposition (5.4) in Theorem 8 in [25]. The resulting series expansions for these functions v_k and $\tilde{b}^j_{m,p}$ themselves can be splitted into a finite sum of terms belonging to countably normed spaces plus a remainder term in $H^t(Q)$, for t sufficiently large. These terms in the finite sum are of the form

$$r^\lambda \theta^\nu \in B^l_\beta, \quad l = 1 \text{ or } 2.$$

It is shown in Lemma 3.1 that even infinite series of those terms belong to countably normed spaces provided λ, ν grow sufficiently fast. Now, those additional singularities can again be approximated exponentially fast due to Theorem 3.3. Thus in view of such an improved decomposition it is left to approximate globally a remainder term in $H^t(Q)$ where $t > 0$ is arbitrarily large. With Theorem 3.2 we obtain that this remainder term can be approximated with $\mathcal{O}(N^{-\alpha})$ with $\alpha = (t-1)/4$.

Remark 3.3. For the Neumann problem we get the analogous result using the spaces $\tilde{H}^{1/2}$ and B^2_β instead of $\tilde{H}^{-1/2}$ and B^1_β, respectively (see [9, 14, 19]).

Remark 3.4. The analogous results hold also for Dirichlet and Neumann problems of the Lamé- and Helmholtz-equations (see [10, 17]).

4 Numerical Results

Here we present numerical results (performed on the CDC 990 and Siemens S400/40 at the RRZN, University of Hannover) for the Dirichlet problem of the Laplacian on a screen. We consider the square Γ with sidelength 2. We choose the function 1 as the right hand side, i.e. $f = 1$. For the Dirichlet problem we look at the capacitance C of the screen Γ, which is given by

$$C = \frac{1}{4\pi} \int_\Gamma \Psi \, ds, \tag{4.1}$$

where the charge density Ψ solves the integral equation

$$V\Psi = \frac{1}{4\pi} \int_\Gamma \frac{1}{|x-y|} \Psi(y) \, ds_y = f \tag{4.2}$$

In [1] a numerical value of 0.733576 in case of the square plate is given for the capacitance C. One obtains an approximation C_N to the capacitance C by inserting the boundary element Galerkin solution Ψ_N for (4.2) into (4.1). Because of the special right hand side $f = 1$ there hold the inequalities

$$c_1 \|\Psi - \Psi_N\|^2_{\tilde{H}^{-1/2}(\Gamma)} \leq |C - C_N| \leq c_2 \|\Psi - \Psi_N\|^2_{\tilde{H}^{-1/2}(\Gamma)} \tag{4.3}$$

for constants $c_1, c_2 > 0$ and therefore the error estimates

$$|C - C_N| \leq c_3 \begin{cases} h^{1-\varepsilon} \\ p^{-2+2\varepsilon} \end{cases} \tag{4.4}$$

for the pure h- and p-versions (see Lemma 1.6 in [1] for the h-version and [22] for the p-version).

Due to Theorem 3.3 and Theorem 3.4 we expect for the hp-version with geometric mesh numerically an exponentially fast convergence. In Figure 4.1 we compare the errors for the different versions on the square plate: The error curves are linear for the pure h- and p-versions and curved downward for the hp-version with geometric mesh. The convergence of the latter is indeed exponential as Figure 4.2 indicates where we obtain nearly straight lines by plotting the logarithmic errors in capacitance against the fourth root of the number of unknowns.

Remark 4.1. Corresponding computations have been done for the Neumann problem of the Laplacian on a screen [9], for Dirichlet and Neumann problems on polyhedrons (cube, L-block) [16], for the Lamé equation [17] and the Helmholtz equation [10] with analogous results.

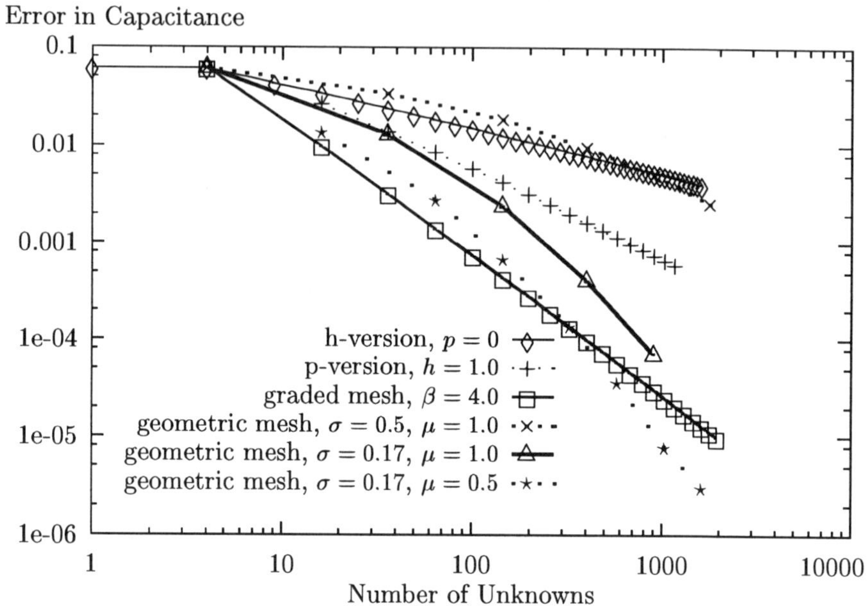

Fig. 4.1. Dirichlet problem on the screen Γ in $\mathbb{R}^3$

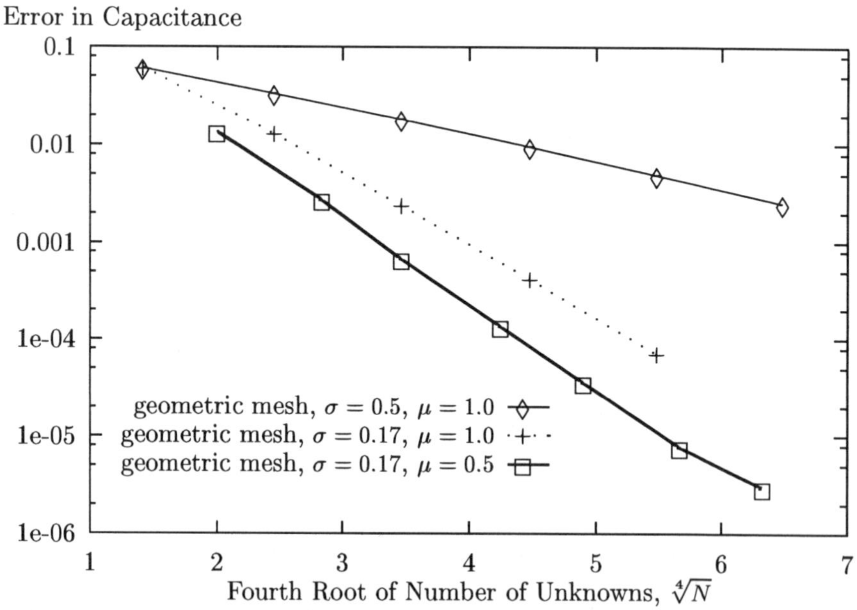

Fig. 4.2. Dirichlet problem on the screen Γ in $\mathbb{R}^3$

Remark 4.2. All numerical investigations have been done using three programs, namely maibem3.f, maibem2.f, maifem2.f . The programs are configurable using a common control language, i.e. all implemented operators, solvers, geometries, etc. are usable in every meaningful combination.

maibem2.f realizes the 2D-BEM on polygonal boundaries, maifem2.f realizes the 2D-FEM on domains with polygonal boundaries and maibem3.f realizes the 3D-BEM on polyhedral boundaries.

We have developed three libraries of subroutines (liblap3.f, liblame3.f, libhelm3.f), which include the analytical computation of the Galerkin elements and potential functions on rectangular and parallelogram elements for an arbitrary hp-distribution (see [12]). Analogously we have developed similar libraries for the 2D-BEM on arbitrary polygonal boundaries (see [13]).

References

1. Abou El-Seoud, S., Ervin, V.J., Stephan, E.P. (1990): An improved boundary element method for the charge density of a thin electrified plate in $\mathbb{R}^3$. Math. Meth. Appl. Sci., 13, 291–303
2. Babuška, I., Guo, B.Q. (1988): Regularity of the solution of elliptic problems with piecewise analytic data, part i. SIAM J. Math. Anal., 19, 172–203
3. Babuška, I., Guo, B.Q. (1994): The h-p version of the finite element method in three dimensions. In: Costabel, M. et al. (eds.) Boundary Value Problems and Integral Equations in Nonsmooth Domains. Dekker, New York
4. Costabel, M. (1988): Boundary integral operators on lipschitz domains: Elementary results. SIAM J. Math. Anal., 19(3), 613–626
5. Dauge, M. (1988): Elliptic Boundary Value Problems on Corner Domains, volume 1341 of Lecture Notes in Mathematics. Springer
6. Elschner, J. (1994): On the exponential convergence of some boundary element methods for Laplace's equation in non-smooth domains. In: Costabel, M. et al. (eds.) Boundary Value Problems and Integral Equations in Nonsmooth Domains. Dekker, New York
7. Guo, B.Q., Babuška, I. (1986): The h-p version of the finite element method, part 1: The basic approximation results. Comp. Mech., 1, 21–41
8. Heuer, N. (1992): hp-Versionen der Randelementmethode. PhD thesis, Hannover
9. Heuer, N., Maischak, M., Stephan, E.P. (1995): The *hp*-version of the boundary element method for screen problems. Preprint, Institut für Angewandte Mathematik, Universität Hannover
10. Holm, H., Maischak, M., Stephan, E.P. (1996): The *hp*-version of the boundary element method for the Helmholtz equation. To appear in Computing
11. Holm, H., Maischak, M., Stephan, E.P. (1996): Exponential convergence of the $h-p$ version bem for mixed bvp's on polyhedrons. Preprint, Institut für Angewandte Mathematik, Universität Hannover
12. Maischak, M. (1995): The analytical computation of the Galerkin elements for the Laplace, Lamé and Helmholtz equation in 3d-bem. Preprint, Institut für Angewandte Mathematik, Universität Hannover

13. Maischak, M. (1995): The analytical computation of the Galerkin elements for the Laplace, Lamé and Helmholtz equation in 2d-bem. Preprint, Institut für Angewandte Mathematik, Universität Hannover
14. Maischak, M. (1995): hp-Methoden für Randintegralgleichungen bei 3D-Problemen, Theorie und Implementierung. PhD thesis, Hannover
15. Maischak, M. (1996): Relations between countably normed spaces for 3D-BEM with weightfunctions defined in different coordinate systems. Preprint, Institut für Angewandte Mathematik, Universität Hannover
16. Maischak, M., Stephan, E.P. (1994): The h–p version of the boundary element method for first kind integral equations on polyhedral surfaces. Preprint, Institut für Angewandte Mathematik, Universität Hannover
17. Maischak, M., Stephan, E.P. (1995): The *hp*-version of the boundary element method for the Lamé equation in 3d. In preparation
18. Maischak, M., Stephan, E.P. (1996): The hp-version of the boundary element method in $\mathbb{R}^3$. The basic approximation results. To appear in Math. Meth. Appl. Sci.
19. Maischak, M., Stephan, E.P. (1995): The hp-version of the boundary element method in $\mathbb{R}^3$. The basic approximation results for the $B^2_\beta(\Gamma)$-space. In preparation
20. Morrison, J.A., Lewis, J.A. (1976): Singularity at the corner of a flat plane. SIAM J. Appl. Math., 31, 233–250
21. Stephan, E.P. (1987): Boundary integral equations for screen problems in $\mathbb{R}^3$. Integral Equations and Operator Theory, 10, 257–263
22. Stephan, E.P. (1988): Improved Galerkin methods for integral equations on polygons and polyhedral surfaces. In: Japan-U.S. Symposium on boundary element methods, pp. 73–80, Tokyo
23. Stephan, E.P., Wendland, W.L. (1976): Remarks to Galerkin and least squares methods with finite elements for general elliptic problems. Manuscripta Geodaetica 1, 93–123
24. von Petersdorff, P. (1989): Randwertprobleme der Elastizitätstheorie für Polyeder – Singularitäten und Approximation mit Randelementmethoden. PhD thesis, Darmstadt
25. von Petersdorff, T., Stephan, E.P. (1990): Decompositions in edge and corner singularities for the solution of the Dirichlet problem of the Laplacian in a polyhedron. Math. Nachrichten 149, 71–104
26. von Petersdorff, T., Stephan, E.P. (1990): Regularity of mixed boundary value problems in $\mathbb{R}^3$ and boundary element methods on graded meshes. Math. Meth. Appl. Sci. 12, 229–249

A Transonic Panel Method for Helicopter Flows

S. Wagner and A. Röttgermann

1 Introduction

1.1 General Remarks

Since the early 1960's, boundary element methods[1] have been an indispensable tool for analyzing the flow field and shaping the surfaces surrounded by flows. After Hess & Smith [17] introduced the panel methods into fluid mechanics, many different methods have been developed [31, 32, 41]; surveys of the existing methods are given by Wagner [45, 46] Hoeijmakers [18] and Hess [16]. With the increasing use of computers, lifting surface theories [43, 44], of which some had been very sophisticated until that time, increasingly lost importance. The scope of application for boundary element methods, however, was extended to include supersonic and unsteady flows.

Reasons for the success of the boundary element methods are:

- the capability of calculating accurate solutions of the linearized potential flow of very complex configurations including the vortex layer
- an acceptable expenditure for the panel discretisation achieved by reducing the problem by one dimension to the body surface

The competing finite–difference, finite–volume and finite–element methods need space networks around the configuration that usually contain many grid cells. Thus, the grid generation and the solution of the flow problem in this grid require a lot of computation time. By novel solution procedures for convergence acceleration and new grid generation strategies, these methods can also be employed on more complex geometries.

To remain competitive, panel methods have to be further developed as well. The boundary element methods will only have future prospects if two of its greatest restrictions can be overcome:

1) The computation expenditures increase quadratically or even cubically according to the number of the used surface elements
2) Panel methods are based on linear integral equations and thus can calculate only incompressible and weak compressible flows

---

[1] In aerodynamics they are usually called **panel methods**.

The computation expenditure can be reduced by several methods: multigrid-methods, clustering by coarse grid estimations [39], panel-clustering techniques [11].

It is getting more and more important to include the interference effects between body and wake vortex layers in calculation, e.g. to study the non-linear lift behavior of delta wings with leading edge vortices [1, 19, 24, 42] or the influence of the blade tip vortex on the circulation distribution of a rotor [8, 22, 35]. In case of rotors, blade–vortex interferences[2] and transonic effects are very important in the area of the blade tip. The separating vortex layers can be especially well described by using vortex–lattice methods. In order to capture the compressible and transonic effects, the second restriction has to be removed and the flow field has to be described by a nonlinear integral equation.

1.2 Objectives

Main emphasis is placed on developing a method that allows to determine the influence of the blade vortex interference and of the compressibility of the fluid on the blades of a rotor.

To capture the development of the free vortex layers and its interferences with bodies, a vortex layer procedure [27, 33, 35] based on a panel method for delta wings by Behr and Urban was developed at the University of the Federal Armed Forces in Munich. For the present work, the method, originally developed only for thin foils and external Neumann boundary condition, was extended to thick blades and extended by the feasibility of using an inner Dirichlet boundary condition. Since the compressible effects of the fluids have to be captured as well, it was necessary to surmount the limitation to the linear integral equation and to employ a nonlinear integral equation. The objective of this work is to prove that transonic effects at the blade tip of helicopter rotors can be described by using an extended integral equation procedure.

In industry, panel methods are very successfully used for the flow computation of complex configurations. However, as a rule, these methods are limited to slightly compressible subsonic flows and pure supersonic flows. When developing a method that includes compressible flows up to the transonic speed domain, the main attention is focused on the easy operation and the combination with already existing panel methods.

The second restriction of the conventional panel methods is the considerable increase of computation time for panel numbers $N > 1000 - 2000$. To facilitate a more efficient computation of the linear part of integral equations, the *panel-clustering method*, introduced by Hackbusch & Nowak [11] and further developed by Sauter [37] is described in order to reduce the expenditures on computation and storage. This method considerably increases

[2] **BVI** — **B**lade–**V**ortex–**I**nterference.

the efficiency of boundary element methods but up to now has not been documented in a form that would be easily comprehensible for most engineers.

The theoretical principles of the field panel method and further results are presented by Röttgermann [34].

2 Basic Equations

2.1 Equations of Motion

For irrotational flows, a velocity potential Φ can be defined, from which the velocity components can be determined by forming the gradient $\mathbf{V} = -\nabla\Phi$, and that complies with the condition of irrotationality due to $\nabla \times \mathbf{V} = \nabla \times \nabla\Phi = 0$. Introducing the velocity potential into the conservation equations yields a formula for the velocity potential Φ at compressible, unsteady flow:

$$\begin{aligned}\left(1-\frac{u^2}{a^2}\right)\frac{\partial^2\Phi}{\partial x^2}+\left(1-\frac{v^2}{a^2}\right)\frac{\partial^2\Phi}{\partial y^2}+\left(1-\frac{w^2}{a^2}\right)\frac{\partial^2\Phi}{\partial z^2}\\ -\frac{2uv}{a^2}\frac{\partial^2\Phi}{\partial x\partial y}-\frac{2vw}{a^2}\frac{\partial^2\Phi}{\partial y\partial z}-\frac{2wu}{a^2}\frac{\partial^2\Phi}{\partial z\partial x}\\ =\frac{1}{a^2}\frac{\partial^2\Phi}{\partial t^2}+\frac{2u}{a^2}\frac{\partial^2\Phi}{\partial x\partial t}+\frac{2v}{a^2}\frac{\partial^2\Phi}{\partial y\partial t}+\frac{2w}{a^2}\frac{\partial^2\Phi}{\partial z\partial t}\,.\end{aligned}$$

With $a = \infty$, the Laplace equation is yielded for incompressible flow that serves as a basis for the boundary element methods in chapter 3:

$$\nabla^2\Phi = \frac{\partial^2\Phi}{\partial x^2}+\frac{\partial^2\Phi}{\partial y^2}+\frac{\partial^2\Phi}{\partial z^2} = 0\,. \tag{2.1}$$

The investigations of the present work are restricted to steady flows so that $\partial/\partial t = 0$ can be set for all time–derivations:

$$\begin{aligned}\left(1-\frac{u^2}{a^2}\right)\frac{\partial^2\Phi}{\partial x^2}+\left(1-\frac{v^2}{a^2}\right)\frac{\partial^2\Phi}{\partial y^2}+\left(1-\frac{w^2}{a^2}\right)\frac{\partial^2\Phi}{\partial z^2}\\ -\frac{2uv}{a^2}\frac{\partial^2\Phi}{\partial x\partial y}-\frac{2vw}{a^2}\frac{\partial^2\Phi}{\partial y\partial z}-\frac{2wu}{a^2}\frac{\partial^2\Phi}{\partial z\partial x}=0\,.\end{aligned} \tag{2.2}$$

The local sound velocity a is a function of Φ: $a^2 = a_\infty^2 + \frac{\kappa-1}{2}(U_\infty^2 - |\nabla\Phi|^2)$. The full potential equation (2.2) is the fundamental mathematical basis for the nonlinear integral equation methods in chapter 4.

2.2 Fundamental Principles of Potential Theory

Conservative fields are completely described by a scalar potential Φ. In agreement with Hess & Smith [17] and Maskew [32], the potential is considered in this work as potential energy, i.e. the field $\mathbf{V}(X,Y,Z)$ is yielded from the negative gradient of the potential Φ.

$$\mathbf{V}(X,Y,Z) = -\nabla\Phi$$

An important tool for the potential theory is *Green*'s third identity:

$$\Phi_P = -\frac{1}{4\pi}\iiint\limits_{\Omega}\frac{\nabla^2\Phi}{r}d\Omega - \frac{1}{4\pi}\iint\limits_{S}\frac{1}{r}\frac{\partial\Phi}{\partial n}dS + \frac{1}{4\pi}\iint\limits_{S}\Phi\frac{\partial}{\partial n}\left(\frac{1}{r}\right)dS\,. \quad (2.3)$$

Equation (2.3) is valid on the conditions that Φ and its partial first- and second–order derivations are steady inside Ω and that the volume integral converges for an unlimited domain [26]. The velocity potential Φ of the full potential equation (2.2) satisfies these conditions so that equation (2.3) can be interpreted as a solution of these differential equations. This formula is the mathematical basis of the field panel method in chapter 4.

In the following it is assumed that the velocity potential Φ is a harmonic function in the whole domain Ω, i.e. $\nabla^2\Phi = 0$ is valid for the Laplace operator. This reduces equation (2.3) to:

$$\Phi_P = -\frac{1}{4\pi}\iint\limits_{S}\frac{1}{r}\frac{\partial\Phi}{\partial n}dS + \frac{1}{4\pi}\iint\limits_{S}\Phi\frac{\partial}{\partial n}\left(\frac{1}{r}\right)dS\,. \quad (2.4)$$

Green's third theorem can also be applied to multi–coherent domains. For instance in Figure 2.1 the space consists of the domains Ω_a exterior and Ω_i interior of S.

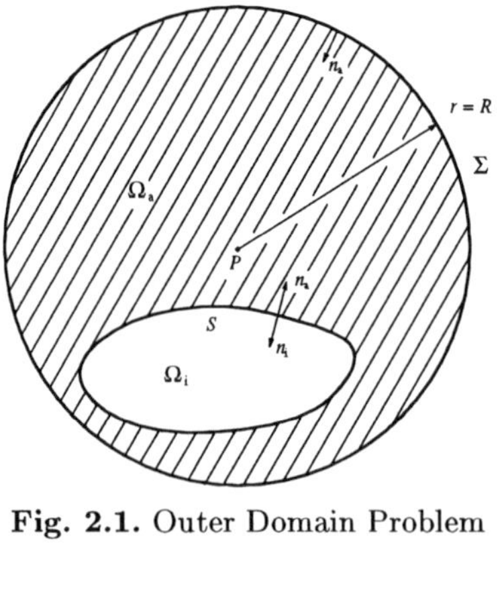

Fig. 2.1. Outer Domain Problem

In these domains the velocity potentials Φ_a and Φ_i are to exist. *Green*'s theorem is applied to both domains. Taking into account that the normals of both domains are of opposite directions ($\mathbf{n}_a = -\mathbf{n}_i$) and adding both terms, the integral over the surface of the sphere Σ is obtained taking the limiting value $R \to \infty$:

$$\Phi_P = \frac{1}{4\pi} \iint_{\mathcal{S}} \left\{ -\frac{1}{r} \mathbf{n}_a (\nabla \Phi_a - \nabla \Phi_i) + (\Phi_a - \Phi_i) \mathbf{n}_a \nabla \left(\frac{1}{r} \right) \right\} dS + \phi_{\infty,P} \, . \tag{2.5}$$

For the potential at an arbitrary point $P \in \mathcal{S}$ [21], equation (2.5) for multi-coherent domains yields:

$$\begin{aligned} \Phi_P &= \phi_{\infty,P} + \frac{1}{2}(\Phi_a - \Phi_i) \\ &+ \frac{1}{4\pi} \iint_{\mathcal{S}} \left\{ -\frac{1}{r} \mathbf{n}_a (\nabla \Phi_a - \nabla \Phi_i) + (\Phi_a - \Phi_i) \mathbf{n}_a \nabla \left(\frac{1}{r} \right) \right\} d\mathcal{S} . \end{aligned} \tag{2.6}$$

For harmonic functions Φ, equation (2.6) gives a solution for the potential at any point P. The equation is the basis for the boundary element method for the description of incompressible flows.

3 Boundary Element Method

If the velocity potential Φ is a harmonic function in the whole space $\Omega = \Omega_a + \Omega_i$, it satisfies the Laplace differential equation (2.1):

$$\nabla^2 \Phi = 0 \, .$$

The velocity potential Φ can also be generated by superimposing the disturbance potential ϕ and the undisturbed initial flow potential ϕ_∞:

$$\Phi = \phi + \phi_\infty \, . \tag{3.1}$$

Since the potential of the undisturbed initial flow is only a function of the space, the following integral equation for the disturbance potential ϕ is obtained:

$$\phi_P = \frac{1}{2}(\phi_a - \phi_i) + \frac{1}{4\pi} \iint_{\mathcal{S}} \left\{ -\frac{1}{r} \mathbf{n}_a (\nabla \phi_a - \nabla \phi_i) + (\phi_a - \phi_i) \mathbf{n}_a \nabla \left(\frac{1}{r} \right) \right\} d\mathcal{S} \, . \tag{3.2}$$

The solution of this potential equation can be interpreted as the superposition of a single and double layer potential.

3.1 Boundary Conditions

In order to explicitly solve the integral equation (3.2) for a domain Ω, potential values have to be prescribed at the two boundary surfaces of the space in Figure 2.1. With increasing distance between the farfield boundary Σ and the surface S, the disturbing influence of the body is getting less significant, the flow being undisturbed at the exterior boundary:

$$\lim_{r\to\infty} \Phi = \lim_{r\to\infty} (\phi + \phi_\infty) = \phi_\infty \,.$$

The potential in equation (3.2) is fixed, if the derivation of the potential and/or the potential itself is known on parts of the boundary or on the whole boundary S. In the following the relationship is valid: $\mathbf{n}_a \equiv \mathbf{n}$.

$$\text{NEUMANN}: \quad \mathbf{n}\nabla\phi = \frac{\partial \phi}{\partial n} = v_n$$

$$\text{DIRICHLET}: \quad \phi = \bar{\phi}$$

In case of the Neumann problem, a value v_n is demanded for the normal derivation of the potential on the boundary. The Dirichlet–boundary condition stipulates a value $\bar{\phi}$ for the potential on the boundary. For each of the two boundary value problems, the fulfillment of the boundary condition can be requested at the interior or exterior side of the border.

In this work, the theoretical presentations will be restricted to the two subsequent formulations:

1. Simultaneous use of single and double layer potential on the body surface. To obtain a definite system of linear equations, the additional conditions $\tau_{1+i} = \tau_{N_T-i}$ and $\mu_{1+i} = -\mu_{N_T-i}$ $(i=0,1,\ldots,N_T/2-1)$ are introduced. The system of equations is solved by using an external Neumann boundary condition.
2. Use of an unknown doublet distribution on the contour surface together with a source distribution, on which the singularity strength is fixed at $\phi_i = 0$ by an inner Dirichlet boundary condition.

Since the surface of the investigated body is to be closed, fluid can neither enter nor leave through the border, i.e. the body is a flow surface and the kinematic flow condition has to be satisfied: $\mathbf{n}\nabla\Phi = 0$. Since the wake layer is assumed to be a flow surface, no fluid can pass through it; the jump in the normal derivation of the velocity over the vortex surface has to disappear: $\mathbf{n}\cdot(\nabla\Phi_o - \nabla\Phi_u) = 0$. The jump in potential between upper and lower side of the vortex layer is the doublet strength of the wake: $\mu_W := -(\Phi_o - \Phi_u)$. In order to simplify this notation, the factor $\frac{1}{4\pi}$ is taken into consideration for the singularity strengths.

	NEUMANN	DIRICHLET
τ	$-\frac{1}{4\pi}\mathbf{n}\,\nabla[\phi]$	$-\frac{1}{4\pi}\mathbf{n}\,\nabla\phi$
μ	$-\frac{1}{4\pi}[\phi] = -\frac{1}{4\pi}[(\Phi - \phi_\infty)]$	$-\frac{1}{4\pi}\phi = -\frac{1}{4\pi}(\Phi - \phi_\infty)$
μ_W	$-\frac{1}{4\pi}(\Phi_o - \Phi_u)$	$-\frac{1}{4\pi}(\Phi_o - \Phi_u)$

With these singularity strengths the basic equations of the boundary element method are obtained for the two boundary conditions .

$$\text{NEUMANN}: \quad \begin{aligned} &\mathbf{n}\,\nabla\iint\limits_{\mathcal{S}} \tau\,\frac{1}{r}\,d\mathcal{S} - 2\pi\,\tau_P - \mathbf{n}\,\nabla\iint\limits_{\mathcal{S}} \mu\,\mathbf{n}\,\nabla\Big(\frac{1}{r}\Big)\,d\mathcal{S} \\ &-\mathbf{n}\,\nabla\iint\limits_{\mathcal{W}} \mu_W\,\mathbf{n}\,\nabla\Big(\frac{1}{r}\Big)\,d\mathcal{W} = -\mathbf{n}\,\nabla\phi_\infty \end{aligned} \tag{3.3}$$

$$\text{DIRICHLET}: \quad \begin{aligned} &\iint\limits_{\mathcal{S}} \tau\,\frac{1}{r}\,d\mathcal{S} - 2\pi\mu_P - \iint\limits_{\mathcal{S}} \mu\,\mathbf{n}\,\nabla\Big(\frac{1}{r}\Big)\,d\mathcal{S} \\ &-\iint\limits_{\mathcal{W}} \mu_W\,\mathbf{n}\,\nabla\Big(\frac{1}{r}\Big)\,d\mathcal{W} = 0 \end{aligned} \tag{3.4}$$

3.2 Discretisation of the Integral Equations

In general, an analytic solution of equations (3.3) and (3.4) on the complete surface of arbitrary body geometries is not possible. For an approximate solution either the Galerkin method can be resumed, satisfying the basic equation on parts of the surface in a weighted mean sense, or the equation can be satisfied exactly by a collocation at a finite number of discrete points — the so–called collocation points or control points. All panel methods used in this work for the description of the incompressible flow for two–dimensional profiles, three–dimensional translatory and rotatory flows use the collocation method.

For the numerical solution of equation (3.2), the integral equations (3.3) and (3.4) have to be discretised. The body surface is divided into individual elements, so–called panels, of which the end or corner points are located on the original contour. Since the determination of the influence coefficients for panel surfaces is much simpler than that for the curved elements, a plane compensation surface, where the collocation points are located, is constructed for the individual surface elements [32]. In the following, equations are derived in case that constant source and doublet distributions are used piecewise on the plane panels.

In this paper, only the discretisation for the Dirichlet boundary problem is indicated. The corresponding formulations for the integral equations (3.3) of the Neumann problem are given in [34, 36].

$$\text{DIRICHLET}: \quad \sum_{\substack{i=1\\i\neq k}}^{N} \tau_i \iint\limits_{\mathcal{S}_i} \frac{1}{r_{ki}}\, d\mathcal{S}_i - 2\pi\,\mu_k - \sum_{\substack{i=1\\i\neq k}}^{N} \mu_i \iint\limits_{\mathcal{S}_i} \mathbf{n}_i \nabla\left(\frac{1}{r_{ki}}\right) d\mathcal{S}_i$$
$$- \sum_{\substack{j=1\\j\neq k}}^{N_W} \mu_{W_j} \iint\limits_{\mathcal{W}_j} \mathbf{n}_j \nabla\left(\frac{1}{r_{kj}}\right) d\mathcal{W}_j = 0 \qquad k = 1,\ldots,N \tag{3.5}$$

Since for the singularity densities τ_i and μ_i constant distributions are set over the individual panels, singularity strengths can be moved in front of the integrals. The following abbreviations are defined for the integrals of the kernel functions:

$$\text{DIRICHLET}: \quad B_{ki} = -\iint\limits_{\mathcal{S}_i} \frac{1}{r_{ki}}\, d\mathcal{S}_i\,,$$
$$C_{ki} = \iint\limits_{\mathcal{S}_i} \mathbf{n}_i \nabla\left(\frac{1}{r_{ki}}\right) d\mathcal{S}_i\,, \tag{3.6}$$
$$C_{kj}^{W} = \iint\limits_{\mathcal{W}_j} \mathbf{n}_j \nabla\left(\frac{1}{r_{kj}}\right) d\mathcal{W}_j\,.$$

The eigen–influences of the panels ($i = k$ and $j = k$ respectively) are included in the source and doublet matrices $\mathbf{B}$ and $\mathbf{C}$. Then the discretised equations can be expressed for both boundary conditions in a unified way:

$$\sum_{i=1}^{N} \tau_i B_{ki} + \sum_{i=1}^{N} \mu_i C_{ki} + \sum_{j=1}^{N_W} \mu_{W_j} C_{kj}^{W} = 0 \qquad k = 1,\ldots,N\,. \tag{3.7}$$

Term (3.7) forms a system of linear equations with N equations for the determination of the unknown singularity distributions. In this presentation, the influence coefficient matrices $\mathbf{B}, \mathbf{C}$ and $\mathbf{C}^W$ contain different values for the Neumann and Dirichlet boundary problem. In case of the Neumann problem, the matrix elements represent the normal components of the velocity influences, whereas in case of the Dirichlet problem, the standard influences of the velocity potential are stored in the matrices.

Putting all values known in the relevant time step to the right–hand side yields:

$$\text{DIRICHLET}: \quad \mathcal{R}_k = -\sum_{i=1}^{N} \tau_i B_{ki} - \sum_{j=1}^{N_W} \mu_{W_j} C^W_{kj} \, .$$

Thus, the system of equations to be established in the program are:

$$\text{DIRICHLET}: \quad \sum_{i=1}^{N} \mu_i C_{ki} = \mathcal{R}_k \qquad k = 1, \ldots, N \, . \tag{3.8}$$

In matrix notation this results in the formula $\mathbf{A} \cdot \boldsymbol{\mu} = \mathcal{R}$:

$$\begin{bmatrix} a_{1,1} & a_{1,2} & \cdots & a_{1,N} \\ a_{2,1} & a_{2,2} & \cdots & a_{2,N} \\ a_{3,1} & a_{3,2} & \cdots & a_{3,N} \\ \vdots & \vdots & \ddots & \vdots \\ a_{N,1} & a_{N,2} & \cdots & a_{N,N} \end{bmatrix} \begin{bmatrix} x_1 \\ x_2 \\ x_3 \\ \vdots \\ x_N \end{bmatrix} = \begin{bmatrix} \mathcal{R}_1 \\ \mathcal{R}_2 \\ \mathcal{R}_3 \\ \vdots \\ \mathcal{R}_N \end{bmatrix} . \tag{3.9}$$

The elements of the influence coefficient matrix[3] **A** correspond to the coefficients of the matrix **C** from equation (3.6).

If the singularity strengths of all panels are known after the solution of the system of linear equations (3.9), the velocity can either be determined at any arbitrary point P in space from the negative gradient of the velocity potential Φ_P:

$$\Phi_P = \sum_{i=1}^{N} \tau_i B_{ki} + \sum_{i=1}^{N} \mu_i C_{ki} + \sum_{j=1}^{N_W} \mu_{W_j} C^W_{kj} + \phi_\infty \, , \tag{3.10}$$

or be calculated directly from:

$$\mathbf{V}_P = -\nabla_p \Phi_P = \sum_{i=1}^{N} \tau_i \widehat{B}_{ki} + \sum_{i=1}^{N} \mu_i \widehat{C}_{ki} + \sum_{j=1}^{N_W} \mu_{W_j} \widehat{C}^W_{kj} + \mathbf{V}_\infty \, . \tag{3.11}$$

The matrices for the velocity influences are defined as follows [32]:

$$\widehat{B}_{ki} = -\nabla_p \iint_{\mathcal{S}_i} \frac{1}{r_{ki}} \, d\mathcal{S}_i \, ,$$

$$\widehat{C}_{ki} = -\nabla_p \iint_{\mathcal{S}_i} \mathbf{n}_i \nabla\left(\frac{1}{r_{ki}}\right) d\mathcal{S}_i \, ,$$

$$\widehat{C}^W_{kj} = -\nabla_p \iint_{\mathcal{W}_j} \mathbf{n}_j \nabla\left(\frac{1}{r_{kj}}\right) d\mathcal{W}_j \, .$$

[3] **AIC — Aerodynamic Influence Coefficient.**

3.3 Vortex–Lattice Method

3.3.1 Doublet Distribution and Vortex–Lattice. In the special case of a constant singularity strength μ_S, the doublet distribution, according to B. Hunt [20], is equivalent to a vortex ring of the strength $\boldsymbol{\Gamma} = \mu\mathrm{t} = constant$ along the boundary of surface (Figure 3.1).

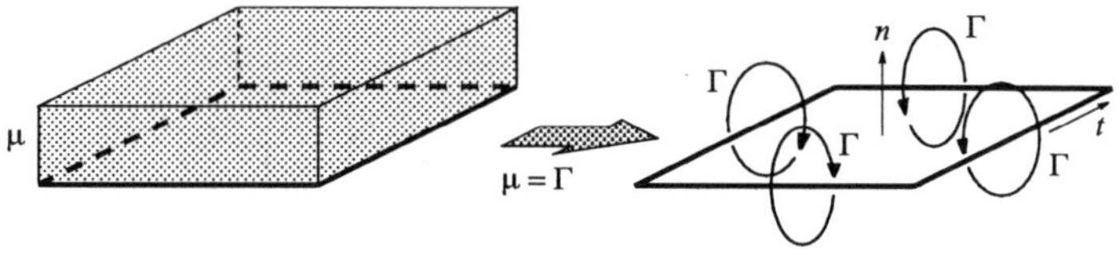

Fig. 3.1. Equivalence between doublet panel and vortex ring

In the vortex–lattice method, the wake layers are composed of adjacent, closed vortex rings. The circulation of individual vortex pieces results form the differences of the doublet strengths of adjoining rings. All wake elements consist of straight finite vortex filaments. In case of a straight, finite vortex filament of the length $\overline{AB}$,the velocity induction is given by:

$$v = \frac{\Gamma}{4\pi r}(\cos\alpha_1 - \cos\alpha_2) . \tag{3.12}$$

For $r \to 0$ the velocity field has a singular point. According to equation (3.12), this results in induced velocities already in its surrounding which yield unrealistic high amounts and directions. Therefore, different corrective procedures have to be introduced in order to correctly determine the velocities in the nearfield of a vortex filament. The effects of the different nearfield methods on the velocity field and the shape of the vortex layers were extensively investigated and described by Urban [42] and Behr et. al. [5].

3.3.2 Discretisation of the Vortex Layer and its Development with Time. In this chapter, the vortex layer procedure ROVLM[4] is discussed, that was developed by Urban [42], Wagner et. al. [47] and Behr & Wagner [1, 2, 3, 4] for delta wings with leading edge separation and first used by Kramer [27], then by Röttgerman et. al. [33, 35] for the aerodynamic simulation of multi–blade rotors.

The vortex–lattice method describes the development of the free wake vortex layer in case of an impulsive start of a wing or a multi-blade rotor. The developed method deals with thick wings with round leading edge and thus allows a separation of the vortex layer only at the trailing edges.

At the beginning of the calculation, blade and airfoil are given without wake, i.e. the number of the wake panels is $N_W = 0$. As basic solution ($t = 0$), the system of equations (3.8) is solved with the right–hand sides being formed

[4] **ROVLM** — **RO**tor **V**ortex **L**attice **M**ethod.

only from the normal components of the velocities of the undisturbed initial flow. For rotors, the initial flow consists of the translatory velocity $\mathbf{V}_\infty$ and a component $\boldsymbol{\Omega} \times \mathbf{r}$ from the rotation of the blade:

$$\mathcal{R}_k(t=0) = -\frac{1}{4\pi} \sum_{i=1}^{N} \left[\mathbf{n}_k(t) \cdot (\mathbf{V}_{\infty,k}(t) + \boldsymbol{\Omega}(t) \times \mathbf{r}_k(t))\right] B_{ki} \,. \tag{3.13}$$

These right–hand sides yield the unknown distributions τ_k and μ_k of the surface panels. With the singularity strengths of the starting solution, equation (3.11) yields the velocities $\mathbf{V}^{BW}$ at the points of the separating lines.

The position of the wake points l shown in Figure 3.2a and 3.2b in the first row of the separated wake vortex layer result from the multiplication of the total velocities at the separating points with the time step Δt.

$$\begin{aligned} \mathbf{r}_l(t=1) &= \mathbf{r}_{HK,l}(t=0) \\ &+ \left(\mathbf{V}_{\infty,l}(t=0) + \boldsymbol{\Omega}(t=0) \times \mathbf{r}_{HK,l}(t=0) + \mathbf{V}_l^{BW}(t=0)\right) \cdot \Delta t \end{aligned} \tag{3.14}$$

The that way resulting points, together with the corner points of the panels on the trailing and lateral edges, form the edges of the vortex rings, from which the first row of the wake vortex layer is developed (Figure 3.2b).

From the Kutta condition follows that the doublet distribution has to be steady at an edge, since any unsteadiness in μ would cause a discrete vortex, resulting in flow around the edge. To avoid that a vortex is originated at the trailing edge, the total vortex strength at the trailing edge has to disappear. The size of the circulation $\Gamma_l = \mu_{W_j}$ of the newly developed vortex rings is determined from the difference of the doublet strengths at the trailing edge panel of the sheet j in Figure 3.3:

$$\mu_{W_j} = \mu_{HK_o,j} - \mu_{HK_u,j} - \frac{1}{4\pi}\left(\phi_{\infty HK_o,j} - \phi_{\infty HK_u,j}\right) \,. \tag{3.15}$$

In the next time step, the newly developed wake row influences the total velocity at the control points of the blades which requires modification of the right–hand sides:

$$\mathcal{R}_k(t) = -\frac{1}{4\pi} \sum_{i=1}^{N} \left[\mathbf{n}_k(t) \cdot (\mathbf{V}_{\infty,k}(t) + \boldsymbol{\Omega}(t) \times \mathbf{r}_k(t))\right] B_{ki} - \sum_{j=1}^{N_W} \mu_{W_j} C_{kj}^W \,. \tag{3.16}$$

The coefficients of the matrices $\mathbf{C}^W$ and $\widehat{\mathbf{C}}^W$ are computed every time for each time step since the wake geometry is constantly changing during the iteration process, which is caused by the interferences between the wings and the vortex layers.

By solving the system of linear equations with the modified right–hand side, the new strengths of the singularity τ_k, μ_k are determined in time step $t + \Delta t$. The new coordinates of the corner points of all wake panels $\mathbf{r}_l$ ($l =$

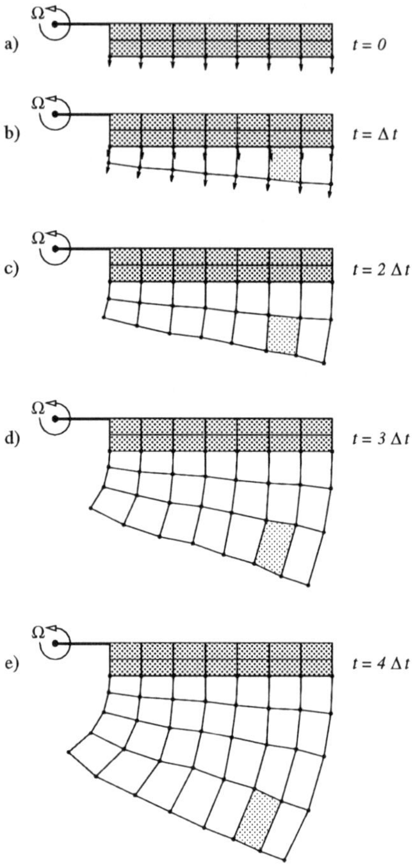

Fig. 3.2. Developement of wake structure

$1, \ldots, N_W$) are yielded from the induced velocities of the airfoil panels $\mathbf{V}_l^{BW}$, the vortex layer $\mathbf{V}_l^{WW}$ and the initial flow:

$$\mathbf{r}_l(t + \Delta t) = \mathbf{r}_l(t) + \left(\mathbf{V}_{\infty,l}(t) + \boldsymbol{\Omega}(t) \times \mathbf{r}_l(t) + \mathbf{V}_l^{BW}(t) + \mathbf{V}_l^{WW}(t)\right) \cdot \Delta t\,. \tag{3.17}$$

where:

$$\mathbf{V}_l^{BW}(t) = \sum_{i=1}^{N} \tau_i(t)\ \widehat{B}_{li} + \sum_{i=1}^{N} \mu_i(t)\ \widehat{C}_{li}\,,, \quad \mathbf{V}_l^{WW}(t) = \sum_{j=1}^{N_W(t)} \mu_{W_j}\ \widehat{C}_{lj}^W\,.$$

The number of wake points N_W is increasing in each time step Δt. The elements of the coefficient matrices $\widehat{\mathbf{B}}$ and $\widehat{\mathbf{C}}$, representing the influence of the wing on the wake, depend on the geometry of the vortex layer and therefore like the matrix $\widehat{\mathbf{C}}^W$ have to be calculated every time in each time step.

In the following time steps (Figs. 3.2c to e), the velocity vectors are determined at all existing wake points with a new wake row arising in each case.

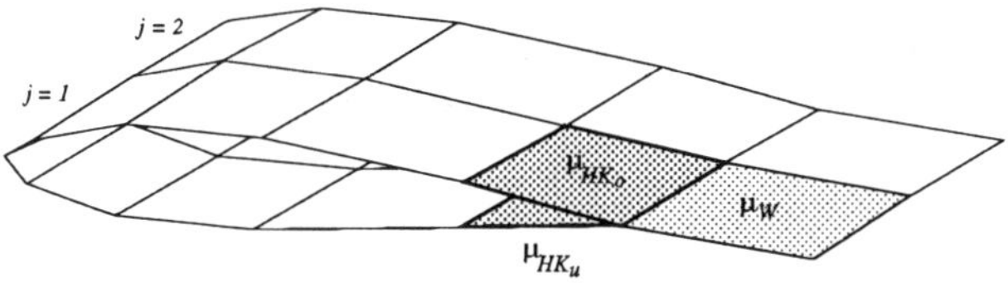

Fig. 3.3. Doublet strength at the trailing edge of the profile

Since the position of the wake layer is determined from the induced velocities of the total system in each time step, the wake is always free of forces. The doublet strengths of already existing vortex rings remain constant and are transported throughout the wake. The shaded wake element in Figure 3.2 indicates, how the circulation is pushed throughout the vortex layer.

The wake vortex layers arise from a series of quasi–steady solutions of the potential equation. This procedure is being continued, until the circulation or drag distribution reaches an approximately steady or cyclic steady state. A more detailed description of the development of the wake vortex layer for thin wings with leading edge separation is documented by Behr et. al. [1, 3, 4, 6] and for thin rotor blades by Röttgermann [33] and Röttgermann et. al. [35].

3.4 Panel Clustering Method

Computing the flow and the aerodynamic coefficients around three–dimensional bodies (e.g. wings) by using a boundary element method, requires to divide the surface of the configuration into individual panels. In case of the chosen method, the discretisation has to be refined to solve high velocity gradients. Here, the orders of magnitude are quickly rising up to some thousand panels. Since all surface elements influence all other panels, the influence coefficients matrices (3.9) are getting very large. Computation of the influence coefficients, their storage, as well as inversion of the matrices require very much computation time and storage capacity.

The panel clustering method, developed by Hackbush & Nowak [10, 11] and further developed by Sauter [37, 38], offers a possibility of reducing these expenses.

The basic idea of the method is to combine panels π with an adequate big distance from the collocation point $\mathbf{x}_k$ in clusters τ. A detailed description of the panel clustering method is presented in [34, 36].

4 Field Panel Method

Coupling of a boundary element method and a field method, e.g. finite–difference method yielded the program ROFPM[5], facilitating the computation of compressible flows without generating a field grid that has to be

[5] **ROFPM** — **RO**tor **F**ield **P**anel **M**ethod.

adapted to the contour. The sole task of the panel method in this compound is to fulfill the Laplace equation and the boundary condition on the body; the field method is only used to capture the compressibility.

The idea is to divide the nonlinear differential equation (4.1) into a homogeneous and a particular part. The homogeneous, linear part is treated by using the well-known panel methods, the nonlinear residual operator is solved by applying the field panel method.

4.1 Basic Equations

The non-conservative potential equation (2.2) is transformed so that all derivations with a coefficient are put on one side of the equation:

$$\Phi_{xx} + \Phi_{yy} + \Phi_{zz} = \frac{u^2}{a^2}\Phi_{xx} + \frac{v^2}{a^2}\Phi_{yy} + \frac{w^2}{a^2}\Phi_{zz} + \frac{2uv}{a^2}\Phi_{xy} + \frac{2vw}{a^2}\Phi_{yz} + \frac{2wu}{a^2}\Phi_{zx} \,. \tag{4.1}$$

For the right-hand side the abbreviation:

$$\sigma := \frac{u^2}{a^2}\Phi_{xx} + \frac{v^2}{a^2}\Phi_{yy} + \frac{w^2}{a^2}\Phi_{zz} + \frac{2uv}{a^2}\Phi_{xy} + \frac{2vw}{a^2}\Phi_{yz} + \frac{2wu}{a^2}\Phi_{zx} \,. \tag{4.2}$$

is introduced; equation (4.2) is a non-conservative formulation of the right-hand side the term σ is used for the right-hand side, the complete potential equation is obtained:

$$\nabla^2 \Phi = \sigma \,. \tag{4.3}$$

The value σ is a function of the derivations of the total velocity potential Φ. The complete potential equation (4.3) is a nonlinear partial differential equation, that generally has to be solved iteratively.

In an iteration procedure computing the right-hand side from the potential values Φ of the preceding iteration step, σ can be considered as constant in the relevant iteration step:

$$\nabla^2 \Phi^n = \sigma^{n-1} = \mathit{constant} \,. \tag{4.4}$$

Equation (4.4) is a Poisson equation with given constant right-hand side σ. This differential equation can be transformed with *Green*'s third identity (2.3) to an integral equation for the velocity potential.
If the total potential Φ is reduced to a component for the undisturbed initial flow ϕ_∞ and a disturbance term ϕ, equation (2.3) yields an integral equation for the disturbance potential ϕ:

$$\phi = \frac{1}{4\pi} \iint_{\mathcal{S}} \tau \, \frac{1}{r} \, d\mathcal{S} - \frac{1}{4\pi} \iint_{\mathcal{S}} \mu \, \mathbf{n} \cdot \nabla\Big(\frac{1}{r}\Big) \, d\mathcal{S} - \frac{1}{4\pi} \iiint_{\mathcal{V}} \sigma \, \frac{1}{r} \, d\mathcal{V} \,. \tag{4.5}$$

The kernel of the volume integral in equation (4.5) corresponds to the kernel function of a single layer potential. This value σ is described as a strength of the source distribution in the flow field or as a field source strength.

Splitting the surface integral into the components for the body $\mathcal{S}$ and the vortex layer $\mathcal{W}$, yields a nonlinear equation for the definition of the disturbance potential:

$$\begin{aligned}\phi &= \frac{1}{4\pi}\iint\limits_{\mathcal{S}} \tau\,\frac{1}{r}\,d\mathcal{S} - \frac{1}{4\pi}\iint\limits_{\mathcal{S}} \mu\,\mathbf{n}\cdot\nabla\Big(\frac{1}{r}\Big)\,d\mathcal{S} \\ &- \frac{1}{4\pi}\iint\limits_{\mathcal{W}} \mu_W\,\mathbf{n}\cdot\nabla\Big(\frac{1}{r}\Big)\,d\mathcal{W} - \frac{1}{4\pi}\iiint\limits_{\mathcal{V}} \sigma\,\frac{1}{r}\,d\mathcal{V}\,.\end{aligned} \tag{4.6}$$

4.2 Boundary Conditions and Field Grid

To include nonlinear effects, the volume internal in equation (4.6) has to be evaluated. According to the break–down into a boundary element method and a field method, the disturbance potential is split up into the components from the panel method ϕ^P and those of the field method ϕ^F:

$$\phi = \phi^P + \phi^F\,. \tag{4.7}$$

The kinematic boundary condition on the body surface is satisfied by the boundary element method so that at the exterior boundary of the field only the farfield boundary condition has to be satisfied for the disturbance potential:

$$\lim_{r\to\infty} \phi = \lim_{r\to\infty} \phi^P + \lim_{r\to\infty} \phi^F = 0\,.$$

As shown in chapter 3.1, the farfield boundary condition for ϕ^P is satisfied implicitly by the singularity strength. Likewise, the condition $\lim\limits_{r\to\infty} \phi^F = 0$ is satisfied due to the $1/r$ – singularity of the field sources.

Poisson equation (4.3) in connection with equation (4.7):

$$\nabla^2\Phi = \nabla^2(\phi^P + \phi^F + \phi_\infty) = \sigma\,.$$

yields with $\nabla^2\phi_\infty = 0$ a condition for the disturbance potential ϕ^F inside the field grid:

$$\nabla^2\phi^F = \sigma - \nabla^2\phi^P\,. \tag{4.8}$$

Since the boundary conditions at the body and at the exterior boundary have already been satisfied, condition (4.8) is the only remaining condition to be satisfied explicitly. A Cartesian field grid can be used that can be generated without requiring much expenditure; it may penetrate the body arbitrarily. The grid boundaries run close to the contour. Since the farfield boundary condition is already satisfied at these boundaries, the grid size can be kept very small.

4.3 Iterative Solution of the System of Equations

The nonlinear differential equation (4.4) is solved by an iterative procedure.

Firstly, in iteration step $n=1$, the homogeneous part of differential equation (2.1) is solved by the panel method described in chapter 3. From the kinematic flow condition $\mathbf{n}\,\nabla\phi = -\mathbf{n}\,\nabla\phi_\infty$ together with equation (4.7) the relation is yielded:

$$\mathbf{n}\,\nabla(\phi^P + \phi^F) = -\mathbf{n}\,\nabla\phi_\infty\,.$$

Thus the potential $\phi^{P,n}$ in the iterative step n is determined by:

$$\mathbf{n}\,\nabla\phi^{P,n} = -\mathbf{n}\,\nabla\phi_\infty - \mathbf{n}\,\nabla\phi^{F,n-1}\,. \tag{4.9}$$

In the first iteration step $n = 1$, the field does not provide a contribution to the disturbance potential, i.e. $\phi^{F,n=0} = 0$. From the boundary condition (4.9), a solution for the singularity strengths of the surface distributions of the body is obtained. These distributions yield the disturbance potential ϕ^P for all points in space. The components of the disturbance potentials of the panel method, the field and the initial flow together constitute the velocity potential:

$$\Phi^n = \phi^{P,n} + \phi^{F,n-1} + \phi^n_\infty\,.$$

The distribution of the field source strength σ, according to equation (4.2), is a function of the potential Φ so that the values of σ in iteration step n are determined from the values of the disturbance potential ϕ^P of the same iteration step and from ϕ^F of the preceding step:

$$\sigma^n = \sigma(\phi^{P,n}, \phi^{F,n-1}, \phi^n_\infty)\,.$$

When the disturbance potential ϕ^P and the field source strengths σ are known, a new solution for the disturbance potential ϕ^F has to be calculated. Equation (4.8) yields the rule for calculating the disturbance potential of the field:

$$\nabla^2\phi^{F,n} = \sigma^n - \nabla^2\phi^{P,n}\,.$$

In the following iteration step, the field potential ϕ^F modifies the boundary condition in equation (4.9). Thus, the contribution of the field to the total potential is determined independently of the boundary element method and iteratively coupled with the panel method via the boundary condition. The iteration process is being continued until the difference from the Laplace operator for the velocity potential and field source strength is below a given limit ε:

$$|\nabla^2\Phi - \sigma| < \varepsilon\,.$$

Field values (ϕ^F,σ) or values of the boundary element methods such as ϕ^P, τ_k and μ_k are considered as convergence criteria. Since the present method captures the compressible influence over the discretised external space and therefore requires an evaluation of the field source strength in each iteration step, the program uses the modification of σ as break–off criterion:

$$\max_i |\sigma_i^n - \sigma_i^{n-1}| \leq \varepsilon_\sigma \,. \tag{4.10}$$

Typically, values in the domain $\varepsilon_\sigma = 0.01 \ldots 0.1$ are used as convergence criteria.

4.4 Discretisation

To facilitate the computation of value σ of the field source strengths, the total velocity has to be known for all field control points k. The contributions of the body and the vortex wake are determined in analogy to equations (3.11). The contribution of the field to this velocity is:

$$\mathbf{V}_k^{FF} = -\nabla_p \phi_k^F = \frac{1}{4\pi} \nabla_p \iiint\limits_{\mathcal{V}} \sigma \frac{1}{r} \, d\mathcal{V} \,. \tag{4.11}$$

After exchanging differentiation and integration, the derivations of the kernel function have to be integrated over $\mathcal{V}$. The space surrounding the contour is divided into M grid cells covered by a constant source distribution in space of the strength σ. The whole cell has the value of σ that results in the center of the volume element. The velocity at a control point k is obtained from the partial inductions of all field panels M:

$$\mathbf{V}_k^{FF} = \frac{1}{4\pi} \sum_{i=1}^{M} \sigma_i \iiint\limits_{\mathcal{V}_i} \nabla_p \left(\frac{1}{r_{ki}} \right) d\mathcal{V}_i \,. \tag{4.12}$$

The velocities for two–dimensional plane field panels are determined from an analogous equation. The following kernel functions have to be developed for the individual velocity components:

$$u_k^{FF} = \frac{1}{4\pi} \sum_{i=1}^{M} \sigma_i \iiint\limits_{\mathcal{V}_i} \frac{\partial}{\partial x_p} \left(\frac{1}{r_{ki}} \right) dx\, dy\, dz = \sum_{i=1}^{M} \sigma_i D_{ki}^x \,, \tag{4.13}$$

$$v_k^{FF} = \frac{1}{4\pi} \sum_{i=1}^{M} \sigma_i \iiint\limits_{\mathcal{V}_i} \frac{\partial}{\partial y_p} \left(\frac{1}{r_{ki}} \right) dx\, dy\, dz = \sum_{i=1}^{M} \sigma_i D_{ki}^y \,, \tag{4.14}$$

$$w_k^{FF} = \frac{1}{4\pi} \sum_{i=1}^{M} \sigma_i \iiint\limits_{\mathcal{V}_i} \frac{\partial}{\partial z_p} \left(\frac{1}{r_{ki}} \right) dx\, dy\, dz = \sum_{i=1}^{M} \sigma_i D_{ki}^z \,. \tag{4.15}$$

Equations (4.15) for the induced velocities result in the standard influence coefficient matrices of the field sources:

$$\widehat{D}_{ki} = \begin{pmatrix} D_{ki}^x \\ D_{ki}^y \\ D_{ki}^z \end{pmatrix} = \frac{1}{4\pi} \iiint\limits_{\mathcal{V}_i} \nabla_p \left(\frac{1}{r_{ki}} \right) d\mathcal{V}_i \,. \tag{4.16}$$

Together with the influence coefficient matrices of the boundary element method from chapter 3.2, all integrals of equation (4.6) are now available in discretised form.

When establishing the iteration scheme, it has to be noted that all three components — body, wake and field — are influencing each other and also themselves. Thus, in comparison to the panel method, additional interactions ocur between field and body and also between field and wake. In Figure 4.1 these interactions and the pertaining matrices are presented.

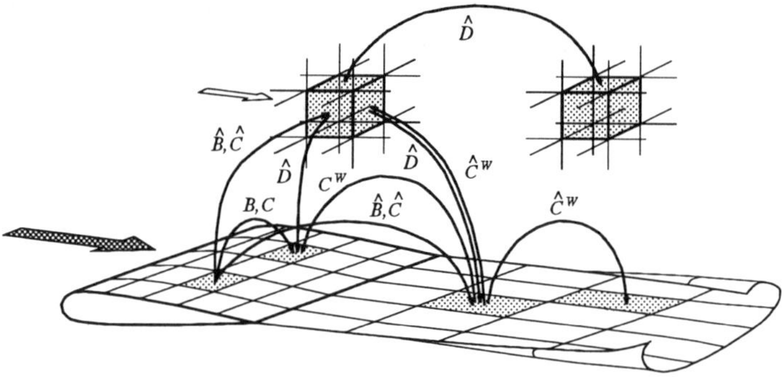

Fig. 4.1. Influences in the Field Panel Method

After the grid generation, at first the influence matrices are calculated, of which the coefficients are not changing during the iteration process and can therefore be stored.

The beginning of the iteration process presents the solution of the homogeneous, system of linear equations for the panel method. By using the methods from section 3, the relevant singularity strengths of the surface distributions are obtained.

The total velocity $\mathbf{V}^F$ required for the determination of the field distribution σ in iteration step n at a point in space k consists of the components of the undisturbed free stream and the induced velocities of the body, the wake, and the field panel:

$$\mathbf{V}_k^{F,n} = \mathbf{V}_{\infty,k}^n + \boldsymbol{\Omega}^n \times \mathbf{r}_k^n + \sum_{i=1}^{N} \tau_i^n \widehat{B}_{ki} + \sum_{i=1}^{N} \mu_i^n \widehat{C}_{ki} + \sum_{j=1}^{N_W} \mu_{W_j}^n \widehat{C}_{kj}^{W,n} + \sum_{i=1}^{M} \sigma_i^{n-1} \widehat{D}_{ki} .$$

Introducing the abbreviations:

$$\mathbf{V}_k^{BF} = \sum_{i=1}^{N} \tau_i \widehat{B}_{ki} + \sum_{i=1}^{N} \mu_i \widehat{C}_{ki} , \tag{4.17}$$

$$\mathbf{V}_k^{WF} = \sum_{j=1}^{N_W} \mu_{W_j} \widehat{C}_{kj}^W , \tag{4.18}$$

$$\mathbf{V}_k^{FF} = \sum_{i=1}^{M} \sigma_i \widehat{D}_{ki} \tag{4.19}$$

yields the equation for the field velocity:

$$\mathbf{V}_k^{F,n} = \mathbf{V}_{\infty,k}^n + \boldsymbol{\Omega}^n \times \mathbf{r}_k^n + \mathbf{V}_k^{BF,n} + \mathbf{V}_k^{WF,n} + \mathbf{V}_k^{FF,n-1} \,. \tag{4.20}$$

In the first step ($n=1$) the field does not contribute to the total velocity since $\sigma^{n=0}=0$ has been described as starting value of the field source strength. In the following, the iteration index n is omitted to simplify the formulation, if all values of the variable come from the same iteration step.

If the velocities in the field — and thus also the source strengths σ of the field panel — are known, equation (4.19) yields the velocities $\mathbf{V}^{FF}$, inducing the field at the field control points.

From the known spatial singularity distribution it is possible to compute the influences of the field on the body and on the wake. The velocity $\mathbf{V}^W$ at the wake points l calculated with equation (3.17) have to be extended by the induced velocity $\mathbf{V}^{FW}$ from the field at the wake points:

$$\mathbf{V}_l^W = \mathbf{V}_{\infty,l} + \boldsymbol{\Omega} \times \mathbf{r}_l + \mathbf{V}_l^{BW} + \mathbf{V}_l^{WW} + \mathbf{V}_l^{FW}$$

with the components defined as follows:

$$\mathbf{V}_l^{BW} = \sum_{i=1}^{N} \tau_i \widehat{B}_{li} + \sum_{i=1}^{N} \mu_i \widehat{C}_{li} \,, \quad \mathbf{V}_l^{WW} = \sum_{j=1}^{N_W} \mu_{W_j} \widehat{C}_{lj}^{W} \,, \quad \mathbf{V}_l^{FW} = \sum_{i=1}^{M} \sigma_i \widehat{D}_{li} \,. \tag{4.21}$$

If the wake shape is allowed to change during the iteration, the influence matrices have to be modified for each iteration step. The new location of the corner points of the wake panels are computed equivalent to equation (3.17):

$$\mathbf{r}_l^{n+1} = \mathbf{r}_l^n + \mathbf{V}_l^{W,n} \cdot \Delta t \,.$$

The field has also an influence on the velocity at the collocation points:

$$\mathbf{V}_k^{FB} = \sum_{i=1}^{M} \sigma_i \widehat{D}_{ki} \,.$$

These induced velocities can be interpreted as an additional free stream velocity and will be accounted for in the right–hand sides of the system of equations. Therefore, the term in equation (3.16) has to be modified:

$$\mathcal{R}_k^n = -\frac{1}{4\pi} \sum_{i=1}^{N} \left[\mathbf{n}_k \cdot \left(\mathbf{V}_{\infty,k} + \boldsymbol{\Omega} \times \mathbf{r}_k + \sum_{i=1}^{M} \sigma_i^n \widehat{D}_{ki} \right) \right] B_{ki} - \sum_{j=1}^{N_W} \mu_{W_j}^n C_{kj}^{W,n} \,. \tag{4.22}$$

The new right–hand sides change the singularity distribution on the surface of the body which also results in new velocities in the outer domain. After adjusting the strengths σ of the field sources, new values of the right–hand sides $\mathcal{R}$ can be determined. This procedure is being continued until the strengths of the field sources fulfill the convergence criterion (4.10).

4.5 Calculation of the Field Source Strengths

The strengths σ of the field sources are determined from the velocities at the field control points which are calculated from the gradient of the potential Φ:

$$\mathbf{V}^F = \begin{pmatrix} u \\ v \\ w \end{pmatrix} = -\begin{pmatrix} \Phi_x \\ \Phi_y \\ \Phi_z \end{pmatrix} .$$

Writing equation (4.2) with the derivatives of the velocities, the strengths σ of the field sources are determined from the velocities and their first derivatives:

$$\sigma = \frac{u^2}{a^2}u_x + \frac{v^2}{a^2}v_y + \frac{w^2}{a^2}w_z + \frac{2uv}{a^2}u_y + \frac{2vw}{a^2}w_y + \frac{2wu}{a^2}u_z . \tag{4.23}$$

4.6 Artificial Viscosity

Since the full potential equation is obtained on the assumption of an adiabatic and irrotational flow, it follows from the energy equation that the flow about a contour has to be isentropic. The artificial viscosity T stabilizes the solution:

$$T = \left(1 - \frac{q^2}{a^2}\right)\frac{a^2}{q^2}\left[\frac{u}{q}\sigma_x\,\Delta x + \frac{v}{q}\sigma_y\,\Delta y + \frac{w}{q}\sigma_z\,\Delta z\right] . \tag{4.24}$$

The artificial viscosity T is added to the strengths σ of the filed sources at supersonic points:

$$\bar{\sigma}_{i,j,k} = \sigma_{i,j,k} + T_{i,j,k} .$$

Using this modified Poisson equation (4.8) in the iterative procedure it is possible to treat flow problems with weak shocks. Test cases have shown that the artificial viscosity can stabilize the solution as long as the Mach numbers in front of the shock are Ma < 1.3. At higher velocities, the flow can no longer be considered to be isentropic. Therefore, the full potential equation is not suitable for flows with strong compression shocks.

4.7 Coupling of the Boundary–Element and the Field Method

The vortex–lattice method for rotors (ROVLM) mentioned in chapter 3 is coupled with the field panel method (FPM) to calculate compressibility effects at the rotor blade tips.

The combined code includes two iteration loops — one is responsible for the calculation of the wake shape, and the second is introduced, to correct the compressibility of the flow. Building up the vortex–lattice for circulatory flows requires about 50 up to 100 time steps. The nonlinear potential equation has to be completely iterated for each time step.

To limit the computational effort of the coupled method, Kandil & Yates [23, 25], propose a procedure where the wake iteration is computed with fixed field source strengths σ and the compressibility iteration is performed with fixed circulation. This procedure continues until the strengths of the field sources and the doublet distribution in the wake are converged.

Firstly in an incompressible iteration a force free wake is built up for a steady blade solution. Gennaretti et. al. [9] and Caradonna & Tung [7] have proven in a numerical and an experimental form that the influence of compressibility on the geometry of the wake shape is very limited and can be neglected. Therefore the wake is treated as a prescribed wake during the compressibility iteration. The singularity strengths on the surface $\mathcal{S}$ change during the iteration steps, just as the doublet strengths μ_{HK} at the trailing edge of the blade. The new wake doublet strength $\mu_W = \mu_{HK_o} - \mu_{HK_u}$ is shed into the wake and one obtains a solution which just differs slightly from the exact solution. Since only the doublet strengths of the wake are modified but the wake geometry remains unchanged, the wake isn't anymore force free. This could be corrected by a few iteration steps to remodel the vortex sheet. The resulting vortices at the trailing edge are very small so that this modification of the wake shape is neglected in this method.

A further advantage of the prescribed wake geometry is that one does not have to calculate the influences and the influence coefficients of the blades, of the field grid and of the vortex sheet on the corner points of the wake vortex rings.

4.8 Reduction of Computational Costs

Integral equation methods can compute boundary values on the body's surface $\mathcal{S}$ very efficiently, but need a lot of computation time and storage when calculating velocities in the outer domain $\mathcal{V}$.

The major part of cpu–time and storage requirement is used for the

- calculation of the influences between the body and the field $\mathbf{V}^{BF}$
- calculation of the influences between the field and the field $\mathbf{V}^{FF}$
- calculation of the influences between the field and the body $\mathbf{V}^{FB}$

In order to limit the computational effort, it is necessary to couple the field panel method with finite–difference methods which are more efficient in calculating stream values in the outer domain than integral methods. Therefore, an AF–scheme, a finite–difference method for calculating the velocities in the outer domain and a trilinear interpolation for calculating the velocities at the collocation points were used [34]. The reduction in cpu–time and storage requirement is shown in Figure 4.2.

5 Results of Rotor Flows

The theoretical background described in the previous chapters has been realized in a two–dimensional and a three–dimensional code.

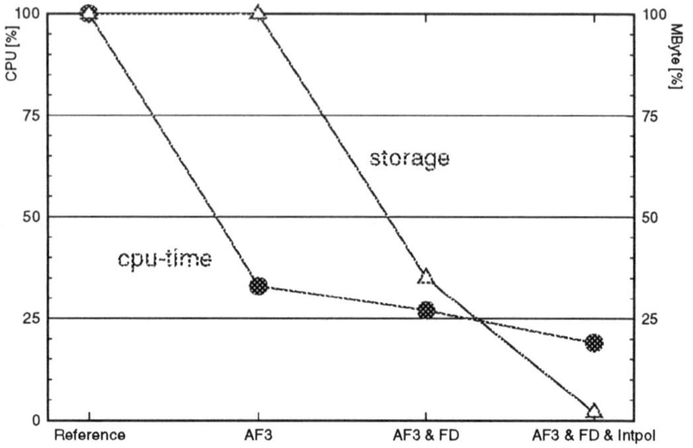

Fig. 4.2. Reduction of cpu–time and storage [34]

In the present paper, only incompressible and compressible calculations of the three–dimensional code are presented. These calculations are limited to steady flows of a hovering rotor.

To validate the program ROFPM, several test cases of the model rotor of Caradonna & Tung [7] were calculated. The CARADONNA–Rotor consists of two untwisted and untapered blades with NACA 0012–sections (aspect ratio $\Lambda = 6$). Since the program should investigate the blade–vortex interference, the discretisation has to be refined at the blade tip. Although the root vortex influences the velocity distribution in the inflow plane of the rotor disk [48], it has no effect on the pressure distribution at the tip. For this reason, the discretisation at the root is not refined so the total number of surface panels can be kept small. In Figure 5.1 a blade of the CARADONNA–rotor [7] is presented.

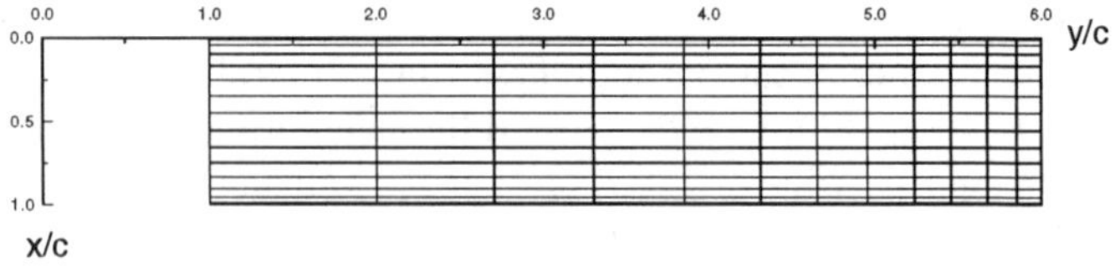

Fig. 5.1. Discretisation of the CARADONNA–testrotor

Both blades are discretised into $N = 360$ surface panels with 12 strips of 30 panels each in chordwise direction. The panel edges are located at 16.7,

33.3, 45, 55, 64.2, 71.8, 77.5, 82.5, 87.2, 90.8, 94.5 and 97.5 % of the blade radius.

Figure 5.2 shows the vortex sheets emanating from the trailing edges of the blades for a collective pitch angle, $\Theta_c = 8°$. The vortex sheets are calculated by the incompressible vortex–lattice method ROVLM [33, 35], see section 3.3.

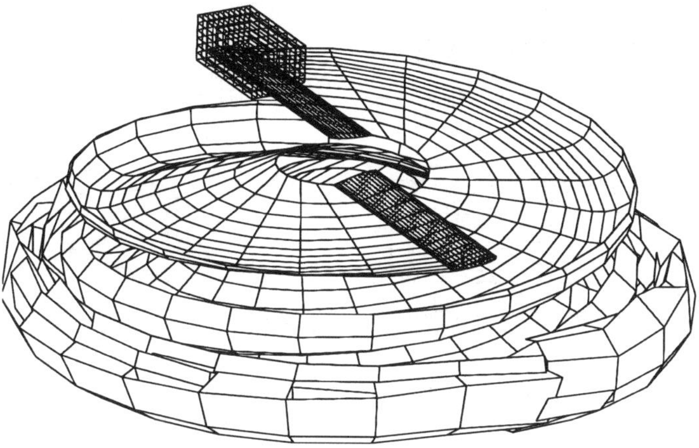

Fig. 5.2. Vortex sheets of CARADONNA–rotor ($\Theta_c = 8°$)

The blades rotate by an azimuthal increment $\Delta\Psi = 15°$ in each iteration cycle. The emanating vortex sheet builds a tip vortex, whereas the root vortex cannot be simulated because of the coarse discretisation at the root. The wake of the first revolution melts to the so–called starting vortex [33] of a diameter larger than the rotor disc. If rotor and wake produce enough induced velocity, the starting vortex declines and the following vortex sheet forms a helical surface which can be assigned to each blade from which it is emanated, see Figure 5.2.

The cartesian field grid shown in Figure 5.3 contains 24 cells in chordwise direction, 14 cells in spanwise direction and 16 elements in the direction of the rotor axis ($M = 5376$ field cells).

The side view in Figure 5.3 illustrates that the field grid does not have to be adapted to the body and may penetrate the contour at an arbitrary position. Due to the collective pitch, areas of nonlinear flow are expected to appear on the upper side of the blades. Therefore, the grid is shifted upwards ($z_{min}/c = -0.25, z_{max}/c = 0.65$). The top view in the same Figure shows that it is sufficient to stretch the boundaries of the field grid 0.1 chord length in front of the leading edge and behind the trailing edge ($x_{min}/c = -0.1, x_{max}/c = 1.1$). In hover, nonlinear effects are due to occur in the tip region. The spatial field is restricted to a relatively small area at the tip of the blade ($y_{min}/c = 4.0, y_{max}/c = 6.3$). The three–dimensional view illustrates the extension of the field grid and the blade.

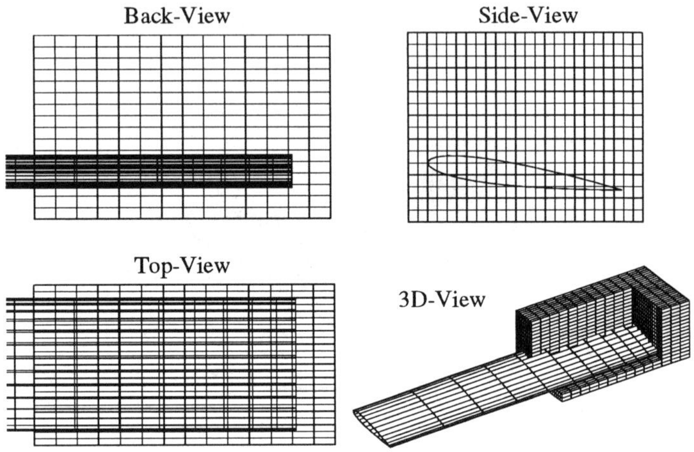

Fig. 5.3. Field grid at the tip of blade
($24 \times 14 \times 16$ fieldpanels, $-0.1/1.1; 4.0/6.3; -0.25/0.65$)

The grid shown in Figure 5.2 and 5.3 was used for all calculations concerning the CARADONNA test rotor.

The wake structure of the incompressible vortex–lattice method is used as prescribed wake in the compressible iteration. During each iteration cycle the shape is kept constant, and the calculation of the induced velocities at the wake corner points can be neglected.

The first test case is a lifting hovering rotor with a tip mach number $\mathrm{Ma}_{tip} = 0.439$ and a collective pitch setting of $\Theta_c = 8°$. The comparison between the calculated pressure distribution c_p of the field panel method and an Euler code with measured data is presented in Figure 5.4 at different radial positions.

Although this is a subsonic case, compressible effects are clearly visible. When comparing the incompressible vortex–lattice method and the field panel method, it can be seen that the pressure distribution on the upper side of the blade is underestimated by the vortex–lattice method. The field panel method is in good agreement with measurement up to the suction peak, which is underestimated. A reason could be that the discretisation of the leading edge is too coarse to resolve the pressure gradient. The Euler code also underestimates the suction peak[6]. Compressibility effects occur at a radial position greater than $r/R \approx 0.70$. The field panel method corrects the incompressible

[6] Equivalent results for Euler codes can be found as well by Hertel [15], Hertel et. al. [12, 13, 14], Krämer et al. [28, 29], Krämer [30] and Stangl et. al. [40].

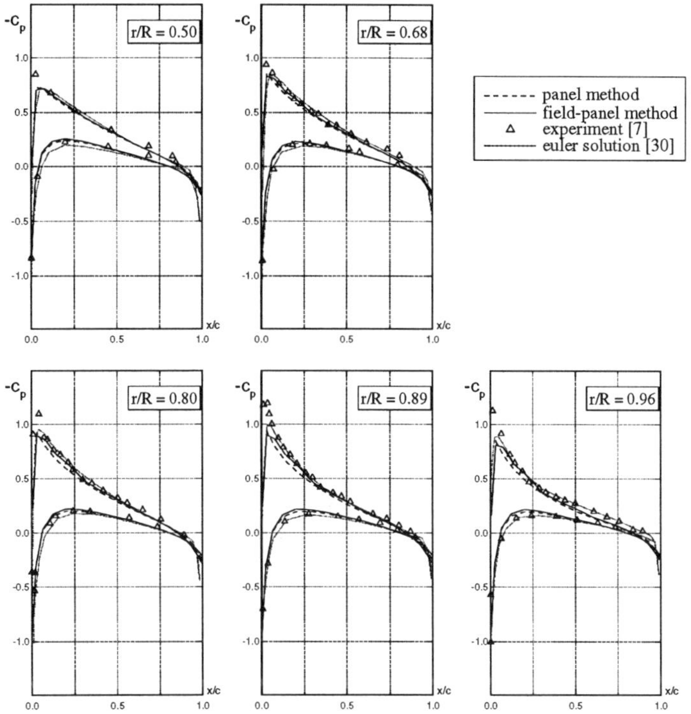

Fig. 5.4. Pressure distribution of the CARADONNA–rotor for $\Theta_c = 8°$, $Ma_{tip} = 0.439$

vortex–lattice results at the outer radial positions, whereas at the inner sections the compressible iteration of the field panel method has no remarkable influence on the results. The comparison of all pressure distributions shows that the field panel method overestimates the pressure on the lower side of the blades.

To prove that the field panel method allows to predict transonic flow, the CARADONNA–rotor was calculated for the collective pitch $\Theta_c = 8°$ and the tip mach number $Ma_{tip} = 0.794$. Near the tip of the blades, supersonic regions with shock waves appear. The pressure distribution c_p at specified radial positions is plotted in Figure 5.5.

The results of ROFPM are in good agreement with the measured data. In particular, the magnitude and location of the shock are represented very well. At the outer most radial position, $(r/R = 0.96)$ the suction peak on the upper side is underestimated. Likewise a difference in the suction peak occurs at the radial position $r/R = 0.80$. The fact the Euler method and the field panel method underestimate the suction peak, could be a hint, that the measured data is wrong[7]. Comparison with the Euler solutions of Krämer

[7] See Krämer [30] for a detailed discussion of the measured data.

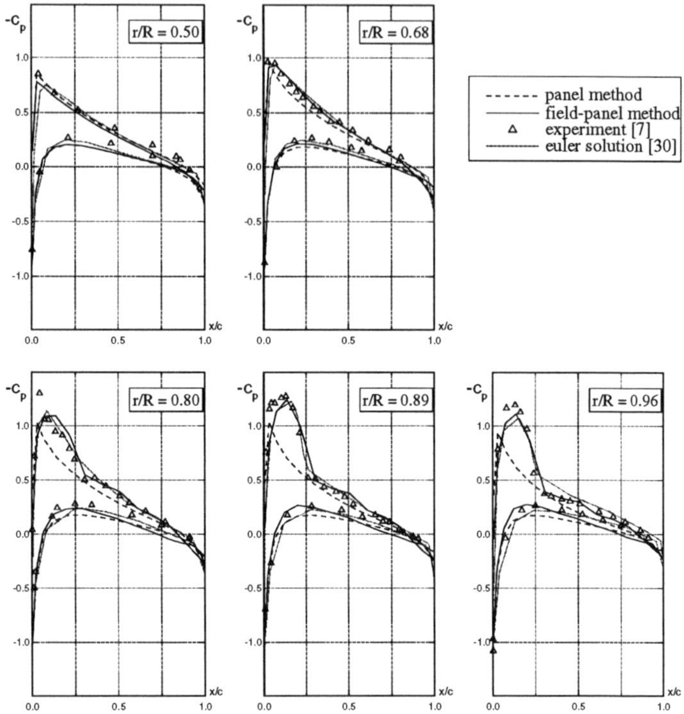

Fig. 5.5. Pressure distribution of the CARADONNA–rotor for $\Theta_c = 8°$, $\mathrm{Ma}_{tip} = 0.794$

[30] shows better agreement of the field panel method with measured data in pressure distribution at the leading edge and prediction of the shock at the blade tip.

The residuum of the field panel method needs 18 cycles of the compressible iteration to reach the convergence criterion $\varepsilon_\sigma = 0.01$. The program needs 90 minutes cpu–time for these 18 cycles on an IBM RISC/6000–320. Adding the cpu–time to determine the wake structure, we need 135 minutes, corresponding to 5% of the computation time for the Euler solution of Krämer [30] on the same machine.

6 Conclusions and Outlook

A *field panel method* has been developed, which takes into account compressible, transonic effects in panel methods which have been developed for linearized potential equation. Basis of the nonlinear extension is the steady full potential equation which is formulated as a Poisson equation $\nabla\Phi = \sigma$. The right–hand side of the Poisson equation includes all nonlinear terms of the potential equation and is interpreted as the singularity strengths of a field source distribution. The Poisson equation is solved by an iterative procedure, where the Laplace operator is solved with a linear panel method and the influence of the field sources σ is included in the right–hand side $\mathcal{R}$ of the linear system of equations.

The field panel method allows to treat two–dimensional and three–dimensional flow problems. In case of three–dimensional calculations, ROVLM is used as the incompressible boundary element method which calculates the free wake of translatory or rotational flows. The extensions for the consideration of the compressible effects are included in ROVLM.

Several advantages of this method are listed below:

- The field panel method uses a Cartesian grid which does not have to be surface fitted, thus avoiding the major task of grid generation.
- The boundary conditions are fulfilled by the panel method alone so that the grid is used only to calculate the source strengths. In practical cases, the field source strengths, which represent the compressibility, decay away from the contour so that the computational grids can be restricted to small regions, where the nonlinear effects cannot be neglected.
- The integral equation method can be combined with other finite–difference methods to reduce computation time and storage requirement.

The method is limited due to the theory and to numerical simplifications:

- The unknown singularity strengths of the boundary element method are taken as constant distribution over the panels
- To reduce storage requirement and computation time for the calculation of the influence coefficients of the field, the field grid is modeled by a Cartesian grid with equidistant length in all directions.
- Up to now, it has not been possible to treat unsteady flows.

A rotor encounters transonic effects at the advancing blade and at the tip. Since there are only subsonic velocities or weak shocks at a rotor, the full potential equation is suitable for treating helicopter flow problems. Since the basis of this method is the steady full potential equation, the investigation is restricted to steady hover flight cases. The field grid is limited to the tip region, but computes also the influence on the neighboring subsonic regions. At the root and in the major part of the surroundings the flow can be considered to be incompressible so that no field grid is required.

The ROFPM code was tested with the CARADONNA test rotor for subsonic and transonic cases. The results were compared with measurements from Caradonna & Tung [7] and with Euler code calculations Hertel [15] and Krämer [30]. Although the code was developed as a preliminary design code, the results show the same quality as the solutions of the Euler code but requiered only 5% of the cpu–time used with the method by Krämer [30].

Since the program ROFPM is able to calculate the shape of the free wake, the code could be used to investigate the blade vortex interference as well as the influence of the body on the rotary flow. The possibility of calculating complex geometries and interference with little effort makes the code a good tool for computing complete helicopter configurations in the preliminary design phase.

Future developments of the ROFPM code and the field panel method could be in the following fields:

- Extending of the program to unsteady forward flight using the unsteady full potential equation. Including flapping and pitching motion of the blades.
- Investigating of blade–vortex interferences in the transonic tip region and analyzing the influence of the body on the rotor flow.
- Using more efficient field methods to reduce storage requirements and cpu–time.
- Modifying the panel clustering method to calculate the influence coefficients of the wake elements in a more efficient way.

References

1. Behr, R., Wagner, S. (1988): Ein nichtlineares Wirbelgitterverfahren zur Berechnung von Interferenzeffekten zwischen freien Wirbelschichten und Tragflügeln. 6. DGLR–Fachsymposium über Strömung mit Ablösung. DGLR–Bericht 88-5, pp. 169 – 180, Braunschweig
2. Behr, R., Wagner, S. (1988): A Low Order Panel Method for the Calculation of Vortex Sheet Roll–Up and Wing–Vortex Interaction. In: Brebbia, C. (ed.) Proceedings Boundary Elements X, Vol. I. Springer, pp. 259 – 274, Southampton UK
3. Behr, R., Wagner, S. (1989): A Vortex–Lattice Method for the Calculation of Vortex Sheet Roll-Up and Wing–Vortex Interaction. In: Hirschel, E.H. (ed.) Notes on Numerical Fluid Mechanics, Vol. 25: Finite Approximations in Fluid Mechanics II. Vieweg, Braunschweig Wiesbaden, pp. 1 – 14
4. Behr, R., Wagner, S. (1990): Anwendung eines Wirbelgitterverfahrens auf die Umströmung schlanker Deltaflügel mit Ablösung. DGLR–Fachsymposium über Strömungen mit Ablösung. DGLR–Bericht 90-06, pp. 166 – 170, RWTH Aachen
5. Behr, R., Wagner, S. (1993): Entwicklung eines Panelverfahrens zur Berechnung der subsonischen Umströmung von endlich dicken Tragflügeln mit Strömungsablösung und freier Wirbelschleppe in einer Flugzeugkonfiguration. Institutsbericht 91/2, Universität der Bundeswehr München
6. Behr, R., Wagner, S. (1990): Application of a Low Order Panel Method to Slender Delta–Wings at High Angles of Attack. Symposium of the International Association for Boundary Element Methods (IABEM 90), Rome. In: Morino, L., Piva, R. (eds.) Boundary Integral Methods — Theory and Application. Springer 1991, pp. 105 – 114
7. Caradonna, F.X., Tung, C. (1980): Experimental and Analytical Studies of a Model Helicopter Rotor in Hover. Sixth European Rotorcraft and Powered Lift Aircraft Forum. pp. 25-1 – 25-19, Bristol, England
8. Egolf, A. (1988): Helicopter Free Wake Prediction of Complex Wake Structures under Blade-Vortex Interaction Operating Conditions

9. Gennaretti, M., Morino, L. (1992): A Boundary Element Method for the Potential, Compressible Aerodynamics of Bodies in Arbitrary Motion. Aeronautical Journal Vol. 96(951), pp. 15 – 19
10. Hackbusch, W. (1989): Integralgleichungen. Teubner, Stuttgart
11. Hackbusch, W., Nowak, Z. (1989): On the Fast Matrix Multiplication in the Boundary Element Method by Panel Clustering. Numerische Mathematik 54, 463 – 491
12. Hertel, J., Krämer, E., Wagner, S. (1988): Berechnung des Strömungsfeldes eines mehrblättrigen Hubschrauberrotors mit Hilfe eines Euler–Verfahrens unter Einbeziehung des Nachlaufes. 6. DGLR-Fach–Symposium über Strömungen mit Ablösung. DGLR–Bericht 88-5, pp. 243 – 254, Braunschweig
13. Hertel, J., Krämer, E., Wagner, S. (1988): Euler Solutions for Steady Flow of a Helicopter Rotor. 2. International Conference on Rotorcraft Basic Research of the American Helicopter Society. Proceedings, pp. H-1 – H-15, Maryland
14. Hertel, J., Krämer, E., Wagner, S. (1990): Complete Euler–Solution for a Rotor in Hover and a Propeller in Forward Flight. 16th European Rotorcraft Forum. Proceedings, Paper No. 2.6, Glasgow, Scotland
15. Hertel, J. (1991): Euler–Lösungen der stationären Rotorströmung für Schwebe– und den axialen Vorwärtsflug mit Einbeziehung linearer Methoden. Dissertation, Universität der Bundeswehr München. Forschungsberichte VDI, Reihe 7: Strömungsmechanik Nr. 197
16. Hess, J. (1990): Panel Methods in Computational Fluid Dynamics. In: Lumley, J., van Dyke, M. (eds.) Annual Review of Fluid Mechanics, pp. 255 – 274. Annual Reviews Inc., Palo Alto, California
17. Hess, J., Smith, A. (1964): Calculation of Nonlifting Potential Flow About Arbitrary Three–Dimensional Bodies. Journal of Ship Research, pp. 22 – 44
18. Hoeijmakers, H. (1991): Panel Methods for Aerodynamic Analysis and Design. Special Course on Engineering Methods in Aerodynamic Analysis and Design of Aircraft, AGARD–R–783, pp. 5-1 – 5-47, Neuilly sur Seine, France
19. Hoeijmakers, H., Bennekers, B. (1978): A Computational Model for the Calculation of the Flow Form about Wings with Leading–Edge Vortices. Technical Report, National Aerospace Laboratory NLR, Amsterdam, the Netherlands
20. Hunt, B. (1977): Relationships between Volume, Surface and Line Distributions of Vorticity, Source and Doublicity. Report Ae/384, British Aircraft Corporation, Military Aircraft Division, Warton Aerodrome, Preston, PR4 IAX
21. Hunt, B. (1980): The Mathematical Basis and Numerical Principles of the Boundary Integral Method for Incompressible Potential Flow over 3–D Aerodynamic Configurations. In: Hunt, B. (ed.) Numerical Methods in Applied Fluid Dynamics, pp. 49 – 135. Academic Press, London New York Toronto Sydney San Francisco
22. Jones, H., Caradonna, F.X. (1986): Full Potential Modelling of Blade Vortex Interactions. 12th European Rotorcraft Forum. DGLR, September 1986
23. Kandil, O. (1985): Computational Technique for Compressible Vortex Flows Past Wings at Large Incidence. Journal of Aircraft, 22(9), 750 – 755
24. Kandil, O., Chu, L., Tureaud, T. (1984): A Nonlinear Hybrid Vortex Method for Wings at Large Angle of Attack. AIAA Journal, 22(3), 329 – 336
25. Kandil, O., Yates, E. (1986): Transonic Vortex Flows Past Delta Wings: Integral Equation Approach. AIAA Journal, 24(11), 1729 – 1736
26. Kellogg, O. (1967): Foundations of Potential Theory. Die Grundlehren der mathematischen Wissenschaften in Einzeldarstellungen Band 31. Springer, Berlin Heidelberg New York

27. Kramer, C. (1988): Erstellen eines mathematischen Modells zur Berechnung der freien Wirbelschicht im Nachlauf von Rotoren. Semesterarbeit, Institut für Luftfahrttechnik und Leichtbau, Universität der Bundeswehr
28. Krämer, E., Hertel, J., Wagner, S. (1987): Computation of Subsonic and Transonic Helicopter Rotor Flow Using Euler Equations. 13th European Rotorcraft Forum, Proceedings, Paper No. 2-14, Arles, France. Vertica, 12(3), 279 – 291, 1988
29. Krämer, E., Hertel, J., Wagner, S. (1990): Euler Procedure for Calculation of the Steady Rotor Flow with Emphasis on Wake Evolution. 8th Applied Aerodynamics Conference, Proceedings, AIAA–paper 90–3007, Portland, Oregon
30. Krämer, E. (1991): Theoretische Untersuchungen der stationären Rotorblattumströmung mit Hilfe eines Euler–Verfahrens. Dissertation, Universität der Bundeswehr München. Forschungsberichte VDI, Reihe 7: Strömungsmechanik Nr. 197
31. Maskew, B., Woodward, F. (1976): Symmetrical Singularity Model for Lifting Potential Flow Analysis. Journal of Aircraft, 13(9), 733 – 734, September 1976
32. Maskew, B. (1987): Program VSAERO Theory Document – A Computer Program for Calculating Nonlinear Aerodynamic Characteristics of Arbitrary Configurations. NASA Contractor Report 4023, NASA National Aeronautics and Space Administration, Ames Research Center
33. Röttgermann, A. (1988): Ein Singularitätenverfahren zur Berechnung des Nachlaufs von Mehrblattrotorsystemen. Diplomarbeit, Institut für Luftfahrttechnik und Leichtbau, Universität der Bundeswehr München
34. Röttgermann, A. (1995): Eine Methode zur Berücksichtigung kompressibler und transsonischer Effekte in Randelementmethoden. Dissertation, Universität Stuttgart
35. Röttgermann, A., Behr, R., Schöttl, Ch., Wagner, S. (1991): Calculation of Blade–Vortex Interaction of Rotary Wings in Incompressible Flow by an Unsteady Vortex–Lattice Method Including Free Wake Analysis. In: Hackbusch, W. (ed.) Notes on Numerical Fluid Mechanics Vol. 33: Numerical Techniques for Boundary Element Methods. Vieweg, Braunschweig, pp. 153 – 166
36. Röttgermann, A., Wagner, S. (1995): ROFPM — Eine transsonische Panelmethode: Wirbelgitterverfahren und Feldpanelverfahren. Institutsbericht IB 95-4, Institut für Aerodynamik und Gasdynamik, Universität Stuttgart
37. Sauter, St. (1991): Der Aufwand der Panel–Clustering–Methode für Integralgleichungen. Institutsbericht 9115, Institut für Informatik und Praktische Mathematik, Christian–Albrechts–Universität Kiel
38. Sauter, St. (1992): Über die effiziente Verwendung des Galerkinverfahrens zur Lösung Fredholmscher Integralgleichungen. Dissertation, Christian–Albrechts–Universität Kiel
39. Slooff, J. (1985): Some New Developments in Exact Integral Equation Formulations for Sub– or Transonic Compressible Potential Flow. In: Morino, L. (ed.) Computational Methods in Potential Aerodynamics, pp. 703 – 725. Springer, Berlin Heidelberg New York Tokyo
40. Stangl, R., Wagner, S. (1992): Calculation of the Steady Rotor Flow Using an Overlapping–Embedded Grid Technique. 18th European Rotorcraft Forum Proceedings, Paper No. ERF 92-71, Avignon, France. Paper No. 78.
41. Tinoco, E., Ball, D., Rice, F. (1987): PAN AIR Analysis of a Transport High Lift Configuration. Journal of Aircraft, 24(3), 181 – 187
42. Urban, Ch. (1988): Ein Wirbelgitterverfahren zur Berechnung von Interferenzeffekten zwischen Tragflächen und freien Wirbelschichten bei Unterschallströmung. Dissertation, Universität der Bundeswehr München

43. Wagner, S. (1967): Beitrag zum Singularitätenverfahren der Tragflächentheorie bei inkompressibler Strömung. Dissertation, Technische Hochschule München. auch Ingenieur–Archiv, 36, 403 – 420, 1968
44. Wagner, S. (1969): On the Singularity Method of Subsonic Lifting-Surface Theory. Journal of Aircraft, 6(6), 549 – 558
45. Wagner, S. (1987): The Panel Method - A Unique Tool for the Aerodynamic Analysis of Complex Configurations in Subsonic and Supersonic Flow. In: Brebbia, C.A., Wendland, W.L., Kuhn, G. (eds.) 9th International Conference – Boundary Element Methods in Engineering – Boundary Elements IX, pp. 519 – 538. Springer, Berlin Heidelberg New York London Paris Tokyo
46. Wagner, S., Urban, Ch. (1986): Current Activities in Basic Research Work on Panel Methods in Germany. In: Hirschel, E.H. (ed.) Notes on Numerical Fluid Mechanics, Vol. 14: Finite Approximations in Fluid Mechanics. Vieweg, Braunschweig Wiesbaden, pp. 273 – 294
47. Wagner, S., Urban, Ch., Behr, R. (1987): A Vortex–Lattice Method for the Calculation of Wing–Vortex Interaction in Subsonic Flows. In: Ballman, J., Eppler, R., Hackbusch, W. (eds.) GAMM-Symposium, Notes on Numerical Fluid Mechanics, Vol. 21: Panel Methods in Fluid Mechanics with Emphasis on Aerodynamics. Vieweg, Kiel, pp. 243 – 251
48. Zerle, L., Wagner, S. (1992): Development and Validation of a Vortex Lattice Method to Calculate the Flowfield of a Helicopter Rotor Including Free Wake Development. 18th European Rotorcraft Forum, Proceedings, Paper No. ERF 92-72, Avignon, France

Interpolation, Triangulation and Numerical Integration on Closed Manifolds

K. Kalik, R. Quatember, and W. L. Wendland

1 Introduction

The first part of the paper deals with the interpolation and triangulation of closed manifolds. Let Γ be a closed surface in $\mathbb{R}^3$, which is implicitly given by a function $H(x) = 0$. We present a new algorithm which creates a "nearly" uniform triangulation of this manifold. Our ideas are based on the procedures by E. L. Allgower [1] and on local geometry properties [2]. The algorithm allows a parallel discretization and was tested for several examples.

In the second part of the paper we study the numerical computation of weakly singular integrals. We develop an adaptive algorithm for the calculation of a class of boundary integrals. Furthermore, a method is presented which allows the construction of C^1-smooth boundary and finite elements on a given closed manifold $\Gamma \subset \mathbb{R}^3$. Finally, Hadamard's regularization technique is used to reduce the computation of singular boundary integrals to the computation of weakly singular integrals.

2 The algorithm for the approximation of a closed surface in $\mathbb{R}^3$ by triangulations

Here we describe a parallel algorithm for eight transputers which yields a nearly uniform triangulation of a closed surface implicitly given by $H(x) = 0$ where $x = (x_1, x_2, x_3)$. We assume that for all corner points p^i of the triangles there holds $H(p^i) = 0$.

The following input data are required (see figure 2.1):

- a surface Γ of a body Ω given by $H(x) = 0$, where $H(x) < 0$ in the interior, $H(x) > 0$ in the exterior of Ω and $\nabla H(x) \neq 0$ in a vicinity of Γ;
- the number N of partitions;
- a starting cube.

The starting cube of edge length $2\,l$ must contain Γ completely (see figure 2.2). This cube is divided into eight subcubes associated with the eight transputers. Each subcube is then divided into N^3 cubical subdomains. In the algorithm only the so-called suitable cubical subdomains are used, where suitable means that in the subcube there exist points p^i with $H(p^i) > 0$ and

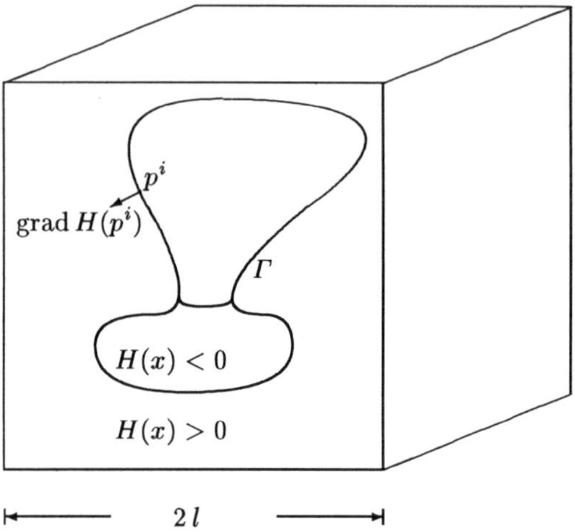

Fig. 2.1. surface $\Gamma : H(x) = 0$ to be triangulated

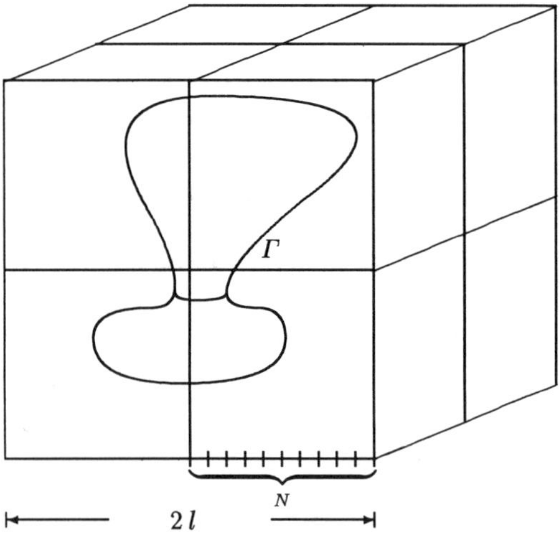

Fig. 2.2. Partition of the surrounding cube with respect to the number of processors

p^j with $H(p^j) \le 0$. The points p^i which are situated at the outer faces of a suitable cubical subdomain are projected in the direction $\nabla H(p^i)$ onto the surface Γ (see figure 2.3). The following algorithm determines this projection:

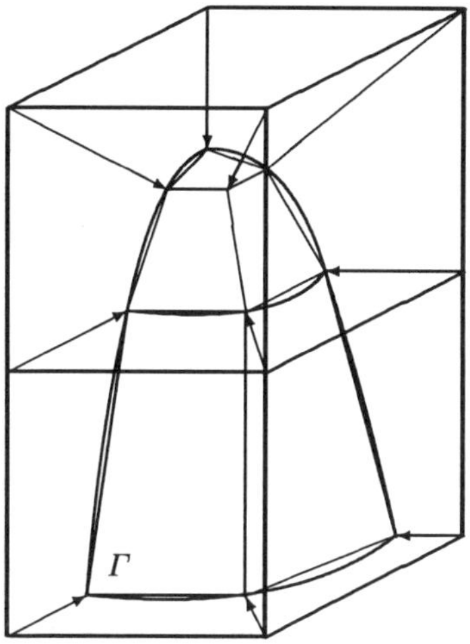

Fig. 2.3. Projection of the outer points onto the surface Γ.

1st step: Define $p^+ := p^i$ and choose

$$p^- := p^i + t\,\nabla H(p^i) \quad \text{with} \quad t \in \mathbb{R} \quad \text{such that} \quad H(p^-) < 0.$$

2nd step: Calculate

$$p^* := \frac{p^+ + p^-}{2}.$$

3rd step: If $|H(p^*)| < \varepsilon$ then $p^i := p^*$ and proceed to the next point p^i.

4th step: If $H(p^*) > 0$, then $p^+ := p^*$; otherwise $p^- := p^*$ and return to step 2.

In the examples below we chose $\varepsilon = 10^{-6}$.

Next, we divide the generated quadrangles on the surface into two triangles along the shortest diagonal.

Unfortunately, in general these triangles are nonuniform. Therefore a correction of the triangulation is necessary. We take for each point the arithmetical mean value of his adjacent neighboring points, replace the point by the mean value and map it back onto the surface. Sweeping twice through the surface points, we obtain a more uniform grid (figure 2.4).

The subgrids generated by the eight transputers will be fit together.

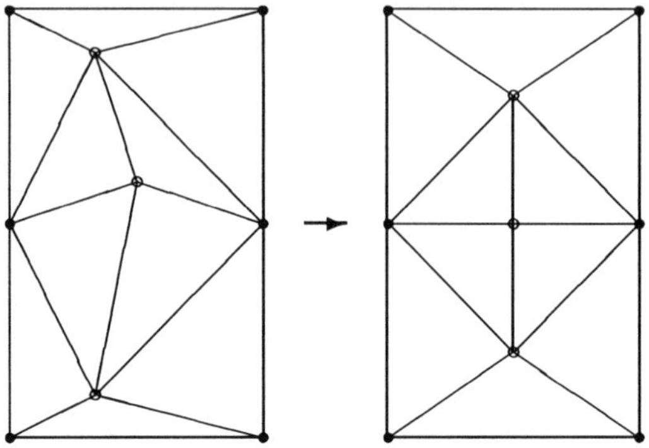

Fig. 2.4. correction of the grid.

2.1 Examples

The following four examples are considered:

1. the sphere: $H(x) = x_1^2 + x_2^2 + x_3^2 - 1$;
2. the ellipsoidal surface: $H(x) = \frac{1}{2}x_1^2 + \frac{1}{2}x_2^2 + x_3^2 - 1$;
3. the torus: $H(x) = (x_1^2 + x_2^2 + x_3^2 + (\frac{1}{2})^2 - (\frac{1}{4})^2)^2 - 4(\frac{1}{2})^2(x_2^2 + x_3^2)$;
4. the double-sphere: $H(x) = \begin{cases} (x_1 - \frac{1}{4})^2 + x_2^2 + x_3^2 - \frac{1}{4} & : \quad x_1 \geq 0, \\ (x_1 + \frac{1}{4})^2 + x_2^2 + x_3^2 - \frac{1}{4} & : \quad x_1 < 0. \end{cases}$

For the starting cube we choose $l = 1.0$, and for all examples the number of partitions is chosen as $N = 20$. The resulting meshwidth is $\delta \leq 0.05$. The generated triangulations are corrected three times. The computations were executed on a transputer-board with eight transputers T 800 and with an accuracy of 10^{-10}. The results are demonstrated in the following table:

	number of		time/number processors; Speedup			
	triangles	points	1	2	4	8
sphere figure 2.5	9072	4538	211 sec 1,0	129 sec 1,6	91 sec 2,3	72 sec 2,9
ellipsoid figure 2.6	6656	3330	145 sec 1,0	90 sec 1,6	64 sec 2,3	51 sec 2,8
torus figure 2.7	7120	3560	168 sec 1,0	103 sec 1,6	72 sec 2,3	57 sec 2,9
double-sphere figure 2.8	6000	2974	127 sec 1,0	80 sec 1,6	57 sec 2,2	46 sec 2,8

Remark:
The results show that the speedup is small. The reason for that is that large parts of the algorithm are running in succession and the loadbalancing is not solved very well. To avoid these problems, we should use one processor as "master" where the main part of the algorithm is done and the remaining processors as "slaves" which do the necesarry computations. This is explained more detailed in [3].

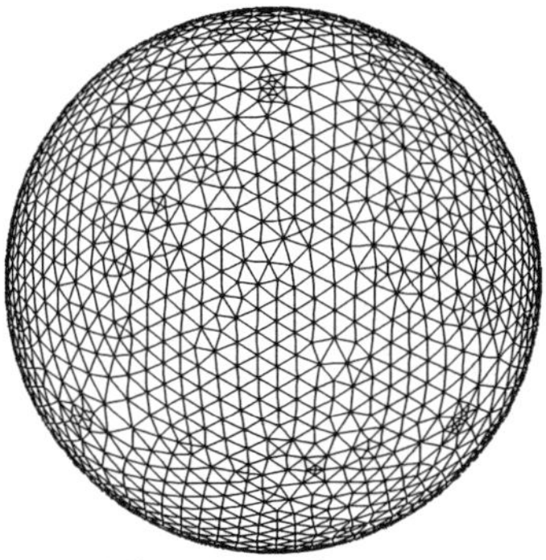

Fig. 2.5. sphere

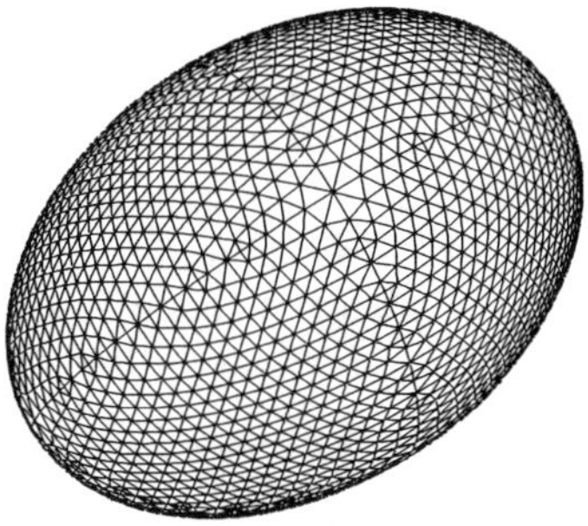

Fig. 2.6. ellipsoid

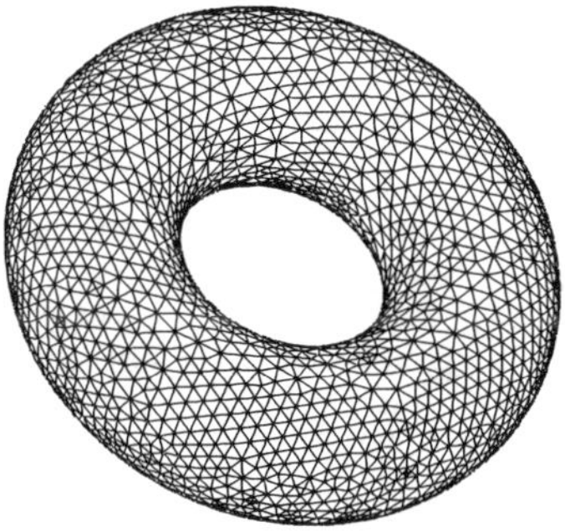

Fig. 2.7. torus

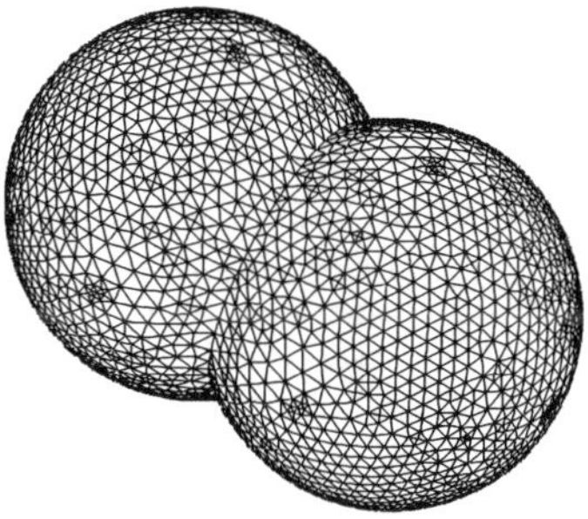

Fig. 2.8. double-sphere

2.2 Extension on surfaces with edges.

The surfaces considered in the examples of the last section were smooth except the double-sphere, i.e. the gradient was defined at each point of the surface. In order to circumvent the problem caused by the non-existence of the gradient along the edge of the double-sphere, two "tricks" are applied. In this example it is possible to choose the starting cube in such a way that the curve of intersection of both spheres is situated in the common face of two adjacent subcubes. Furthermore, we define a "modified gradient":

$$\nabla H(x) := \begin{cases} (2x_1 - \frac{1}{2},\, 2x_2,\, 2x_3) & \text{for } x_1 > 10^{-6}, \\ (2x_1 + \frac{1}{2},\, 2x_2,\, 2x_3) & \text{for } x_1 < -10^{-6}, \\ (0,\, 2x_2,\, 2x_3) & \text{otherwise.} \end{cases}$$

Clearly, this approach is only possible if we consider a grid where the curves of intersection of the composed bodies lie in a plane.

In addition we considered bodies with more general edges:

1. A cylinder with edges (see figure 2.9).
2. Two cones glued together (see figure 2.10).
3. A triple torus and a satellite (see figures 2.11 and 2.12).
4. A cube where three cylinders are cut out. This cube has the following geometrical singularities:
   a) corner points,
   b) edges at the exterior faces, and
   c) complicated edges in the interior of the cube.

The difficulties occuring here can be classified as follows:

1. The corner points will be cut by the algorithm and replaced by one triangle face. A modification is necessary where several triangles with the given corner point as common vertex are used.
2. The algorithm produces a fringed grid at the exterior edges. This can be avoided by an appropriate choice of the starting cube and modification of the gradient.
3. Our approach does not work for edges which appear at non-plane interfaces of composite bodies.
4. The exact calculation of the gradient for surfaces of complicated structure is expensive, hence it should be approximated numerically.

2.2.1 Numerical approximation of the gradient. We compute the gradient at the point x^0 approximately with the help of centered differences:

$$\nabla H(x^0) := \begin{pmatrix} \frac{H(x^0+\varepsilon\, e^1)-H(x^0-\varepsilon\, e^1)}{2\,\varepsilon} \\ \frac{H(x^0+\varepsilon\, e^2)-H(x^0-\varepsilon\, e^2)}{2\,\varepsilon} \\ \frac{H(x^0+\varepsilon\, e^3)-H(x^0-\varepsilon\, e^3)}{2\,\varepsilon} \end{pmatrix},$$

where e^i is the i-th unit vector in $\mathbb{R}^3$ and $\varepsilon = 10^{-6}$. In order to decide wether the gradient is well-defined or has a jump at x^0, we calculate the gradient at all points adjacent to x^0:

$$x^{i,j,k} := x^0 + i\, e^1 + j\, e^2 + k\, e^3 \quad \text{with} \quad i, j, k \in \{-1, 0, 1\}.$$

If

$$\nabla H(x^{i,j,k}) \neq 0 \quad \text{and} \quad \nabla H(x^0) \neq 0$$

and therefore

$$\frac{\langle \nabla H(x^{i,j,k}), \nabla H(x^0)\rangle}{|\nabla H(x^{i,j,k})|\, |\nabla H(x^0)|} > \delta \quad \text{with} \quad \delta = 0.8 \quad \text{for all} \quad x^{i,j,k},$$

the gradient is well-defined. If this inequality is not satisfied for at least one of the points $x^{i,j,k}$, then the gradient has a jump in the vicinity of x^0. In this case x^0 is an edge or corner point of the surface Γ.

However, our algorithm needs a well-defined gradient. Therefore we replace x_0 in the discontinuous case by a point close to x^0 on the initial triangle where the gradient exists, then this replaced point is projected onto the surface.

We now assume that a triangulation with grid points on the surface Γ is available. As described before we correct the triangulation in order to obtain a more uniform grid. Unfortunately, in the neighborhood of edges and corners this method fails and a special correction for these geometrical singularities is necessary.

2.2.2 Correction of the corner points. The corner points, which are known as such, have to be included into the grid. This is realized by replacing the grid point which is one of the closest to the corner point, by the corner point itself. Grid points which are corner points will be marked; and they are fixed in the mean value correction procedure described before.

Applying this correction procedure we end up with a nearly uniform grid with correct corner points.

2.2.3 Correction of the edges. The algorithm for the correction of the edges will be described in seven steps:

1st step: Determine the triangular line $(p^1; p^2)$ which satisfies

$$\frac{\langle \nabla H(p^1), \nabla H(p^2) \rangle}{|\nabla H(p^1)|\,|\nabla H(p^2)|} \leq \delta \quad \text{for} \quad \delta = 0.8\,;$$

i.e. one edge point lies between p^1 and p^2. Points with $\nabla H(p) = 0$ are edge points which are marked and fixed in the following steps.

2nd step: Define

$$p_0^1 := p^1 \quad \text{and} \quad p_0^2 := p^2 .$$

3rd step: Determine the mid-point

$$\tilde{p}_k^{12} := \frac{p_k^1 + p_k^2}{2}$$

and project $\tilde{p}_k^{12}$ onto Γ such that

$$H(p_k^{12}) = 0 \quad \text{with} \quad p_k^{12} := \tilde{p}_k^{12} + \lambda \nabla H(\tilde{p}_k^{12}) .$$

4th step: If

$$\frac{\langle \nabla H(p_k^{12}), \nabla H(p_k^1) \rangle}{|\nabla H(p_k^{12})|\,|\nabla H(p_k^1)|} > \frac{\langle \nabla H(p_k^{12}), \nabla H(p_k^2) \rangle}{|\nabla H(p_k^{12})|\,|\nabla H(p_k^2)|},$$

then put

$$p_{k+1}^1 := p_k^{12} \quad \text{und} \quad p_{k+1}^2 := p_k^2\,;$$

otherwise put

$$p_{k+1}^1 := p_k^1 \quad \text{und} \quad p_{k+1}^2 := p_k^{12} .$$

5th step: If

$$|p_{k+1}^1 - p_{k+1}^2| > \varepsilon$$

then return to step 3. (In the examples above we chose $\varepsilon = 10^{-6}$.)

6th step: If

$$|p^1 - p_k^{12}| < |p^2 - p_k^{12}|,$$

then put

$$p^1 := p_k^{12} \text{ and mark } p^1 \text{ as an edge point;}$$

otherwise put

$$p^2 := p_k^{12} \text{ and mark } p^2 \text{ as an edge point.}$$

7th step: If there is another triangular line with an edge point then start with step 1 again.

The resulting grid (except the corner and edge points) is corrected anew by the mean value correction procedure described at the beginning of this section. The following examples show that the resulting triangulations are nearly uniform.

2.3 Examples with edges

1. cylinder: $H(x) = \max\left\{|x_3| - 1,\, x_1^2 + x_2^2 - 1\right\}$
2. two cones: $H(x) = \max\left\{|x_3| - 1,\, x_1^2 + x_2^2 - (1 - |x_3|)^2\right\}$
3. triple torus: $H(x) = (x_1^2+x_2^2+x_3^2+1.9)^2 - 8\max\left\{x_1^2+x_2^2,\, x_2^2+x_3^2,\, x_1^2+x_3^2\right\}$
4. satellite: $H(x) = \min\left\{H_1(x),\, H_2(x),\, H_3(x)\right\}$ with

$$H_1(x) = \begin{cases} \min\left\{x_1^2 + x_2^2 - 0.05,\, x_1^2 + x_2^2 + (|x_3| - 4)^2 - \frac{1}{2}\right\} & \text{if } |x_3| \le 3, \\ x_1^2 + x_2^2 + (|x_3| - 4)^2 - \frac{1}{2} & \text{else.} \end{cases}$$

$$H_2(x) = \begin{cases} \min\left\{x_2^2 + x_3^2 - 0.05,\, x_2^2 + x_3^2 + (|x_1| - 4)^2 - \frac{1}{2}\right\} & \text{if } |x_1| \le 3, \\ x_2^2 + x_3^2 + (|x_1| - 4)^2 - \frac{1}{2} & \text{else.} \end{cases}$$

$$H_3(x) = \begin{cases} \min\left\{x_1^2 + x_3^2 - 0.05,\, x_1^2 + x_3^2 + (|x_2| - 4)^2 - \frac{1}{2}\right\} & \text{if } |x_2| \le 3, \\ x_1^2 + x_3^2 + (|x_2| - 4)^2 - \frac{1}{2} & \text{else.} \end{cases}$$

5. cube: $H(x) = \max\left\{H_1(x),\, H_2(x)\right\}$ with

$$H_1(x) = \max\left\{|x_1| - 1,\, |x_2| - 1,\, |x_3| - 1\right\}$$

$$H_2(x) = \max\left\{\frac{1}{2} - x_1^2 - x_2^2,\, \frac{1}{2} - x_2^2 - x_3^2,\, \frac{1}{2} - x_1^2 - x_3^2\right\}$$

With the modified algorithm we obtained the following results:

	number of triangles	points	CPU-time for the mesh generation	correction
cylinder (Fig. 2.9)	2672	1338	12,0 sec	140,0 sec
two cones (Fig. 2.10)	12752	6378	225,8 sec	459,9 sec
triple torus (Fig. 2.11)	14976	7452	304,6 sec	896,9 sec
satellite (Fig. 2.12)	19056	9530	495,3 sec	6078,6 sec
cube (Fig. 2.13)	10032	5008	142,6 sec	500,6 sec

2.4 Convergence

In this section we investigate the convergence of the algorithm for a polyhedral domain. To this end, we introduce barycentric coordinates:

Definition 2.1. *Let $v^0, \dots, v^n$ be $n+1$ points which are not all situated on a hyperplane of $\mathbb{R}^n$. Then the values $\lambda_0(x), \dots, \lambda_n(x) \in \mathbb{R}$ associated with the point $x \in \mathbb{R}^n$ with*

$$\sum_{i=0}^{n} \lambda_i(x) = 1 \quad \textit{and} \quad x = \sum_{i=0}^{n} \lambda_i(x)\, v^i$$

are called the **barycentric coordinates of the point x**.

The barycentric coordinates are defined by the points $v^0, \dots, v^n$; in $\mathbb{R}^3$ we need four points.

Since we need the barycentric coordinates with respect to a plane triangle in $\mathbb{R}^3$, we choose an auxiliary point v_0 for the fourth barycentric coordinate in $\mathbb{R}^3$. Then $\lambda_0(x) = 0$ for all x lying in the plane defined by the triangle.

The corresponding procedure is given as follows: Choose the point v^0, e. g.

$$v^0 := \frac{v^1 + v^2 + v^3}{3} + \frac{v^2 - v^1}{|v^2 - v^1|} \times \frac{v^3 - v^1}{|v^3 - v^1|};$$

calculate the $\lambda_i(x)$:

$$\lambda_0(x) := \frac{1}{D} \det \begin{pmatrix} 1 & 1 & 1 & 1 \\ x_1 & v_1^1 & v_1^2 & v_1^3 \\ x_2 & v_2^1 & v_2^2 & v_2^3 \\ x_2 & v_2^1 & v_2^2 & v_2^3 \end{pmatrix}, \quad \lambda_1(x) := \frac{1}{D} \det \begin{pmatrix} 1 & 1 & 1 & 1 \\ v_1^0 & x_1 & v_1^2 & v_1^3 \\ v_2^0 & x_2 & v_2^2 & v_2^3 \\ v_2^0 & x_2 & v_2^2 & v_2^3 \end{pmatrix}$$

$$\lambda_2(x) := \frac{1}{D} \det \begin{pmatrix} 1 & 1 & 1 & 1 \\ v_1^0 & v_1^1 & x_1 & v_1^3 \\ v_2^0 & v_2^1 & x_2 & v_2^3 \\ v_2^0 & v_2^1 & x_2 & v_2^3 \end{pmatrix}, \quad \lambda_3(x) := \frac{1}{D} \det \begin{pmatrix} 1 & 1 & 1 & 1 \\ v_1^0 & v_1^1 & v_1^2 & x_1 \\ v_2^0 & v_2^1 & v_2^2 & x_2 \\ v_2^0 & v_2^1 & v_2^2 & x_2 \end{pmatrix}$$

where

$$D = \det \begin{pmatrix} 1 & 1 & 1 & 1 \\ v_1^0 & v_1^1 & v_1^2 & v_1^3 \\ v_2^0 & v_2^1 & v_2^2 & v_2^3 \\ v_2^0 & v_2^1 & v_2^2 & v_2^3 \end{pmatrix}.$$

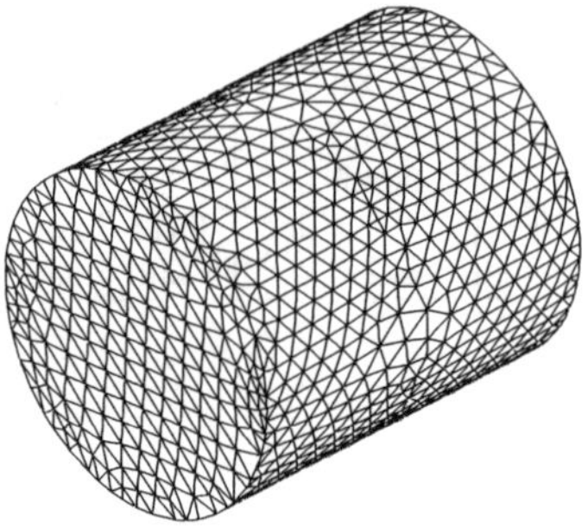

Fig. 2.9. cylinder

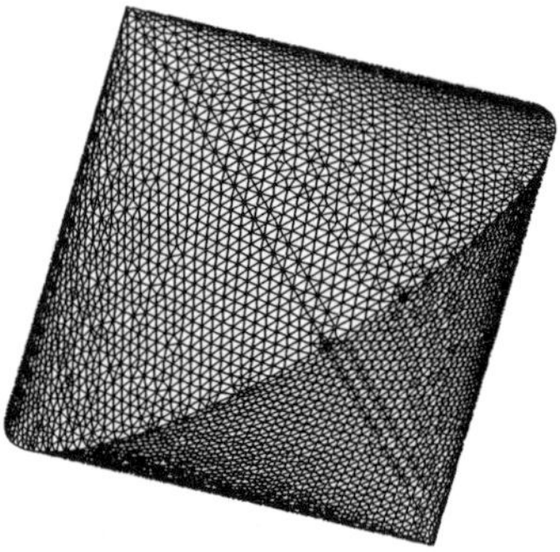

Fig. 2.10. two cones

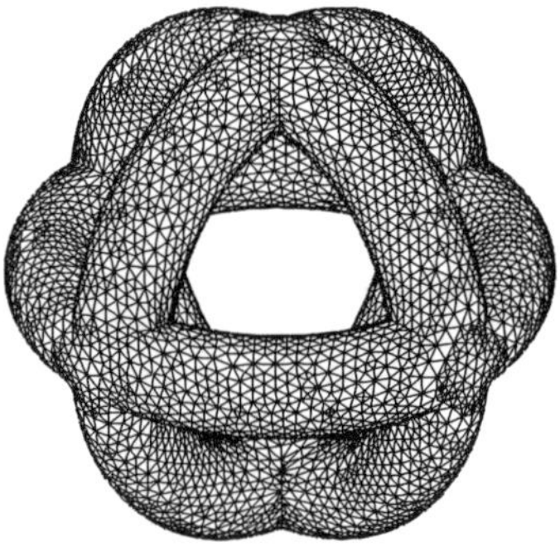

Fig. 2.11. triple torus

Fig. 2.12. satellite

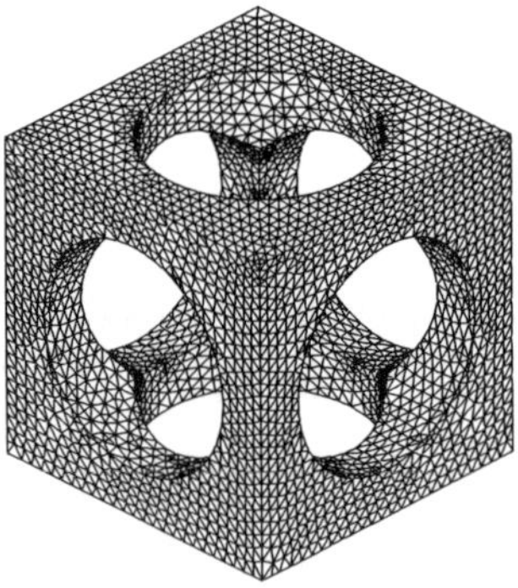

Fig. 2.13. cube

Now we show the convergence of our algorithm:

We assume that the surface Γ given by $H(y) = 0$ is approximated by a polyhedron Γ_h which consists of triangles Δ^i $(i = 1, \dots, N)$ of maximal diameter

$$h := \max\left\{|x-y| \;\middle|\; x, y \in \Delta^i, \quad i = 1, \dots, N\right\}.$$

Let us consider a triangle Δ^i which is spanned by the points $v^{i,1}$, $v^{i,2}$ and $v^{i,3}$. At the same time these points generate a curvilinear triangle Γ^i on the surface Γ. In short, we write

$$\Delta \text{ for } \Delta^i, \qquad v^1 \text{ for } v^{i,1} \text{ (analogously, } v^2 \text{ and } v^3) \qquad \text{and } \Omega \text{ for } \Gamma^i.$$

Let y be an arbitrary point of Ω and y' its projection in the direction of gradient $\nabla H(y)$ onto Δ. Our aim is to show that

$$|y - y'| \to 0 \text{ converges as } h \to 0.$$

For any point $x \in \mathbb{R}^3$ we have

$$H(x) = H(y) + \nabla H(y)(x-y) + \frac{1}{2} D^2\, H(y^*)(x-y)^2,$$

where $y^* \in \Omega$. We assume that H is piecewise $C^2(\Omega)$-smooth.

Then, for the points $v^1, \dots, v^3$ we have

$$H(v^i) = H(y) + \nabla H(y)(v^i - y) + \frac{1}{2} D^2\, H(y^*)(v^i - y)^2. \tag{2.1}$$

With the help of the barycentric coordinates, the point $y' \in \Delta$ can be represented as

$$y' = \sum_{i=1}^{3} \lambda_i(y')\, v^i.$$

Insert this formula into (2.1) to obtain

$$\sum_{i=1}^{3} \lambda_i(y')\, H(v^i) = \sum_{i=1}^{3} \lambda_i(y') \left[H(y) + \nabla H(y)(v^i - y) + \frac{1}{2} D^2\, H(y^*)(v^i - y)^2 \right].$$

With $H(v^i) = H(y) = 0$, this yields

$$\begin{aligned} 0 &= \sum_{i=1}^{3} \lambda_i(y') \left[\nabla H(y)(v^i - y) + \frac{1}{2}\, D^2\, H(y^*)(v^i - y)^2 \right] \\ &= \nabla H(y)(y' - y) + \sum_{i=1}^{3} \lambda_i(y')\, \frac{1}{2}\, D^2\, H(y^*)(v^i - y)^2. \end{aligned} \tag{2.2}$$

With the estimate

$$\left|\frac{1}{2}D^2 H(y^*)(v^i-y)^2\right| = \left|\frac{1}{2}|v^i-y|^2 D^2 H(y^*)\frac{(v^i-y)^2}{|v^i-y|^2}\right|$$
$$\leq \frac{h^2}{2}\max_{1\leq i,j\leq 3}\left\{\max_{x\in\Omega}\left|\frac{\partial^2 H}{\partial y_i\,\partial y_j}(x)\right|\right\}$$

because of $|v^i - y| \leq h$, we obtain from equation (2.2) the second order estimate

$$|\nabla H(y)(y'-y)| \leq \left|\sum_{i=1}^{3}\lambda_i(y')\frac{h^2}{2}\max_{1\leq i,j\leq 3}\left\{\max_{x\in\Omega}\left|\frac{\partial^2 H}{\partial y_i\,\partial y_j}(x)\right|\right\}\right|$$
$$\leq \frac{h^2}{2}\max_{1\leq i,j\leq 3}\left\{\max_{x\in\Omega}\left|\frac{\partial^2 H}{\partial y_i\,\partial y_j}(x)\right|\right\} =: \frac{h^2}{2}M. \quad (2.3)$$

Since $\nabla H(y)$ is parallel to $(y-y')$, i. e. $\nabla H(y) = c\,(y-y')$, it follows that

$$|\nabla H(y)(y'-y)| = |\nabla H(y)|\,|y'-y|.$$

Furthermore, since $\nabla H(y) \neq 0$ for all $y \in \Omega$, equation (2.3) implies

$$|y'-y| \leq \frac{1}{|\nabla H(y)|}\frac{h^2}{2}M =: h^2 M^*. \quad (2.4)$$

Hence, we have shown that the polyhedron Γ_h converges to the surface Γ : $H(y) = 0$ for $h \to 0$ quadratically.

3 An adaptive method for the computation of surface integrals

Our aim is to calculate integrals of the form

$$\int_\Gamma \varphi\,d\sigma \quad \text{and} \quad \int_\Gamma \frac{\varphi(x)}{|x-x^0|}\,d\sigma.$$

First, a mesh $\mathcal{N} \subset \Gamma$ is generated. Then a small positive number $\delta > 0$ is chosen as a threshold which is given by the overall accuracy required. The mesh $\mathcal{N}$ corresponds to a polyhedron $\mathcal{P} := \{\Delta_i\}_{1\leq i\leq M}$, where the Δ_i, $1 \leq i \leq M$ are created by the grid points of $\mathcal{N}$. The corner points of the polyhedron $\mathcal{P}$ are located on Γ. We use once more the projector $P_\Gamma(x) := x + \lambda(x)\nabla H(x)$, which is described in section 2 Let

$$\gamma_i := P_\Gamma(\Delta_i) \text{ for } 1 \leq i \leq M.$$

The $\gamma_i \subset \Gamma$ are obviously curved triangles. If the mesh $\mathcal{N} \subset \Gamma$ is sufficiently fine, then the family $\{\gamma_i\}_{1\leq i\leq M}$ becomes a triangulation of the surface Γ, i. e. we may require the following conditions:

1. For all $1 \leq i, j \leq M$ with $i \neq j$, the intersection $\gamma_i \cap \gamma_j$ is either an empty set or a common line or a common corner point, and
2. $\Gamma \subset \bigcup_{i=1}^{M} \gamma_i$.

Under these conditions, we have

$$\int_{\Gamma} \varphi \, d\sigma = \sum_{i=1}^{M} \int_{\gamma_i} \varphi \, d\sigma.$$

The integrals

$$I(\gamma_i) = \int_{\gamma_i} \varphi \, d\sigma$$

will be evaluated separately. Therefore we introduce a local coordinate transformation

$$y = Ax + b,$$

in accordance with figure 3.1

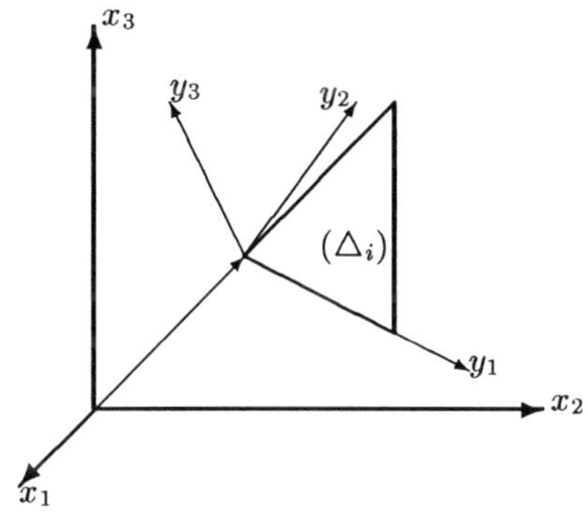

Fig. 3.1. transformation to local coordinates

whereby the triangle $\Delta_i = [x^1, x^2, x^3]$ is mapped into the coordinate plane (y_1, y_2). For simplicity we denote the corner points of the triangle Δ_i by $x^1, x^2, x^3 \in \mathcal{N}$. We further denote the transformed function of H by H as well. Let

$$y^j := Ax^j + b \quad \text{for} \quad 1 \leq j \leq 3.$$

Finally we introduce the projector $Q : \mathbb{R}^3 \longrightarrow \mathbb{R}^2$ defined by

$$Q(y_1, y_2, y_3) := (y_1, y_2) \quad \text{for all} \quad (y_1, y_2, y_3) \in \mathbb{R}^3.$$

Now let

$$\omega_i := Q(\gamma_i),$$

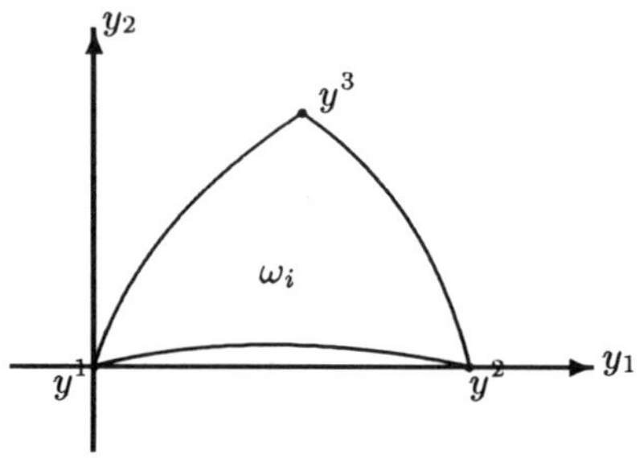

Fig. 3.2. projection of γ_i onto the plane (y_1, y_2)

i. e. ω_i is the projection of γ_i into the plane (y_1, y_2) as depicted in figure 3.2.

Corresponding to the local transformation, the integral $I(\gamma_i)$ is transformed as well and replaced by:

$$I(\gamma_i) = \int_{\omega_i} \psi(y_1, y_2)\, dy_1\, dy_2 \approx I_1(\gamma_i) := \frac{\psi(y^1) + \psi(y^2) + \psi(y^3)}{3} |\Delta_i|,$$

where $|\Delta_i|$ is the measure of the triangle Δ_i. The next, second approximation $I_2(\gamma_1)$ will be computed according to the following refinement procedure:

First, introduce four auxiliary triangles

$$\Delta_{i,k_1} := \left[y^{1;k_1}, y^{2;k_1}, y^{3;k_1}\right], \; i \leq k_1 \leq 4,$$

where the corner points $y^{1;k_1}, y^{2;k_2}$ and $y^{3;k_3}$ are determined as follows:

1. For all $1 \leq k_1 \leq 3$:

$$\left.\begin{aligned} y^{1;k_1} &:= y^{k_1}, \\ y^{2;k_1} &:= Q \circ P_\Gamma\left(\tfrac{y^{k_1}+y^{\mu}}{2}\right) \text{ with } \mu = (k+1) \pmod 3, \\ y^{3;k_1} &:= Q \circ P_\Gamma\left(\tfrac{y^{k_1}+y^{\mu}}{2}\right) \text{ with } \mu = (k_1+2) \pmod 3, \end{aligned}\right\} \quad (3.1)$$

and

2.

$$\left.\begin{aligned} y^{1;4} &:= Q \circ P_\Gamma\left(\tfrac{y^2+y^3}{2}\right), \\ y^{2;4} &:= Q \circ P_\Gamma\left(\tfrac{y^3+y^4}{2}\right), \\ y^{3;4} &:= Q \circ P_\Gamma\left(\tfrac{y^1+y^2}{2}\right). \end{aligned}\right\} \quad (3.2)$$

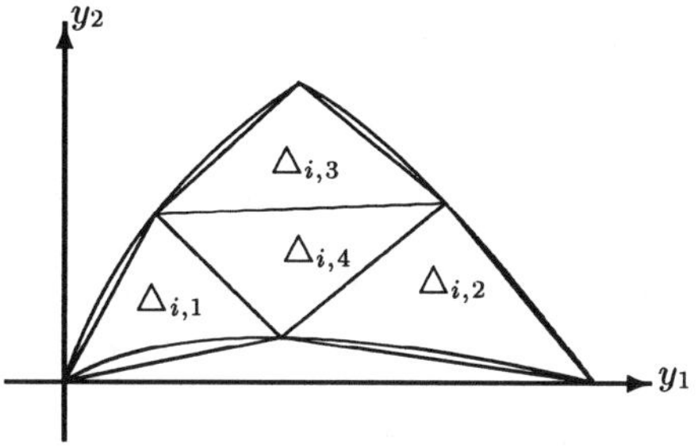

Fig. 3.3. refinement of ω_i

These four new triangles represent a triangulation of ω_i as shown in figure 3.3. We calculate

$$I_2(\Delta_{i,k_1}) := \frac{\psi(y^{1;k_1}) + \psi(y^{2;k_1}) + \psi(y^{3;k_1})}{3} |\Delta_{i,k_1}| \tag{3.3}$$

and obtain the second approximation of the integral $I(\gamma_i)$ by

$$I_2(\gamma_i) := \sum_{k_1=1}^{4} I_2(\Delta_{i,k_1}).$$

Next, we check the inequality

$$|I_1(\gamma_i) - I_2(\gamma_i)| \leq \delta$$

with the given tolerance $\delta > 0$. If this inequality is satisfied then we choose $I(\gamma_i) := I_2(\gamma_i)$, otherwise we compute the next approximation $I_3(\gamma_i)$ of $I(\gamma_i)$ and proceed with the next subdivision by replacing one triangle by four new ones, e.g. the triangle $\Delta_{i,1}$ by

$$\Delta_{i,1,k_2} = \left[y_1^{1,k_2}, y_2^{1,k_2}, y_3^{1,k_2}\right], \quad 1 \leq k_2 \leq 4.$$

The corner points $y^{1;1,k_2}, y^{2;1,k_2}, y^{3;1,k_2}$ are calculated by using (3.1) and (3.2), after replacing y^1, y^2, y^3 by $y^{1;1}$, $y^{2;1}$, $y^{3;1}$ and $y^{1;k_1}$, $y^{2;k_1}$, $y^{3;k_1}$ by $y^{1;1,k_2}$, $y^{2;1,k_2}$, $y^{4;1,k_2}$.

Then one computes the numbers $I_3(\Delta_{i,1,k_2})$ with the help of formula (3.3) by modifying the indices correspondingly and obtain

$$I_3(\gamma_i) = \sum_{k_2=1}^{4} I_3(\Delta_{i,1,k_2}) + \sum_{k_1=2}^{4} I_2(\Delta_{i,k_1}).$$

Now we check the inequality

$$|I_2(\gamma_i) - I_3(\gamma_i)| \leq \delta.$$

The further steps are obvious.

It can be shown that the error is of the order $\mathcal{O}\left(\frac{h^4}{4^{n_\sigma+1}}\right)$ if $\varphi \in C^2$ and of the order $\mathcal{O}\left(\frac{h}{4^{n_\sigma+1}}\right)$, if the function φ possesses a weak singularity. n_σ denotes the degree of refinement.

With this algorithm, the integrals appearing in boundary element methods can be computed numerically. Figure 3.4 shows the needed computational time depending on the number of original triangles on the surface Γ and figure 3.5 shows the accuracy of the computation of the double layer potential on an ellipsoid where the singularity is situated on the surface.

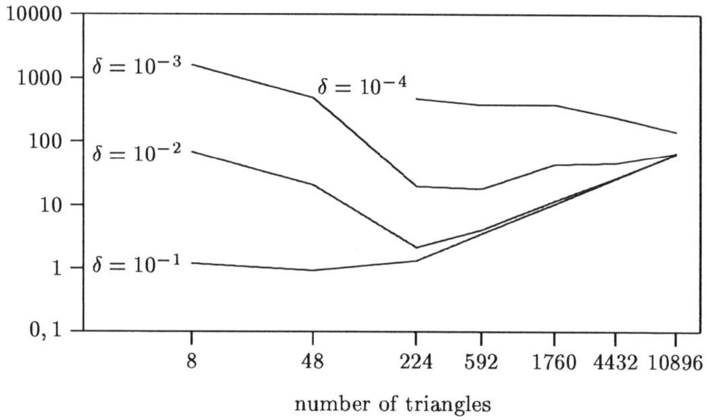

Fig. 3.4. CPU-time for the computation of the double layer potential on an ellipsoid with the singular point situated on the surface

4 Boundary elements of the class $\mathcal{C}^1$ and finite-element bases of the class $\mathcal{C}^1$.

Again, we consider an implicitly given surface $\Gamma := \{x \in \mathbb{R}^3 \mid H(x) = 0\}$ and a mesh $\mathcal{N} = \{x^i\}_{1 \leq i \leq N} \subset \Gamma$. We assume that $\mathcal{N}$ is dense enough in order to determine a triangulation of Γ which approximates Γ well enough. Let $1 \leq i \leq N$ be a fixed index.

We introduce local coordinates (y_1, y_2, y_3) with respect to $x^i \in \mathcal{N}$ by the transformation

$$y = A(x - x^i), \quad \text{for} \quad x \in \mathbb{R}^3,$$

where the matrix A is chosen such that the plane (y_1, y_2) coincides with the tangential plane of Γ in x^i. A is a rotation and $A^{-1} = A^*$.

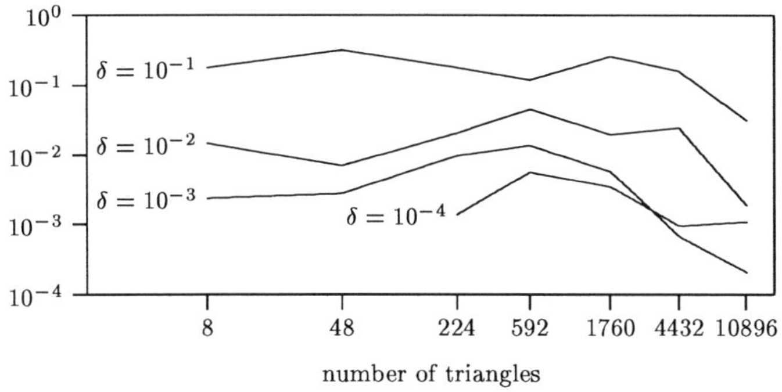

Fig. 3.5. Accuracy for the computation of the double layer potential on an ellipsoid with the singular point situated on the surface

Let $\mathcal{N}_i := \{x^{i+1}, \ldots, x^{i+p}\} \subset \mathcal{N}$ be the neighboring points of x^i given in local coordinates by

$$(y_1^j, y_2^j, y_3^j) := A(x^{i+j} - x^i) \quad \text{for} \quad 1 \leq j \leq p$$

where

$$y^j := (y_1^j, y_2^j)$$

denotes the projection of the grid point $(y_1^j, y_2^j, y_3^j) \in \Gamma$ onto the tangential plane (y_1, y_2). Without loss of generality we assume $y^0 := (0,0)$.

Now we consider the triangles

$$\Delta_j := [y^0, y^j, y^{j+1}] \text{ for } 1 \leq j \leq p-1 \text{ and } \Delta_p := [y^0, y^p, y^1]$$

as shown in figure 4.1, together with the closed region $\overline{\omega_i} := \bigcup_{j=1}^{p} \overline{\Delta_j}$.

Now we are in the position to formulate and to determine the desired interpolation.

For simplicity, we denote the function H after the transformation $x \longmapsto y$ again by the symbol H. Let

1. $h : \overline{\omega_i} \longrightarrow \mathbb{R}$ be the solution of

$$H\left(y_1, y_2, h(y_1, y_2)\right) = 0 \quad \text{for all} \quad (y_1, y_2) \in \overline{\omega_i}$$

and

2. $\gamma_j := \left\{(y_1, y_2, y_3) \in \mathbb{R}^3 \;\middle|\; (y_1, y_2) \in \Delta_j, \text{ for } 1 \leq j \leq p \text{ and } y_3 = h(y_1, y_2)\right\}.$

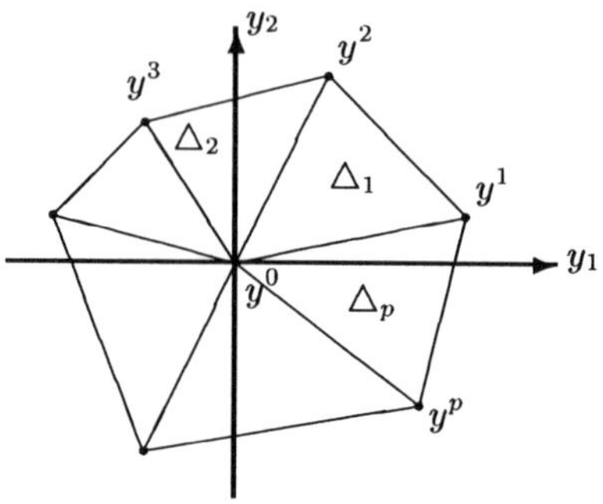

Fig. 4.1. closed region $\overline{\omega_i} := \bigcup_{j=1}^{p} \overline{\Delta_j}$

We assume that the mesh is already dense enough such that the condition $\Gamma \subset \bigcup_{i=1}^{N} \left(\bigcup_{j=1}^{p} \gamma_j \right)$ is satisfied.

Now we formulate the following interpolation problem: Let

$$\left\{ p_{j,\alpha} \in \mathbb{R} \mid 0 \leq j \leq p, \quad \text{and} \quad \alpha \in \mathbb{N}_0^2 \quad \text{with} \quad |\alpha| \leq 1 \right\}$$

be given. Then we have to find the function $w_i : \overline{\omega_i} \longrightarrow \mathbb{R}$ with the following properties:

1. $w_i \in C^1(\overline{\omega_i})$,
2. $\partial^\alpha \, w_i(y_j) = p_{j,\alpha}$ for $0 \leq j \leq p$ and for all $\alpha \in \mathbb{N}_0^2$ with $|\alpha| \leq 1$,
3. w_i consists of piecewise polynomials of degree at most three.

We solve this interpolation problem by determining w_i on each triangle Δ_i separately. For doing so, we choose one triangle, e. g. Δ_1, and determine w_i on Δ_1 with the help of a Clough-Toucher element.

By the aim of the center of gravity

$$z := \frac{y^0 + y^1 + y^2}{3}$$

we obtain three triangles

$$\Delta_{1,0} = \left[z, y^1, y^2 \right], \quad \Delta_{1,1} = \left[y^0, z, y^2 \right], \quad \text{and} \quad \Delta_{1,2} = \left[y^0, y^1, z \right]$$

as shown in figure 4.2.

Now we introduce barycentric coordinates with respect to these triangles:

1. $\lambda = (\lambda_1, \lambda_2, \lambda_3)$ for $\Delta_{1,0}$ with $y^1 = (1,0,0)$, $y^2 = (0,1,0)$, $z = (0,0,1)$;
2. $\mu = (\mu_1, \mu_2, \mu_3)$ for $\Delta_{1,1}$ with $y^0 = (1,0,0)$, $z = (0,1,0)$, $y^2 = (0,0,1)$;
3. $\nu = (\nu_1, \nu_2, \nu_3)$ for $\Delta_{1,2}$ with $y^0 = (1,0,0)$, $y^1 = (0,1,0)$, $z = (0,0,1)$.

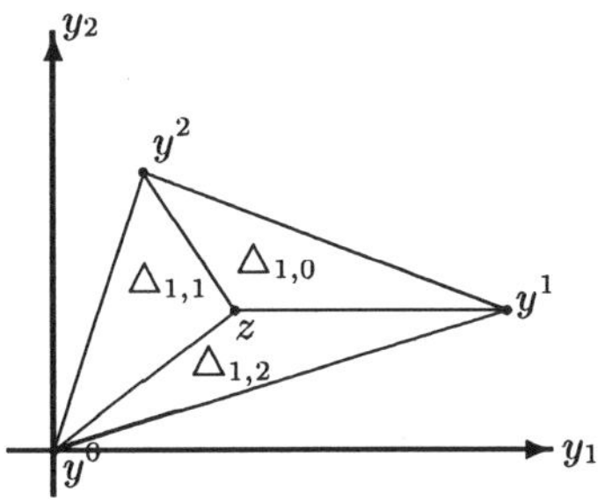

Fig. 4.2. subdivision of Δ_1 by the center of gravity

and determine corresponding Bézier polynomials

$$p_0(\lambda_1, \lambda_2, \lambda_3) = \sum_{\alpha_1+\alpha_2+\alpha_3=3} \frac{3!}{\alpha_1!\alpha_2!\alpha_3!} a_\alpha \lambda_1^{\alpha_1} \cdot \lambda_2^{\alpha_2} \cdot \lambda_3^{\alpha_3},$$

$$p_1(\mu_1, \mu_2, \mu_3) = \sum_{\alpha_1+\alpha_2+\alpha_3=3} \frac{3!}{\alpha_1!\alpha_2!\alpha_3!} b_\alpha \mu_1^{\alpha_1} \cdot \mu_2^{\alpha_2} \cdot \mu_3^{\alpha_3},$$

$$p_2(\nu_1, \nu_2, \nu_3) = \sum_{\alpha_1+\alpha_2+\alpha_3=3} \frac{3!}{\alpha_1!\alpha_2!\alpha_3!} c_\alpha \nu_1^{\alpha_1} \cdot \nu_2^{\alpha_2} \cdot \nu_3^{\alpha_3}.$$

Figure 4.3 illustrates the calculcation of the coefficients a_α, b_α and c_α which are given by the following formulae. The $r_{j,\alpha}$ are parameters which are to be chosen later.

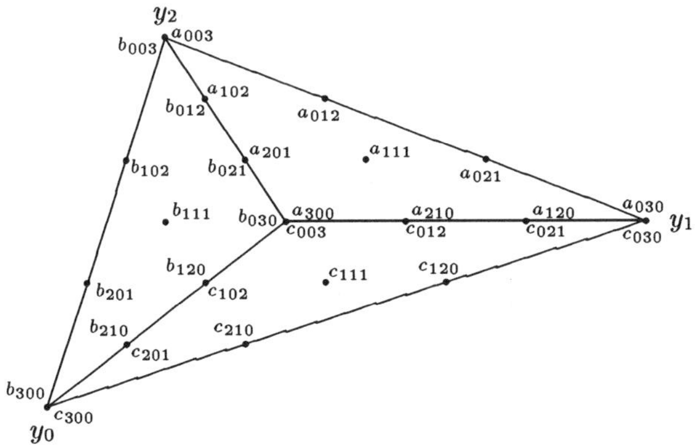

Fig. 4.3. Bézier-control-points

1. $a_{003} = b_{003} = r_{2,(0,0)}, \quad b_{300} = c_{300} = r_{0,(0,0)}, \quad c_{030} = a_{030} = r_{1,(0,0)},$

2. $$a_{021} = r_{1,(0,0)} + \frac{1}{3}\left[r_{1,(1,0)} \cdot (y_{1,2} - y_{1,1}) + r_{1,(0,1)} \cdot (y_{2,2} - y_{1,2})\right],$$

$$a_{012} = r_{2,(0,0)} + \frac{1}{3}\left[r_{2,(1,0)} \cdot (y_{1,1} - y_{2,1}) + r_{2,(0,1)} \cdot (y_{1,2} - y_{2,2})\right],$$

$$b_{102} = r_{2,(0,0)} + \frac{1}{3}\left[r_{2,(1,0)} \cdot y_{2,1} + r_{2,(0,1)} \cdot y_{2,2}\right],$$

$$b_{201} = r_{0,(0,0)} + \frac{1}{3}\left[r_{0,(1,0)} \cdot y_{2,1} + r_{0,(0,1)} \cdot y_{2,2}\right],$$

$$c_{210} = r_{0,(0,0)} + \frac{1}{3}\left[r_{0,(1,0)} \cdot y_{1,1} + r_{0,(0,1)} \cdot y_{1,2}\right],$$

$$c_{120} = r_{1,(0,0)} - \frac{1}{3}\left[r_{1,(1,0)} \cdot y_{1,1} + r_{1,(0,1)} \cdot y_{1,2}\right],$$

3. $$a_{120} = c_{021} = \frac{a_{021} + r_{1,(0,0)} + c_{120}}{3},$$

$$a_{102} = b_{012} = \frac{a_{012} + r_{2,(0,0)} + b_{102}}{3},$$

$$b_{210} = c_{201} = \frac{b_{201} + r_{0,(0,0)} + c_{210}}{3},$$

4. $$a_{111} = \frac{a_{021} + a_{012} + a_{102} + a_{120}}{2} - \frac{r_{1,(0,0)} + r_{2,(0,0)}}{4},$$

$$b_{111} = \frac{b_{102} + b_{201} + b_{210} + b_{012}}{2} - \frac{r_{2,(0,0)} + r_{0,(0,0)}}{4},$$

$$c_{111} = \frac{c_{210} + c_{120} + c_{021} + c_{201}}{2} - \frac{r_{0,(0,0)} + r_{1,(0,0)}}{4},$$

5. $$a_{210} = c_{012} = \frac{c_{111} + c_{021} + a_{111}}{3},$$

$$a_{201} = c_{021} = \frac{a_{111} + a_{102} + b_{111}}{3},$$

$$b_{120} = c_{102} = \frac{b_{111} + b_{210} + c_{111}}{3},$$

6. $$a_{300} = b_{030} = c_{003} = \frac{b_{021} + b_{120} + a_{210}}{3}.$$

In the same way one determines the corresponding polynomials p_0, p_1 and p_2 for the triangles $\Delta_2, \dots, \Delta_p$. The desired interpolating function is finally given by

$$w_i(y) := \begin{cases} p_0(\lambda) & \text{for all } \ y \in \Delta_{j,0} \ \text{ with } \ 1 \le j \le p, \\ p_1(\mu) & \text{for all } \ y \in \Delta_{j,1} \ \text{ with } \ 1 \le j \le p, \\ p_2(\nu) & \text{for all } \ y \in \Delta_{j,2} \ \text{ with } \ 1 \le j \le p. \end{cases}$$

Now let $\Gamma_i := \bigcup\limits_{j=1}^{p} \gamma_j$ and

$$W_i(y_1, y_2, y_3) := w_i(y_1, y_2) \ \text{ for all } \ (y_1, y_2) \in \overline{\omega}_i, \quad y_3 = h(y_1, y_2).$$

With the help of these functions, both, boundary elements of the class C^1 and finite-element-bases of the class C^1 can be generated.

First we investigate the boundary elements. Here we define

$$r_{j,\alpha} := \partial^\alpha h(y^j), \ \text{ for } \ 0 \le j \le p, \ \text{ and } \ \alpha \in \mathbb{N}_0^2 \ \text{ with } \ |\alpha| \le 1, \tag{4.1}$$

and

$$\sigma_j := \left\{ (y_1, y_2, y_3) \in \mathbb{R}^3 \mid (y_1, y_2) \in \Delta \ \text{ and } \ y_3 = w_i(y_1, y_2) \right\}.$$

This set of surface patches σ_j generates a set of boundary elements of the class C^1 because of the following properties:

- For $j_1 \neq j_2$, the set $\sigma_{j_1} \cap \sigma_{j_2}$ is either empty or a common line or a common corner point,
- the crossing of two arbitrary adjacent elements is continuous and their tangential planes coincide on the crossing line,
- the corner points of any element σ_i are situated on Γ and the tangential planes of Γ and σ_j coincide at the corner point.

In order to obtain the basis functions for the finite element spaces we solve the interpolational problem for all $\beta \in \mathbb{N}_0^2$ with $|\beta| \le 1$ and

$$r_{j,\alpha}^{\beta} := \begin{cases} \delta_{\alpha,\beta} & \text{if } j = 0 \\ 0 & \text{if } 1 \le j \le p \end{cases} \qquad \text{for all } \ \alpha \in \mathbb{N}_0^2 \ \text{ with } \ |\alpha| \le 1.$$

With the help of the determined interpolation $w_{i,\beta} : \overline{\omega}_i \longrightarrow \mathbb{R}$ we generate the functions

$$W_{i,\beta}(y_1, y_2, y_3) := \begin{cases} w_{i,\beta}(y_1, y_2) & \text{for all } (y_1, y_2) \in \overline{\omega}_i \ \text{ and } \ y_3 = h(y_1, y_2) \\ 0 & \text{for all } (y_1, y_2, y_3) \in \Gamma \backslash \Gamma_i. \end{cases}$$

The functions $W_{i,\beta}$ are defined everywhere on Γ belonging to the class C^1 on Γ; and they are linearly independent. The linear span of these functions is the desired finite element space on Γ called boundary elements.

5 On Hadamard's regularization of derivatives of potentials

We consider a singular integral of the form

$$\int_{\Gamma_i} D_\mu E(x, x^i) W_j(x)\, d\sigma_x \tag{5.1}$$

where $E(x, x^i)$ is the fundamental solution of an elliptic equation, $\mu = (\mu_1, \mu_2, \mu_3)$ is a given direction at the point $x \in \Gamma$ and

$$D_\mu := \mu_1 \frac{\partial}{\partial x_1} + \mu_2 \frac{\partial}{\partial x_2} + \mu_3 \frac{\partial}{\partial x_3},$$

for $x^i \in \mathrm{N} \subset \Gamma$ and $\Gamma_i = \bigcup\limits_{j=1}^{p} \gamma_j$. As a model we consider

$$E(x, x^i) = \frac{1}{|x - x^i|},$$

where the regularization can be performed explicitly. With the help of the coordinate-transformation $y = A \cdot (x - x^i)$ from the previous section, one obtains

$$D_\mu \frac{1}{|x - x_i|} = D_\nu \frac{1}{\sqrt{y_1^2 + y_2^2 + y_3^2}} = -\frac{\nu_1 \cdot y_1 + \nu_2 \cdot y_2 + \nu_3 \cdot y_3}{(y_1^2 + y_2^2 + y_3^2)^{\frac{3}{2}}}$$

and

$$d\sigma_x = \sqrt{1 + \left(\frac{\partial h}{\partial y_1}\right)^2 + \left(\frac{\partial h}{\partial y_2}\right)^2}\, dy_1\, dy_2.$$

Let $\varrho := \sqrt{y_1^2 + y_2^2}$ then the function h satisfies

$$h(0,0) = 0, \quad \frac{\partial h}{\partial y_1}(0,0) = 0, \quad \text{and } \frac{\partial h}{\partial y_2}(0,0) = 0;$$

hence h is of the form

$$h(y_1, y) = \varrho^2 \tilde{h}(y_1, y_2)$$

with a smooth function $\tilde{h}$. This yields

$$(y_1^2 + y_2^2 + y_3^2)^{\frac{3}{2}} = \varrho^3 \left(1 + \varrho^2 \tilde{h}(y_1, y_2)\right)^{\frac{3}{2}}$$

and

$$D_\mu \frac{1}{|x - x^i|} = -\left(\nu_1 \frac{y_1}{\varrho^3} + \nu_2 \frac{y_2}{\varrho^3} + \nu_3 \frac{\tilde{h}}{\varrho}\right) \frac{1}{(1 + \varrho^2 \tilde{h}^2)^{\frac{3}{2}}}.$$

With the notation

$$\varphi_j := \frac{\sqrt{1+\left(\frac{\partial h}{\partial y_1}\right)^2+\left(\frac{\partial h}{\partial y_2}\right)^2}}{(1+\varrho^2\tilde{h}^2)^{\frac{3}{2}}} \cdot W_j,$$

we obtain

$$\int_{\Gamma_i} D_\mu \frac{1}{|x-x^i|} \cdot W_j(x)\, d\sigma_x$$
$$= -\int_{\omega_i} \left[\nu_1 \frac{y_1}{\varrho^3} + \nu_2 \frac{y_2}{\varrho^3}\right] \varphi_j\, dy_1 dy_2 - \int_{\omega_i} \nu_3 \frac{\tilde{h}}{\varrho}\, \varphi_j\, dy_1 dy_2,$$

which implies the formula

$$p.f. \int_{\omega_i} \left[\nu_1 \frac{y_1}{\varrho^3} + \nu_2 \frac{y_2}{\varrho^3}\right] \varphi_i\, dy_1 dy_2 = \int_{\omega_i} \left[\nu_1' \frac{1}{\varrho}\frac{\partial \varphi_i}{\partial y_1} + \nu_2 \frac{1}{\varrho}\frac{\partial \varphi_i}{\partial y_2}\right] dy_1 dy_2.$$

This formula describes nothing but Hadamard's regularization of the singular integral with the help of weakly singular integrals and hence makes its numerical computation possible. For more details on the numerical integration of Hadamard finite part integrals see also [4].

References

1. Allgower, E. L., Gnutzmann, S. (1987): An algorithm for piecewise linear approximation of implicitly defined two-dimensional surfaces. SIAM J. Numer. Anal. **24**, 425–469
2. Kalik, K., Wendland, W. L. (1992): The approximation of closed manifolds by triangulated manifolds and the triangulation of closed manifolds. Computing **47**, 255–275
3. Kalik, K., Quatember, R., Wendland, W. L. (1993): Load-Balancer unter Helios. Preprint 93–2, Universität Stuttgart
4. Kieser, R., Schwab, C., Wendland, W. L. (1992): Numerical evaluation of singular and finite-part surface integrals on curved surfaces using symbolic manipulation. Computing **49**, 279–301

Asymptotic Expansions of Elastic Fields in Domains with Boundary and Structural Singularities

A.–M. Sändig and W.L. Wendland

1 Introduction

It is well known that the solutions of elliptic boundary value problems in domains with corners, edges and interfaces develop local singularities at these geometrical and structural peculiarities. Based on V.A. Kondratiev's fundamental work [31], G. Fichera [23, 24], V.A. Kondratiev and O.A. Oleinik [33, 34], V.G. Maz'ya and B.A. Plamenevski [41, 44], M. Dauge [18], V. Kozlov, V.G. Maz'ya and C. Schwab [37], T.v.Petersdorff [53], H.Schmitz, K.Volk and W.L.Wendland [62] and M. Costabel [14] have analyzed the local asymptotic expansion for the elliptic system of elasticity. This local expansion can be obtained by using Mellin transformation with respect to the distance to boundary or structural singularities. The Mellin transformation yields pencils of differential operators depending on a complex parameter. The resulting holomorphic eigenvalue problem provides the basic information on the local asymptotic expansion determined by the distribution of the corresponding eigenvalues and the associated Jordan chains of generalized eigenfunctions.

Instead of treating the boundary value problem directly, one can also use equivalent pseudodifferential equations on the boundary. One reduces the problem to appropriate boundary integral equations for the missing Cauchy data. Using a Wiener–Hopf–technique or the Mellin transformation, one can obtain a local asymptotic expansion of the unknown Cauchy data about the boundary and structural singularities. In particular, this technique is used for cracks and interface cracks between different materials in [16, 20, 21] and [62]. The singular expansions are needed for criterions of crack propagation and its modelling on one hand and, on the other hand, for the compensation of the deterioration of numerical stress analysis due to the well-known global pollution effect of singularities. The improvement of any numerical stress computation can be achieved either by mesh grading based on the a priori knowledge of the form of the singularities [2, 3, 6, 55, 57, 54, 39] or by adaption. The latter is accomplished either by automatic local mesh refinement and increase of approximation order based on local error indicators [1] or via model adaption by incorporating explicit singular expansion terms into the trial spaces as in the augmented Fix method [10, 64]. All these techniques are used in finite difference, finite element and boundary element discretizations.

This paper is devoted to the analysis of three–dimensional elastic fields involving singularities and some related boundary and finite element methods. In Section 2 we derive boundary integral equations for three–dimensional crack and interface problems. We discuss the direct formulations, where the relations between the Cauchy data via the Calderon projection are used, as well as indirect methods, where the solutions are reproduced through chosen potentials with unknown densities. In Section 3 we present results on the asymptotic expansions of the unknown Cauchy data (stress or displacement fields) in the framework of boundary integral equations for domains with cracks, interface cracks and polyhedral corners [20, 21, 62]. In Section 4 we describe asymptotic expansions of the displacement fields in domains with edges or polyhedral corners. We formulate some new results on Neumann problems and give modified coefficient formulae in domains with edges [56]. In Section 5 we present the relations between the two asymptotic expansions, one generated by Mellin techniques for boundary integral equations and the other by Mellin techniques applied to the original Dirichlet problem as derived in [38]. For three-dimensional problems it can be shown that the Jordan chains of an associated one–dimensional boundary integral equation on some curve γ on the unit sphere is related to corresponding chains belonging to an operator pencil on the sphere, and their jumps across γ. Section 6 is concerned with mesh grading for h–versions of finite element domain methods based on a priori knowledge on the form of the singularities and asymptotic expansions about corners and edges [5]. This general method can be extended to boundary element methods as it is done in [54] for a mixed boundary value problem for the Laplacian in a polyhedral domain.

2 Boundary integral equations for three-dimensional crack and interface problems

Let Ω be a three–dimensional domain consisting of two domains Ω_1 and Ω_2 occupied by possibly different anisotropic materials where either both domains are bounded or Ω_2 is the unbounded exterior $\mathbb{R}^3 \setminus \overline{\Omega_1}$. The boundaries $\partial\Omega_i, i = 1, 2$, are divided into two parts; one part is the interface $\Gamma = \partial\Omega_1 \cap \partial\Omega_2$ with transmission conditions, the other part Γ_i consists of the boundary pieces where boundary conditions are imposed (see Fig. 2.1 and Fig. 2.2). If Ω_2 is unbounded, we assume that appropriate radiation conditions are satisfied at infinity. Cracks between two different materials, the so–called interface cracks as well as cracks in an elastic homogeneous body Ω of one material are included; the lips of the crack are supposed to belong to the boundary pieces Γ_i.

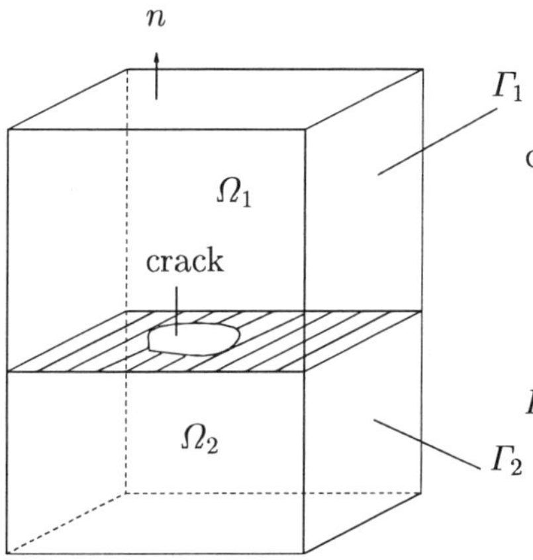

Fig. 2.1. A bounded coupled structure

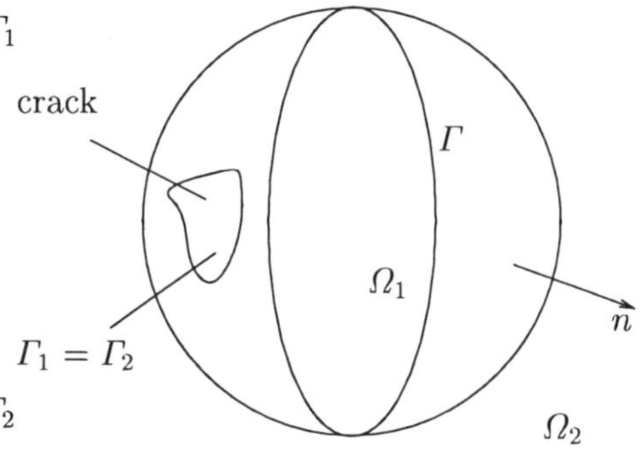

Fig. 2.2. A coupled structure with unbounded exterior

2.1 Crack problems

Let the domain $\Omega \subset \mathbb{R}^3$ describe a homogeneous anisotropic elastic body containing a crack surface S in the interior. The closure $\overline{S}$ can be represented as a manifold with a C^k–smooth boundary ∂S ($k \geq 3$). The crack-faces are denoted by $S^\pm$ and the corresponding restrictions of a function u by $u^\pm = u|_{S^\pm}$. The equations of linear elasticity read

$$Lu = D^\top C D u = 0 \qquad \text{in } \Omega, \tag{2.1}$$

where u is the vector–valued displacement field, C is the 6×6 symmetric, positive definite matrix defined by the 21 elastic moduli (see [23]) and

$$D = \begin{pmatrix} \partial_1 & 0 & 0 & 0 & \partial_3 & \partial_2 \\ 0 & \partial_2 & 0 & \partial_3 & 0 & \partial_1 \\ 0 & 0 & \partial_3 & \partial_2 & \partial_1 & 0 \end{pmatrix}^\top, \quad \partial_j = \frac{\partial}{\partial x_j}.$$

The Dirichlet and Neumann conditions for the crack read

$$u|_{S^\pm} = g^\pm \tag{2.2}$$

and

$$T(\partial_x, n(x))u|_{S^\pm} := N^\top C D u|_{S^\pm} := t^\pm, \tag{2.3}$$

respectively, where

$$N = \begin{pmatrix} n_1 & 0 & 0 & 0 & n_3 & n_2 \\ 0 & n_2 & 0 & n_3 & 0 & n_1 \\ 0 & 0 & n_3 & n_2 & n_1 & 0 \end{pmatrix}^\top .$$

For the isotropic case, the Dirichlet and Neumann cracks have been investigated in [16]. In the following we will present the results from [20]. For brevity we assume $\Omega = \mathbb{R}^3 \setminus \overline{S}$ and the radiation condition

$$u(x) = o(1) \quad \text{for } |x| \to \infty. \tag{2.4}$$

Theorem 2.1. *[20]: Let $g^\pm \in H_2^{\frac{1}{2}}(S)$ with $g_0 := (g^+ - g^-) \in \widetilde{H}_2^{\frac{1}{2}}(S)$ (see below). Then $u \in H^1_{loc}(\mathbb{R}^3\setminus\overline{S})$ is the solution of the Dirichlet crack problem (2.1), (2.2) and (2.4) if and only if*

$$u(z) = \mathbf{V}\varphi(z) - \mathbf{U}g_0(z) \quad \textit{for } z \in \mathbb{R}^3\setminus\overline{S},$$

where $\varphi \in \widetilde{H}_2^{-\frac{1}{2}}(S)$ is the solution to the pseudodifferential equation

$$V\varphi(x) = \left[\frac{1}{2}(g^+ + g^-) + K^* g_0\right](x) \textit{ for } x \in S. \tag{2.5}$$

Here $\widetilde{H}_2^{\frac{1}{2}}(S) := \{u \in H_2^{\frac{1}{2}}(\widetilde{S}) : \text{supp } u \in \overline{S}\}$, where $\widetilde{S}$ is a closed compact C^k–surface with $\overline{S} \Subset \widetilde{S}$ and $\widetilde{H_2}^{-\frac{1}{2}}(S) = \left[H_2^{\frac{1}{2}}(S)\right]'$. The condition $g_0 \in \widetilde{H}_2^{\frac{1}{2}}(S)$ ensures the compatibility between g^+ and g^- on ∂S.
Furthermore, we have used the notations

$$\begin{aligned}
\mathbf{V}\,\psi(z) &:= \int_S E(z-y)\psi(y)\,d_yS \quad \text{for } z \in \mathbb{R}^3\setminus\overline{S}, \\
\mathbf{U}\,\varphi(z) &:= \int_S [T(\partial_y, n(y))E(z-y)]^\top \varphi(y)\,d_yS \quad \text{for } z \in \mathbb{R}^3\setminus\overline{S}, \\
K^*\psi(x) &:= \int_S T(\partial_x, n(x))E(z-y)\psi(y)\,d_yS \quad \text{for } x \in S, \\
V\psi(x) &:= \int_S E(x-y)\psi(y)\,d_yS \quad \text{for } x \in S
\end{aligned}$$

with a fundamental solution E for the operator $-L$.

Theorem 2.2. *[20]: Let $t^\pm \in H_2^{-\frac{1}{2}}(S)$ with $t_0 := (t^+ - t^-) \in \widetilde{H}_2^{-\frac{1}{2}}(S)$. Then $u \in H^1_{loc}(\mathbb{R}^3\setminus\overline{S})$ is the variational solution of the Neumann crack problem (2.1), (2.3) and (2.4) if and only if*

$$u(z) = \mathbf{V}t_0(z) - \mathbf{U}\psi(z) \quad \textit{for } z \in \mathbb{R}^3\setminus\overline{S},$$

where $\psi \in \widetilde{H}_2^{\frac{1}{2}}(S)$ solves the pseudodifferential equation

$$D\psi(x) = \left[\frac{1}{2}(t^+ + t^-) + K^* t_0\right](x) \quad \textit{for } x \in S. \tag{2.6}$$

The hypersingular integral operator D is defined by

$$D\psi(x) = T(\partial_x, n(x)) \int_S [T(\partial_y, n(y))E(x-y)]^\top \psi(y)\, d_yS.$$

2.2 Interface problems

Let $\Omega = \Omega_1 \cup \Omega_2 \cup \Gamma$ be a composite as introduced before (see Fig. 2.1 and Fig. 2.2). The interface or boundary transmission problem then reads:
Find functions $u_i, i = 1, 2$, such that

$$\begin{aligned} L_i u_i &= 0 \quad \text{in } \Omega_i, \\ B_i u_i &= g_i \quad \text{on } \Gamma_i, \\ u_1 - u_2 &= h_1 \quad \text{on } \Gamma, \\ T_1 u_1 - T_2 u_2 &= h_2 \quad \text{on } \Gamma. \end{aligned} \tag{2.7}$$

The operators L_i and T_i are defined by (2.1) and (2.2); the boundary operators B_i describe Neumann or Dirichlet conditions on Γ_i. In short, we say that (2.7) is a **D–D interface problem** if on Γ_1 and Γ_2 the Dirichlet data g_i are given; (2.7) is a **N–N interface problem** if on Γ_1 and Γ_2 the Neumann data $g_i = t_i$ are given; and (2.7) is an **N–D interface problem** if on Γ_1 the Dirichlet datum and on Γ_2 the Neumann datum are given.
The derivation of the corresponding boundary integral equations for these problems requires a few particular steps which we sketch in the following:

(i) [21, 12, 58]
 - Consider extensions G_i of the boundary data g_i onto all of Γ.
 - Solve the decoupled boundary value problems and write the solution as a solution operator acting on G_i.
 - Determine the extensions G_i in such a way, that the transmission conditions on Γ are satisfied.

(ii) [47, 12, 19]
 Assume that the boundary pieces Γ_i coincide with the crack faces $S^{\pm}$, i.e. we consider interface cracks (see Fig. 2.2).
 - Consider extensions H_i of h_i onto $S = \Gamma \cup \partial\Gamma \cup \Gamma_1$ and identify $\Gamma_1 = \Gamma_2$.
 - Solve the "pure" transmission problem and write the solution as a solution operator operating on H_i.
 - Determine the extensions H_i in such a way, that the boundary conditions on Γ_i are satisfied.

Both solution procedures can be realized with direct methods [67, 65, 16, 58], where the relations between the Cauchy data, described by the Calderon projector, are used. They were also carried out with indirect methods [47,

12, 20, 29], where the solutions are represented by chosen potentials with unknown densities. We shall formulate results for both methods:

Interface Neumann cracks (direct method):
We present here an extension of the approach in [16, 65] to coupled structures, developed in [58]. Let us consider the domain $\Omega = \mathbb{R}^3 = \Omega_1 \cup \Omega_2 \cup S$ (see Fig. 2.2), where $S = \partial\Omega_1 = \Gamma \cup \partial\Gamma \cup \Gamma_1$ is the common C^μ–boundary consisting of the interface (or contact zone) Γ and the crack faces $\Gamma_1 = \Gamma_2$. As before, we introduce the potentials

$$\mathbf{V_i}\psi(z) = \int\limits_S E_i(z-y)\psi(y)\,d_yS, \quad \text{and}$$

$$\mathbf{U_i}\chi(z) = \int\limits_S \left[T(\partial_y, n(y))E_i(z-y)\right]^\top \chi(y)\,d_yS \quad \text{for } z \in \mathbb{R}^3\backslash\overline{S}.$$

E_i is the fundamental solution of the operator $-L_i, i = 1, 2$. The corresponding boundary integral operators (pseudodifferential operators) are denoted by

$$V_i\psi(x) = \int\limits_S E_i(x-y)\psi(y)\,d_yS,$$

$$K_i^*\psi(x) = \int\limits_S T_i(\partial_x, n(x))E_i(x-y)\psi(y)\,d_yS \quad \text{for } x \in S.$$

We use the approach (i) and consider some chosen extensions $\widehat{G}_i \in H^{-\frac{1}{2}}(S)$ of the Neumann data $g_i \in H^{-\frac{1}{2}}(\Gamma_i)$. Every further extension $G_i \in H^{-\frac{1}{2}}(S)$ of g_i can then be written as $G_i = \widehat{G}_i + \mathring{G}_i$ with $\mathring{G}_i \in \widetilde{H}^{-\frac{1}{2}}(\Gamma)$. Assume that the solvability conditions for the decoupled Neumann problem for the bounded domain Ω_1 is satisfied, i.e.

$$\left\langle \widehat{G}_1 + \mathring{G}_1, e_j \right\rangle := \int\limits_S (\widehat{G}_1 + \mathring{G}_1)e_j\,dS = 0 \quad \text{for } j = 1, \dots, 6, \tag{2.8}$$

where e_j are the traces of the linearly independent rigid motions.
Based on the concept (i) and the relations between the Cauchy data on $\partial\Omega_i$ which are given by the Calderon projections [17],

$$\begin{pmatrix} u_i \\ T_iu_i \end{pmatrix} = \begin{pmatrix} \frac{1}{2}I + (-1)^iK_i & (-1)^{i+1}V_i \\ (-1)^iD_i & \frac{1}{2}I + (-1)^{i+1}K_i^* \end{pmatrix} \begin{pmatrix} u_i \\ T_iu_i \end{pmatrix} \tag{2.9}$$

we define the corresponding Steklov–Poincaré operators

$$S_i = V_i^{-1}\left(\frac{(-1)^{i+1}}{2}\, I + K_i\right).$$

The inverse Poincaré–Steklov operator which maps the boundary tractions into the boundary displacements is well defined for the Neumann problem in the exterior domain Ω_2 satisfying (2.4) and with the projection properties of the Calderon projection one gets $S_2^{-1} = V_2(-\frac{1}{2}I + K_2^*)^{-1}$. For the bounded domain Ω_1 we use the pseudoinverse operator $S_1^{(-1)} = V_1(\frac{1}{2}I + K_1^*)^{(-1)}$ acting on functions satisfying (2.8) (see [21]).
If the exterior domain is replaced by a bounded domain Ω_2 (see Fig. 2.1), then we use the pseudoinverse $S_2^{(-1)}$, correspondingly.
For the Dirichlet, the Neumann and the mixed problem we have the following results obtained in [58]:

Theorem 2.3. *(D–D interface problem): Let* $\widehat{G}_i \in H^{\frac{1}{2}}(\partial\Omega_i), i = 1, 2$, *be some extensions of the Dirichlet data* g_i *of the D–D interface problem (2.7), which satisfy the compatibility condition*

$$h_1 - r|_\Gamma(\widehat{G}_1 - \widehat{G}_2) = \widehat{h}_1 \in \widetilde{H}^{\frac{1}{2}}(\Gamma).$$

Let $G_i = \widehat{G}_i + \mathring{G}_i$ *be the desired boundary trace with* $\mathring{G}_i \in \widetilde{H}^{\frac{1}{2}}(\Gamma)$. *Then*

$$u_i(z) = (-1)^{i+1}[\boldsymbol{V_i}(S_iG_i) - \boldsymbol{U_i}(G_i)](z) \quad \textit{for } z \in \Omega_i, i = 1, 2,$$

are the uniquely determined solutions $u_i \in H^1(\Omega_i)$ *of the D–D interface problem (2.7), if and only if the functions* $\mathring{G}_i$ *satisfy the following system of boundary pseudodifferential equations on the interface* Γ*:*

$$\begin{pmatrix} I & -I \\ r|_\Gamma S_1 & -r|_\Gamma S_2 \end{pmatrix} \begin{pmatrix} \mathring{G}_1 \\ \mathring{G}_2 \end{pmatrix} = \begin{pmatrix} h_1 - r|_\Gamma(\widehat{G}_1 - \widehat{G}_2) \\ h_2 - r|_\Gamma(S_1\widehat{G}_1 - S_2\widehat{G}_2) \end{pmatrix}.$$

Theorem 2.4. *(N–N interface problem): Let* $\widehat{G}_i \in H^{-\frac{1}{2}}(\partial\Omega_i)$ *be some extensions of the Neumann data* $g_i = t_i$ *of the N–N interface problem (2.7) which satisfy the solvability conditions*

$$\left\langle \widehat{G}_i, e_j \right\rangle_{\partial\Omega_i} = 0, \quad \textit{for } j = 1, \ldots, 6, \quad i = 1, 2, \tag{2.10}$$

where e_j *are the traces of the rigid motions of the form* $[a \times x] + b$ *with* $a, b \in \mathbb{R}^3$. *Let* $G_i = \widehat{G}_i + \mathring{G}_i \in H^{-\frac{1}{2}}(\partial\Omega_i)$ *be the desired boundary stresses satisfying (2.10). Then*

$$u_i(z) = (-1)^{i+1}[\boldsymbol{V_i}(G_i) - \boldsymbol{U_i}(S_i^{(-1)}G_i)](z), \quad i = 1, 2$$

are solutions $u_i \in H^1(\Omega_i)$ *of the N–N interface problem (2.7), if and only if* $\mathring{G}_i$ *are solutions of the system of boundary pseudodifferential equations on the interface* Γ*:*

$$\begin{pmatrix} r|_\Gamma S_1^{(-1)} & -r|_\Gamma S_2^{(-1)} \\ r|_\Gamma & -r|_\Gamma \end{pmatrix} \begin{pmatrix} \mathring{G}_1 \\ \mathring{G}_2 \end{pmatrix} = \begin{pmatrix} h_1 - r|_\Gamma(S_1^{(-1)}\widehat{G}_1 - S_2^{(-1)}\widehat{G}_2 \\ h_2 - r|_\Gamma(\widehat{G}_1 - \widehat{G}_2) \end{pmatrix}.$$

Theorem 2.5. *(N–D interface problem): Let* $\widehat{G}_1 \in H^{-\frac{1}{2}}(\partial\Omega_i)$ *and* $\widehat{G}_2 \in H^{\frac{1}{2}}(\partial\Omega_i)$ *be some extensions of the Neumann datum* $g_1 = t_1$ *and of the Dirichlet datum* g_2*, respectively, so that (2.10) holds for* $i = 1$*. Let* $G_i = \widehat{G}_i + \overset{\circ}{G}_i \in H^{i-1-\frac{1}{2}}(\partial\Omega_i)$ *be any other extensions of* g_i *where* G_1 *also satisfies (2.10). Then*

$$u_1(x) = [\boldsymbol{V_1}(G_1) - \boldsymbol{U_1}(S_1^{(-1)}(G_1)](x),$$

$$u_2(x) = -[\boldsymbol{V_2}(S_2G_2) - \boldsymbol{U_2}(G_2)](x)$$

are the solutions $u_i \in H^1(\Omega_i)$ *of the N–D interface problem (2.7), if and only if* $\overset{\circ}{G}_i$ *are the solutions of the system of boundary pseudodifferential equations on the interface* Γ

$$\begin{pmatrix} r|_\Gamma S_1^{(-1)} & -r|_\Gamma \\ r|_\Gamma & -r|_\Gamma S_2 \end{pmatrix} \begin{pmatrix} \overset{\circ}{G}_1 \\ \overset{\circ}{G}_2 \end{pmatrix} = \begin{pmatrix} \widehat{h}_1 \\ \widehat{h}_2 \end{pmatrix}.$$

Interface Neumann cracks (indirect method) [21]:
In the indirect approach we seek the solution u_i as a single layer potential of the Neumann interface crack problem (2.7) in the form

$$u_i(z) = \mathbf{V_i}\psi_i(z) \quad \text{for } z \in \Omega_i, \tag{2.11}$$

where ψ_i on S are still to be determined. Here one has the following result.

Theorem 2.6. *: The function* $u \in H^1_{loc}(\mathbb{R}^3 \setminus \overline{S})$ *given by*

$$u(z) = \begin{cases} \mathbf{V_1}\left[\sum_{j=1}^{6} c_j V_1^{-1} e_j + (\frac{1}{2}I + K_1^*)^{(-1)}(\widehat{G}_1 + \overset{\circ}{G}_1)\right](z), & \text{for } z \in \Omega_1 \\ \mathbf{V_2}\left[(-\frac{1}{2}I + K_2^*)^{-1}(\widehat{G}_2 + \overset{\circ}{G}_2)\right](z), & \text{for } z \in \Omega_2 \end{cases}$$

is a solution of the Neumann interface crack problem (2.7), if

$$\widetilde{G} = (\overset{\circ}{G}_1, c_1, \ldots, c_6) \in \widetilde{H}^{-\frac{1}{2}}(\Gamma) \times \mathbb{R}^6$$

is a solution of the system of pseudodifferential equations

$$\widetilde{S}^{-1}\widetilde{G} = \widetilde{f} = (f, d_1, \ldots, d_6), \tag{2.12}$$

provided the conditions (2.8) together with the compatibility conditions

$$h_1 - \left(S_1^{(-1)}\widehat{G}_1 - S_2^{-1}\widehat{G}_2\right)|_\Gamma = \widehat{h}_1 \in \widetilde{H}_2^{\frac{1}{2}}(\Gamma),$$

$$h_2 - \left(\widehat{G}_1 - \widehat{G}_2)\right)|_\Gamma = \widehat{h}_2 \in \widetilde{H}_2^{-\frac{1}{2}}(\Gamma)$$

are satisfied. Here

$$\widetilde{S}^{-1} = \begin{pmatrix} r|_\Gamma(S_1^{(-1)} - S_2^{-1}) & A^* \\ A & 0 \end{pmatrix} \tag{2.13}$$

is a matrix of operators, where $A^*(\{c_1, \ldots, c_6\}) = \sum_{j=1}^6 c_j e_j$, $A\mathring{G}_1 = \left\{\left\langle \mathring{G}_1, e_j \right\rangle_S\right\}_{j=1,\ldots,6}$, $f = \widehat{h}_1 - r|_\Gamma S_2^{(-1)}\widehat{h}_2$, $d_i = -\left\langle \widehat{G}_1, e_j \right\rangle_S$ *and* $\mathring{G}_2 = \mathring{G}_1 - \widehat{h}_2$. *The restriction or trace on* Γ *is denoted by* $r|_\Gamma$.

3 Asymptotic expansions on the boundary

Here we present results on asymptotic expansions of the Cauchy data (stress or displacement fields) near corners and edges (crack fronts included) in the framework of integral equations. Two approaches are applied, the Wiener–Hopf method [22, 16, 20, 21] and Mellin techniques [62].

3.1 Crack problems

In [22, Section 9], the Wiener–Hopf method is adapted to the investigation of scalar multi–dimensional pseudodifferential equations in the half space $\mathbb{R}^n_+$. This method is generalized to systems of pseudodifferential equations in [22, Section 26] and [9]. The Wiener–Hopf technique is used by M. Costabel and E. P. Stephan in [16] in order to get the explicit asymptotic expansions for the displacement and traction fields at the crack surface in three–dimensional homogeneous isotropic bodies. They have derived the explicit factorization of 3×3 matrix symbols of the corresponding integral equations. These results were generalized in [20] to cracks in homogeneous anisotropic materials by using an implicit factorization of symbols. In [21] interface cracks between two anisotropic materials are investigated. In this case the corresponding symbols of $\widetilde{S}^{-1}$ (see 2.13) are not explicitly factorizable and do not have the generalized transmission property anymore. The explicit form of the leading exponents of the asymptotic expansion of the surface traction near the crack-front is then derived by suitable adaption of the Wiener–Hopf method. These results read as follows.

Theorem 3.1. *(Dirichlet crack) [20]: Assume that the right–hand sides of the Dirichlet crack problem (2.1), (2.2) and (2.4),* $g^\pm$ *belong to* $H_2^{\frac{3}{2}+\sigma}(S)$, $-\frac{1}{2} < \sigma < \frac{1}{2}$. *Then the boundary traction field* τ *is the solution of the boundary pseudodifferential equation (2.5) and has the following asymptotic expansion:*

$$\tau(s,r) = \eta(r)\tau_0(s)r^{-\frac{1}{2}} + \tau_1(s,r),$$

where $r = dist(x, \partial S)$ *and* s *denotes the arclength on* ∂S, $\tau_0 \in H^{\frac{1}{2}+\sigma}(\partial S)$, $\tau_1 \in \widetilde{H}_2^{\frac{1}{2}+\sigma}(S)$. *The function* η *is an appropriate cut–off function.*

Theorem 3.2. *(Neumann crack) [20]: Assume that the right-hand sides $t^\pm$ of the Neumann crack problem (2.1), (2.3) and (2.4) belong to $H_2^{\frac{1}{2}+\sigma}(S)$ with $-\frac{1}{2} < \sigma < \frac{1}{2}$. Then the boundary displacement field ψ is the solution of the boundary pseudodifferential equation (2.6) and has the following asymptotic expansion:*

$$\psi(s,r) = \eta(r)\psi_0(s)r^{\frac{1}{2}} + \psi_1(s,r),$$

where $\psi_0 \in H_2^{\frac{3}{2}+\sigma}(\partial S)$ and $\psi_1 \in \widetilde{H}_2^{\frac{3}{2}+\sigma}(S)$.

Theorem 3.3. *(Neumann interface crack) [21]: Assume that the right-hand sides of the Neumann interface crack problem $g_i = t_i \in H_p^{\frac{1}{p}+\sigma}(\Gamma_i)$ and $h_i \in H_p^{\frac{1}{p}+2-i+\sigma}(\Gamma)$ with $-\frac{1}{2} < \sigma < \frac{1}{2}$. Then $f \in H_p^{\frac{1}{p}+1+\sigma}(\Gamma)$ in (2.12) and the solution $\mathring{G}_1$ of (2.12) has the asymptotic expansion*

$$\mathring{G}_1(s,r) = \eta(s)K(s)\begin{pmatrix} r^{\frac{1}{2}} & 0 & 0 \\ 0 & r^{\frac{1}{2}+i\delta} & 0 \\ 0 & 0 & r^{\frac{1}{2}-i\delta} \end{pmatrix} K^{-1}(s)\begin{pmatrix} c_1(s) \\ c_2(s) \\ c_3(s) \end{pmatrix} + W(s,r)\,, \tag{3.1}$$

where the coefficients in the first term depend on the arclength s and belong to $H_p^{\frac{1}{p}+\sigma'}(\partial\Gamma)$ and $W(s,r) \in \widetilde{H}_p^{\frac{1}{p}+\sigma}(\Gamma)$ for $\sigma' < \sigma$, respectively. $\delta = \delta(s) = \frac{\ln\lambda(s)}{2\pi}$ is an eigenvalue of the matrix $\mathcal{S}^{-1}(s,0,-1)\,\mathcal{S}(s,0,1)$, where $\mathcal{S}$ denotes the symbol of $\widetilde{\mathcal{S}}^{-1}$. The matrix $K(s)$ is given by the principal symbol matrix of $\widetilde{\mathcal{S}}^{-1}$.

A calculation of the exponent δ in the expansion (3.1) has already been done by M. L. Williams [69] and T. C. T. Ting [68], based on results of the two-dimensional case. In [21] we present a rigorous justification for the three-dimensional case. As an example we calculate the eigenvalues $\lambda_j(s), j = 1,2,3$ for two isotropic materials with the Lamé constants λ_1, μ_1 and λ_2, μ_2, respectively, with our method (see [21]). With the help of local coordinates $\widetilde{\kappa}_j$ we have

$$\mathcal{S}(s,0,\pm 1) = \left[\widetilde{\kappa}_j'(0,x)^\top\right]^{-1} \mathcal{S}(\pm 1)\widetilde{\kappa}_j'(0,x)^\top$$

with

$$\mathcal{S}(\pm 1) = 2\begin{pmatrix} \gamma_4 & 0 & 0 \\ 0 & \gamma_5 & \pm i\gamma_6 \\ 0 & \mp i\gamma_6 & \gamma_5 \end{pmatrix},$$

where

$$\gamma_{1,k} = \frac{\lambda_k + 3\mu_k}{2\mu_k(\lambda_k + 2\mu_k)}, \quad \gamma_{2,k} = \frac{\lambda_k + \mu_k}{\lambda_k + 3\mu_k}, \quad \gamma_{3,k} = \frac{\mu_k}{\lambda_k + 2\mu_k}$$

and

$$\gamma_4 = \sum_{k=1}^{2} \gamma_{1,k}(1+\gamma_{2,k}),\ \gamma_5 = \sum_{k=1}^{2} \frac{\gamma_{1,k}}{1-(\gamma_{3,k})^2},\ \gamma_6 = \frac{\gamma_{1,1}\gamma_{3,1}}{1-(\gamma_{3,1})^2} - \frac{\gamma_{1,2}\gamma_{3,2}}{1-(\gamma_{3,2})^2}.$$

The eigenvalues of $\mathcal{S}^{-1}(-1)\mathcal{S}(+1)$ are the zeros of
$\det[\mathcal{S}(1) - \lambda\mathcal{S}(-1)] = 8(1-\lambda)[\gamma_5^2(1-\lambda)^2 - \gamma_6^2(1+\lambda)^2] = 0.$
Thus we get

$$\lambda_1 = 1,\ \lambda_2 = \frac{\gamma_5 - \gamma_6}{\gamma_5 + \gamma_6},\ \lambda_3 = \frac{1}{\lambda_2} \text{ and } \delta = \delta_2 = \frac{1}{2\pi} \ln \lambda_2 .$$

Corollary 1. There are no oscillation terms in (3.1), i.e. each δ_i vanishes if and only if $\gamma_6 = 0$, i.e. if and only if $\frac{\mu_1}{\mu_2} = \frac{1-2\nu_1}{1-2\nu_2}$, where $\nu_i = \frac{\lambda_i}{2(\lambda_i+\mu_i)}$ are the Poisson ratios for $i = 1, 2$.

This result coincides with results by Williams [69] and Ting [68].

3.2 Corner problems

The singular expansions of edge singularities can locally be reduced to two–dimensional problems [42, 48, 45] and will be described in Section 4. Here we consider the three–dimensional singularities of elastic fields near polyhedral vertices where edges meet. This approach is based on the papers [62, 38], where Mellin techniques for boundary integral equations are used for the reduction of the boundary value problem to a one–dimensional boundary integral equation on a closed curve on the sphere.
Asymptotic expansions near vertices of rotationally symmetric cones under symmetric [8] and non–symmetric loads [60, 35] are special cases and are calculated by applying the Mellin transformation to the original boundary value problems.
Let Ω be a polyhedral domain and Ω^c its complement. Let us consider the Dirichlet problem

$$\begin{aligned} Lu &= 0 \quad \text{in } \Omega \quad \text{or in } \Omega^c, \\ u &= g \quad \text{on } \partial\Omega, \end{aligned} \tag{3.2}$$

where L is introduced in (2.1).

Since we are interested in local representations near a vertex, we associate with (3.2) the Dirichlet problem in a cone

$$K_+ = K = \left\{(r,w), r \in \mathbb{R}^+, w \in S_+\right\},$$

or in $K_- = \mathbb{R}^3 \setminus \overline{K}$, respectively, where $S_+ = K \cap S^2$ is a bounded domain on the unit sphere S^2 with a piecewise smooth boundary curve $\gamma = \partial K \cap S_+$ and where $S_- = S^2 \setminus \overline{S}_+$. We consider the equivalent boundary integral equation of the first kind for the stress field t on ∂K:

$$Vt(x) := \int_{\partial K} E(x-y)t(y)\, d_yS = f_\pm(x) \quad \text{for } x \in \partial K. \tag{3.3}$$

The sign + is used for the interior, the sign − for the exterior problem. V is considered as a mapping between weighted Sobolev spaces

$$V : V_0^{-\frac{1}{2}}(\partial K) \to V_0^{\frac{1}{2}}(\partial K),$$

where $V_0^{\frac{1}{2}}(\partial K)$ is the trace space of the Sobolev space $V_0^1(K)$. The Sobolev space $V_0^1(K)$ is defined via the norm

$$||v||_{V_0^1(K)} := \left(\int_K |\nabla v|^2 + r^{-2}|v|^2 \, dx \right)^{\frac{1}{2}}.$$

By $V_0^{-\frac{1}{2}}(\partial K) = \left[V_0^{\frac{1}{2}}(\partial K) \right]'$ we denote the dual space.

Lemma 3.1. *[38]: The operator V is $V_0^{-\frac{1}{2}}(\partial K)$-elliptic, i. e. there exists a positive constant c such that*

$$(V\varphi, \varphi)_{L_2(\partial K)} \geq c||\varphi||^2_{V_0^{-\frac{1}{2}}(\partial K)} \quad \textit{for all } \varphi \in V_0^{-\frac{1}{2}}(\partial K).$$

Moreover, V defines an isomorphism

$$V : V_0^{-\frac{1}{2}}(\partial K) \to V_0^{\frac{1}{2}}(\partial K).$$

The Mellin transformation

$$\widehat{h}(\alpha) = \int_0^\infty r^{-\alpha-1} h(r)\, dr \quad \text{for } \alpha \in \mathbb{C}. \tag{3.4}$$

will be applied to (3.3) after rewriting (3.3) in terms of polar coordinates $x = (r, \omega)$ and $y = (\rho, \vartheta)$. Then the Mellin transformed equation reads:

$$\widehat{Vt}(\alpha, \omega) = \int_\gamma a(\alpha, \omega, \vartheta)\widehat{t}(\alpha - 1, \vartheta)\, d_\vartheta\gamma = \widehat{f}(\alpha, \omega),$$

where the Mellin kernel is given by

$$a(\alpha, \omega, \vartheta) := \int_0^\infty E(\rho\omega - \vartheta)\rho^{\alpha-1}\, d\rho.$$

Let us denote by $\widehat{V}(\alpha)$ the operator

$$\widehat{V}(\alpha)\varphi(\omega) = \int_\gamma a(\alpha, \omega, \vartheta)\varphi(\vartheta)\, d_\vartheta\gamma,$$

$$\widehat{V}(\alpha) : H^{-\frac{1}{2}}(\gamma) \to H^{\frac{1}{2}}(\gamma). \tag{3.5}$$

Lemma 3.2. *[38]: The meromorphic operator function $\widehat{V}(\alpha)$ is Fredholm of index zero. The spectrum of $\widehat{V}(\alpha)$ consists only of isolated points α_j which are the eigenvalues of $\widehat{V}(\alpha)$ having finite algebraic multiplicity ν_j.*

For the formulation of the asymptotic results we need the definition of operator pencils $\mathcal{A}_\pm(\alpha)$, which are obtained by applying the Mellin transformation to the interior and exterior Dirichlet problem in K_+ (or K_-). Furthermore, we assume that K has m edges which meet at the vertex O.

Theorem 3.4. *[38]: Let $h \in \mathbb{N}_0, \beta \in \mathbb{R}^1$ and $\boldsymbol{\sigma} \in \mathbb{R}^m$ satisfy condition (4.15) of section 4. Suppose that the line $Re\,\alpha = H := h + \frac{1}{2} - \beta - \sum_{1\le j\le m}\sigma_j > -\frac{1}{2}$ is free of eigenvalues of $\mathcal{A}_+(\alpha)$ and $\mathcal{A}_-(\alpha)$. Let the right–hand side $f_\pm$ in (3.3) be given in $V_0^{\frac{1}{2}}(\partial K) \cap V_{\beta,\boldsymbol{\sigma}}^{h+\frac{3}{2}}(\partial K)$, where this space will be defined in (3.7).*
Then the solution $t \in V_0^{\frac{1}{2}}(\partial K)$ of the boundary integral equation (3.3) admits the expansion

$$t(x) = \eta(r) \sum_{-\frac{1}{2}<Re\,\alpha_j<H} r^{\alpha_j - 1} \sum_{i=0}^{I_j} \sum_{s=0}^{m_{ji}-1} c_{ijs} \sum_{k=0}^{s} \frac{(\log r)^{s-k}}{(s-k)!} t_{jik}(\omega) + \widetilde{t}(x), \quad (3.6)$$

where I_j and m_{ji} are the geometric and partial multiplicities of the eigenvalues α_j and where $\{t_{jik}\}_{k=0}^{m_{ji}-1}(\omega)$ is a canonical set of Jordan chains corresponding to α_j. The remainder satisfies $\widetilde{t} \in V_{\beta,\boldsymbol{\sigma}}^{h+\frac{1}{2}}(\partial K)$.

The weighted trace space $V_{\beta,\boldsymbol{\sigma}}^l(\partial K)$ is defined as follows: The curve $\gamma = \partial S_+$ has m corner points q_j. For $\rho_j = \rho_j(w) = |w - q_j|$ and $k \in \mathbb{N}_0$ let $V_{\boldsymbol{\sigma}}^k(S_\pm)$ be the space of functions with the finite norm

$$||v||_{V_{\boldsymbol{\sigma}}^k(S_\pm)} = \left(\sum_{s=0}^{k} || \prod_{j=1}^{m} \rho_j^{\sigma_j + s - k} v||_{H^s(S_\pm)}^2 \right)^{\frac{1}{2}}.$$

$V_{\beta,\boldsymbol{\sigma}}^k(K_\pm)$ with $\beta \in \mathbb{R}$ is the space of functions with the norm

$$||v||_{V_{\beta,\boldsymbol{\sigma}}^k}(K_\pm) = \left(\int_0^\infty r^{2(\beta + \sum_{j=1}^m \sigma_j - k)} \sum_{s=0}^{k} ||(r\partial_r)^s v(r,\cdot)||_{V_{\boldsymbol{\sigma}}^{k-s}(S_\pm)}^2 r^2\, dr \right)^{\frac{1}{2}}. \quad (3.7)$$

Finally, $V_{\beta,\boldsymbol{\sigma}}^l(\partial K)$ is for $l = k - \frac{1}{2}$ the trace space to $V_{\beta,\boldsymbol{\sigma}}^k(K_+)$ (or $V_{\beta,\boldsymbol{\sigma}}^k(K_-)$.

A similar asymptotic expansion is derived in [61] in usual Sobolev spaces, where the spectral properties of $\widehat{V}(\alpha)$ are used directly. The connection between these two operator pencils will be described in Section 5.

4 Asymptotic expansions in the domain

4.1 The edge-problem

The general form of edge-singularities for solutions of elliptic boundary value problems has been described in many papers (see e.g. [18, 15, 32, 45, 48]). Moreover, in [11, 43, 46, 45] formulae for the coefficients in the asymptotic expansion are derived. However, the papers [11, 43, 46, 45] contain certain conditions for the boundary operators which exclude e.g. the Neumann problem. Therefore, we investigated variational solutions not only of the Dirichlet, but also of the Neumann and mixed problem for the equations of linear elasticity, based on ideas of the paper [56].

Let G be a bounded domain in $\mathbb{R}^n$ with piecewise smooth boundary ∂G. We assume that there is a smooth one-dimensional submanifold $M \subset \partial G$, such that $\partial G \backslash M$ is smooth and G is locally diffeomorphic to a dihedron in some neighbourhood of any point $z \in M$. Assume that $\partial G \backslash M$ is decomposed into two smooth parts Γ_D and Γ_N and define the subspace

$$\mathcal{V} = \{u \in W_2^1(G) : u = 0 \text{ on } \Gamma_D\}$$

of $W_2^1(G)$. Then we consider the following variational problem:

Find a vector-function $u \in \mathcal{V}$, such that

$$\begin{aligned} b(u,v) &:= \int_G \sum_{i,h,k,l=1}^{3} a_{ihkl}\varepsilon_{kl}(u)\overline{\varepsilon_{ih}(v)}\,dx \\ &= \langle F, v\rangle = \int_G f\bar{v}\,dx + \int_{\Gamma_N} g\bar{v}\,d\sigma,\ \forall v \in \mathcal{V}, \end{aligned} \tag{4.1}$$

where $F \in \mathcal{V}'$, $f \in L_2(G)$, $g \in W_2^{1/2}(\Gamma_N)$.

Here $\varepsilon_{kl}(u) = \varepsilon_{lk}(u) = (\partial_{x_k} u_l + \partial_{x_l} u_k)/2$ denotes the components of the strain tensor, a_{ihkl} are the given elasticities, f is the given volume force and g is the surface traction.

If $u \in W_2^2(G)$ then the variational solution $u \in \mathcal{V}$ of (4.1) is also a classical solution of the boundary value problem

$$\Big(L(\partial_x)u\Big)_i = a_{ihkl}\partial_{x_h}\partial_{x_k}u_l = f_i,$$

in short

$$L(\partial_x)u = f \quad \text{in } G,$$

satisfying

$$Tu = \sigma(u)\cdot n = g \quad \text{on } \Gamma_N \tag{4.2}$$

with $\sigma_{ij}(u) = a_{ihkl}\varepsilon_{kl}(u)$ and the exterior normal vector n.

In this paper we consider the following situations: $\Gamma_D = \emptyset$ (Neumann problem), $\Gamma_N = \emptyset$ (Dirichlet problem), $\Gamma_D, \Gamma_N \neq \emptyset$ (mixed problem).

We now assume for the sesquilinear form b that a Gårding inequality

$$\mathrm{Re}\; b(u,u) + \gamma \|u\|^2_{L_2(G)} \geq c\|u\|^2_{\mathcal{V}} \tag{4.3}$$

with positive c is satisfied for all $u \in \mathcal{V}$. In elasticity, (4.3) corresponds to Korn's inequality.

Operator pencils
For the description of the behaviour of the solution of the variational problem (4.1) in a neighbourhood of a point $x_0 = (0, 0, z_0) \subset M$ we introduce the following two operator pencils:

The operator pencil $\mathbf{A}(\cdot)$:
Let $D_{x_0} = K_{x_0} \times \mathbb{R}$ be the local dihedron at the edge point x_0 with the faces $\Gamma^{\pm}_{x_0}$, where $K_{x_0} = \{(r, \varphi) : 0 < r < \infty, 0 < \varphi < \varphi_0\}$ is the plane wedge, $r = |x - x_0|$. Then $M = \mathbb{R}$ is the edge of D_{x_0}. We discard the derivative ∂_{x_3} in $b(u, v)$ and get the 2D–sesquilinear form

$$\widetilde{b}(u,v) = \int_{K_{x_0}} \sum_{i,h,k,l=1}^{2} a_{ihkl}\varepsilon_{kl}(u)\overline{\varepsilon_{ih}(v)}\, dx_1\, dx_2 .$$

Introducing the polar coordinates (r, φ) and using the Mellin transformation with respect to r $(r\,\partial/\partial r \to \alpha)$ we get finally a parameter-depending sesquilinear form

$$\begin{aligned} \widehat{a}(\widehat{u}, \widehat{v}, \alpha) &= \langle \mathbf{A}(\alpha)\widehat{u}, \widehat{v}\rangle_{(0,\varphi_0)}, \text{ where} \\ \mathbf{A}(\alpha) &: \widehat{\mathcal{V}} \to \widehat{\mathcal{V}}'; \end{aligned} \tag{4.4}$$

$\widehat{\mathcal{V}} = \{\widehat{u} \in W^1_2(0, \varphi_0) : \gamma^{\pm}\widehat{u}(\varphi^{\pm}) = 0\}$, $\varphi^+ = 0$, $\varphi^- = \varphi_0$, $\gamma^{\pm} = 1$ if $\Gamma^{\pm}_{x_0} \cap \Gamma_{D_{x_0}} = \emptyset$ and $\gamma^{\pm} = 0$ if $\Gamma^{\pm}_{x_0} \cap \Gamma_N = \emptyset$.

The operator pencil $\mathcal{A}(\cdot)$:
We denote by $Q = Q_{x_0}$ and $q^{\pm}_{x_0} = q^{\pm}$ the intersection of D_{x_0} and $\Gamma^{\pm}_{x_0}$, respectively, with the unit sphere $S^2_{x_0} = \{x : |x - x_0| = \varrho = 1\}$. Introducing in (4.1) the spherical coordinates $(\varrho, \boldsymbol{\omega}) = (\varrho, \varphi, \vartheta)$ and using the Mellin transformation $(\varrho\,\partial/\partial\varrho \to \alpha)$ we get finally a parameter-depending sesquilinear form

$$\widetilde{a}(\widetilde{u}, \widetilde{v}, \alpha) = \langle \mathcal{A}(\alpha)\widetilde{u}, \widetilde{v}\rangle, \tag{4.5}$$

where

$$\mathcal{A}(\alpha) : \widetilde{\mathcal{V}} \to \widetilde{\mathcal{V}}', \ \widetilde{\mathcal{V}} = \{\widetilde{u} \in W_2^1(Q) : \gamma^{\pm}\widetilde{u} = 0 \text{ on } q^{\pm}\}. \tag{4.6}$$

$\mathcal{A}(\alpha) = \{\mathcal{L}(\alpha), \mathcal{B}^{\pm}(\alpha)\}$ can be considered as the operator pencil defined by the boundary value problem

$$\begin{aligned} \mathcal{L}(\omega, \partial_\omega, \alpha)\widetilde{u} &= \widetilde{f} \text{ in } Q \\ \mathcal{B}^{\pm}(\omega, \partial_\omega, \alpha)\widetilde{u} &= \widetilde{g}_{\pm} \text{ on } q^{\pm}, \end{aligned} \tag{4.7}$$

where $L(\partial_x) = \varrho^{-2}\mathcal{L}(\omega, \partial_\omega, \varrho\,\partial_\varrho)$, $\mathcal{B}^{\pm} = 1$ and $\widetilde{g}_{\pm} = 0$ if $\gamma^{\pm} = 1$, $\mathcal{B}^{\pm} = t^{\pm}$ if $\gamma^{\pm} = 0$ with $t^{\pm}(\partial_x) = \varrho^{-1}t^{\pm}(\omega, \partial_\omega, \varrho\,\partial_\varrho)$.

Lemma 4.1. *[30, 31]: The spectra of* $\mathbf{A}(\cdot)$ *and* $\mathcal{A}(\cdot)$ *consist of isolated eigenvalues having finite algebraic multiplicities. For an eigenvalue* α_j *of* $\mathbf{A}(\cdot)$ *there exist Jordan chains*

$$\{\widetilde{u}_j^{k,s}(\varphi)\}, k = 1, \dots p_j \, (\textit{geometric multiplicity}) \tag{4.8}$$

$$s = 0, \dots \kappa_{j,k} - 1 (\kappa_{j,k} \textit{ - Riesz index}) \tag{4.9}$$

or $\{\widetilde{u}_j^{k,s}(\omega)\}$ *for* $\mathcal{A}(\cdot)$, *respectively.*

Asymptotic decompositions

For the local characterization of solutions we need the weighted Sobolev spaces

$$W^k_{2,\beta}(D) = \{u \in L_2(D) : \sum_{|\gamma| \le k} \int_D r^{2\beta} |D^\gamma u|^2 \, dx = \|u\|^2_{W^k_{2,\beta}} < \infty\},$$

where $r = r(x) = \operatorname{dist}(x, M)$ and $\beta \in \mathbb{R}$.

Theorem 4.1. *[56]: Suppose that the sesquilinear form (4.1) provides a Gårding inequality (4.3) and that* $u \in \mathcal{V}$ *is a solution of (4.1) with the right-hand side*

$$\langle \eta F, \overline{v} \rangle = \int_D f \eta \overline{v} \, dx + \sum_{\pm} \int_\Gamma g_{\pm} \eta \overline{v} \, d\sigma,$$

where $f \in W^o_{2,\beta}(D)$, $g_{\pm} \in W^{1/2}_{2,\beta}(\Gamma^{\pm})$, $0 < \beta \le 1$, *and where* $\eta = \eta(r)$ *is a* C^∞ *cut-off function. Furthermore, we assume, that the line* $\operatorname{Re} \alpha = 1 - \beta$ *is free of eigenvalues of the operator pencil* $\mathbf{A}(\cdot)$. *Then the solution* u *has the local representation*

$$\chi u = \chi \left[c_0(z) + \sum_{\substack{j,k,s \\ 0 < \operatorname{Re} \alpha_j < 1 - \beta}} c_{j,k,s}(z) r^{\alpha_j} \sum_{\sigma=1}^{s} \frac{1}{\sigma!} (\ln r)^\sigma \widehat{u}_j^{k,s-\sigma}(\varphi) + w(x) \right], \tag{4.10}$$

where $z \in M$, $c_0 \in W_2^{1-\beta-\epsilon}(M)$, $c_{j,k,s} \in W_2^{1-\beta-\operatorname{Re}\alpha_j-\epsilon}(M)$ *(*ϵ *is an arbitrary small positive number),* $w \in W^0_{2,\beta-2}(D)$, $\partial w/\partial x_l \in W^0_{2,\beta-1}(D)$ *for* $l = 1, 2$. χ *is a cut-off function with* $\chi\eta = \chi$. *The eigenvalues of* $\mathbf{A}(\alpha)$ *are denoted by* α_j; *and* $\{\widehat{u}_j^{k,s}(\varphi)\}$ *are the corresponding Jordan chains.*

Coefficient formulae

In analogy to [43, 46] it is shown in [56] that

$$c_{j,k,\kappa_{j,k}-1-s}(z_0) = \int_G f\bar{v}_j^{k,s}(z_0)\,dx + \int_{\Gamma_N} g\bar{v}_j^{k,s}(z_0)\,d\sigma,$$

if the solution u of (4.1) vanishes outside $G \cap D_{x_0}$. Here $x_0 = (0,0,z_0)$ and the $v_j^{k,s}$ are the so-called dual singular functions. These are singular solutions of the homogeneous adjoint problem, which have the form

$$v_j^{k,s}(\varrho, w) = -\varrho^{-1-\bar{\alpha}_j} \sum_{\varsigma=0}^{s} \frac{1}{\sigma!}(-\ln\varrho)^{\varsigma}\widetilde{v}_j^{k,s-\varsigma}(\omega),$$

where $\{\widetilde{v}_j^{k,s}\}$ are Jordan chains of the adjoint operator $\mathcal{A}^*(\alpha)$ belonging to the eigenvalue $-1-\bar{\alpha}_j$ and $\widetilde{v}_j^{k,s} \in \widetilde{\mathcal{V}}$. Unfortunately the calculation of the vector functions $\widetilde{v}_j^{k,s}$ is difficult. Therefore we weaken the condition $\widetilde{v}_j^{k,s} \in \widetilde{\mathcal{V}}$ by assuming $s = 0$ and

$$\widetilde{v}_j^k = (\sin\vartheta)^{-\alpha_j} P_{j,k}(\varphi, \cos\vartheta) + w_j^k \text{ with } w_j^k \in W_2^1(Q), \tag{4.11}$$

where $P_j^k(\varphi, \cos\vartheta)$ are polynomials in $\cos\vartheta$ whose coefficients are infinitely differentiable with respect to φ and satisfy the biorthonormality condition

$$\left(\frac{\partial\mathcal{L}}{\partial\alpha}(\alpha_j)\widetilde{u}_j^{(m)}, \widetilde{v}_j^k\right)_Q - \sum_{\pm}\left(\frac{\partial t^{\pm}}{\partial\alpha}(\alpha_j)\widetilde{u}_j^m, \widetilde{v}_j^k\right)_{q^{\pm}} = \delta_{k,m} \tag{4.12}$$

for $0 < \operatorname{Re}\alpha_j < 1-\beta$, $k, m = 1, \ldots p_j$.

Theorem 4.2. *[56]: Let $\widetilde{v}_j^k$ be eigenfunctions of $\mathcal{A}^*(\cdot)$ to the eigenvalue $-1-\overline{\alpha_j}$ which have the form (4.11). Then the following formula holds for the coefficients in (4.10):*

$$\begin{aligned} c_{j,k}(z_0) &+ \sum_{m=1}^{p_j} d_j^{k,m} \int_0^\infty \frac{c_{j,m}(z_0+\varrho) - c_{j,m}(z_0-\varrho)}{2\varrho}\,d\varrho \\ &= \int_D f\bar{\widetilde{v}}_j^k\,dx + \sum_{\pm}\int_{\Gamma^{\pm}} g_{\pm}\bar{\widetilde{v}}_j^k\,d\sigma \end{aligned}$$

where

$$d_j^{k,m} = -\langle\mathcal{A}(\alpha_j)\cos\vartheta\,\widetilde{u}_j^m, \widetilde{v}_j^k\rangle_Q.$$

For some boundary value problems of the Laplace operator and the Lamé system, the weight functions $\widetilde{v}_j^{k,s}$ have been calculated with the help of a Computer Algebra Program [40].

Remarks:

The calculation of the eigenvalues α_j of the operator pencil $\mathbf{A}(\alpha)$ and of the corresponding Jordan chains can be found in different papers (e.g. [14, 59, 28]). Here we recall some results from [50] for a half-space interface crack between two isotropic elastic materials. In this case, the displacement fields have the following decomposition near the crack front, $j = 1, 2$:

$$\chi u_j(x) = \chi \left[c_1(z) r^{\frac{1}{2}} \begin{pmatrix} 0 \\ 0 \\ u^1_{3j}(\varphi) \end{pmatrix} \right.$$

$$\left. + \; c_2(z) r^{\frac{1}{2}+i\epsilon} \begin{pmatrix} u^2_{1j}(\varphi) \\ u^2_{2j}(\varphi) \\ 0 \end{pmatrix} + c_3(z) r^{\frac{1}{2}-i\epsilon} \begin{pmatrix} u^3_{1j}(\varphi) \\ u^3_{2j}(\varphi) \\ 0 \end{pmatrix} + w_j \right]. \quad (4.13)$$

Here

$$\epsilon = \frac{1}{2\pi} \ln \frac{1+\beta}{1-\beta}, \; \beta = \frac{\mu_2(1-2\nu_1) - \mu_1(1-2\nu_2)}{2[\mu_2(1-\nu_1) + \mu_1(1-\nu_2)]}$$

(compare [69, 68]). μ_j denote the shear moduli and ν_j the Poisson ratios. It follows that

$$\epsilon = 0 \Leftrightarrow \beta = 0 \Leftrightarrow \frac{\mu_2}{\mu_1} = \frac{1-2\nu_2}{1-2\nu_1}$$

(compare Corollary 1, where we recall results of [69] and [68] for the singular expansion of the boundary stress fields). The decoupling in planar and antiplanar states is for anisotropic materials not always possible. Nevertheless, the displacement field $u(x)$ has a decomposition analogous to (4.13) where nonzero components occur, in general.

4.2 The corner problem

The behaviour of displacement fields near polyhedral corner points has already been discussed in Section 2.2 in the framework of boundary integral equations. Especially, in Theorem 3.4 we have referred to spectral properties, which now will be explained in more detail.

For simplicity we consider the boundary value problems in the polyhedral cones $K_\pm$. Assume that $\partial K = \partial K_+ = \cup_{j=1}^m \Gamma_j \cup M_j$, where Γ_j are the faces and M_j the edges meeting at the vertex O.

We consider the boundary value problem:

$$\begin{aligned} L(x, D_x)u_\pm = L_\pm(x, D_x)u_\pm &= f_\pm \quad && \text{in } K_+ \text{ (or } K_-) \quad (4.14) \\ u_\pm|_{\Gamma_j} &= g_j^\pm && \text{for } j \in N_1 \subset \{1, \dots, m\}, \\ T(\partial_x, n_\pm(x))u_\pm = T_\pm u_\pm|_{\Gamma_k} &= t_k^\pm && \text{for } k \in N_2 \subset \{1, \dots, m\}, \end{aligned}$$

$N_1 \cup N_2 = \{1, \dots, m\}, N_1 \cap N_2 = \emptyset$. Clearly, the Dirichlet and Neumann problems are included.

Introducing the spherical coordinates $x = (r, \omega)$ in (4.14) and applying the Mellin transformation we get in analogy to (4.5), (4.6) and (4.7) the operator pencil

$$\mathcal{A}_\pm(\alpha) := \left\{\mathcal{L}_\pm(\alpha), \mathcal{B}_j^\pm(\alpha)\right\}_{j=1\dots m} :$$
$$H^1(S_\pm) \to H^{-1}(S_\pm) \times \prod_{j=1}^{m} H^{1-m_j-\frac{1}{2}}(\gamma_j), \tag{4.15}$$

where $S_\pm = K_\pm \cap S^2, \gamma_j = \Gamma_j \cap S^2$ and where m_j is the order of the corresponding boundary operator.
For every edge M_j we use local coordinates (y_1, y_2, y_3) centered at the corner points q_j of S_+. The y_3–direction is normal to S^2, (y_1, y_2) is tangential to S^2. Similar to the operator pencil $\mathbf{A}(\alpha)$ in Section 3.1 we introduce the pencil $\mathbf{A}_j^\pm(\alpha)$ for every edge.
Denote by $h_j^\pm$ the maximal positive number such that the strip $|Re\,\alpha| < h_j^\pm$ does not contain any eigenvalues of $\mathbf{A}_j^\pm(\alpha)$.
For $h \in \mathbb{N}_0$ and $\boldsymbol{\sigma} \in \mathbb{R}^m$ we assume that

$$|h + 1 - \sigma_j| < \min\{h_j^+, h_j^-\} \quad \text{for } j = 1, \dots, m \tag{4.16}$$

or, less restrictive

$$|h + 1 - \sigma_j| < h_j^\pm \qquad \text{for } j = 1, \dots, m. \tag{4.17}$$

Theorem 4.3. *[48, 38, 49, 36, 35]: Suppose that $f \in V_0^{-1}(K_\pm) \cap V_{\beta,\sigma}^h(K_\pm)$ (see (3.7)) and $g_j \in V_0^{1-m_j-\frac{1}{2}}(\Gamma_j) \cap V_{\beta,\sigma}^{h+\frac{3}{2}-m_j}(\Gamma_j)$ are given where h and σ_j satisfy condition (4.16). Furthermore, assume that the line $Re\,\alpha = H = h + \frac{1}{2} - \beta - \sum_{1\le j\le m} \sigma_j > -\frac{1}{2}$ is free of eigenvalues of $\mathcal{A}_\pm(\cdot)$.*
Then the solution $u_\pm \in V_0^1(K_\pm)$ of the problem (4.14) has the following asymptotic representation:

$$u_\pm(x) = \eta(r) \sum_{-\frac{1}{2}<Re\,\alpha_j^\pm<H} r^{\alpha_j^\pm} \sum_{i=1}^{I_j^\pm} \sum_{s=0}^{m_{ji}^\pm-1} c_{ijs}^\pm u_{jis}^\pm(x) + w^\pm(x), \tag{4.18}$$

where $w^\pm \in V_{\beta,\sigma}^{h+2}(K^\pm)$, η is a C^∞ cut–off function and

$$u_{ijs}^\pm = r^{\alpha_j^\pm} \sum_{k=0}^{s} \frac{(\ln r)^{(s-k)}}{(s-k)!} e_{jik}^\pm(w). \tag{4.19}$$

Here $\{e_{jik}^\pm(w)\}_{k=0}^{m_{ji}^\pm-1}$ is a canonical system of Jordan chains of $\mathcal{A}_\pm(\cdot)$ belonging to the eigenvalue $\alpha_j^\pm$ where $I_j^\pm$ and $m_{ji}^\pm$ denote the respective geometrical and partial multiplicities.

Note that this theorem also holds if x_0 is an edge point since the wedge through x_0 can be considered as a vertex to a specific corner.

5 Relations between the corner asymptotics in the domain and on the boundary

We discuss the connection between the spectral properties of the operator pencil $\widehat{V}(\alpha)$ (see (3.5)) and the operator pencils $\mathcal{A}_\pm(\alpha)$ (see (4.15)) belonging to the exterior and interior Dirichlet problem (i.e. $N_2 = \emptyset$ in (4.14)). In analogy to (4.7) and (4.15) we denote the Mellin transformed surface tractions by $\mathcal{T}_\pm(\alpha)$.

Theorem 5.1. *[38]: The following equivalence holds between the eigenvalue problems belonging to $\widehat{V}(\alpha)$ and $\mathcal{A}_\pm(\alpha)$, respectively: The number α_0 is an eigenvalue of $\widehat{V}(\alpha)$ and $t_o, \ldots, t_\kappa$ is a corresponding Jordan chain if and only if the following two conditions are valid:*

(i) α_0 is an eigenvalue of at least one of the operator pencils $\mathcal{A}_\pm(\alpha)$.
(ii) There exist chains $e_0^\pm, \ldots, e_\kappa^\pm$ of the operator pencils $\mathcal{A}_\pm(\alpha)$ corresponding to α_0; and at least one of them is a Jordan chain, i.e. the first element does not vanish. Moreover

$$t_j = \mathcal{T}_-(\alpha_0)e_j^- - \mathcal{T}_+(\alpha_0)e_j^+ \qquad \text{for } j = 0, \ldots, \kappa. \tag{5.1}$$

Denote by μ_j for $j \in N_0$ the eigenvalues of $\widehat{V}(\alpha)$ lying in the half plane $R(\alpha) > -\frac{1}{2}$ and by ν_j' the algebraic multiplicity of μ_j. To each eigenvalue μ_j we accompany the ν_j'-dimensional space $\mathcal{N}_j$ consisting of the functions

$$\left\{ r^{\mu_j - 1} \sum_{i=0}^{\kappa} \frac{(\ln r)^{\kappa - i}}{(\kappa - i)!} t_i(w) \right\},$$

where $t_0, \ldots, t_\kappa$ is any Jordan chain of $\widehat{V}(\mu_j)$.
We construct a generating system of $\mathcal{N}_j$ with the help of Theorem 5.1. There exist integers $n, l \geq 0$ such that $\mu_j = \alpha_n^-$ or $\mu_j = \alpha_l^+$ and $\nu_j' \leq \kappa_n^- + \kappa_l^+ =: \nu_j$. Furthermore, (5.1) yields that $\mathcal{N}_j$ is generated by the sum of the normal stresses of the chains corresponding to the eigenvalues α_n^- or α_l^+ of the operator pencil $\mathcal{A}_\pm(\alpha)$:

$$\mathcal{T}_- u_{nks}^-, k = 1, \ldots, I_n^-, s = 0, \ldots, m_{nk}^- - 1$$
$$\mathcal{T}_+ u_{lpq}^+, p = 1, \ldots, I_l^+, q = 0, \ldots, m_{lp}^+ - 1$$

(see (4.18)) We enumerate the elements of this system such that

$$\mathcal{T}_{js} = \mathcal{T}_- u_{nks}^- \quad (\text{or } \mathcal{T}_+ u_{lpq}^+)$$

In this way we get again a justification of Theorem 3.4 by a different approach.

Theorem 5.2. *[38]: Suppose that the assumptions of Theorem 3.4 are satisfied. Then the solution $t \in V_0^{\frac{1}{2}}(\partial K)$ of the boundary integral equation (3.3) admits the expansion*

$$t(x) = \eta(r) \sum_{-\frac{1}{2} < Re\,\mu_j < H} \; \sum_{1 \le k \le \nu_j} c_{jk} T_{jk}(r,\omega) + \widetilde{t}(x)$$

with $\widetilde{t} \in V_0^{-\frac{1}{2}}(\partial K) \cap V_{\beta,\sigma}^{h+\frac{1}{2}}(\partial K)$.

6 Graded Mesh Refinement

It is well known that standard numerical techniques lose accuracy as a result of the nonsmooth boundary, which causes loss of regularity of the solution in comparison to smooth problems. Therefore a variety of adapted numerical methods has been developed, e.g. the augmented Fix method [7, 13, 10, 64] based on the knowledge of the singular terms in the asymptotics or graded mesh refinements near corners or edges based on the local behaviour of the solution [2, 3, 6, 55, 57, 54, 39]. In [5] one finds finite element schemes on locally graded meshes for a large class of second order boundary value problems in polygonal and polyhedral domains. The grading and the corresponding error estimates are based on regularity results in weighted Sobolev spaces and on the local asymptotics as described in Section 3.
The **graded partitions** are defined as follows:
Let Ω be a polygonal domain in $\mathbb{R}^2$ or a polyhedral domain in $\mathbb{R}^3$. We consider a family of finite element partitions $\mathcal{T}_h$ of $\overline{\Omega}$ with the usual regularity properties:

(i) $\overline{\Omega} = \bigcup_{\Omega_e \in \mathcal{T}_h} \overline{\Omega_e}$, where $\overline{\Omega_e}$ are polygons in $\mathbb{R}^2$ (triangles or quadrilaterals) or polyhedra in $\mathbb{R}^3$ (for instance tetrahedra or bricks).
(ii) $\Omega_{e_1} \cap \Omega_{e_2} = \emptyset$ for $e_1 \neq e_2$.
(iii) Any edge (for $n = 2,3$) or face (for $n = 3$) of Ω_{e_1} is either a subset of $\partial\Omega$ or an edge or the face of another Ω_{e_2}.

Denote by h_e the diameter of Ω_e and by ρ_e the diameter of the largest inner ball of Ω_e, then we assume that there is a constant σ independent of $\mathcal{T}_h$ with

(iv) $\frac{h_e}{\rho_e} < \sigma$ for all e with $\Omega_e \in \mathcal{T}_h$.

Further denotes $h = \max_{\Omega_e \in \mathcal{T}_h} h_e$ the meshwidth parameter of the family. In order to treat the singularities of the solution near the irregular part M of the boundary, we assume that the partition $\mathcal{T}_h$ is graded in the following way:

(v) if $\overline{\Omega}_e \cap M \neq \emptyset$ then $\underline{C}_1 h^{1/\mu} \le h_e \le \overline{C}_1 h^{1/\mu}$,
if $\overline{\Omega}_e \cap M = \emptyset$ then $\underline{C}_2 h r_e^{1-\mu} \le h_e \le \overline{C}_2 h r_e^{1-\mu}$,

where $r_e = \text{dist}\,(\Omega_e, M)$, and $\mu \in (0,1]$ is a control parameter of the grading. Note that for $\mu = 1$ we have an unrefined partition.
Such refinements were studied for the Poisson problem in polygonal domain in [27, 52, 55] and in three–dimensional domains with edges in [4]. Approximation results can also be found in [25, 26]. We here give a generalization of these results to polyhedral domains and to more general differential operators.
Let $u \in \mathcal{V}$ be a weak solution of the variational problem (4.1), where Ω is a polygonal or polyhedral domain

$$b(u,v) =< F, v > \qquad \forall v \in \mathcal{V}, \text{ and for a given } F \in \mathcal{V}'. \tag{6.1}$$

We consider the Galerkin equations

$$b(u_h, v_h) =< F, v_h > \qquad \forall v_h \in S_h \subset \mathcal{V}. \tag{6.2}$$

Here $\{S_h\}$ is a family of piecewise polynomial functions with respect to the mesh parameter h with the following properties:

(i) *Conformity*, i. e. $S_h \subset \mathcal{V}$.
(ii) *Transformability*, in the sense that the elements shall belong to an affine or isoparametric family.
(iii) *Approximability*, in the sense that $\lim\limits_{h\to 0} \inf\limits_{v_h \in S_h} \|u - v_h\|_{W^{1,2}(\Omega)} = 0$.

Theorem 6.1. *[5]: Let Ω be a polygonal domain in $\mathbb{R}^2$ or a polyhedral domain in $\mathbb{R}^3$ and $\mathcal{T}_h$ a family of graded partitions. Then the following error estimates for the GALERKIN solution u_h of (6.2) holds*

$$\|u - u_h\|_{W_2^1(\Omega)} \le ch^\gamma \|f\|_{L_2(\Omega)}, \quad \textit{where } \gamma = \begin{cases} 1 & \textit{for } \mu < H_0 \\ \frac{H_0 - \varepsilon}{\mu} & \textit{for } \mu \ge H_0 \end{cases}$$

and ε is a small positive number.

The number H_0 is defined

- for a polygonal domain near a corner point:
  $H_0 = Re\,\alpha_0 - \varepsilon > 0$,
  where α_0 is that eigenvalue of the operator $\mathcal{A}(\alpha)$ (which is defined in the plane case analogously to (4.5)) such that the strip $0 < Re\,\alpha < Re\,\alpha_0$ is free of eigenvalues.
- for a polyhedral domain near a vertex, where m edges meet:
  $H_0 = \min\{Re\,\alpha_0 + \frac{1}{2}, h_j^+, j = 1, \ldots, m\} - \varepsilon$,
  where α_0 is an eigenvalue of the operator $\mathcal{A}(\alpha)$, defined in (4.5), such that the strip $-\frac{1}{2} < Re\,\alpha < Re\,\alpha_0$ is free of eigenvalues. h_j^+ is defined in (4.16). (Here we consider only the interior boundary value problem.)

– for a polyhedral domain near an edge:
$H_0 = Re\,\alpha_0 - \varepsilon$,
where α_0 is an eigenvalue of the operator $\mathcal{A}(\alpha)$, defined in (4.4), such that the strip $0 < Re\,\alpha < Re\,\alpha_0$ is free of eigenvalues.

Numerical experiments have been made in [5] for mixed problems for the Lamé equations in an L–shaped domain and for the Poisson equation in a polyhedral domain with a Fichera corner.
We underline that the general ideas on graded mesh refinements for FEM presented here can be also formulated for Galerkin schemes in BEM. We refer especially to [54] where boundary element methods on graded meshes are studied for the mixed boundary value problem with the Laplacian in polyhedral domains.

References

1. Andersson, B., Babuška, I., Falk, U., v.Petersdorff, T. (1995): Reliable stress and fracture mechanics analysis of complex components using a h-p version of FEM. Intern.J.for Numer. Methods in Engrg. 38, 2135–2163
2. Apel, Th. (1991): Finite–Elemente–Methoden über lokal verfeinerten Netzen für elliptische Probleme in Gebieten mit Kanten. Thesis, TU Chemnitz
3. Apel, Th., Dobrowolski, M. (1992): Anisotropic interpolation with applications to the finite element method. Computing 47, 277–293
4. Apel, Th., Heinrich, B. (1995): Mesh refinement and windowing near edges for some elliptic problems. To appear in: SIAM J. Numer. Anal.
5. Apel, Th., Sändig, A.–M., Whiteman, J.R. (1996): Graded Mesh Refinement and Error Estimates for Finite Element Solutions of Elliptic Boundary Value Problems in Non–Smooth Domains. Math. Meth. in the Appl. Sci. 19, 63–85
6. Babuška, I., Kellogs, R.B., Pitkäranta, J. (1979): Direct and inverse error estimates for finite elements with mesh refinements. Numer. Math 33, 447–471
7. Babuška, I., Rheinboldt, W.C. (1979): On the reliability and optimality of the finite element method. Comput. Struct. 10, 87–94
8. Beagles, A.E., Sändig, A.–M. (1991): Singularities of rotationally symmetric solutions of boundary value problems for the Lamé equations. ZAMM 71, 423–431
9. Bennish, J. (1993): Asymptotics for elliptic boundary value problems for systems of pseudo–differential equations. Journal of Mathematical Analysis and Applications 179, 417–445
10. Blum, H., Dobrowolski, M. (1982): On finite element methods for elliptic equations on domains with corners. Computing 28, 53–63
11. Bourlard, M., Dauge, M., Lubuma, M.S., Nicaise, S. (1988): Coefficients des Singularites pour des Problemes aux limites elliptiques sur une Domaine a Points coniques. Equations aux Derivees partielles, Univ. Nantes 1, 1–53
12. Chkadua, O. (1995): The nonclassical boundary–contact problems of elasticity for homogeneous anisotropic media. Submitted to: Mathematische Nachrichten
13. Ciarlet, Ph. (1978): The finite element method for elliptic problems. Amsterdam

14. Costabel, M., Dauge, M. (1995): Computation of corner singularities in linear elasticity. In: Boundary value problems and integral equations in nonsmooth domains. Lecture Notes in Pure and Applied Mathematics 167, 59–69. Marcel Dekker, New York
15. Costabel, M., Dauge, M. (1993): General edge asymptotics of solutions of second order elliptic boundary value problems. Proc. of Royal Soc. of Edinbourgh 123A, 109–155,157–184
16. Costabel, M., Stephan, E. (1987): An improved boundary element Galerkin method for three–dimensional crack problems. Integral Equations and Operator Theory 10, 467–504
17. Costabel, M., Wendland, W.L. (1986): Strong ellipticity of boundary integral operators. Crelle's Journal für die Reine und Angewandte Mathematik 372, 34–63
18. Dauge, M. (1988): Elliptic Boundary Value Problems in Corner Domains. Smoothness and Asymptotics of Solutions. Lecture Notes in Math. 1341. Springer, Berlin
19. Duduchava, R., Natroshvili, D., Shargorodsky, E. (1990): Boundary value problems of the mathematical theory of cracks. Proceedings of I. Vekua Institute of applied mathematics, some problems of elasticity theory 39, 68–84
20. Duduchava, R., Wendland, W.L. (1995): The Wiener–Hopf method for systems of pseudodifferential equations with an application to crack problems. Integral Equations and Operator Theory 123,294–335
21. Duduchava, R., Sändig, A.–M., Wendland, W.L. (1995): Asymptotics of solutions to boundary equations with applications to interface cracks in anisotropic composites. Preprint 95–9, University Stuttgart
22. Eskin, G. (1981): Boundary Value Problems for Elliptic Pseudodifferential Equations. AMS, Rhode Island
23. Fichera, G. (1972): Existence theorems in elasticity. In: Handbuch der Physik Vol.VIa/2,Festk"orpermechanik II, Springer, Berlin, pp.437–389
24. Fichera, G. (1978): Numerical and Quantitative Analysis. Pitman, London
25. Fritsch, R. (1990): Optimale Finite–Elemente–Approximationen für Funktionen mit Singularitäten. Thesis, TU Dresden
26. Fritsch, R., Oswald, P. (1988): Zur optimalen Gitterwahl bei Finite–Elemente–Approximationen. Wiss. Zeitschr. TU Dresden 37, 155–158
27. Grisvard, P. (1985): Elliptic problems in non–smooth domains. Pitman, Boston
28. Grisvard, P. (1986): Problèmes aux limites dans les polygons, mode d'emploi. EDF Bus Direction Etudes Rech. Sér. C. Math. Inform 1, 21–59
29. Jentsch, L., Natroshvili, D., Wendland, W.L. (1996): General transmission problems in the theory of elastic oscillations of anisotropic bodies. Preprint 96, University Stuttgart
30. Keldysh, M.V. (1971): On the completeness of the eigenfunctions of certain classes of non–selfadjoint operators. Uspechi Matem. Nauk 26, 4,15–41
31. Kondratiev, V.A. (1967): Boundary value problems for elliptic equations in domains with conical or angular points. Transactions Moskow Math. Soc. 10, 227–313
32. Kondratiev, V.A. (1977): Singularities of the solution to the Dirichlet problem for a second order elliptic equation in a neighbourhood of an edge. Diff. Uravn. 13, 11,2026–2032
33. Kondratiev, V.A., Oleinik, O.A. (1983): Boundary value problems for partial differential equations in non–smooth domains. Russian Math. Surveys 38, 1–86
34. Kondratiev, V.A., Oleinik, O.A. (1988): Boundary value problems for a system in elasticity theory in unbounded domains. Korn's inequalities. Uspekhi Math. Nauk 43, 55–99

35. Kozlov, V., Maz'ya, V. (1992): On the spectrum of an operator pencil generated by the Neumann problem in a cone. St. Petersburg Math. J. 3, 2,333-353
36. Kozlov, V., Maz'ya, V. (1988): Spectral properties of operator pencils generated by elliptic boundary value problems in a cone. Funktsional. Anal. i Prilozhen. 22, 38–46
37. Kozlov, V., Maz'ya, V., Schwab, C. (1992): On singularities of solutions of the Dirichlet problem for the Lamé system near the vertex of a cone. Arch. Rat. Mech. Anal. 119, 197–227
38. Kozlov, V., Wendland, W.L., Goldberg, H. (1996): The behaviour of elastic fields and boundary integral Mellin techniques near conical points. Math.Nachrichten 179,95–133
39. Maischak, M., Stephan, E.P. (1995): The hp–Version of the Boundary Element Method in $\mathbb{R}^3$. The basic approximation results. To appear in Math. Math. Appl. Sci.
40. Maple V Release 3 Copyright (c) 1981–1994 by Waterloo Maple Software and the University of Waterloo.
41. Maz'ya, V.G., Plamenevski, B.A. (1977): Elliptic boundary value problems on manifolds with singularities. Probl. Math. Anal. 6, 85–142
42. Maz'ya, V.G., Plamenevski, B.A. (1978): L_p–estimates for solutions of elliptic boundary value problems in domains with edges. Trudy Mosk. Mat. Obshch 37, 49–93
43. Maz'ya, V.G., Plamenevski, B.A. (1976): On the coefficients in the asymptotics of solutions of elliptic boundary value problems near the edge. Soviet Math. Dokl. 17, 4,974
44. Maz'ya, V.G., Plamenevski, B.A. (1983): The first boundary value problem for classical equations of mathematical physics in domains with piecewise smooth boundaries. Zeitschr. f. Anal. Anw. 2, 335–359,523–551
45. Maz'ya, V.G., Rossmann, J. (1988): Über die Asymptotik der Lösungen elliptischer Randwertaufgaben in der Umgebung von Kanten. Math. Nach. 138, 27–53
46. Maz'ya, V.G., Rossmann, J. (1984): Über die Lösbarkeit und die Asymptotik der Lösungen elliptischer Randwertaufgaben in Gebieten mit Kanten III. Preprint 31/84, pp.1–44, University of Berlin
47. Natroshvili, N. (1994): Mathematical problems of anisotropic elasticity for piecewise homogeneous bodies. In: Problems and Methods in Mathematical Physics. Teubner–Texte zur Mathematik 134, 100–110
48. Nazarov, S.A., Plamenevski, B.A. (1994): Elliptic problems in domains with piecewise smooth boundary. Walter de Gruyter, Berlin
49. Nicaise, S. (1992): About the Lamé System in a Polygonal or a Polyhedral Domain and a coupled Problem between the Lamé System and the Plate Equation. I: Regularity of the Solutions. Ann. Scuola Norm. Sup. Pisa IV,XIX, 327–361
50. Nicaise, S., Sändig, A.–M. (1994): General Interface Problems I/II. Math. Meth. in the Appl. Sci. 17, 395–450
51. Nikishkin, V.A. (1979): Singularities of solutions to the Dirichlet problem for a second order equation in a neighbourhood of an edge. Vestnik Moskow Univ. Math. Mech. 2, 51–62
52. Oganesyan, L.A., Ruhkhovets, L.A. (1979): Variational–difference methods for solving elliptic equations. Izdatel'stvo Akad. Nauk. Arm. SSR, Jerevan
53. von Petersdorff, T. (1989): Randwertprobleme der Elastizitätstheorie für Polyeder–Singularitäten und Approximation mit Randelementmethoden. Doctoral Thesis, Techn. University Darmstadt

54. von Petersdorff, T., Stephan, E.P. (1990): Regularity of mixed boundary value problems in $\mathbb{R}^3$ and boundary element methods on graded meshes. Math. Meth. Appl. Sci. 12, 229–249
55. Raugel, G. (1978): Résolution numérique par une méthode d'élements finis du probléme Dirichlet pour le Laplacien dans un polygone. C. R. Acad. Sci. Paris. Sér A 286, 791–794
56. Rossmann, J., Sändig, A.-M. (1995): Formulas for the coefficients in the asymptotics of solutions of boundary value problems for second order systems near edges. Preprint 95-6, University Stuttgart
57. Sändig, A.-M. (1990): Error Estimates for Finite Element Solutions of Elliptic Boundary Value Problems in Non Smooth Domains. ZAA 9, 2,133–153
58. Sändig, A.-M. (1995): Boundary integral equations for coupled structures in solid mechanics — The direct method. Preprint 95–20, University Stuttgart
59. Sändig, A.-M., Richter, U., Sändig, R. (1989): The Regularity of Boundary Value Problems for the Lamé Equations in a Polygonal Domain. Rostocker Math. Kolloquium 36, 21–50
60. Sändig, A.-M., Sändig, R. (1992): Singularities of Non-Rotationally Symmetric Solutions of Boundary Value Problems for the Lamé Equations in a 3 dimensional Domain with Conical Points. Teubner-Texte zur Mathematik 131, 181–193
61. Schmitz, H. (1989): Über das singuläre Verhalten der Lösungen von Integralgleichungen auf Flächen mit Ecken. Ph.D. thesis, University Stuttgart
62. Schmitz, H., Volk, K., Wendland, W.L. (1993): On three-dimentional singularities of elastic fields near vertices. Numerical Methods for Partial Differential Equations 9, 323–337
63. Schulze, B.-W. (1988): Regularity with continuous and branching asymptotics for elliptic operators on manifolds with edges. Integral Equations, Operator Theory 11, 557–602
64. Stephan, E.P., Wendland, W.L. (1984): An augmented Galerkin procedure for the boundary integral method applied to two-dimensional screen and crack problems. Appl. Anal. 18, 183–220
65. Stephan, E.P., Wendland, W.L. (1990): A hypersingular boundary integral method for two-dimensional screen and crack problems. Archive Rat.Mech. Anal. 112 363–390
66. Wendland, W.L. (1987): Strongly elliptic boundary integral equations. In: The state of the Art in Numerical Analysis, pp.511–561. Clarendon, Oxford
67. Wendland, W.L., Stephan, E.P., Hsiao, G.C. (1979): On the integral equation method for the plane mixed boundary value problem with the Laplacian. Math.Methods Appl. Sci. 1, 265–321
68. Ting, T.C.T. (1986): Explicit solution and invariance of the singularities at an interface crack in anisotropic composites. Int. J. Solids Structures 22, 965–983
69. Williams, M.L. (1959): Stresses Around a Fault of Crack in Dissimilar Media. Bull. Seismol. Sec. America 49

Local, Residual-Based A Posteriori Error Estimates Forcing Adaptive Boundary Element Methods

H. Schulz and W. L. Wendland

1 Introduction

The aim of this contribution is to prove local residual-based a posteriori error estimates for a wide class of *nonlocal* operators arising in the boundary element method corresponding to the Galerkin discretization scheme. First results of this type were proved by Saranen [14, 15], Saranen and Wendland [16] and Wendland and Yu [25]. These results provide estimates only on fixed parts of the boundary surface, which is not appropriate for adaptive methods. Here we present a new result to overcome this disadvantage and present for the first time *local* a posteriori error bounds. Furthermore we present a new technique to calculate Sobolev norms with non-integer Sobolev indices based on a window technique.

A different approach to adaptivity was used by Carstensen & Stephan who localize in [2] the standard a posteriori residual error bounds to drive adaptivity and control the *global* error. Faermann shows in [5, 6] that the local Babuška-Rheinboldt error indicator can also be used to drive adaptivity for boundary element methods for boundary integral operators of orders $2\alpha > -1$ and permits in this case *global* error inclusions. There it is also shown that, for collocation methods with continuous splines of odd order and boundary integral equations on curves, her approach also works for $2\alpha = -1$.

In the first section of this paper we give a short introduction into the problem and the background of boundary value problems. We then introduce the Galerkin method and revoke some well known global a priori and a posteriori error estimates [9].

In general the operators arising in boundary element methods are *nonlocal* pseudodifferential operators, which will be considered in the sections 2, 3 and 4. The crucial point of section 4 will be the continuity property of pseudodifferential operators as a mapping between Sobolev spaces. This continuity is well known [7, 11, 23], but for the local a posteriori estimates in section 5 we need the corresponding norms in an *explicit form*. For this purpose we revisit this well known result in theorem 4.1 in section 4. The proof will be given in the appendix.

In section 5 we first present in theorem 5.1 a result similar to the result in [16] but having the same disadvantage as mentioned above. The main tool for the proof of theorem 5.1 is the *commutator property* of pseudodifferential operators. That means: If A is an operator of order $\alpha \in \mathbb{R}$ and $\omega \in C^\infty$, then the operator $[\omega, A] := (\omega A - A\omega)$ is again a pseudodifferential operator, now of order $2\alpha - 1$. The function ω plays the rôle of a truncation function for localizing a posteriori error estimates. If the measure of the support of ω is proportional to the meshsize then the localizing effects are destroyed if the meshsize parameter h tends to zero. Therefore we carefully control the norm of the pseudodifferential operator $[\omega, A]$ with respect to decreasing support of ω. This is done in subsection 5.2. Using that result, one obtains a localized a posteriori error estimate on *mesh-dependent* parts of the boundary surface, as presented in theorem 5.2. They are based on the localized residual.

In section 6 we present a new method for the computation of Sobolev norms with non-integer Sobolev index. The method is based on a certain window technique which is related to the ideas presented in [19, 20] and leads to the solution of quasi-local problems i.e. that we have to solve local Galerkin equations with a globally defined right-hand side.

In section 7 we present an adaptive feedback algorithm based on that window technique and show its efficiency for two simple numerical examples.

2 Pseudodifferential Operators

Here we recall definitions and properties of pseudodifferential operators following the presentations in [11] and [23]. For the following well known definitions we consider a bounded domain $X \subset \mathbb{R}^n$.

Definition 2.1. *Let* $0 \le \rho, \delta_1, \delta_2 \in \mathbb{R}$ *be given numbers.* $p(x, y, \xi)$ *is called a* symbol of order $m \in \mathbb{R}$ *in the class of symbols* $S^m_{\rho,\delta_1,\delta_2}(X \times X \times \mathbb{R}^n)$*, iff for every multi-index* $(\alpha_1, \alpha_2, \alpha_3)$ *there exists a constant* $C_{\alpha_1,\alpha_2,\alpha_3}$ *such that the estimate*

$$|D_x^{\alpha_1} D_y^{\alpha_2} D_\xi^{\alpha_3} p(x, y, \xi)| \le C_{\alpha_1,\alpha_2,\alpha_3}(1 + |\xi|)^{m-\rho|\alpha_3|+\delta_1|\alpha_1|+\delta_2|\alpha_2|} \tag{2.1}$$

holds for all $x, y \in X$. *The corresponding* pseudodifferential operator B of order m *is defined by the oscillatory integral*

$$(Bu)(x) := \frac{1}{(2\pi)^n} \int_{\mathbb{R}^n}\int_{\mathbb{R}^n} e^{i(x-y)^\top \xi} p(x, y, \xi) u(y) dy d\xi. \tag{2.2}$$

For the following we consider classical pseudodifferential operators, i.e. we assume $\rho = 1$ and $\delta_1 = \delta_2 = 0$, that means

$$|D_x^{\alpha_1} D_y^{\alpha_2} D_\xi^{\alpha_3} p(x, y, \xi)| \le C_{\alpha_1,\alpha_2,\alpha_3}(1 + |\xi|)^{m-|\alpha_3|}. \tag{2.3}$$

We denote the corresponding symbol and operator classes by $S^m(X \times X \times \mathbb{R}^n)$ and $\mathrm{Op}S^m$, respectively. Furthermore we set $S^\infty := \cup S^m$.

Sometimes an additional regularization is necessary:

Definition 2.2. *We define the space of* pseudodifferential operators $\Psi^m(X)$ *as the space of operators* B *of the form*

$$B = B_0 + R \tag{2.4}$$

with $B_0 \in OpS^m$ *and* $R \in OpS^{\infty}$.

Next we define pseudodifferential operators on the boundary surface Γ. For simplicity we consider a manifold $\Gamma \in C^{\infty}$.

Definition 2.3. *Let* $\Gamma \in C^{\infty}$ *be a compact manifold and* $\{(\mathcal{U}_i, \Phi_i), i \in I\}$ *(with a finite index set* I*) a corresponding atlas (see e.g. [26] for definitions). Let*

$$B: \quad C_0^{\infty}(\Gamma) \longrightarrow C^{\infty}(\Gamma)$$

be a linear continuous operator. Then the operator B *is called a* pseudodifferential operator of order m on Γ, *iff for every local chart* $(\mathcal{U}_i, \Phi_i)$, *the operator*

$$\tilde{B}_i u := (B(u \circ \Phi_i)) \circ \Phi_i^{-1} \tag{2.5}$$

is a pseudodifferential operator of order m *as in definition 2.1, over* $\Phi_i(\mathcal{U}_i \cap \Gamma)$.

For simplicity we shall identify the manifold Γ with a domain $X \subset \mathbb{R}^n$.

For solving equations with pseudodifferential operators by using Galerkin methods with boundary elements it turned out, that the property of *strong ellipticity* [22] is essential (see [24]).

Definition 2.4. *A pseudodifferential operator* A *is called* strongly elliptic *iff there exist a positive constant* γ_0 *and a complex function* $\Theta \in C^{\infty}(\Gamma)$ *such that for all* $x \in \Gamma$ *and all* $\eta \in \mathbb{R}^n$ *with* $|\eta| = 1$, *the principal symbol* $a_0(x, \eta)$ *of* A *satisfies*

$$Re\, \Theta(x) a_0(x, \eta) \geq \gamma_0. \tag{2.6}$$

This definition is a simplification of strong ellipticity of a $p \times p$ matrix of pseudodifferential operators, given e.g. in [17].

3 Pseudodifferential Equations

Let us consider a domain Ω in $\mathbb{R}^N$, $N = 2, 3$, with (at least) Lipschitz continuous boundary surface $\Gamma = \partial\Omega$ of dimension $n := N - 1$; and the equation

$$Au = f \tag{3.1}$$

with a given pseudodifferential operator A of order 2α and a given right-hand side $f \in H^{-\alpha}(\Gamma)$. For the unknown solution u we define an approximation by the Galerkin-Bubnov method: Find $u_h \in S_h^{(r,d)} \subset H^{\alpha}(\Gamma)$ such that

$$\langle \varphi_h, Au_h \rangle = \langle \varphi_h, f \rangle \tag{3.2}$$

holds for all $\varphi_h \in S_h^{(r,d)}$. The finite element space $S_h^{(r,d)}$ on Γ is defined in the sense of [1] with piecewise polynomial splines with polynomial degree d and smoothness r (here $r = -1$ corresponds to the discontinuous splines) on a mesh family with maximal meshsize parameter h.

Example 3.1. We consider the *Dirichlet* boundary value problem:

$$\begin{aligned} P(x, D)U &= f_0 \quad \text{for } x \in \Omega, \\ \gamma_0 U &= g_0 \quad \text{for } x \in \Gamma \end{aligned}$$

where $P(x, D)$ is a given strongly elliptic linear differential operator of second order and f_0 and g_0 are given functions defined on Ω and Γ, respectively. γ_0 is the trace operator. Then it is well known [6, 9] that the solution U can be represented with the given trace g_0 on the boundary surface and the conormal derivative $\partial_\nu U$ by the sum of a single layer and a double layer boundary potential [8] or by a Taylor expansion [19, 20]. One possibility to compute the unknown Cauchy data $\partial_\nu U$ consists of solving a boundary integral equation of the form of equation (3.1) where the *single layer potential operator* plays the rôle of A ($2\alpha = -1$) and f is given by the boundary values of the double layer potential applied to the given trace function g_0 and a Newton potential operator applied to f_0.

Example 3.2. We consider the *Neumann* boundary value problem:

$$\begin{aligned} P(x, D)U &= f_1 \quad \text{for } x \in \Omega, \\ \partial_\nu U &= g_1 \quad \text{for } x \in \Gamma \end{aligned}$$

Now the unknown trace $\gamma_0 U = g_0$ can be computed by an integral equation of the second kind where the sum of the identity and the double layer potential operator plays the rôle of A ($2\alpha = 0$) in (3.1). f is given by the single layer potential operator applied to g_1 and a Newton potential operator applied to f_1.

For mixed boundary value problems see e.g. [21].

If (3.1) is solved via Galerkin's method then, for the Galerkin solution u_h there holds an asymptotic optimal global a priori error estimate [9]:

Proposition 3.1. *Let A be a bijective strongly elliptic pseudodifferential operator of order 2α. Then we have the optimal asymptotic a priori error estimate:*

$$||u - u_h||_{H^s(\Gamma)} \leq c\,h^{t-s}\,||u||_{H^t(\Gamma)} \tag{3.3}$$

with $2\alpha - (d+1) \leq s \leq \alpha \leq t \leq d+1$ and $\alpha < r + 1/2$ ($N = 2$) or $\alpha \leq r$ ($N = 3$). h is the maximal stepsize. The constant c is independent of the exact solution u and the meshsize h.

Remark 3.1. In [4] a modification of proposition 3.1 for Galerkin-Petrov schemes is proved. The extension to systems is done in [24].

For simplicity we assume that the operator A^{-1} exists. Then A^{-1} is again a pseudodifferential operator, but of order -2α. From the continuity of pseudodifferential operators (see [7, 11, 23] and section 4) there follows the *global a posteriori error estimate*

$$c_l \|r_h\|_{H^{s-2\alpha}(\Gamma)} \leq \|e_h\|_{H^s(\Gamma)} \leq c_r \|r_h\|_{H^{s-2\alpha}(\Gamma)} \tag{3.4}$$

where $e_h := u - u_h$ is the error of the Galerkin discretization and $r_h := f - Au_h$ is the residual. The residual is computable, so estimate (3.4) gives a *computable asymptotic bound for the error.* The constants c_l and c_r are the continuity constants of A and A^{-1} and in general not known explicitly. Desirable is now

1. A *localized* version of (3.4). First results of this type were proved in [14, 15, 16] and [25], but on *fixed* parts of the boundary surface which do not change if $h \to 0$.
2. Methods to *compute* H^α-norms (with $\alpha \notin \mathbb{N}$ e.g. $\alpha = 1/2$ or $\alpha = -1/2$) for the localized residual. One possibility to compute H^α-norms with $\alpha \notin \mathbb{N}$ of the residual is the use of the explicit Sobolev-Slobodecki norm as in [5, 6] (see e.g. [26] for the definition). With an uniform mesh and B-splines, the norms can be expressed via circulant matrices which are used in [12, 13]. A modification leading to sparse matricies was developed in [10]. For all these computations, the residual is needed at some appropriate points $x \in \Gamma$. By the definition of the residual we have

$$\begin{aligned} r_h(x) &= f(x) - Au_h(x) = f(x) - \sum_{i=1}^{N_0} u_i A\varphi_i(x) \\ &= f(x) - \sum_{i=1}^{N_0} u_i \int_{\text{supp } \varphi_i} K(y, y-x)\, \varphi_i(y) ds_y, \end{aligned}$$

   where N_0 is the dimension of $S_h^{(r,d)}$ and $K(x, z)$ denotes the Schwartz kernel of A. For each point x we have to compute N_0 integrals numerically (including singular or hypersingular and some nearly singular integrals). This means, that for the approximation of the residual in N_0 points we need at least the computation of N_0^2 integrals.
   Another possibility for computing $H^\alpha(\Gamma)$-norms of the localized residual is the *window technique.* There we will obtain that the computation of the norm of the localized residual is equivalent to the solution of a certain local problem.

4 Continuity of Pseudodifferential Operators

Theorem 4.1. *Let $p(x,y,\xi) \in S^m(\Gamma \times \Gamma \times \mathbb{R}^n)$ be a symbol which satisfies the decay conditions (2.3). Then the corresponding pseudodifferential operator B is a continuous mapping from $H^s(\Gamma)$ to $H^{s-m}(\Gamma)$ for any $s \in \mathbb{R}$ and there holds the estimate*

$$||B||_{H^s,H^{s-m}} \leq c(s,m)\, \hat{C}_{\ell_0,k_0;0} \cdot M_{|s|-2\ell_0}(\delta) \cdot M_{|s-m|-2k_0}(\delta) \qquad (4.1)$$

with $\mathbb{N} \ni \ell_0 > \frac{1}{2}(n+|s|)$, $\mathbb{N} \ni k_0 > \frac{1}{2}(n+|s-m|)$ and (see [11])

$$M_t(u) := (2\pi)^{-n/2} \int\limits_{\mathbb{R}^n} (1+|\xi|)^t \, |\hat{u}(\xi)| \, d\xi. \qquad (4.2)$$

δ is the Dirac measure at the origin. The constant $C_{\ell_0,k_0;0}$ can be computed from the constants $C_{\alpha_1,\alpha_2,\alpha_3}$ of the decay conditions (2.3) by

$$\hat{C}_{\ell_0,k_0;\alpha_3} := \sum_{i=0}^{\ell_0} \sum_{j=0}^{k_0} \binom{\ell_0}{i} \binom{k_0}{j} \sum_{|\gamma_1|=2i} \sum_{|\gamma_2|=2j} C_{\gamma_1,\gamma_2,\alpha_3}. \qquad (4.3)$$

The constant $c(s,m)$ is independent *of the symbol $p(x,y,\xi)$.*

The **proof** of theorem 4.1 is given in the appendix.

Consequence: To control the norm of a pseudodifferential operator B : $H^s(\Gamma) \to H^{s-m}(\Gamma)$ it is necessary to control the derivatives of the symbol with respect to x up to an order $\ell_0 > \frac{1}{2}(n+|s|)$ and with respect to y up to an order $k_0 > \frac{1}{2}(n+|s-m|)$. This knowledge will be decisive in section 5.2.

5 Localized A Posteriori Error Estimates

5.1 Localized A Posteriori Error Estimates on Fixed Parts of the Boundary Surface

In this subsection we show a simple error estimate similar to the estimates in [14, 15, 16, 25] by using the commutator property [7, 23]

$$||[\omega, A]v||_{H^{s-2\alpha+1}(\Gamma)} \leq C(\omega, A)\, ||v||_{H^s(\Gamma)} \quad \text{for all } v \in H^s(\Gamma) \qquad (C)$$

and for any function $\omega \in C^\infty(\Gamma)$. The operator $[\omega, A]$ denotes the commutator

$$[\omega, A]v := (\omega A - A\omega)v = \omega(x) \cdot (Av)(x) - (A(\omega v))(x).$$

The fundamental relation for the development of a localized a posteriori error estimate is the *error equation*:

$$Ae_h = r_h \tag{5.1}$$

and the reformulation of (5.1) into

$$A\omega e_h + [\omega, A]e_h = \omega r_h \tag{5.2}$$

for arbitrary ω. Based on this formulation and the commutator property we obtain the following theorem:

Theorem 5.1. *Let A be a strongly elliptic pseudodifferential operator of order 2α with bounded inverse A^{-1}. Let Γ_i be an element of the decomposition of the boundary surface in to finite elements and let $\tilde{\Gamma}_i$ be a vicinity of Γ_i, i.e. $\Gamma_i \Subset \tilde{\Gamma}_i \subset \Gamma$. We consider truncation functions $\omega_i \in C^\infty(\Gamma)$ with the following properties:*

1. *$\omega_i(x) = 1$ for all $x \in \Gamma_i$,*
2. *$\operatorname{supp}\omega_i \subset \tilde{\Gamma}_i$,*
3. *$\omega_i(x) \geq 0$ for all $x \in \Gamma$.*

Then there holds the following estimate for the localized error:

$$c_l||\omega_i r_h||_{H^{s-2\alpha}(\Gamma)} - \varepsilon_h^{(l)}(u) \leq ||\omega_i e_h||_{H^s(\Gamma)} \leq c_r||\omega_i r_h||_{H^{s-2\alpha}(\Gamma)} + \varepsilon_h^{(r)}(u) \tag{5.3}$$

with the perturbation terms

$$\varepsilon_h^{(l,r)}(u) = \tilde{c}_{l,r} h^{t-s+1}||u||_{H^t(\Gamma)}$$

and $2\alpha - (d+1) \leq s-1 \leq \alpha \leq t \leq d+1$, $\alpha < r + 1/2$ ($N = 2$) or $\alpha \leq r$ ($N = 3$). h is the maximal meshsize.

Proof. For the estimate of the localized error from above we apply the operator A^{-1} to equation (5.2) and obtain

$$\omega_i e_h = A^{-1}\omega_i r_h - A^{-1}[\omega_i, A]e_h. \tag{5.4}$$

From (5.4) there follows with the commutator property

$$\begin{aligned} ||\omega_i e_h||_{H^s(\Gamma)} &\leq ||A^{-1}\omega_i r_h||_{H^s(\Gamma)} + ||A^{-1}[\omega_i, A]e_h||_{H^s(\Gamma)} \\ &\leq c_r \cdot \left(||\omega_i r_h||_{H^{s-2\alpha}(\Gamma)} + ||[\omega_i, A]||_{H^{s-1},H^{s-2\alpha}}||e_h||_{H^{s-1}(\Gamma)}\right) \end{aligned}$$

with $c_r = ||A^{-1}||_{H^{s-2\alpha},H^s}$. Proposition 3.1 with $2\alpha - (d+1) \leq s-1 \leq \alpha \leq t \leq d+1$ leads to

$$||e_h||_{H^{s-1}(\Gamma)} \leq c_G h^{t-s+1}||u||_{H^t(\Gamma)},$$

which concludes the proof for the estimate from above. We obtain for $\tilde{c}_r$

$$\tilde{c}_r = c_G \cdot ||A^{-1}||_{H^{s-2\alpha},H^s} \cdot ||[\omega_i, A]||_{H^{s-1},H^{s-2\alpha}}. \tag{5.5}$$

For the estimate of the localized error from below we obtain in a similar way

$$\begin{aligned}||\omega_i r_h||_{H^{s-2\alpha}(\Gamma)} &\leq ||A\omega_i e_h||_{H^{s-2\alpha}(\Gamma)} + ||[\omega_i, A]e_h||_{H^{s-2\alpha}(\Gamma)} \\ &\leq ||A||_{H^s,H^{s-2\alpha}}||\omega_i e_h||_{H^s(\Gamma)} + ||[\omega_i, A]||_{H^{s-1},H^{s-2\alpha}}||e_h||_{H^{s-1}(\Gamma)}.\end{aligned}$$

This leads for c_l to the expression $c_l = 1/||A||_{H^s,H^{s-2\alpha}}$. The norm $||e_h||_{H^{s-1}(\Gamma)}$ we estimate again by using proposition 3.1. Then we obtain

$$\tilde{c}_l = c_G \cdot \frac{1}{||A||_{H^s,H^{s-2\alpha}}} \cdot ||[\omega_i, A]||_{H^{s-1},H^{s-2\alpha}}. \tag{5.6}$$

Remark 5.1. The behaviour of the commutator norm $||[\omega_i, A]||_{H^{s-1},H^{s-2\alpha}}$ is essential for the behaviour of the perturbation terms in theorem 5.1. Using the "inverse" error equation

$$A^{-1}r_h = e_h$$

one obtains another commutator and different Sobolev indices. Based on the equation

$$A^{-1}\omega r_h + [\omega, A^{-1}]r_h = \omega e_h \tag{5.7}$$

one obtains essentially the same inequality as in (5.3). Using this approach for the behaviour of the perturbation terms, the behaviour of the commutator norm $||[\omega_i, A^{-1}]||_{H^{s-2\alpha-1},H^s}$ is decisive. In the context of the generalization of the local error estimate (5.3) in section 5.2 to h-dependent parts of the boundary surface this fact will be particulary important.

Remark 5.2. In theorem 5.1, for the estimate of the localized error in the $H^s(\Gamma)$-norm, we have used the fact that the error can optimally be estimated in the $H^{s-1}(\Gamma)$-norm in order to obtain perturbation terms which are of one order less than the error itself. We have not used the optimal error estimate up to the order $2\alpha - (d+1)$. There are two ways to obtain error estimates with much smaller perturbation terms:

- Recursive application of the commutator property.
- Using the *product property:* The operator $\omega_1 B \omega_2$ is a pseudodifferential operator of order $-\infty$, provided $\operatorname{supp}\omega_1 \cap \operatorname{supp}\omega_2 = \emptyset$.

Both ways have the disadvantage that the constants c_l and c_r depend on the truncation function ω.

5.2 Generalization of the Commutator Property

Definition 5.1. *Let Γ_i be an element of the decomposition of the boundary surface in to finite elements and let $\varepsilon_i > 0$ be a given number. We define an associated vicinity of Γ_i by:*

$$\Gamma_i^{(\varepsilon_i)} := \{x \in \Gamma : \ \exists y \in \Gamma_i \ \textit{with} \ \operatorname{dist}(x,y) \leq \varepsilon_i\} \subset \Gamma \tag{5.8}$$

We call a function $\omega_i(x)$ an admissible truncation function, *iff*

1. $supp\, \omega_i(x) \subset \Gamma_i^{(\varepsilon_i)}$,
2. $\omega_i(x) = 1$ *on* Γ_i, $0 \le \omega_i(x) \le 1$ *for all* $x \in \Gamma$,
3. $|D^\alpha \omega_i(x)| \le c_D\, \varepsilon_i^{-|\alpha|}$ *for all* $x \in \Gamma$.

For brevity we write in the following ε *instead of* ε_i.

Lemma 5.1. *Let* B *be a pseudodifferential operator of order* $m = 2\beta$ *with symbol* $p(x, y, \xi)$ *satisfying the decay condition (2.3). Furthermore let* ω_i *be an admissible truncation function. Then, for fixed* ε, *the commutator*

$$(B_i u)(x) := [\omega_i, B]u(x) = \omega_i(x)(Bu)(x) - (B\omega_i u)(x)$$

is a pseudodifferential operator of order $2\beta - 1$. *Furthermore, for the symbol of* $(B_i u)(x)$ *one obtains the decay condition*

$$|D_x^{\alpha_1} D_y^{\alpha_2} D_\xi^{\alpha_3} \tilde{p}_i(x, y, \xi)| \le \tilde{C}'_{\alpha_1,\alpha_2,\alpha_3} \cdot \varepsilon^{-|\alpha_1+\alpha_2|-1} \cdot (1 + |\xi|)^{2\beta-1-|\alpha_3|}, \quad (5.9)$$

where the constant $\tilde{C}'_{\alpha_1,\alpha_2,\alpha_3}$ is independent *of the behavior of the truncation function* ω_i.

Proof. From the definition (2.2) of the pseudodifferential operator B there follows

$$(B_i v)(x) = (2\pi)^{-n} \int_{\mathbb{R}^n} \int_X e^{\mathrm{i}(x-y)^\top \xi} [\omega_i(x) - \omega_i(y)] p(x, y, \xi) v(y) dy d\xi. \quad (5.10)$$

The difference $\omega_i(x) - \omega_i(y)$ can be written in the following manner:

$$\omega_i(x) - \omega_i(y) = \sum_{j=1}^{n} (\delta_j \omega_i)(x, y) \qquad \text{with} \qquad (5.11)$$

$$(\delta_j \omega_i)(x, y) := \omega_i(y_1, ..., y_{j-1}, x_j, x_{j+1}, ..., x_n) - \omega_i(y_1, ..., y_{j-1}, y_j, x_{j+1}, ..., x_n).$$

Using (5.11) and the relation $D_{\xi_j} e^{\mathrm{i}(x-y)^\top \xi} = \mathrm{i}(x_j - y_j) e^{\mathrm{i}(x-y)^\top \xi}$ one obtains from (5.10)

$$(B_i v)(x) = (2\pi)^{-n} (-\mathrm{i}) \sum_{j=1}^{n} \int_{\mathbb{R}^n} \int_X D_{\xi_j} e^{\mathrm{i}(x-y)^\top \xi} h_j^{(i)}(x, y) p(x, y, \xi)\, v(y) dy d\xi.$$

where

$$h_j^{(i)}(x, y) := \frac{(\delta_j \omega_i)(x, y)}{x_j - y_j}.$$

Partial integration with respect to ξ_j then yields

$$(B_i v)(x) = (2\pi)^{-n}\, \mathrm{i} \sum_{j=1}^{n} \int_{\mathbb{R}^n} \int_X e^{\mathrm{i}(x-y)^\top \xi}\, h_j^{(i)}(x, y) D_{\xi_j} p(x, y, \xi)\, v(y)\, dy d\xi.$$

This leads for the symbol of $B_i(x)$ to the representation

$$\tilde{p}_i(x,y,\xi) := \mathrm{i}\cdot\sum_{j=1}^{n} h_j^{(i)}(x,y) D_{\xi_j} p(x,y,\xi). \tag{5.12}$$

Now we have to justify an estimate of the form

$$\left|D_x^{\alpha_1} D_y^{\alpha_2} D_\xi^{\alpha_3} \tilde{p}_i(x,y,\xi)\right| \le \tilde{C}_{\alpha_1,\alpha_2,\alpha_3} (1+|\xi|)^{2\beta-1-|\alpha_3|}. \tag{5.13}$$

The Leibniz rule gives

$$D_x^{\alpha_1} D_y^{\alpha_2} D_\xi^{\alpha_3} \tilde{p}_i(x,y,\xi) = \mathrm{i}\sum_{j=1}^{n} D_x^{\alpha_1} D_y^{\alpha_2} \left\{ h_j^{(i)}(x,y) D_\xi^{\alpha_3} D_{\xi_j} p(x,y,\xi)\right\}$$

$$= \mathrm{i}\sum_{\substack{\beta_1\le\alpha_1\\ \beta_2\le\alpha_2}} \binom{\alpha_2}{\beta_2}\binom{\alpha_1}{\beta_1} \sum_{j=1}^{n} D_x^{\alpha_1-\beta_1} D_y^{\alpha_2-\beta_2} h_j^{(i)}(x,y)\cdot D_x^{\beta_1} D_y^{\beta_2} D_\xi^{\alpha_3} D_{\xi_j} p(x,y,\xi) \tag{5.14}$$

Elementary computations show that

$$\left|D_x^{\gamma_1} D_y^{\gamma_2} h_j^{(i)}(x,y)\right| \le c_{\mathrm{D}}\, \varepsilon^{-|\gamma_1+\gamma_2|-1}, \quad j=1,\dots n, \tag{5.15}$$

holds for arbitrary multi-indices γ_1 and γ_2. Using (5.15) in (5.14) we obtain for the decay of $\tilde{p}_i(x,y,\xi)$ the estimate

$$\left|D_x^{\alpha_1} D_y^{\alpha_2} D_\xi^{\alpha_3} \tilde{p}_i(x,y,\xi)\right| \le c_{\mathrm{D}}\cdot \sum_{\substack{\beta_1\le\alpha_1\\ \beta_2\le\alpha_2}} \binom{\alpha_2}{\beta_2}\binom{\alpha_1}{\beta_1} \varepsilon^{-|\alpha_1-\beta_1+\alpha_2-\beta_2|-1}\cdot$$
$$\cdot\sum_{j=1}^{n} \left|D_x^{\beta_1} D_y^{\beta_2} D_\xi^{\alpha_3} D_{\xi_j} p(x,y,\xi)\right|.$$

Using the decay condition (2.3) of $p(x,y,\xi)$ we obtain

$$\left|D_x^{\alpha_1} D_y^{\alpha_2} D_\xi^{\alpha_3} \tilde{p}_i(x,y,\xi)\right| \le c_{\mathrm{D}}\cdot \sum_{\substack{\beta_1\le\alpha_1\\ \beta_2\le\alpha_2}} \binom{\alpha_2}{\beta_2}\binom{\alpha_1}{\beta_1} \varepsilon^{-|\alpha_1-\beta_1+\alpha_2-\beta_2|-1}\cdot$$
$$\cdot\bar{C}_{\beta_1,\beta_2,\alpha_3} (1+|\xi|)^{2\beta-1-|\alpha_3|},$$

where

$$\bar{C}_{\beta_1,\beta_2,\alpha_3} := \sum_{j=1}^{n} C_{\beta_1,\beta_2,\alpha_3+1_{(j)}}$$

and $C_{\beta_1,\beta_2,\alpha_3+1_{(j)}}$ are the constants from inequality (2.3). Here $1_{(j)}$ means the multi-index with 1 in the j-th position and otherwise 0. So we obtain for the decay of $\tilde{p}_i(x,y,\xi)$

$$\left| D_x^{\alpha_1} D_y^{\alpha_2} D_\xi^{\alpha_3} \tilde{p}_i(x,y,\xi) \right| \leq \tilde{C}_{\alpha_1,\alpha_2,\alpha_3} \cdot (1+|\xi|)^{2\beta-1-|\alpha_3|} \tag{5.16}$$

where

$$\tilde{C}_{\alpha_1,\alpha_2,\alpha_3} := c_{\mathrm{D}} \sum_{\substack{\beta_1 \leq \alpha_1 \\ \beta_2 \leq \alpha_2}} \binom{\alpha_2}{\beta_2} \binom{\alpha_1}{\beta_1} \varepsilon^{-|\alpha_1-\beta_1+\alpha_2-\beta_2|-1} \cdot \bar{C}_{\beta_1,\beta_2,\alpha_3}.$$

For $\varepsilon < 1$ and $0 \leq \beta_i \leq \alpha_i$, $i = 1,2$, there holds

$$\varepsilon^{-|\alpha_1-\beta_1+\alpha_2-\beta_2|-1} \leq \varepsilon^{-|\alpha_1+\alpha_2|-1}. \tag{5.17}$$

Using (5.17) we obtain from (5.16) the desired estimate (5.9).

Using lemma 5.1 we are able to estimate the norms of the commutators $[\omega_i, A]$ and $[\omega_i, A^{-1}]$ (to control the perturbation terms in theorem 5.1) with respect to the decay of the truncation functions in dependence on ε.

Lemma 5.2. *For the norm of the commutator $[\omega_i, A]$ there holds the estimate*

$$|||[\omega_i, A]|||_{H^{s-1},H^{s-2\alpha}} \leq C\,\varepsilon^{-\ell-k-1} \tag{5.18}$$

with $\ell > n + |s-1|$ and $k > n + |s-2\alpha|$. For the norm of $[\omega_i, A^{-1}]$ we obtain

$$|||[\omega_i, A^{-1}]|||_{H^{s-2\alpha-1},H^s} \leq C'\,\varepsilon^{-\ell-k-1} \tag{5.19}$$

with $\ell > n + |s-1-2\alpha|$ and $k > n + |s|$. The constants C and C' are independent *of ε.*

Proof. We only show (5.18). Then estimate (5.19) follows in the same manner. theorem 4.1 gives, with $|s-1| - 2\ell_0 < -n$ and $|s-2\alpha| - 2k_0 < -n$,

$$|||[\omega_i, A]|||_{H^{s-1},H^{s-2\alpha}} \leq c\,\hat{C}_{\ell_0,k_0;0} \cdot M_{|s-1|-2\ell_0}(\delta) \cdot M_{|s-2\alpha|-2k_0}(\delta)$$

with $\hat{C}_{\ell_0,k_0;0}$ defined in (4.3), but using there $\tilde{C}_{\gamma_1,\gamma_2,\alpha_3}$ (from (5.13)) instead of $C_{\gamma_1,\gamma_2,\alpha_3}$ (from (2.3)). From (5.16) and (5.17) there follows

$$\tilde{C}_{\alpha_1,\alpha_2,\alpha_3} \leq \tilde{C}'_{\alpha_1,\alpha_2,\alpha_3} \cdot \varepsilon^{-|\alpha_1+\alpha_2|+1} \tag{5.20}$$

and, hence, it is possible to estimate $\hat{C}_{\ell_0,k_0;0}$ (defined in (4.3)) in the following way:

$$\begin{aligned}
\hat{C}_{\ell_0,k_0;0} &= \sum_{i=0}^{\ell_0} \sum_{j=0}^{k_0} \binom{\ell_0}{i} \binom{k_0}{j} \sum_{|\gamma_1|=2i} \sum_{|\gamma_2|=2j} \tilde{C}_{\gamma_1,\gamma_2,0} \\
&\leq \sum_{i=0}^{\ell_0} \sum_{j=0}^{k_0} \binom{\ell_0}{i} \binom{k_0}{j} \sum_{|\gamma_1|=2i} \sum_{|\gamma_2|=2j} \tilde{C}'_{\gamma_1,\gamma_2,0} \cdot \varepsilon^{-|\gamma_1+\gamma_2|-1} \\
&\leq \varepsilon^{-2\ell_0-2k_0-1} \cdot \sum_{i=0}^{\ell_0} \sum_{j=0}^{k_0} \binom{\ell_0}{i} \binom{k_0}{j} \sum_{|\gamma_1|=2i} \sum_{|\gamma_2|=2j} \tilde{C}'_{\gamma_1,\gamma_2,0},
\end{aligned}$$

where the constant $C'_{\gamma_1,\gamma_2,0}$ is independent of ε. Setting $\ell := 2\ell_0 > n + |s - 1 - 2\alpha|$ and $k := 2k_0 > n + |s-2\alpha|$ concludes the proof.

5.3 Localized A Posteriori Error Estimates on Mesh-Dependent Parts of the Boundary Surface

Theorem 5.2. *Let the assumptions of theorem 5.1 be fulfilled. Furthermore let $0 < \delta \leq 1$ be a given number and ω_i an admissible truncation function with respect to a boundary element Γ_i. In addition we assume that the mesh is β-regular, i.e. there exist constants $\beta \geq 1$ and $c_R > 0$ such that $h_{\min} \geq c_R h^\beta$. The decay parameter ε_i of the truncation function ω_i we specify by the relation*

$$\varepsilon_i \geq c_0 \, h_i^\gamma \quad \text{with } h_i := |\Gamma_i| \tag{5.21}$$

and

$$\gamma \leq \frac{1-\delta}{\beta(\ell+k+1)}. \tag{5.22}$$

ℓ and k have to satisfy either

a) $\ell > n + |s-1|$ and $k > n + |s-2\alpha|$ or
b) $\ell > n + |s-1-2\alpha|$ and $k > n + |s|$

and we choose that case where $\ell + k$ is minimal. Then for the localized error there holds

$$c_l ||\omega_i r_h||_{H^{s-2\alpha}(\Gamma)} - \varepsilon_h^{(l)}(u) \leq ||\omega_i e_h||_{H^s(\Gamma)} \leq c_r ||\omega_i r_h||_{H^{s-2\alpha}(\Gamma)} + \varepsilon_h^{(r)}(u) \tag{5.23}$$

with perturbation terms

$$\varepsilon_h^{(l,r)}(u) = \bar{c}_{l,r} h^{t-s+\delta} ||u||_{H^t(\Gamma)}$$

and $2\alpha - (d+1) \leq s-1 \leq \alpha \leq t \leq d+1$, $\alpha < r + 1/2$ ($N=2$) or $\alpha \leq r$ ($N=3$).

Proof. We assume that a) is satisfied. From the proof of theorem 5.1 (see (5.5) and (5.6)) we conclude for the perturbation terms

$$\varepsilon_h^{(l,r)}(u) \leq \tilde{c}_{l,r} \cdot h^{t-s+1} ||u||_{H^t(\Gamma)} \leq \bar{c} \cdot ||[\omega_i, A]||_{H^{s-1}, H^{s-2\alpha}} \cdot h \cdot h^{t-s} ||u||_{H^t(\Gamma)}.$$

From assumption (5.21) with (5.22) one obtains

$$\varepsilon_i^{-\ell-k-1} \cdot h \leq c \, h_i^{\gamma(\ell+k+1)} \cdot h \leq c \, h^{-\gamma\beta(\ell+k+1)} \cdot h \leq c \, h^{\delta-1} \cdot h = c \, h^\delta .$$

Using lemma 5.2 on the norm of the commutator $[\omega_i, A]$ we find

$$\varepsilon_h^{(l,r)}(u) \leq \bar{c}' \cdot \varepsilon^{-\ell-k-1} \cdot h \cdot h^{t-s} ||u||_{H^t(\Gamma)} \leq \bar{c}'' \cdot h^\delta \cdot h^{t-s} ||u||_{H^t(\Gamma)}.$$

The case of condition b) follows in the same manner.

Remark 5.3. Assume $s = \alpha$. Then the conditions a) and b) for the numbers ℓ and k from theorem 5.2 are

a) $\ell > n + |\alpha - 1|$ and $k > n + |\alpha|$ or

b) $\ell > n + |\alpha + 1|$ and $k > n + |\alpha|$.

To minimize ℓ we have to distinguish between the case that A is a pseudodifferential operator of positive order and the case of negative order. In the case of positive order it is obvious that condition a) requires a smaller ℓ (that means a smaller support of ω_i) than condition b); and vice versa in the case of an operator of negative order.

6 Window Techniques for the Computation of Norms of the Localized Residual

Let ω_i be an admissible truncation function. We define the following family of problems: Find functions $g^{(i)} \in H^\alpha(\Gamma)$ such that

$$Ag^{(i)} = \omega_i f - [\omega_i, A]u_h \tag{6.1}$$

holds for $i = 1, ..., N_1$, where N_1 is the number of boundary elements. Furthermore we define

$$\eta^{(i)} := g^{(i)} - \omega_i u_h. \tag{6.2}$$

Lemma 6.1. *Let the assumptions of theorem 5.2 and $A = A^*$ be fulfilled. Then there holds*

$$||\omega_i u - g^{(i)}||_{H^s(\Gamma)} \leq ||u - u_h||_{H^{s-1}(\Gamma)} \leq ch^{t-s+\delta}||u||_{H^t(\Gamma)} \tag{6.3}$$

with $2\alpha - (d+1) \leq s - 1 \leq \alpha \leq t \leq d+1$ *and*

$$||\eta^{(i)}||_E = ||\omega_i r_h||_{E^-}. \tag{6.4}$$

Here $||\cdot||_E := \langle \cdot, A(\cdot)\rangle$ *is the energy norm with respect to the operator* A*;* $||\cdot||_{E^-}$ *is the energy norm with respect to* A^{-1}.

Proof. For the proof of (6.3) we first consider case a) (see the assumptions of theorem 5.2). From (6.1) and the equation

$$A\omega_i u = \omega_i f - [\omega_i, A]u$$

there follows

$$A(\omega_i u - g^{(i)}) = -[\omega_i, A](u - u_h). \tag{6.5}$$

Using lemma 5.2 we obtain

$$\begin{aligned} ||\omega_i u - g^{(i)}||_{H^s(\Gamma)} &\leq ||[\omega_i, A]||_{H^{s-1}, H^{s-2\alpha}} ||u - u_h||_{H^{s-1}(\Gamma)} \\ &\leq c\varepsilon_i^{-2\ell-2k-1} ||u - u_h||_{H^{s-1}(\Gamma)}. \end{aligned}$$

Applying proposition 3.1 one gets (6.3). In the case of condition b) we use the relation

$$[\omega_i, A] = -A[\omega_i, A^{-1}]A,$$

then estimate the norm of the commutator $[\omega_i, A]$ by

$$||[\omega_i, A]||_{H^{s-1},H^{s-2\alpha}} \le ||A||_{H^s,H^{s-2\alpha}}||[\omega_i, A^{-1}]||_{H^{s-1-2\alpha},H^s}||A||_{H^{s-1},H^{s-1-2\alpha}}$$

and apply lemma 5.2 for $||[\omega_i, A^{-1}]||_{H^{s-1-2\alpha},H^s}$.

Now we prove equation (6.4). From the definition (6.1) of the functions $g^{(i)}$ there follows

$$Ag^{(i)} = \omega_i f - (\omega_i A - A\omega_i)u_h = \omega_i r_h + A\omega_i u_h,$$

therefore we have

$$A(g^{(i)} - \omega_i u_h) = \omega_i r_h. \tag{6.6}$$

With the help of (6.6) it is easy to obtain (6.4), since

$$\begin{aligned} ||\eta^{(i)}||_E^2 &= \langle \eta^{(i)}, A\eta^{(i)}\rangle = \langle g^{(i)} - \omega_i u_h, A(g^{(i)} - \omega_i u_h)\rangle \\ &= \langle g^{(i)} - \omega_i u_h, \omega_i r_h\rangle = \langle A(g^{(i)} - \omega_i u_h), A^{-1}\omega_i r_h\rangle \\ &= \langle \omega_i r_h, A^{-1}\omega_i r_h\rangle = ||\omega_i r_h||_{E^-}^2. \end{aligned}$$

Remark 6.1. The identity (6.4) is *independent* of special properties of pseudodifferential operators (e.g. the commutator property).

Consequence: By theorem 5.2, the last lemma and the norm equivalences

$$||\cdot||_E \sim ||\cdot||_{H^\alpha(\Gamma)} \text{ and } ||\cdot||_{E^-} \sim ||\cdot||_{H^{-\alpha}(\Gamma)},$$

we obtain an estimate of the error from below and from above

$$d_l||\eta^{(i)}||_E - \varepsilon_h^{(l)}(u) \le ||\omega_i e_h||_{H^\alpha(\Gamma)} \le d_r||\eta^{(i)}||_E + \varepsilon_h^{(r)}(u)$$

with the perturbation terms

$$\varepsilon_h^{(l,r)}(u) = \hat{c}_{l,r}h^{t-s+\delta}||u||_{H^t(\Gamma)}.$$

Now let us define local approximations of the functions $g^{(i)}$ and $\eta^{(i)}$. Let

$$S_i^{(2)} \subset \tilde{H}^\alpha(\tilde{\Gamma}_i)$$

be a local space of ansatz functions of degree $d_2^{(i)}$, smoothness $r_2^{(i)}$, local meshsize $h_2^{(i)}$ and support in $\tilde{\Gamma}_i$. We define local functions $g_{h_2}^{(i)} \in S_i^{(2)}$ by

$$\langle \varphi, Ag_{h_2}^{(i)}\rangle = \langle \varphi, \omega_i f\rangle - \langle \varphi, ([\omega_i, A]u_h\rangle \qquad \forall \varphi \in S_i^{(2)} \tag{6.7}$$

and

$$\eta_{h_2}^{(i)} := g_{h_2}^{(i)} - \omega_i u_h. \tag{6.8}$$

Lemma 6.2. *Let us assume $h_2 \leq c_{\mathrm{loc}} h_i^{\varrho}$ with $\varrho \geq 1$. Then we have for the differences between the global and the local functions the estimate*

$$||g^{(i)} - g_{h_2}^{(i)}||_E \leq c \left(h^{t-\alpha+\delta} + h_2^{(1-\frac{\gamma}{\varrho})t-\alpha} \right) ||u||_{H^t(\Gamma)} \tag{6.9}$$

and

$$\Big| ||\eta^{(i)}||_E - ||\eta_{h_2}^{(i)}||_E \Big| \leq c \left(h^{t-\alpha+\delta} + h_2^{(1-\frac{\gamma}{\varrho})t-\alpha} \right) ||u||_{H^t(\Gamma)}, \tag{6.10}$$

provided $\alpha \leq t \leq \min\{d, d_2^{(i)}\} + 1$, $t \in \mathbb{N}$, and with δ, γ defined in theorem 5.2.

Proof. Equation (6.1) is equivalent to

$$\langle \varphi, A g^{(i)} \rangle = \langle \varphi, \omega_i f \rangle - \langle \varphi, ([\omega_i, A] u_h \rangle \quad \forall \varphi \in H^{\alpha}(\Gamma). \tag{6.11}$$

From equation (6.11) with $\varphi \in S_i^{(2)}$ and (6.7) one obtains

$$0 = \langle \varphi, A(g^{(i)} - g_{h_2}^{(i)}) \rangle \qquad \forall \varphi \in S_i^{(2)}. \tag{6.12}$$

Using this orthogonality and the definition of the norm $||\cdot||_E$ we find

$$\begin{aligned} ||g^{(i)} - g_{h_2}^{(i)}||_E^2 &= \langle g^{(i)} - g_{h_2}^{(i)}, A(g^{(i)} - g_{h_2}^{(i)}) \rangle \\ &= \langle g^{(i)}, A(g^{(i)} - g_{h_2}^{(i)}) \rangle \\ &= \langle g^{(i)} - \varphi, A(g^{(i)} - g_{h_2}^{(i)}) \rangle \quad \forall \varphi \in S_i^{(2)}. \end{aligned}$$

Applying the Cauchy-Schwarz inequality and dividing by $||g^{(i)} - g_{h_2}^{(i)}||_E$, this gives

$$||g^{(i)} - g_{h_2}^{(i)}||_E \leq ||g^{(i)} - \varphi||_E \quad \forall \varphi \in S_i^{(2)}. \tag{6.13}$$

With the triangle inequality and taking the minimum with respect to $\varphi \in S_i^{(2)}$ one obtains

$$||g^{(i)} - g_{h_2}^{(i)}||_E \leq ||g^{(i)} - \omega_i u||_E + \min_{\varphi \in S_i^{(2)}} ||\omega_i u - \varphi||_E. \tag{6.14}$$

With (6.3) and the norm equivalence between $||\cdot||_E$ and $||\cdot||_{H^\alpha(\Gamma)}$ we estimate the first term of the right-hand side of (6.14) by

$$||g^{(i)} - \omega_i u||_E \leq c\, ||g^{(i)} - \omega_i u||_{H^\alpha(\Gamma)} \leq c h^{t-\alpha+\delta} ||u||_{H^t(\Gamma)} \text{ with } \alpha \leq t \leq d+1.$$

For the second term in (6.14) there holds with some $\varphi^* \in S_i^{(2)}$,

$$\min_{\varphi \in S_i^{(2)}} ||\omega_i u - \varphi||_E = ||\omega_i u - \varphi^*||_E \leq c\, h_2^{t-\alpha}\, ||\omega_i u||_{H^t(\Gamma)} \text{ for } \alpha \leq t \leq d_2^{(i)} + 1.$$

From definition 5.1 we obtain

$$||\omega_i u||_{H^t(\Gamma)} \leq c\varepsilon_i^{-t}||u||_{H^t(\Gamma)}.$$

The estimate

$$\varepsilon_i^{-t} \leq ch_i^{-t\gamma} \leq c''h_2^{-t\frac{\gamma}{e}}$$

concludes the proof for inequality (6.9).

Inequality (6.10) follows directly from (6.9) and the definition of $\eta^{(i)}$ and $\eta_{h_2}^{(i)}$:

$$\begin{aligned}\left| ||\eta^{(i)}||_E - ||\eta_{h_2}^{(i)}||_E \right| &\leq ||\eta^{(i)} - \eta_{h_2}^{(i)}||_E \leq ||g^{(i)} - g_{h_2}^{(i)}||_E \\ &\leq c\left(h^{t-\alpha+\delta} + h_2^{(1-\frac{\gamma}{e})t-\alpha}\right)||u||_{H^t(\Gamma)},\end{aligned}$$

provided $\alpha \leq t \leq \min\{d, d_2^{(i)}\} + 1$.

Finally we collect our result in the following theorem:

Theorem 6.1. *Under the assumptions of theorem 5.2, lemma 6.2 and* $d \leq d_2$ *there holds for the localized error the estimate*

$$d_l||\eta_{h_2}^{(i)}||_E - \varepsilon_h^{(l)}(u) \leq ||\omega_i e_h||_{H^\alpha(\Gamma)} \leq d_r||\eta_{h_2}^{(i)}||_E + \varepsilon_h^{(r)}(u)$$

with the perturbation terms

$$\varepsilon_h^{(l,r)}(u) = \tilde{d}_{l,r}\left(h^{t-s+\delta} + h_2^{(1-\frac{\gamma}{e})t-\alpha}\right)||u||_{H^t(\Gamma)}.$$

The **proof** follows immediately from theorem 5.2 and lemma 6.2.

Remark 6.2. For a global error estimation, the result of theorem 6.1 is not useful. There are other possibilities, for instance an approximation of the error based on the error equation and elements with higher polynomial degree [18].

7 Adaptive Methods, Numerical Examples

7.1 The Adaptive Algorithm

Let $TOL > 0$ and $\theta \in [0,1]$ are given numbers. For the following two numerical examples we have realized the algorithm:

1. Compute a Galerkin solution.
2. Compute the local error estimators $\tau_i := ||\eta_{h_2}^{(i)}||_E$ for every element Γ_i, compute η as the sum of the local values τ_i.
3. If $\eta \leq TOL$ is fulfilled, stop. Otherwise refine all elements Γ_i with

   $$\tau_i \geq \theta \cdot \tau_{\max}$$

   and proceed with step 1.

In the following examples we have chosen $\theta = 0.01$.

7.2 Two Numerical Examples

Example 7.1. We consider the boundary integral equation

$$-\frac{1}{2\pi}\int_{\Gamma} \ln|x-y|u(y)\,ds_y = f(x), \quad x\in\Gamma \tag{7.1}$$

on the boundary Γ of a circle Ω with radius 1/2. The right-hand side is chosen in such a way that the solution of (7.1) is given by

$$u(x) = \frac{\partial U}{\partial n} \text{ with } U(x) = c_1 \ln|x - x_{\varepsilon_1}| + c_2 \ln|x - x_{\varepsilon_2}| + c_3 x_1 + c_4 x_2.$$

x_{ε_1} and x_{ε_2} are points outside of the circle, the parameters ε_1 and ε_2 represent the distance to the boundary. Here we have chosen $x_{\varepsilon_1} = (0, 0.501)^\top$, $x_{\varepsilon_2} = (-0.501, 0)^\top$ ($\varepsilon_1 = \varepsilon_2 = 0.001$) and $c_1 = c_2 = 1$, $c_3 = c_4 = 20$. To solve (7.1) we consider a Galerkin method with piecewise constant trial- and test-functions.

Example 7.2. We consider the same equation and the same method, but on the boundary of an L-shape like domain:

$$\Omega := \{x\in\mathbb{R}^2 : \ x = x(r,\varphi) = (r\cos\varphi, r\sin\varphi)^\top, \ r\in[0,\frac{1}{2}), \ \varphi\in(0,\varphi_0)\}$$

with $\varphi_0 < 2\pi$. The right-hand side is chosen in such a way that the solution of (7.1), given by

$$u(x) = \frac{\partial U}{\partial n} \quad \text{with} \quad U(x) = U(r,\varphi) = r^{\frac{\pi}{\varphi_0}} \sin(\pi\frac{\varphi}{\varphi_0}),$$

has a singularity at the origin. We choose $\varphi_0 = 7/4\cdot\pi$.

Figures 7.1 and 7.2 display the errors with respect to the number of degrees of freedom of the adaptive method in comparison with uniform meshes (Example 7.1):

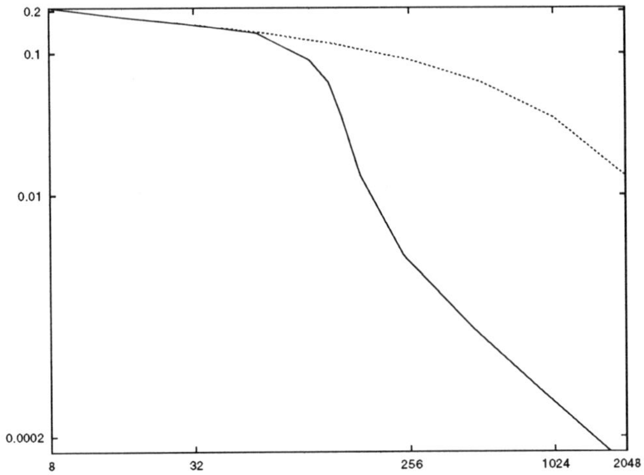

Figure 7.1: Energy error

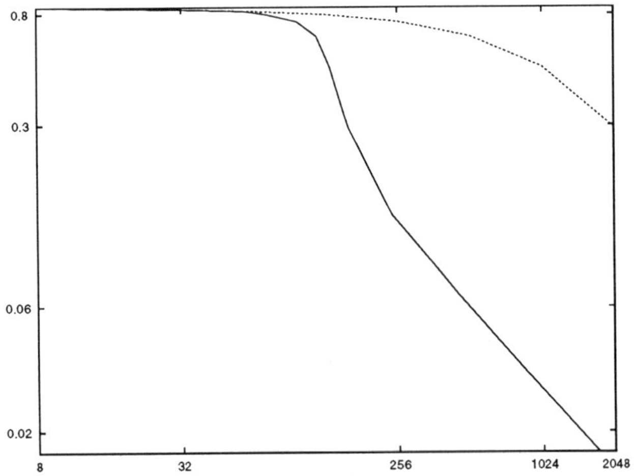

Figure 7.2: L^2-error

Remarks: In Example 7.1, the solution is nearly singular, but smooth. For that reason the error of the uniform grid refinements converges optimally, but there is a long pre-asymptotic region. The adaptive method gives optimal convergence with a significantly smaller pre-asymptotic region.

Figures 7.3 and 7.4 show the errors with respect to the computing times (sec.) of the adaptive method in comparison with uniform meshes (Example 7.1):

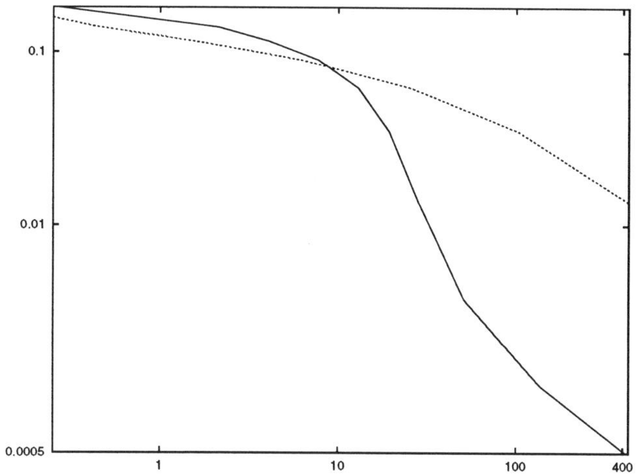

Figure 7.3: Energy error

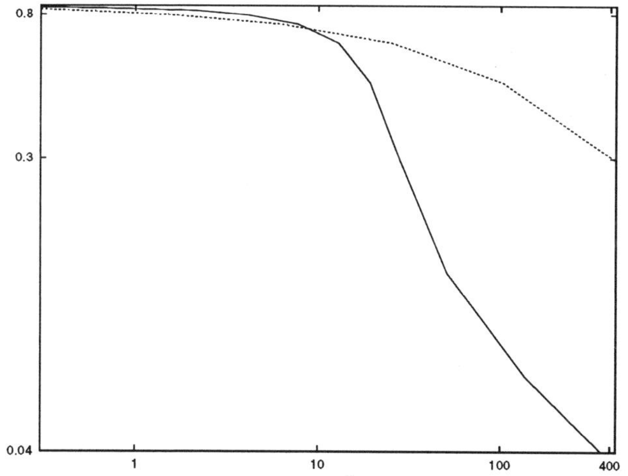

Figure 7.4: L^2-error

Remarks: The behaviour with respect to the time is in principal the same as the behaviour with respect to the number of degrees of freedom. Only in the first adaptive steps one observes that the adaptive method is a little more time consuming.

Figures 7.5 and 7.6 display the errors with respect to the number of degrees of freedom of the adaptive method in comparison with uniform meshes (Example 7.2):

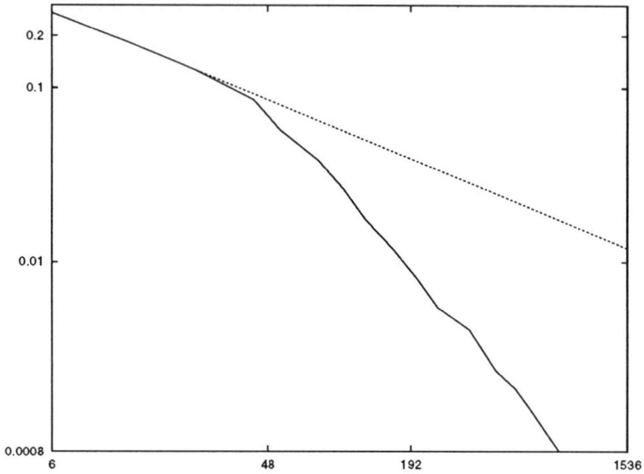

Figure 7.5: Energy error

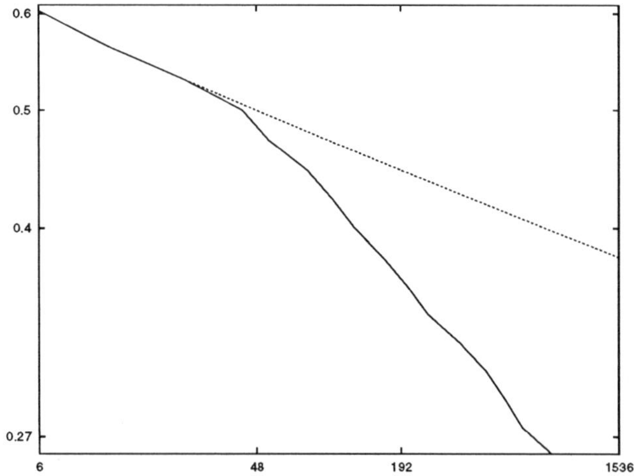

Figure 7.6: L^2-error

Remarks: In Example 7.2 the solution is nonsmooth. Here the uniform meshes cannot produce the optimal order of convergence. For the error in the energy norm the adaptive method detects the optimal convergence order. Also the error in the L^2-norm convergences much faster than on the uniformly refined grids.

Figures 7.7 and 7.8 show the errors with respect to the computing times (sec.) of the adaptive method in comparison with uniform meshes (Example 7.2):

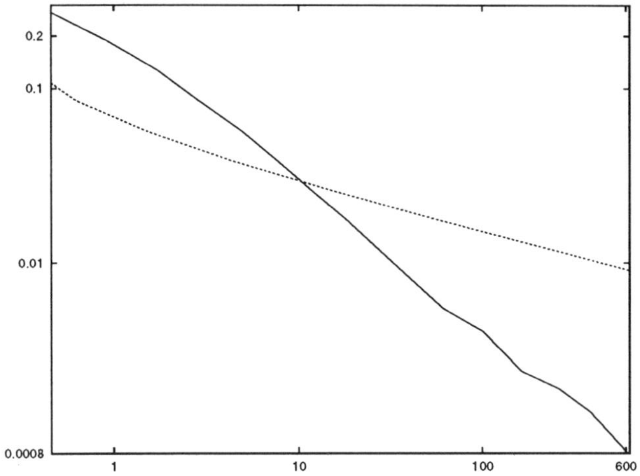

Figure 7.7: Energy error

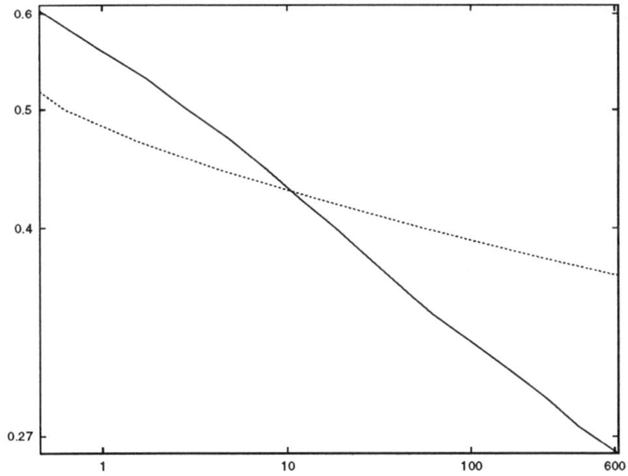

Figure 7.8: L^2-error

Remarks: The behaviour with respect to the time is in principal the same as the behaviour with respect to the number of degrees of freedom. Only in the first adaptive steps one observes that the adaptive method is more costly. A numerical comparison between $||\eta_{h_2}^{(i)}||_E$ and the real localized error in the energy norm is given in [18].

A Proof of theorem 4.1

The following proof resembles the standard proof in [23], however we will stress the explicit interaction between the constants in the decay condition of the symbol (2.3) and the operator norm in (4.1).

For the proof of theorem 4.1 we need the following basic inequalities:

- Young's inequality:

$$||f * g||_{L^r(\mathbb{R}^n)} \leq ||f||_{L^p(\mathbb{R}^n)} \, ||g||_{L^q(\mathbb{R}^n)}, \tag{A.1}$$

  where $f * g$ is the *convolution* between functions $f \in L^p$ and $g \in L^q$.
- Peetre's inequality:

$$\left(1+|x|^2\right)^t \leq (1+|x-y|)^{2|t|} \left(1+|y|^2\right)^t, \tag{A.2}$$

  where t is any real number and $x, y \in \mathbb{R}^n$.

Furthermore we shall need

$$\left(1+|x|^2\right)^t \leq (1+|x|)^{2t} \leq 2^t \left(1+|x|^2\right)^t \quad \text{for} \quad t \geq 0 \tag{A.3}$$

and

$$2^t \left(1+|x|^2\right)^t \leq (1+|x|)^{2t} \leq \left(1+|x|^2\right)^t \quad \text{for} \quad t < 0. \tag{A.4}$$

It is necessary to show that there holds:

$$||Bu||_{H^{s-m}(\Gamma)} \leq \hat{C}_{\ell_0,k_0;0} \cdot M_{|s|-2\ell_0}(\delta) \cdot M_{|s-m|-2k_0}(\delta) \cdot ||u||_{H^s(\Gamma)} \quad \forall u \in C^\infty(\Gamma), \tag{A.5}$$

where δ is the Dircac measure at the origin. We denote by $\hat{p}$ the Fourier transform of the symbol $p(x,y,\xi)$ with respect to x and to y:

$$\hat{p}(\xi', \eta', \xi) = (2\pi)^{-n} \int\limits_X e^{-\mathrm{i}y^\top \eta'} \int\limits_X e^{-\mathrm{i}x^\top \xi'} p(x,y,\xi) \, dx dy \tag{A.6}$$

In the first step we prove an estimate for the derivatives $D_\xi^{\alpha_3} \hat{p}(\xi', \eta', \xi)$. With

$$(1-\Delta_x)^{k_0} e^{-\mathrm{i}x^\top \xi'} = \left(1+|\xi'|^2\right)^{k_0} e^{-\mathrm{i}x^\top \xi'}$$

and

$$(1-\Delta_y)^{\ell_0} e^{-\mathrm{i}y^\top \eta'} = \left(1+|\eta'|^2\right)^{\ell_0} e^{-\mathrm{i}y^\top \eta'},$$

there follows by integration by parts

$$\hat{p}(\xi', \eta', \xi) = (2\pi)^{-n} \left(1+|\xi'|^2\right)^{-k_0} \left(1+|\eta'|^2\right)^{-\ell_0} \cdot$$
$$\cdot \int\limits_X e^{-\mathrm{i}y^\top \eta'} \int\limits_X e^{-\mathrm{i}x^\top \xi'} (1-\Delta_y)^{\ell_0} (1-\Delta_x)^{k_0} p(x,y,\xi) \, dx dy. \tag{A.7}$$

This gives

$$\left| D_\xi^{\alpha_3} \hat{p}(\xi', \eta', \xi) \right| \le c\,(2\pi)^{-n} \left(1 + |\xi'|^2\right)^{-k_0} \left(1 + |\eta'|^2\right)^{-\ell_0} \cdot$$

$$\cdot \sup_{x,y \in X} \left| D_\xi^{\alpha_3} \left(1 - \Delta_y\right)^{\ell_0} \left(1 - \Delta_x\right)^{k_0} p(x,y,\xi) \right| . \quad \text{(A.8)}$$

The Leibniz formula

$$(1 - \Delta)^{\ell_0} g(x) = \sum_{j=0}^{\ell_0} \binom{\ell_0}{j} (-1)^j \Delta^j g(x)$$

implice

$$\left| D_\xi^{\alpha_3} \left(1 - \Delta_y\right)^{\ell_0} \left(1 - \Delta_x\right)^{k_0} p(x,y,\xi) \right| \le$$

$$\sum_{i=0}^{\ell} \sum_{j=0}^{k_0} \binom{\ell_0}{i} \binom{k_0}{j} \left| D_\xi^{\alpha_3} \Delta_y^i \Delta_x^j p(x,y,\xi) \right| .$$

Hence, we obtain by the definition of the constant $\hat{C}_{\ell_0,k_0,\alpha_3}$ (4.3)

$$\left| D_\xi^{\alpha_3} \left(1 - \Delta_y\right)^{\ell_0} \left(1 - \Delta_x\right)^{k_0} p(x,y,\xi) \right| \le \hat{C}_{\ell_0,k_0,\alpha_3} \cdot (1 + |\xi|)^{m - |\alpha_3|} .$$

Using (A.4) we end up with the estimate

$$\left| D_\xi^{\alpha_3} \hat{p}(\xi', \eta', \xi) \right| \le c \hat{C}_{\ell_0,k_0;\alpha_3} (2\pi)^{-n} \left(1 + |\xi'|\right)^{-2k_0} \left(1 + |\eta'|\right)^{-2\ell_0} \left(1 + |\xi|\right)^{m - |\alpha_3|} . \quad \text{(A.9)}$$

In the next step we show

$$(\widehat{Bu})(\xi') = (2\pi)^{-n} \int_{\mathbb{R}^n} \int_{\mathbb{R}^n} \hat{p}(\xi' - \xi, \xi - \eta', \xi) \hat{u}(\eta')\, d\xi d\eta' . \quad \text{(A.10)}$$

We begin with the representation (2.2). Using the Fourier transformation we obtain

$$(\widehat{Bu})(\xi') = (2\pi)^{-\frac{3}{2}n} \int_X e^{-\mathrm{i}x^\top \xi'} \int_{\mathbb{R}^n} \int_X e^{\mathrm{i}(x-y)^\top \xi} p(x,y,\xi) u(y)\, dy d\xi dx$$

$$= (2\pi)^{-\frac{3}{2}n} \int_{\mathbb{R}^n} \int_X e^{-\mathrm{i}y^\top \xi} \int_X e^{-\mathrm{i}(\xi' - \xi)^\top x} p(x,y,\xi)\, dx\, u(y)\, dy d\xi$$

$$= (2\pi)^{-n} \int_{\mathbb{R}^n} \int_X e^{-\mathrm{i}y^\top \xi} \tilde{p}(\xi' - \xi, y, \xi) u(y)\, dy d\xi ,$$

where $\tilde{p}$ denotes the Fourier transformed of p with respect to x. Using the inverse Fourier transformation, this implies

$$(\widehat{Bu})(\xi') = (2\pi)^{-\frac{3}{2}n} \int\limits_{\mathbb{R}^n}\int\limits_{\mathbb{R}^n}\int\limits_{X} e^{-i(\xi-\eta')^\top y}\, \tilde{p}(\xi'-\xi, y, \xi)\, dy \hat{u}(\eta')\, d\xi d\eta' .$$

The integral with respect to y is the Fourier transformed of $\tilde{p}$ with respect to y. This concludes the proof of formula (A.10).

From (A.10), Peetre's inequality (A.2) and the inequalities (A.3) and (A.4) there follows

$$\left(1+|\xi'|^2\right)^{(s-m)/2} \left|(\widehat{Bu})(\xi')\right| \le (2\pi)^{-2n} \int\limits_{\mathbb{R}^n}\int\limits_{\mathbb{R}^n} (1+|\xi-\xi'|)^{|s-m|} \cdot$$

$$\cdot \left(1+|\xi|^2\right)^{(s-m)/2} |\hat{p}(\xi'-\xi, \xi-\eta', \xi)|\, |\hat{u}(\eta')|\, d\xi d\eta'$$

$$\le c(s,m)(2\pi)^{-2n} \int\limits_{\mathbb{R}^n}\int\limits_{\mathbb{R}^n} (1+|\xi-\xi'|)^{|s-m|} \cdot$$

$$\cdot \left(1+|\xi|^2\right)^{s/2} (1+|\xi|)^{-m} |\hat{p}(\xi'-\xi, \xi-\eta', \xi)|\, |\hat{u}(\eta')|\, d\xi d\eta' .$$

Applying Peetre's inequality again yields

$$\left(1+|\xi'|^2\right)^{(s-m)/2} \left|(\widehat{Bu})(\xi')\right| \le c(s,m)(2\pi)^{-2n} \int\limits_{\mathbb{R}^n}\int\limits_{\mathbb{R}^n} (1+|\xi-\xi'|)^{|s-m|} \cdot$$

$$\cdot (1+|\xi-\eta'|)^{|s|} (1+|\xi|)^{-m} \left(1+|\eta'|^2\right)^{s/2} |\hat{p}(\xi'-\xi, \xi-\eta', \xi)|\, |\hat{u}(\eta')|\, d\xi d\eta' .$$

Using the decay condition (A.9) of $\hat{p}(\cdot,\cdot,\cdot)$ (with $\alpha_3 = 0$) we obtain

$$(1+|\xi'|)^{(s-m)/2} \left|(\widehat{Bu})(\xi')\right| \le c(s,m)(2\pi)^{-2n}\, \hat{C}_{\ell_0,k_0;0} \int\limits_{\mathbb{R}^n}\int\limits_{\mathbb{R}^n} (1+|\xi-\eta'|)^{|s|-2\ell_0}$$

$$(1+|\xi-\xi'|)^{|s-m|-2k_0} \left(1+|\eta'|^2\right)^{s/2} |\hat{u}(\eta')|\, d\xi d\eta' . \qquad \text{(A.11)}$$

The right-hand side of (A.11) can be written as a three-fold convolution of three functions h_1, h_2 and h_3:

$$(h_3 * h_2 * h_1)(\xi') = \int\limits_{\mathbb{R}^n} (h_2 * h_1)(\xi) \cdot h_3(\xi'-\xi)\, d\xi$$

$$= \int\limits_{\mathbb{R}^n}\int\limits_{\mathbb{R}^n} h_2(\xi-\eta') \cdot h_3(\xi'-\xi)\, d\xi h_1(\eta')\, d\eta' ,$$

Using Young's inequality (A.1) with $r=2$, $p=1$ and $q=2$ two times gives

$$||h_3 * h_2 * h_1||_{L^2(\mathbb{R}^n)} \le ||h_3||_{L^1(\mathbb{R}^n)} \cdot ||h_2||_{L^1(\mathbb{R}^n)} \cdot ||h_1||_{L^2(\mathbb{R}^n)} . \qquad \text{(A.12)}$$

With $h_1(\cdot) := \left(1+|\cdot|^2\right)^{s/2} |\hat{u}(\cdot)|$, $h_2(\cdot) := (1+|\cdot|)^{|s|-2\ell_0}$ and $h_3(\cdot) := (1+|\cdot|)^{|s-m|-2k_0}$ we obtain from (A.11) and (A.12)

$$\| \left(1+|\cdot|^2\right)^{(s-m)/2} (\widehat{Bu})(\cdot)\|_{L^2(\mathbb{R}^n)}$$
$$\leq c(s,m)(2\pi)^{-2n}\hat{C}_{\ell_0,k_0;0}\|h_3\|_{L^1(\mathbb{R}^n)}\|h_2\|_{L^1(\mathbb{R}^n)}\| \left(1+|\cdot|^2\right)^{s/2} \hat{u}(\cdot)\|_{L^2(\mathbb{R}^n)}.$$

By the definition of the Sobolev norm this is equivalent to

$$\|Bu\|_{H^{s-m}(\mathbb{R}^n)} \leq c(s,m)(2\pi)^{-2n}\hat{C}_{\ell_0,k_0;0} \cdot \|h_3\|_{L^1(\mathbb{R}^n)}\|h_2\|_{L^1(\mathbb{R}^n)}\|u\|_{H^s(\mathbb{R}^n)}. \tag{A.13}$$

Using

$$\|h_2\|_{L^1(\mathbb{R}^n)} = \int\limits_{\mathbb{R}^n} (1+|\eta|)^{|s|-2\ell_0} \, d\eta = (2\pi)^n \, M_{|s|-2\ell_0}(\delta)$$

for $2\ell_0 > n + |s|$ and

$$\|h_3\|_{L^1(\mathbb{R}^n)} = \int\limits_{\mathbb{R}^n} (1+|\eta|)^{|s-m|-2k_0} \, d\eta = (2\pi)^n \, M_{|s-m|-2k_0}(\delta)$$

for $2k_0 > n + |s-m|$, we obtain from (A.13) the estimate for the operator norm in (4.1). □

References

1. Babuška, I., Aziz, A.K. (1972): Survey lectures on the mathematical foundations of the finite element method. In: Aziz, A.K. (ed) The Mathematical Foundation of the Finite Element Method with Applications to Partial Differential Equations. Academic Press, New York, pp.3-359
2. Carstensen, C., Stephan, E.P. (1995): A posteriori error estimates for boundary element methods. Math. Comput. 64 No. 210, 483-500
3. Costabel, M. (1987): Boundary integral operators on Lipschitz domains – elementary results. SIAM J. Math. Anal. 19, 613–626
4. Costabel, M., Stephan, E.P. (1988): Duality estimate for the numerical solution of integral equations. Numer. Math. 54, 339-353.
5. Faermann, B. (1993): Lokale a–posteriori–Fehlerschätzer bei der Diskretisierung von Randintegralgleichungen. Thesis, Christian–Albrechts–Universität Kiel
6. Faermann, B.: Local a posteriori error estimators for the Galerkin discretization of boundary integral equations. To appear in: Numer. Math.
7. Hörmander, L. (1985): The Analysis of Linear Partial Differential Operators III. Springer, Berlin Heidelberg New York Tokyo
8. Hsiao, G.C., Wendland, W.L. (1977): A finite element method for some integral equations of the first kind. J. Math. Anal. Appl. 58, 449-481
9. Hsiao, G.C., Wendland, W.L. (1981): The Aubin-Nitsche lemma for integral equations. Journal of Integral equations 3, 299-315
10. Kiss, B., Krebsz, A. (1996): On the matrix representation of the Sobolev norms of non-integer order and applications. Report 1996/4, Department of Mathematics, Széchenyi István College, Györ, Hungary
11. Petersen, B.E. (1983): Introduction to the Fourier Transform and Pseudo-Differential Operators. Pitman, Boston London Melbourne

12. Rjasanow, S. (1990): Vorkonditionierte iterative Auflösung von Randelementgleichungen für die Dirichlet-Aufgabe. Wissenschaftliche Schriftenreihe 7, TU Chemnitz
13. Rjasanow, S. (1995): Optimal preconditioner for boundary element formulation of the Dirichlet problem in elasticity. Math. Meth. Appl. Sci. 18, 603-613
14. Saranen, J. (1987): Local error estimates for some Petrov-Galerkin methods applied to strongly elliptic equations on curves. Math. Comp. 48, 485-502
15. Saranen, J. (1993): On local residual-type error estimates for boundary element methods. Preprint, Department of Mathematics, University of Oulu
16. Saranen, J., Wendland, W.L. (1993): Local residual-type error estimates for adaptive boundary element methods on closed curves. Appl. Anal. 48, 37-50
17. Schatz, A.H., Thomé, V., Wendland, W.L. (1990): Mathematical Theory of Finite and Boundary Element Methods. Birkhäuser, Basel Boston Berlin
18. Schulz, H.: Über lokale und globale Fehlerabschätzungen für adaptive Randelementmethoden. Thesis, University of Stuttgart, in preparation
19. Schwab, C., Wendland, W.L. (1996): On the extraction technique in boundary integral equations. Preprint No. 96-3, Mathematics Institute A, Stuttgart University
20. Schulz, H., Schwab, C., Wendland, W.L. (1996): An extraction technique for boundary element methods. In: Hackbusch, W. (ed.) Proceedings of the 12th GAMM-Seminar, Kiel, January 19-21, 1996. Vieweg, Braunschweig, pp.219-231
21. Steinbach, O. (1996): Fast solvers for the symmetric boundary element method. In: Hackbusch, W. (ed.) Proceedings of the 12th GAMM-Seminar, Kiel, January 19-21, 1996. Vieweg, Braunschweig, pp.232-242
22. Stephan, E.P., Wendland, W.L. (1976): Remarks to Galerkin and least squares methods with finite elements for general elliptic problems. Manuscripta Geodaetica 1, 93-123.
23. Treves, F. (1980): Pseudodifferential and Fourier Integral Operators. Plenum Press, New York London
24. Wendland, W.L. (1987): Strongly elliptic boundary integral equations. In: Iserles, A., Powell, M. (eds.) State of the Art in Numerical Analysis. Clarendon Press, Oxford, pp.511-561
25. Wendland, W.L., Yu, D. (1990): Local error estimates of boundary element methods for pseudo-differential of non-negative order on closed curves. J. Comp. Math. 10, 273-289
26. Wloka, J. (1982): Partielle Differentialgleichungen. Teubner, Stuttgart

Domain Decomposition and Preconditioning Techniques in Boundary Element Methods

O. Steinbach and W. L. Wendland

1 Introduction

Domain decomposition methods are powerful tools to parallelize the implementation of boundary element methods, especially if different material parameters in substructures appear. Moreover, such an approach is well suited for the coupling with other discretization schemes such as finite element methods; and for problems in infinite regions.

Boundary element methods are based on equivalent boundary integral equations of corresponding mixed boundary value problems. We restrict ourselves to the direct approach for selfadjoint problems of second order, where the mapping properties of the resulting boundary integral operators are well known [6, 7]. The basic tool for the boundary element domain decomposition formulation is the Calderon projection in terms of two boundary integral equations from which we can derive two representations of the Dirichlet–Neumann map based on the so–called Steklov–Poincaré operator. According to the choice of these operators, we distinguish between the symmetric and the nonsymmetric formulation. The latter is based on an integral equation involving only single and double layer potentials, which allows us to use either Galerkin or collocation methods for their discretization. This approach is used for coupling finite and boundary element methods in a nonuniform way, see e.g. [14, 24].

The symmetric formulation is based on both of the boundary integral equations given by the Calderon projection [17] and is widely used in symmetric coupling of finite and boundary element methods [5, 8, 9].

Both formulations lead to algebraic systems of linear equations of the same type; however, in the nonsymmetric formulation, the resulting system is symmetric only up to a small perturbation. For the solution of these systems one can use different iterative schemes like Schur complement algorithms or transformation–based solvers like the Bramble/Pasciak approach for skew–symmetric positive definite systems [1, 8], which can be generalized to the nonsymmetric case, too. However, these methods require the use of efficient preconditioners to keep the number of iterations small which are needed for a given relative error reduction independent of the discretization parameters. Such optimal preconditioners can be classified into two classes: the local type preconditioners for the single layer potential operator to realize the action of the local Steklov–Poincaré operators, and of global type for the assembled Schur complement, which corresponds to the globally defined discrete Dirichlet–Neumann map.

Preconditioners for discretized boundary integral operators are well known in the two–dimensional case for uniform meshes, where one can use circulant or Toeplitz matrices approximating the boundary integral operators on a circle [13], or one may use fast realizations for the Steklov–Poincaré operator and its inverse by solving finite element schemes for the underlying partial differential equation on uniform meshes in the domain. For both methods there are generalizations to the 3D–case [10, 11], which are restricted to partially uniform meshes and a suitable ordering of the geometrical nodes and elements.

Here we will give a general approach for the preconditioning of discrete boundary integral operators, which is well suited for the symmetric formulation and which has no restrictions to the space dimension and the discretization scheme used [22].

The construction of global preconditioners for the discrete Steklov–Poincaré operator in its symmetric representation is based on the spectral equivalence to the Galerkin discretization of the hypersingular integral operator. This allows us to construct several preconditioners of hierarchical type, where the spectral condition number depends on the relation between the local and the diameter of the substructures. Due to the spectral equivalence described above, we may also take the discrete hypersingular integral operator as an optimal preconditioner of the global Schur complement system. The realization of this preconditioner, i.e. the inversion of the discrete hypersingular operator can be obtained either by multigrid schemes or any other preconditioned iterative methods.

Based on these preconditioners we construct fast algorithms for the iterative solution of linear algebraic systems resulting from mixed boundary value problems in a variational formulation with respect to a given domain decomposition. Numerical examples will demonstrate the efficiency of the proposed methods.

2 Variational Formulations for Mixed Boundary Value Problems

Let us consider a selfadjoint and elliptic boundary value problem of 2nd order in a bounded domain $\Omega \subset \mathbb{R}^n$ with $n = 2, 3$ and a Lipschitz continuous boundary $\Gamma = \partial\Omega$, which is decomposed into two distinct parts Γ_D and Γ_N, where boundary conditions of Dirichlet or Neumann type, respectively, are given:

$$\left.\begin{array}{rcll} L(x)u(x) & = & f(x) & \text{for } x \in \Omega, \\ \ell_0 u(x) & = & g(x) & \text{for } x \in \Gamma_D, \\ \ell_1 u(x) & = & h(x) & \text{for } x \in \Gamma_N. \end{array}\right\} \qquad (2.1)$$

Here, the differential operator is given by

$$L(x)u(x) = -\sum_{k,\ell=1}^{n} \frac{\partial}{\partial x_k}\left(a_{k\ell}(x)\frac{\partial}{\partial x_\ell}u(x)\right) + c(x)u(x), \qquad (2.2)$$

whereas the ℓ_j are the corresponding boundary operators with $\ell_0 u$ the traditional trace and

$$\ell_1 u := \sum_{k,\ell=1}^{n} \nu_k a_{k\ell}\frac{\partial u}{\partial x_\ell}$$

the conormal derivative with $\boldsymbol{\nu} = (\nu_1, \ldots, \nu_n)$ the exterior unit normal vector on Γ. E.g. in linear elasticity $\ell_0 u$ are the boundary displacements while $\ell_1 u$ represents the boundary stress.

For a nonoverlapping domain decomposition

$$\overline{\Omega} = \bigcup_{i=1}^{p} \overline{\Omega}_i, \ \Omega_i \cap \Omega_j = \emptyset \text{ for } i \neq j, \ \Gamma_i := \partial\Omega_i, \ \Gamma_{ij} := \Gamma_i \cap \Gamma_j \qquad (2.3)$$

with the skeleton

$$\Gamma_S := \bigcup_{i=1}^{p} \Gamma_i \qquad (2.4)$$

we suppose, that the coefficients of the differential operator (2.2) are piecewise constant:

$$a_{k\ell}(x) = a^i_{k\ell}, \quad c(x) = c^i \quad \text{for } x \in \Omega_i .$$

Hence, the fundamental solutions of the differential operators exist locally, i.e. for all $i = 1, \ldots, p$ there exist functions $E^i(x,y)$ satisfying

$$L(x)E^i(x,y) = \delta(x-y)\cdot \mathbf{I}(x) \quad \text{for } x, y \in \Omega_i .$$

As in most applications, we choose the local trace operators $\ell^i_j u(\cdot)$ for $j = 0, 1$ and $i = 1, \ldots, p$ as

$$u_i(x) = \ell^i_0 u(x), \quad t_i(x) = \ell^i_1 u(x) \quad \text{for } x \in \Gamma_i.$$

Then the solution $u(\cdot)$ of the mixed boundary value problem (2.1) is given by the local representation formulae

$$u(x) = \int_{\Gamma_i} t_i(y)\cdot \ell^i_0 E^i(x,y)ds_y - \int_{\Gamma_i} u_i(y)\cdot \ell^i_1 E^i(x,y)ds_y + \int_{\Omega_i} f(y)\cdot E^i(x,y)dy$$

for $x \in \Omega_i$ and $i = 1, \ldots, p$. In addition to the boundary conditions

$$u_i(x) = g(x) \quad \text{for } x \in \Gamma_i \cap \Gamma_D,$$
$$t_i(x) = h(x) \quad \text{for } x \in \Gamma_i \cap \Gamma_N,$$

the densities u_i and t_i have to satisfy homogeneous transmission conditions

$$u_i(x) - u_j(x) = 0,\; t_i(x) + t_j(x) = 0 \quad \text{for } x \in \Gamma_{ij}\,, \tag{2.5}$$

which may be generalized to inhomogeneous conditions, too. Associated with the fundamental solutions $E^i(\cdot,\cdot)$, we define the common boundary integral operators by

$$(V_i t_i)(x) = \int_{\Gamma_i} \left(\ell_0^i(y) E^i(x,y)\right) t_i(y)\, ds_y\,,$$

$$(K_i u_i)(x) = \int_{\Gamma_i} \left(\ell_1^i(y) E^i(x,y)\right)^\top u_i(y)\, ds_y\,,$$

$$(K_i' t_i)(x) = \int_{\Gamma_i} \left(\ell_1^i(x) E^i(x,y)\right) t_i(y)\, ds_y\,,$$

$$(D_i u_i)(x) = -\ell_1^i(x) \int_{\Gamma_i} \left(\ell_1^i(y) E^i(x,y)\right)^\top u_i(y)\, ds_y$$

for $x \in \Gamma_i$ and the boundary traces of the volume or Newton potentials by

$$(N_j^i f)(x) \;=\; \ell_j^i \int_{\Omega_i} E^i(x,y) f(y)\, dy \quad \text{for } x \in \Gamma_i \text{ and } j = 0,1.$$

In terms of the Calderon projector the resulting local boundary integral equations can be written as

$$\begin{pmatrix} u_i \\ t_i \end{pmatrix} = \begin{pmatrix} \frac{1}{2}I - K_i & V_i \\ D_i & \frac{1}{2}I + K_i' \end{pmatrix} \begin{pmatrix} u_i \\ t_i \end{pmatrix} + \begin{pmatrix} N_0^i f \\ N_1^i f \end{pmatrix} \tag{2.6}$$

for $i = 1, \ldots, p$. These equations imply the nonsymmetric representation of the local Dirichlet–Neumann maps

$$t_i \;=\; V_i^{-1} \left(\frac{1}{2} I + K_i\right) u_i \;-\; V_i^{-1} N_0^i f \tag{2.7}$$

as well as their symmetric representation

$$\begin{aligned} t_i \;&=\; \left[\left(\tfrac{1}{2}I + K_i'\right) V_i^{-1} \left(\tfrac{1}{2}I + K_i\right) + D_i\right] u_i \\ &\qquad\qquad + N_1^i f - \left(\tfrac{1}{2}I + K_i'\right) V_i^{-1} N_0^i f. \end{aligned} \tag{2.8}$$

This may be written as

$$t_i \;=\; S_i u_i - f_i \tag{2.9}$$

by using the Steklov–Poincaré operator

$$S_i u_i \;=\; V_i^{-1} \left(\frac{1}{2} I + K_i\right) u_i \;=\; \left[\left(\frac{1}{2} I + K_i'\right) V_i^{-1} \left(\frac{1}{2} I + K_i\right) + D_i\right] u_i$$

and

$$f_i = V_i^{-1} N_0^i f = \left(\frac{1}{2} I + K_i'\right) V_i^{-1} N_0^i f - N_1^i f .$$

All the boundary integral operators involved are pseudodifferential operators of orders $-1, 0, +1$, respectively, and their mapping properties are well known, c.f. for instance [4, 6, 7].

2.1 The nonsymmetric formulation

The variational formulation of the mixed boundary value problem (2.1) with respect to the domain decomposition (2.3) satisfying the transmission conditions (2.5) has the following form:

Find $u \in H^{1/2}(\Gamma_S)$ with $\ell_0 u_{|\Gamma_D} = g$, such that

$$\sum_{i=1}^{p} \langle S_i \ell_0^i u, \ell_0^i v \rangle_{L^2(\Gamma_i)} = \int_{\Gamma_N} h \cdot \ell_0 v_{|\Gamma_N} ds_x + \sum_{i=1}^{p} \langle f_i, \ell_0^i v \rangle_{L^2(\Gamma_i)} \tag{2.10}$$

for all $v \in H^{1/2}(\Gamma_S)$ with $\ell_0 v_{|\Gamma_D} = 0$.

Here, the continuity of u across the coupling boundaries Γ_{ij} is satisfied in the strong $H^1(\Omega)$–sense, while the compatibility conditions of the stresses t_i are required in a weak form. This concept corresponds to variational formulations for finite element domain decomposition methods and is therefore also well suited for coupled finite–boundary element methods [8].

In the contrary to (2.10), one can use the inverse Steklov–Poincaré operators to require continuity of the stresses t_i and a weak coupling of the u_i [14]; generally one may take the weak coupling of both components, this would lead to the so–called mortar element methods.

Due to the $H^{1/2}(\Gamma_S)$–ellipticity of the sum of the Steklov–Poincaré operators involved in (2.10), the variational problem has a unique solution. Moreover, for a finite–dimensional subspace

$$V_H = \text{span}\{\varphi_k^\mu\}_{k=1}^M \subset H^{1/2}(\Gamma_S)$$

of spline functions of polynomial degree μ we have the quasi–optimal error estimate by Céas lemma,

$$||u - u_H||_{H^{1/2}(\Gamma_S)} \le c \cdot \inf_{v_H \in V_H} ||u - v_H||_{H^{1/2}(\Gamma_S)}$$

for the solution u_H of the variational problem (2.10) where inf is taken over the subspace V_H with corresponding boundary conditions. However, the Steklov–Poincaré operators involved admit no explicit discretization. Hence we introduce approximations of the local Steklov–Poincaré operators S_i by

$$S_i^h \ell_0^i v_H := t_h^i \in W_h^i = \text{span}\{\varphi_{i,k}^\nu\}_{k=1}^{N_i} \subset H^{-1/2}(\Gamma_i) \tag{2.11}$$

satisfying the local finite–dimensional variational boundary integral equations

$$\langle V_i t_h^i, \tau_h^i \rangle_{L^2(\Gamma_i)} = \langle (\frac{1}{2} I + K_i) \ell_0^i u, \tau_h^i \rangle_{L^2(\Gamma_i)} \tag{2.12}$$

for all $\tau_h^i \in \tilde{W}_h^i \subset H^{-1/2}(\Gamma_i)$. This approach includes Galerkin– as well as collocation schemes depending on the choice of the test spaces $\tilde{W}_h^i$ to compute the approximate Steklov–Poincaré operators S_i^h. If we suppose quasi–optimal error estimates (see e.g. [16]) for the local problems (2.12), i.e.

$$||(S_i^h - S_i)\ell_0^i u_H||_{H^{-1/2}(\Gamma_i)} \leq c \cdot h^{t+\frac{1}{2}} \cdot ||S_i \ell_0^i u_H||_{H^t(\Gamma)} \tag{2.13}$$

with $-\frac{1}{2} \leq t$ and $t < \mu + \frac{1}{2}$ for $n = 2$, $t \leq \mu$ for $n = 3$, then we can derive the following existence and convergence results for the solution of the approximate variational problem:

Theorem 2.1. [14, 24] *Let h, H be the maximal diameters of the finite elements in V_H and W_h respectively. If the condition $h < C \cdot H$ is satisfied with an appropriate constant C depending only on the ellipticity and boundedness of the single layer potential operators V_i in (2.12) and on the constant in the inverse inequality for V_H, then the sum of the operators S_i^h is positive definite and bounded. Hence, there exists a unique solution $\tilde{u}_H \in V_H$ with $\ell_0 \tilde{u}_{H|\Gamma_D} = g$ satisfying the variational problem*

$$\sum_{i=1}^{p} \langle S_i^h \gamma_0^i \tilde{u}_H, \gamma_0^i v_H \rangle_{L^2(\Gamma_i)} = f(v_H) \tag{2.14}$$

for all test functions $v_H \in V_H$ with $\gamma_0 v_{H|\Gamma_D} = 0$ and we have the error estimate

$$||u - \tilde{u}_H||_{H^{1/2}(\Gamma_S)} \leq c \cdot \left\{ ||u - u_H||_{H^{1/2}(\Gamma_S)} + ||(S - S^h)u||_{H^{-1/2}(\Gamma_S)} \right\}. \tag{2.15}$$

The proof is due to the first Strang lemma (c.f. e.g. [3]) based on the local error estimates (2.13) and uses the inverse property of the trial space V_H.

The discretization, either by a Galerkin or collocation method, leads now to the system of linear equations

$$M_h^\top V_h^{-1} \left(\frac{1}{2} \tilde{M}_h + K_h \right) \underline{u} = \underline{f}, \tag{2.16}$$

with $\tilde{M}_h = M_h$ in the Galerkin–Bubnov case. This is algebraically equivalent to the coupled system

$$\begin{pmatrix} V_h & -\frac{1}{2}\tilde{M}_h - K_h \\ M_h^\top & 0 \end{pmatrix} \begin{pmatrix} \underline{v} \\ \underline{u} \end{pmatrix} = \begin{pmatrix} \underline{0} \\ \underline{f} \end{pmatrix}, \tag{2.17}$$

note, that (2.16) is the Schur complement system of (2.17).

2.2 The symmetric formulation

Instead of the approximations of the local Steklov–Poincaré operators S_i one may use the symmetric formulation [5, 17] to discretize the original variational problem (2.10). This approach is based on both boundary integral equations in (2.6) and leads to the following variational formulation [9].

Find $u \in H^{1/2}(\Gamma_S)$ with $\ell_0 u_{|\Gamma_D}$ and $t_i \in H^{-1/2}(\Gamma_i)$ with $t_{i|\Gamma_N} = h$ for all $i = 1, \ldots, p$ such that

$$\sum_{i=1}^{p} \left\{ \langle D\ell_0^i u, \ell_0^i v\rangle_{L^2(\Gamma_i)} + \frac{1}{2}\langle t_i, \ell_0^i v\rangle_{L^2(\Gamma_i)} + \langle t_i, K_i \ell_0^i v\rangle_{L^2(\Gamma)^i)} \right\} = f_2(v)$$

$$\langle V_i t_i, \tau_i\rangle_{L^2(\Gamma_i)} - \frac{1}{2}\langle \ell_0^i u, \tau_i\rangle_{L^2(\Gamma_i)} - \langle K_i \ell_0^i u, \tau_i\rangle_{L^2(\Gamma_i)} = f_1^i(\tau_i)$$

holds for all test functions $v \in H^{1/2}(\Gamma_S)$ with $\ell_0 v_{|\Gamma_D} = 0$ and $\tau_i \in H^{-1/2}(\Gamma_i)$ with $\tau_{i|\Gamma_N} = 0$ for $i = 1, \ldots, p$.

Here, the linear forms on the right hand side are given by

$$f_2(v) = \int_{\Gamma_N} h(x) \cdot \ell_0 v(x)\, ds_x \; - \sum_{i=1}^{p} \langle N_1^i f, \ell_0^i v\rangle_{L^2(\Gamma_i)},$$

$$f_1^i(\tau_i) = -\langle N_0^i f, \tau_i\rangle_{L^2(\Gamma_i)} \,.$$

Since the associated bilinear form is positive definite and bounded, the variational problem is uniquely solvable and we get quasi–optimal error estimates for the corresponding solution in finite–dimensional subspaces without additional conditions. The Galerkin discretization of the variational problem leads to the skew–symmetric system

$$\begin{pmatrix} V_h & -\tilde{K}_h \\ \tilde{K}_h^\top & D_h \end{pmatrix} \begin{pmatrix} \underline{t} \\ \underline{u} \end{pmatrix} = \begin{pmatrix} \underline{f}_1 \\ \underline{f}_2 \end{pmatrix}, \tag{2.18}$$

where $\tilde{K}_h = K_h$ is the Galerkin matrix which couples the unknown Dirichlet datum on Γ_N with the unknown Neumann datum on Γ_D while $\tilde{K}_h = \frac{1}{2}M_h + K_h$ is the Galerkin matrix for the coupling boundaries $\Gamma_C := \Gamma_S \backslash \Gamma$.

Applying a block Gauss–Seidel step to (2.18) we get the Schur complement system with the matrix

$$\left[D_h + \tilde{K}_h^\top V_h^{-1} \tilde{K}_h\right] \underline{u} = \underline{f}_2 - \tilde{K}_h^\top V_h^{-1} \underline{f}_1 = \underline{f} \tag{2.19}$$

which is symmetric and positive definite. Hence we can use the preconditioned method of conjugate gradients to solve (2.19) iteratively. In general, we have to assume, that there are two preconditioners available, i.e. matrices C_V and C_S satisfying the spectral equivalence inequalities

$$c_1^V \cdot (C_V \underline{t}, \underline{t}) \le (V_h \underline{t}, \underline{t}) \le c_2^V \cdot (C_V \underline{t}, \underline{t})$$

for all vectors $\underline{t}$ associated with the (local) trial functions t_h^i and

$$c_1^S \cdot (C_S \underline{u}, \underline{u}) \leq ((D_h + \tilde{K}_h^\top V_h^{-1} \tilde{K}_h)\underline{u}, \underline{u}) \leq c_2^S \cdot (C_S \underline{u}, \underline{u})$$

for all vectors $\underline{u}$ associated with the (global) trial functions u_h, where all constants should not depend on the actual discretization parameters. We will give a general approach to construct such preconditioners in chapter 3. For the construction of preconditioners to the discrete symmetric Steklov–Poincaré operator

$$\tilde{S}_h = D_h + \tilde{K}_h^\top V_h^{-1} \tilde{K}_h , \tag{2.20}$$

which is only an approximation of the Galerkin discretization of the Steklov–Poincaré operator S_i defined in (2.9) given by the matrix entries

$$S_h[\ell, k] = \langle S\varphi_k^\mu, \varphi_\ell^\mu \rangle_{L^2(\Gamma_S)} = \sum_{i=1}^p \langle S_i \varphi_k^\mu, \varphi_\ell^\mu \rangle_{L^2(\Gamma_i)} \tag{2.21}$$

for appropriate trial functions φ_k^μ of degree μ, we invoke the following basic result.

Theorem 2.2. [22] *The matrices defined in* (2.20) *and* (2.21) *are spectrally equivalent, i.e. we have the inequalities*

$$c_1^{\tilde{S}} \cdot (S_h \underline{u}, \underline{u}) \leq (\tilde{S}_h \underline{u}, \underline{u}) \leq c_2^{\tilde{S}} \cdot (S_h \underline{u}, \underline{u})$$

for all vectors $\underline{u}$ associated with trial functions $u_h \in V_h$, where the constants do not depend on the discretization.

Therefore it is only necessary to construct preconditioners for the single layer potentials V_h, and either for the Steklov–Poincaré operator S_h or for D_h, respectively, which are Galerkin approximations to pseudodifferential operators of order ± 1. Obviously, there is a strong relation between these operators, which we will use for the construction of efficient preconditioning techniques in boundary element methods.

3 The Construction of Local Preconditioners

Let us consider strongly elliptic and self–adjoint pseudodifferential operators A, B of orders $\pm 2\alpha$, respectively, in appropriate Sobolev spaces on Γ:

$$A : H^s(\Gamma) \to H^{s-2\alpha}(\Gamma), \quad B : H^{s-2\alpha}(\Gamma) \to H^s(\Gamma).$$

With the admissible quotient spaces

$$V = H^s(\Gamma)_{/\ker A}, \quad W = H^{s-2\alpha}(\Gamma)_{/\ker B}$$

we assume the coerciveness and boundedness inequalities

$$c_1^A \cdot ||t||^2_{H^s(\Gamma)} \leq \langle At, t\rangle_{H^{s-\alpha}(\Gamma)} \leq c_2^A \cdot ||t||^2_{H^s(\Gamma)} \qquad \forall t \in V\,,$$

$$c_1^B \cdot ||u||_{H^{s-2\alpha}(\Gamma)} \leq \langle Bu, u\rangle_{H^{s-\alpha}(\Gamma)} \leq c_2^B \cdot ||u||^2_{H^{s-2\alpha}(\Gamma)} \qquad \forall u \in W.$$

For the nontrivial finite–dimensional nullspace $\ker B$ we use an orthonomal basis representation

$$\ker B \;=\; \left\{v \in H^{s-2\alpha}(\Gamma) \,:\, Bv = 0\right\} \;=\; \operatorname{span}\{v_k\}_{k=1}^m$$

with

$$\langle v_k, v_\ell\rangle_{H^{s-2\alpha}(\Gamma)} \;=\; \delta_{k\ell} \quad \text{for all } k, \ell = 1, \ldots, m.$$

Hence we can introduce the quotient space

$$V^{s,0}(B) \;=\; \left\{t \in H^s(\Gamma) \,:\, \langle t, v_k\rangle_{H^{s-\alpha}(\Gamma)} \;=\; 0 \quad \text{for all } v_k \in \ker B\right\};$$

and the pseudoinverse operator

$$\dot{B}^{-1} \,:\, V^{s,0}(B) \to W$$

is then well defined, i.e. $\dot{B}^{-1}$ is positive definite and bounded for all $t \in V^{s,0}(B)$. Thus we have immediately

Lemma 3.1. *For every $t \in V \cap V^{s,0}(B)$, the inequalities*

$$\gamma_1 \cdot \langle \dot{B}^{-1}t, t\rangle_{H^{s-\alpha}(\Gamma)} \;\leq\; \langle At, t\rangle_{H^{s-\alpha}(\Gamma)} \;\leq\; \gamma_2 \cdot \langle \dot{B}^{-1}t, t\rangle_{H^{s-\alpha}(\Gamma)}$$

are satisfied with positive constants $\gamma_1 = c_1^A \cdot c_1^B$ and $\gamma_1 = c_2^A \cdot c_2^B$.

To construct a preconditioner we have to split the space V into an orthogonal sum

$$V \;=\; V^{s,0}(B) \oplus \left[V^{s,0}(B)\right]^{\perp}. \tag{3.1}$$

To this end, we characterize the Bessel potential operator

$$J \,:\, H^{s-2\alpha}(\Gamma) \to H^s(\Gamma)$$

for $u \in H^{s-2\alpha}(\Gamma)$ by the variational formulation

$$\langle Ju, v\rangle_{H^{s-\alpha}(\Gamma)} \;=\; \langle u, v\rangle_{H^{s-2\alpha}(\Gamma)} \quad \text{for all } v \in H^{s-2\alpha}(\Gamma).$$

Then, for any $t \in V$ there exists the uniquely determined projection

$$t_0(x) = t(x) - \sum_{j=1}^{m} \langle t, v_j\rangle_{H^{s-\alpha}(\Gamma)} (Jv_j)(x) \in V^{s,0}(B).$$

Now we can formulate one of the basic results for our preconditioning technique [22]:

Theorem 3.1. *The bounded linear operator defined by the bilinear form*

$$c(t,\tau) = \langle \dot{B}^{-1} t_0, \tau_0 \rangle_{H^{s-\alpha}(\Gamma)} + \sum_{k=1}^{m} \langle t, v_k \rangle_{H^{s-\alpha}(\Gamma)} \langle \tau, v_k \rangle_{H^{s-\alpha}(\Gamma)} \tag{3.2}$$

for all $t, \tau \in V$ is spectrally equivalent to A, i.e. there hold the inequalities

$$\gamma_1 \cdot c(t,t) \le \langle At, t \rangle_{H^{s-\alpha}(\Gamma)} \le \gamma_2 \cdot c(t,t) \quad \textit{for all } t \in V$$

with positive constants

$$\gamma_1 = c_1^A \cdot \min\{1, c_1^B\}, \quad \gamma_2 = c_2^A \cdot \max\{1, c_2^B\}.$$

Let a finite approximation space $V_h \subset V$ be given by

$$V_h = \mathrm{span}\{\varphi_k^\nu(\cdot)\}_{k=1}^N,$$

e.g. by the smoothest B–splines $\varphi_k^\nu(\cdot)$ of piece–wise polynomial degree ν. Then as a direct consequence of Theorem 3.1, the matrices A_h and C_h defined by their elements

$$A_h[\ell,k] = \langle A\varphi_k^\nu, \varphi_\ell^\nu \rangle_{H^{s-\alpha}(\Gamma)} \text{ and } C_h[\ell,k] = c(\varphi_k^\nu, \varphi_\ell^\nu) \quad \text{for } k,\ell = 1,\dots,N$$

are spectrally equivalent, i.e. the spectral condition number of the preconditioned system $C_h^{-1} A_h$ is bounded by a constant independent of the discretization parameters like mesh adaptivity or the degree of the trial functions used. To use the proposed preconditioner, one has to discretize the bilinear form (3.2) and to realize the multiplication with C_h^{-1} in the most efficient way.

For the orthogonal complement to $V^{s,0}(B)$ we suppose a representation in terms of the above basis of V_h, i.e.

$$(Jv_k)(x) = \sum_{j=1}^{N} v_k^j \cdot \varphi_j^\nu(x).$$

By the partial Gram–Schmidt orthogonalization,

$$\varphi_k^{\nu,0}(x) = \varphi_k^\nu(x) - \sum_{j=1}^{m} \langle \varphi_k^\nu, v_j \rangle_{H^{s-\alpha}(\Gamma)} \cdot (Jv_j)(x) \in V^{s,0}(B),$$

we find the discrete analog to (3.1) by the basis transformation

$$V_h = \mathrm{span}\{\varphi_1^\nu, \dots, \varphi_N^\nu\} = \mathrm{span}\{\varphi_1^{\nu,0}, \dots, \varphi_{N-m}^{\nu,0}, Jv_1, \dots, Jv_m\}.$$

For a given function $t_h \in V_h$ this may be written as

$$t_h(x) = \sum_{k=1}^{N} t_k \cdot \varphi_k^\nu(x) = \sum_{k=1}^{N-m} \tilde{t}_k \cdot \varphi_k^{\nu,0}(x) + \sum_{k=1}^{m} \tilde{t}_{N-m+k} \cdot (Jv_k)(x)$$

with the relations

$$t_k = \alpha_k \cdot \tilde{t}_k + \sum_{\ell=1}^{m} v_\ell^k \left(\tilde{t}_{N-m+\ell} - \sum_{j=1}^{N-m} \langle \varphi_j^\nu, v_\ell \rangle_{H^{s-\alpha}(\Gamma)} \, \tilde{t}_j \right)$$

for $k = 1, \ldots, N$ and

$$\alpha_k = \begin{cases} 1 & \text{for } k = 1, \ldots, N-m, \\ 0 & \text{for } k = N-m+1, \ldots, N. \end{cases}$$

It can be shown [19], that this basis transformation is transposed to a transformation T defined as follows: Let the Galerkin weights $f_k = f(\varphi_k^\nu)$ for a linear form $f(\cdot)$ be given. Using the transformed trial functions we find

$$\tilde{f}_{N-m+\ell} := f(Jv_\ell) = \sum_{j=1}^{N} v_{\ell,j} \cdot f_j \quad \text{for } \ell = 1, \ldots, m \text{ and}$$

$$\tilde{f}_j := f(\varphi_j^{\nu,0}) = f_j - \sum_{\ell=0}^{m} \langle \varphi_j^\nu, v_\ell \rangle_{H^{s-\alpha}(\Gamma)} \cdot \tilde{f}_{N-m+\ell}$$

for $j = 1, \ldots, N-m$; in short $\underline{\tilde{f}} = T\underline{f}$.

Due to the orthogonal decomposition of the trial space V_h, the Galerkin discretization of the bilinear form (3.2) with respect to the transformed trial functions leads to the block diagonal matrix

$$\tilde{C}_h = \begin{pmatrix} C_0 & 0 \\ 0 & I_m \end{pmatrix}$$

with the matrix elements

$$C_0[\ell, k] = \langle \dot{B}^{-1} \varphi_k^{\nu,0}, \varphi_\ell^{\nu,0} \rangle_{H^{s-\alpha}(\Gamma)} \quad \text{for } k, \ell = 1, \ldots, N-m.$$

We still have to compute these Galerkin weights, where the pseudoinverse operator $\dot{B}^{-1}$ is in general not given in explicit form. Hence we have to find an approximation u_h^0 of the exact function $u_0 = \dot{B}^{-1} t_0 \in W$ for any function $t_0 \in V^{s,0}(B)$. This can be done by solving the finite–dimensional variational problem

$$\langle B u_0^h, v_0^h \rangle_{H^{s-\alpha}(\Gamma)} = \langle t_0, v_0^h \rangle_{H^{s-\alpha}(\Gamma)}$$

for all test functions v_0^h in an appropriate subspace $W_h^0 \subset W$ as follows.

For a finite–dimensional subspace $W_h \subset H^{s-2\alpha}(\Gamma)$ with the basis representation

$$W_h = \operatorname{span}\{\varphi_k^\mu(\cdot)\}_{k=1}^{M},$$

we define an orthogonal decomposition

$$W_h = W_h^0 \oplus \{v_k\}_{k=1}^{m} \quad \text{with } W_h^0 = \operatorname{span}\{\varphi_k^{\mu,0}\}_{k=1}^{M-m} \subset W$$

with the transformed trial functions

$$\varphi_k^{\mu,0}(x) = \varphi_k^\mu(x) - \sum_{\ell=1}^m \langle \varphi_k^\mu, v_\ell \rangle_{H^{s-2\alpha}(\Gamma)} \cdot v_\ell(x) .$$

Then the discretization of the pseudoinverse operator $\dot{B}^{-1}$ leads to the matrix representation

$$\tilde{C}_0 = \tilde{M}_h B_h^{-1} \tilde{M}_h^\top$$

with the entries

$$B_h[j,i] = \langle B\varphi_i^\mu, \varphi_j^\mu \rangle_{H^{s-\alpha}(\Gamma)},$$

$$\tilde{M}_h[j,k] = \langle \varphi_k^\nu, \varphi_j^\mu \rangle_{H^{s-\alpha}(\Gamma)} - \sum_{\ell=1}^m \langle \varphi_k^\nu, v_\ell \rangle_{H^{s-\alpha}(\Gamma)} \cdot \langle \varphi_j^\mu, v_\ell \rangle_{H^{s-2\alpha}(\Gamma)}$$

for $i,j = 1,\ldots,M-m, k = 1,\ldots,N-m$. From the additional condition

$$c \cdot ||t_h||_{H^s(\Gamma)} \le \sup_{v_h \in W_h^0} \frac{|\langle t_h, v_h \rangle_{H^{s-\alpha}(\Gamma)}|}{||v_h||_{H^{s-2\alpha}(\Gamma)}} \quad \text{for all } t_h \in V_h^0 \tag{3.3}$$

then immediatly follows the spectral equivalence of C_0 and $\tilde{C}_0$. On the other hand, an efficient preconditioner requires the action of $\tilde{C}_0^{-1}$, which in this case can easily be done by

$$\tilde{C}_0^{-1} = \tilde{M}_h^{-\top} B_h \tilde{M}_h^{-1}$$

if we suppose the invertibility of $\tilde{M}_h$, which corresponds to (3.3) with $N = M$. Due to the basis transformation introduced above, we have:

Lemma 3.2. *The modified mass matrix $\tilde{M}_h$ is invertible if and only if the mass matrix defined by*

$$M_h[\ell,k] = \langle \varphi_k^\nu, \varphi_\ell^\mu \rangle_{H^{s-\alpha}(\Gamma)}$$

for $k,\ell = 1,\ldots,N$ is invertible.

For the special choices $\nu = \mu + 2k$ with $k \in I\!N_0$, the mass matrix M_h is strongly diagonal dominant, hence there exists the inverse independent of h; and condition (3.3) is satisfied. Then the action of the proposed preconditioner is given by

$$C_h^{-1} = T^\top \begin{pmatrix} \tilde{M}_h^{-\top} B_h \tilde{M}_h^{-1} & 0 \\ 0 & I_m \end{pmatrix} T .$$

To demonstrate this technique, let us consider the Laplace equation, where A is the single layer potential operator of order $2\alpha = -1$ given by

$$(Vt)(x) = \int_\Gamma E(x,y)\, t(y) ds_y$$

with the fundamental solution

$$E(x,y) = \frac{1}{2(n-1)\pi} \cdot \begin{cases} -\log|x-y| & \text{for } n=2, \\ |x-y|^{-1} & \text{for } n=3. \end{cases}$$

Let B be the hypersingular integral operator

$$(Du)(x) = -\frac{\partial}{\partial n_x} \int_\Gamma \frac{\partial}{\partial n_y} E(x,y)\, u(y)\, ds_y\,,$$

then both operators satisfy all previous assumptions with

$$\ker V = \{0\} \quad \text{and} \quad \ker D = \{1\}\,.$$

From

$$v_1(x) = \sum_{k=1}^{N} \varphi_k^\nu(x)$$

we find the basis transformation T by

$$u_h(x) = \sum_{k=1}^{N} u_k \cdot \varphi_k^\nu(x) = \sum_{k=1}^{N-1} \tilde{u}_k \cdot \varphi_k^{\nu,0}(x) + \omega \cdot v_1(x)$$

with the modified trial functions

$$\varphi_k^{\nu,0} = \varphi_k^\nu(x) - \frac{m_k^\nu}{|\Gamma|} \cdot v_1(x) \text{ and } m_k^\nu = \int_\Gamma \varphi_k^\nu(x)\, ds_x\,.$$

For $\mu = \nu + 2k$ with $k \in I\!N_0$ the modified mass matrix is given by

$$\tilde{M}_h = M_h - \frac{1}{|\Gamma|} \underline{m}^\nu (\underline{m}^\mu)^\top\,,$$

where the inverse can be computed from the Sherman–Morrison formula [12],

$$\tilde{M}_h^{-1} = M_h^{-1} - \frac{1}{\alpha - |\Gamma|} M_h^{-1} \underline{m}^\nu (\underline{m}^\mu)^\top M_h^{-1}$$

with

$$\alpha = (\underline{m}^\mu)^\top M_h^{-1} \underline{m}^\nu\,.$$

Note, that M_h is sparse and strongly diagonally dominant, hence the inverse can be computed by applying some steps of the Jacobi iteration to get a spectrally equivalent approximation of M_h^{-1}. Moreover, for $n = 2$ and $\nu = \mu = 1$, the mass matrix M_h is tridiagonal due to the basis transformation. Then, M_h^{-1} can be computed directly in optimal order by a Cholesky decomposition.

However, the basis transformation used may produce some difficulties in applying these techniques to more general problems, like in linear elasticity, where we have up to 6 eigensolutions of the hypersingular integral operator. To avoid these complications, we modify our preconditioner in the following way. Since the hypersingular integral operator associated with the Laplacian is positive definite and bounded with respect to the $|\cdot|_{H^{1/2}(\Gamma)}$–semi–norm, i.e.

$$c_1^D \cdot |u|^2_{H^{1/2}(\Gamma)} \leq \langle Du, u\rangle_{L^2(\Gamma)} \leq c_2^D \cdot |u|^2_{H^{1/2}(\Gamma)} \quad \forall u \in H^{1/2}(\Gamma), \tag{3.4}$$

then it follows immediately, that the modified operator

$$B = \alpha \cdot I + D$$

is positive definite and bounded with respect to the norm in $H^{1/2}(\Gamma)$. For $\alpha \in [c_1^D, c_2^D]$ we have the same constants as in (3.4). In this approach, no basis transformation is needed, which clearly can be extended to more general problems without any difficulty. Note, that for more general equations, the above modification only depends on the nontrivial nullspace of the hypersingular integral operator D [22]. The amount of work is of the same order as the multiplication with the original matrix A_h where in the symmetric approach to mixed boundary value problems and in domain decomposition methods both operators are still discretized and therefore available.

In general, this technique can be used to construct efficient preconditioners for boundary integral operators, if pairs of operators of order $\pm 2\alpha$ are available. There are no restrictions to the space dimension. Moreover, this method is also applicable, if Γ is not a closed boundary curve, as it appears in mixed boundary value problems.

At the end of this chapter let us present some numerical examples. For an L–shaped domain without further decomposition we consider the Laplace equation with Dirichlet boundary conditions to check both preconditioners proposed for the single layer potential. The Galerkin discretization is done for $\nu = \mu = 1$ and an uniform mesh, N denotes the number of degrees of freedom, $\kappa(C_h^{-1}V_h)$ is an approximate bound for the spectral condition number and Iter and sec mean the number of iterations and the time needed, respectively, in order to achieve a relative error reduction of 10^{-8} in the preconditioned conjugate gradient iteration.

Table 3.1. Preconditioners for V_h.

	$C_h^{-1} = T^\top \tilde{M}_h^{-\top} D_h \tilde{M}_h^{-1} T$			$C_h^{-1} = M_h^{-\top}(M_h + D_h)M_h^{-1}$		
N	$\kappa(C_h^{-1}V_h)$	Iter	sec	$\kappa(C_h^{-1}V_h)$	Iter	sec
64	1.70	8	0.02	1.85	8	0.01
128	1.70	8	0.07	1.92	8	0.04
256	1.71	8	0.32	1.99	9	0.21
512	1.71	8	1.42	2.04	9	0.96
1024	1.72	8	6.42	2.09	9	3.80
2048	1.72	8	24.18	2.12	9	15.18

From the numerical results given in Table 3.1 we conclude experimentally the optimality of the preconditioners proposed. As we can see, the basis transformation guarantees a sharper estimate of the spectral condition number, however, the numerical amount of work of the modified preconditioner without this basis transformation is significantly less.

Obviously, we can use this technique in the opposite direction to get a preconditioner for the discrete hypersingular integral operator and, due to Theorem 2.2, also for the discrete Steklov–Poincaré operator. Table 3.2 shows the numerical results.

Table 3.2. Preconditioner for D_h.

	$C_h^{-1} = M_h^{-1} V_h M_h^{-\top}$		
N	$\kappa(C_h^{-1} D_h)$	Iter	sec
64	1.77	9	0.01
128	1.80	9	0.04
256	1.82	9	0.19
512	1.82	9	0.93
1024	1.83	9	3.63
2048	1.83	9	14.50

Finally, we conclude this section with the statement, that the proposed preconditioning technique is optimal for the symmetric formulation of boundary integral equations to handle mixed boundary value problems by domain decomposition methods. For $\nu = \mu$, all boundary integral operators needed for preconditioning are discretized anyway, so this method seems to be the most natural one. Moreover, there are no restrictions to the geometrical discretization and to the underlying partial differential equation either.

4 Hierarchical Domain Decomposition Preconditioners

Due to Theorem 2.2 it is sufficient to construct a spectrally equivalent preconditioner for the Galerkin discretization of the Steklov–Poincaré operator given by the bilinear form

$$a(u,v) = \sum_{i=1}^{p} \langle S_i \ell_0^i u, \ell_0^i v \rangle_{L^2(\Gamma_i)} \tag{4.1}$$

for $u, v \in H^{1/2}(\Gamma_S)$. To this end we recall, that a boundary element domain decomposition method can be considered as a parallel realization of a Trefftz formulation. Hence it seems to be natural, to decompose any given function $u \in H^{1/2}(\Gamma)$ by

$$u(x) = u_H(x) + \tilde{u}(x) \quad \text{for } x \in \Gamma_S \tag{4.2}$$

into a coarse grid part $u_H(\cdot)$ and a fine grid function $\tilde{u}(\cdot)$, where the latter vanishes at all global coarse grid nodes $x_C \in \omega$. Here, ω is the q–dimensional set of all coarse grid nodes of the given domain decomposition (2.3). Obviously, this decomposition is only unique, if we choose the ansatz

$$u_H(x) = \sum_{k=1}^{q} u(x_{C,k}) \cdot \varphi_k^H(x), \tag{4.3}$$

where $\varphi_k^H(\cdot)$ are harmonic trial functions, which are piecewise polynomials on Γ_S and satisfy the given partial differential equation with homogeneous right hand side in the subdomains Ω_i.

If we define by

$$V_H^q(\Gamma_S) = \operatorname{span}\left\{\varphi_k^H\right\}_{k=1}^q$$

the trial space of the coarse grid functions, we can define a unique basis transformation

$$T : V_h(\Gamma_S) \to V_H^q(\Gamma_S) \cup V_h^q(\Gamma_S),$$

of the usual complete trial space $V_h(\Gamma_S)$ onto a hierarchical trial space. If we denote by I_h the piecewise linear interpolant of a coarse grid function, this basis transformation is given by

$$T = \begin{pmatrix} I_q & 0 \\ -I_h & I \end{pmatrix} \tag{4.4}$$

provided a corresponding ordering of the coefficients $\underline{u}$ of the original fine grid function u_h.

The Galerkin discretization of the variational problem (2.10) with respect to the hierarchical trial space now leads to the reordered algebraic system with the stiffness matrix

$$S_{hierarchical} = \begin{pmatrix} S_{HH} & S_{hH} \\ S_{hH}^\top & S_{hh} \end{pmatrix}. \tag{4.5}$$

From this representation we derive different preconditioners for the globally defined discrete Steklov–Poincaré operator based on the relation

$$S_h = T^\top \cdot S_{hierarchical} \cdot T. \tag{4.6}$$

If we are given a hierarchical preconditioner $C_{hierarchical}$, which is spectrally equivalent to (4.5), the resulting preconditioner for the discrete Steklov–Poincaré operator is given by

$$C_S = T^{-1} \cdot C_{hierarchical}^{-1} T^{-\top} \tag{4.7}$$

where we use the inverse basis transformation matrix

$$T^{-1} = \begin{pmatrix} I_q & 0 \\ I_h & I \end{pmatrix}.$$

The construction of hierarchical preconditioners to (4.5) is based on the following result [19]:

Theorem 4.1. *Let $v_h \in V_h(\Gamma_S)$ be an arbitrary, given function. The corresponding coarse grid function v_H is obviously the linaer interpolant with respect to the coarse grid, i.e. $v_H = I_H v_h$. Then there holds the stability estimate*

$$||v_h - v_H||_{H^{1/2}(\Gamma_S)} \leq c \cdot K(h,H) \cdot ||v_h||_{H^{1/2}(\Gamma_S)} \tag{4.8}$$

with the positive constant

$$K(h,H) = \gamma^{n-2} \cdot (\log \gamma)^{n-1} \quad \textit{with } \gamma = \frac{H}{h}. \tag{4.9}$$

The proof of Theorem 4.1 is based on the Sobolev embedding theorem,

$$|u(x) - u(y)|^2 \leq c \cdot h_k^{2s-1} \cdot |u|^2_{H^s(\Gamma_k)}, \quad x, y \in \Gamma_k, \quad h_k = \operatorname{diam} h_k$$

for $\frac{n-1}{2} < s < n-1$ $(n = 2, 3)$ with the constant

$$L(s) = (2s - n + 1)^{1-n}$$

describing explicitly the dependence on s and n. The above estimate implies the error estimate for the linear interpolation function u_I,

$$||u - u_I||^2_{H^t(\Gamma)} \leq c \cdot L(s) \cdot h^{2(s-t)} \cdot ||u||^2_{H^s(\Gamma)} \quad \text{for } 0 \leq t < 1,\ t \leq s;$$

and combining this estimate with the inverse inequality for $V_H(\Gamma_S)$ provides the inequality (4.8) with (4.9) in Theorem 4.1.

As a consequence of this theorem we conclude estimates of the spectral condition number of the preconditioned system by using different preconditioning techniques. Here we will summarize the results, for a more detailed representation we refer to a forthcoming paper.

Neglecting the coupling terms in (4.5) we get the BPS–type preconditioner

$$C_{BPS} = \begin{pmatrix} S_{HH} & 0 \\ 0 & S_{hh} \end{pmatrix},$$

which gives the estimate

$$\kappa(C_{BPS}^{-1} S_h) \leq c_1 \cdot [1 + K(h,H)].$$

For a preconditioner of (4.5) which maintains the coupling, we factorize the block matrix (4.5) and choose

$$C_{CGS} = \begin{pmatrix} I & 0 \\ S_{hH}^{\top} S_{HH}^{-1} & I \end{pmatrix} \begin{pmatrix} S_{HH} & 0 \\ 0 & S_{hh} \end{pmatrix} \begin{pmatrix} I & S_{HH}^{-1} S_{hH} \\ 0 & I \end{pmatrix}$$

as preconditioner which provides the estimate

$$\kappa(C_{CGS}^{-1} S_h) \leq c_2 \cdot K(h,H).$$

Both preconditioners need a preconditioning matrix C_{hh} which is spectrally equivalent to S_{hh}. For this preconditioning one may use different approaches such as the ones based on the results in chapter 3 or of Neumann–Neumann type [23]. In any case, however, the spectral condition number of the preconditioned system will depend of the ratio H/h, which ist not optimal. To avoid this dependence, we note that we have the optimal bound

$$\kappa(D_h^{-1} S_h) \leq c_3$$

due to Theorem 2.2. So we can use the explicitly given matrix D_h as a preconditioner of the implicitly given Schur complement matrix S_h. To realize the action of this preconditioner, i.e. to multiply a given vector $\underline{v}$ by D_h^{-1}, one can use either a linear multigrid scheme [8] or any other iterative scheme preconditioned by the hierarchical methods described above. Here we will compare all three preconditioners for a simple model problem with the Laplacian for a domain decomposition of the unit square into 64 subdomains. N denotes the local degree of freedom in the subdomains, while the iteration number and the time correspond to an error reduction of 10^{-6}:

Table 4.1. Preconditioners for S_h.

	BPS		CGS		$CG(D_h)$	
N	Iter	sec	Iter	sec	Iter	sec
64	24	22.52	20	26.95	11	41.28
128	24	26.55	22	33.20	12	47.01
256	25	45.47	23	52.09	13	62.11
512	27	127.39	25	133.02	13	107.80
1024	28	489.38	26	482.16	14	326.77

Note, that the number N of 1024 local degrees of freedom per subdomain corresponds for 64 subdomains to a global number of 4.190.209 inner finite element nodes. The computations documented in Table 4.1, are carried out on an Intel Paragon machine.

Comparing all three preconditioners presented, we conclude, that the application of D_h^{-1} by a hierarchically preconditioned conjugate gradient scheme is asymptotically optimal due to the bounded number of global iterations. Similar results were obtained in the case of linear elasticity in [19].

5 Conclusions

The Galerkin discretization of the symmetric boundary element domain decomposition formulation leads to a skew–symmetric, positive definite system of linear equations, where the Schur complement is positive definite and symmetric. Both local and global preconditioners for the discrete single layer potential operators and the discrete Steklov–Poincaré operators, respectively, may be constructed by related boundary integral operators without any additional work. Moreover, the construction of a preconditioner corresponding to the global Schur complement system can be reduced to the discrete hypersingular integral operator, which is given explicitly in the contrary to the nonsymmetric boundary element formulation as well as to the finite element domain decomposition methods.

References

1. J. H. Bramble, J. E. Pasciak. A preconditioning technique for indefinite systems resulting from mixed approximations of elliptic problems. Math. Comput. 50, pp. 1–17, 1988.
2. J. H. Bramble, J. E. Pasciak, A. H. Schatz. The construction of preconditioners for elliptic problems by substructuring I. Math. Comput. 47, no. 175, pp. 103–134, 1987.
3. P. G. Ciarlet. The Finite Element Method for Elliptic Problems. North–Holland, 1978.
4. M. Costabel. Starke Elliptizität von Randintegraloperatoren erster Art. Habilitation thesis, TH Darmstadt, 1984.
5. M. Costabel. Symmetric methods for the coupling of finite elements and boundary elements. In: Boundary Elements IX (C. A. Brebbia, G. Kuhn, W. L. Wendland eds.), pp. 411–420, Springer, 1987.
6. M. Costabel. Boundary integral operators on Lipschitz domains: Elementary results. SIAM J. Math. Anal. 19, no. 3, pp. 613–626, 1988.
7. M. Costabel, W. L. Wendland. Strong ellipticity of boundary integral operators. Crelle's J. Reine Angew. Math. 372, pp. 34–63, 1986.
8. G. Haase, B. Heise, M. Kuhn, U. Langer. Adaptive Domain Decomposition Methods for Finite and Boundary Element Equations. This book.
9. G. C. Hsiao, W. L. Wendland. Domain Decomposition Methods in Boundary Element Methods. In: Domain Decomposition Methods in Science and Engineering (R. Glowinski et. al. eds.), SIAM, pp. 41–49, 1990.
10. B. N. Khoromskij, W. L. Wendland. Spectrally equivalent preconditioners in substructuring techniques. East–West J. Numer. Math. 1, pp. 1–25, 1992.
11. A. Meyer, S. Rjasanow. An effective direct solution methods for certain boundary element equations in 3D. Math. Meth. Appl. Sci. 13, pp. 43–53, 1990.
12. J. M. Ortega, W. C. Rheinboldt. Iterative Solution of Nonlinear Equations in Several Variables. Academic Press, 1970.
13. S. Rjasanow. Vorkonditionierte iterative Auflösung von Randelementgleichungen für die Dirichlet–Aufgabe. Wissenschaftliche Schriftenreihe 7, TU Karl–Marx–Stadt, 1990.
14. G. C. Hsiao, E. Schnack, W. L. Wendland. A hybrid coupled finite–boundary element method. Preprint, Universität Stuttgart, 1995.
15. R. Quatember, O. Steinbach, W. L. Wendland. Domain decomposition based solvers for industrial stress analysis with boundary elements. In: Progress in Industrial Mathematics at ECMI 94 (H. Neunzert ed.), John Wiley & Sons and B. G. Teubner, pp. 88–95, 1996.
16. A. Schatz, V. Thomée, W. L. Wendland. Mathematical Theory of Finite and Boundary Element Methods. Birkhäuser–Verlag, Basel, 1990.
17. S. Sirtori. General stress analysis by means of integral equations and boundary elements. Meccanica 14, pp. 210–218, 1979.
18. B. F. Smith. Domain decomposition algorithms for the partial differential equations of linear elasticity. PhD–Thesis, New York University, 1991.

19. O. Steinbach. Gebietszerlegungsmethoden mit Randintegralgleichungen und effiziente numerische Lösungsverfahren für gemischte Randwertprobleme. PhD–Thesis, Universität Stuttgart, 1996.
20. O. Steinbach. Fast solvers for boundary integral equations related to mixed boundary value problems. In preparation.
21. O. Steinbach and W. L. Wendland. Efficient Preconditioners for Boundary Element Methods and their Use in Domain Decomposition Methods. In: Proceedings of the Eighth International Conference on Domain Decomposition Methods in Science and Engineering (R. Glowinski et. al. eds.), to appear.
22. O. Steinbach, W. L. Wendland. The construction of optimal preconditioners in the boundary element method. In preperation.
23. P. Le Tallec. Domain decomposition methods in computational mechanics. Comput. Mech. Adv. 1, no. 2, pp. 121–220, 1994.
24. W. L. Wendland. On asymptotic error estimates for combined BEM and FEM. In: Finite Element and Boundary Element Techniques from Mathematical and Engineering Point of View (E. Stein, W. L. Wendland eds.), CISM Courses and Lectures 301, Springer, pp. 273–333, 1988.

Benchmark Problems for Boundary Element Methods

H. Schulz, O. Steinbach, and W. L. Wendland

1 Introduction

Over the last 6 years a lot of new techniques and algorithms for boundary element methods have been developed; e.g. for the numerical integration of singular integrals, for fast matrix times vector realizations as the panel clustering method or wavelet techniques to get sparse stiffness matrices; for domain decomposition methods and parallel algorithms for boundary integral equations including preconditioning techniques; for applications to complicated three dimensional nonlinear problems in structural mechanics.

Related to the theoretical results, computer codes have been developed by different groups, partially in cooperation, to incorporate, confirm and exploit the theory. Due to the analytically involved formulation of boundary element methods, we could not expect to generate an optimal code for solving general boundary value problems by the BEM. Instead, the current goal is now to compare particular components in order to find an optimal strategy for the future design and development of boundary element software.

Such a comparison via benchmark problems should be done with respect to various criteria. With our first proposal we started a significant discussion on the design of benchmark problems, i.e. the problems and criteria. At this stage, only a qualitative comparison can be expected. The ultimate aim will be a quantitative comparison with respect to the computational costs for results within a given tolerance. Benchmark problems should be simple on one hand, but they should also contain typical features of real applications. As model problems we have to include the Laplace equation and problems in linear elasticity in two and three dimensions. A crucial task is the numerical integration of the singular surface integrals for the discretization of the boundary integral operators, i.e. the amount of work for the corresponding computer realization. In general, this has to be related to the accuracy of the numerical solution, so we have to check also the order of convergence by applying academic examples. Another important topic is the solution of the resulting systems of linear equations by preconditioned iterative solvers including wavelet and panel clustering techniques. The discussion of performance is in general based on the formulation of the boundary value problem as a system of boundary integral equations, i.e. the direct or the indirect approach; the discretization of these systems, either by collocation or Galerkin methods, or more general combinations of these. Error estimators, mesh adapting algorithms and hp–methods are needed to get an optimal relation between the amount of work and the required accuracy. Clearly, the costs of an estimator should be low. Due to the fast development of parallel computer architectures, domain decomposition methods and the parallelization of existing codes are useful to handle coupling problems and problems with a large number of degrees of freedom. However, these methods require an adequate formulation and appropriate preconditioning techniques.

The most important criterion for a boundary element realization should be the robustness of the code with respect to singular perturbations either in geometry or in given boundary data. In general we are interested in the numerical amount of work to get a specified accuracy of the approximate solution, and in the way how this result is obtained.

The comparison of existing boundary element codes depends strongly on the individual realization, the programming language and on the computer architecture used. Hence we have to use a simple program including main features of the boundary element methods to scale up all computers used in the comparison.

Note, that the discussion of benchmark problems – not only for boundary element methods – is still in progress. Our main goal is not only to find out and to demonstrate, which method can solve which class of problems, we also want to provide a basis for modern techniques in boundary element methods and their application in professional software codes.

2 A first attempt for defining benchmark problems

As a simple model problem for the comparison of the numerical integrations of the boundary integral operators as well as of the solution procedures for the resulting systems of linear equations we consider first the Laplace equation in a two–dimensional L–shaped domain with either Dirichlet or Neumann boundary conditions. To check the order of convergence we used as an explicit solution the fundamental solution with a source point outside the domain. The main task is then to solve this boundary value problem by computing the remaining Cauchy datum and the solution along a described line inside the domain. In general, there are different possibilities for solving this problem as described in the introduction.

The first benchmark problem was chosen as follows:

$$\Delta u(x) = 0 \quad \text{for } x \in \Omega_L = [-0.25, 0.25]^2 \backslash [-0.25, 0]^2.$$

Dirichlet problem:

$$u(x) = \ln |x - x^*| \quad \text{for } x \in \Gamma.$$

Neumann problem:

$$\frac{\partial u}{\partial n}(x) = \frac{(x - x^*, n(x))}{|x - x^*|^2} \quad \text{for } x \in \Gamma.$$

The given source point is $x^* = (-0.01, -0.01)^\top$.

As a more complicated problem we consider in the domain Ω_L a mixed boundary value problem in two–dimensional linear elasticity with plain strain, where the boundary conditions are given according to the following sketch:

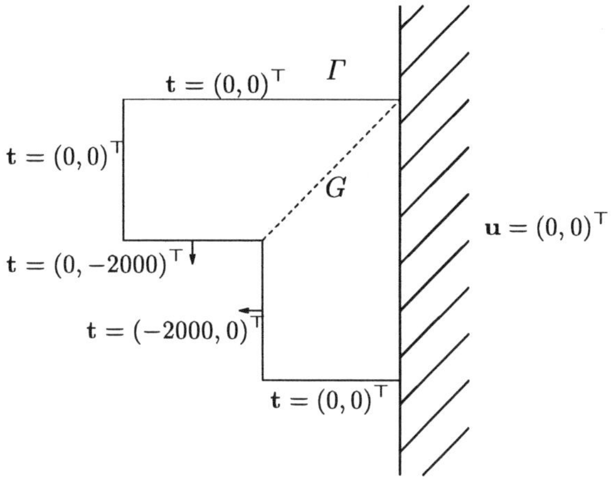

As material parameters we choose for the Young modulus $E = 200.000$ and for the Possion ratio $\nu = 0.3$. In addition to the complete Cauchy data, the displacements and the main stresses along G are required.

As a three–dimensional problem in linear elasticity we consider an L–cube, the so–called Fichera corner as sketched in the following figure:

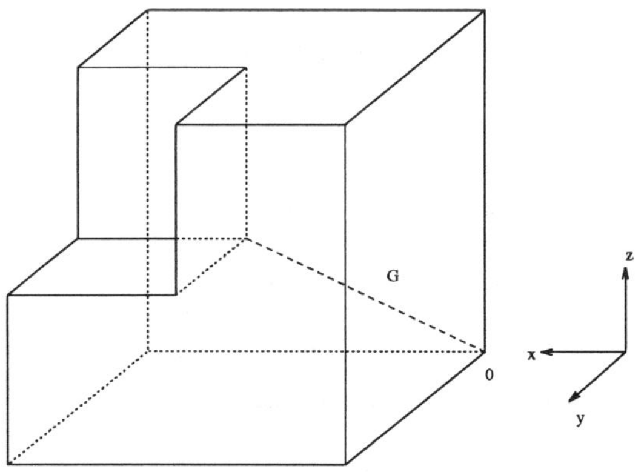

The length of the edges of the cube are two units, while the length of the Fichera edge is one unit. The material parameters E and ν are the same as in the two–dimensional problem. Mixed boundary conditions are given along the faces as

$$\begin{aligned} z = 0 &: t_1 = t_2 = t_3 = 0; \\ z = 2 &: t_1 = t_2 = u_3 = 0; \\ x = 0 &: t_1 = -1000, t_2 = t_3 = 0; \\ y = 2 &: t_1 = u_2 = t_3 = 0; \\ y = 0 &: t_1 = t_2 = t_3 = 0; \\ x = 2 &: u_1 = t_2 = t_3 = 0 \end{aligned}$$

and along the sides of the Fichera edge:

$$u_1 = u_2 = u_3 = 0.$$

Again we require the remaining Cauchy data and the displacements and the principal stresses along G.

3 Results

Here we present some results for the above benchmark problems obtained by different groups in the priority research programme. For a detailed description of the algorithms used we refer to the corresponding contributions in this volume.

The uniform discretization of the Dirichlet problem for the Laplace equation gives for $N = 1024$ boundary elements an absolute L^2–error of the normal derivative of approximately 0.06 for piecewise constant trials. This accuracy was obtained by either a Galerkin discretization of Symm's integral equation or with the method of difference potentials by A. Hommel (Chemnitz). To get moderately better convergence results one may use parallel computers to realize a domain decomposition with $N = 4096$ boundary elements and obtain an error of $1.41 \cdot 10^{-2}$, this result was presented by M. Kuhn (Linz). Due to the large number of degrees of freedom, it seems more preferable to use an adaptive discretization, as by M. Maischak (Hannover). With an adaptive hp–version he got an L^2–error of 10^{-2} by using just 100 boundary elements. However, such an adaptive discretization requires the use of an appropriate preconditioner to limit the number of iterations needed. For the hp–method, the simplest choice is diagonal preconditioning, which brings some improvement only in the adaptive case. Then the number of iterations for the adaptive version grows only as in the non–preconditioned uniform case. For the Dirichlet problem also collocation methods were presented by R. Bialecki (Erlangen). Since he does not use piecewise constant trials, he has to formulate additional equations at the Dirichlet corners or to use non–uniform trial functions. In the first case, the result depends on the special choice of these additional equations. Note, that in the case of a Galerkin formulation there are no such problems.

The mixed boundary value problem in linear elasticity was also solved via different approaches. Collocation methods based on the boundary integral equation including only single and double layer potentials were used by M. Kreienmeyer (Hannover) for an uniform mesh comparing a sequentieal with a parallel implementation as well as an adaptive mesh refinement. E. Schnack and his crew (Karlsruhe) used a macro boundary element coupled with a finite element method where the Neumann–Dirichlet map is realized by a Neumann series based on the discrete double layer potential. The program package BEAT was used by G. Kuhn and his group (Erlangen) to solve the two–dimensional problem in linear elasticity. Due to the use of only one boundary integral equation for collocation one cannot expect a particularly efficient solution method. More suitable is the symmetric formulation discretized by some Galerkin–Bubnov method. This method was used by U. Brink (Hannover) based on the code by M. Maischak (Hannover), as well as in the coupled finite–boundary element software package by U. Langer and his co–workers (Linz).

The situation for the three–dimensional problem in linear elasticity is quite similar. Here codes based on collocation were used by G. Kuhn (Erlangen), L. Gaul (Stuttgart), E. Schnack (Karlsruhe) and R. Quatember (Stuttgart). Most of these codes are based on adaptive numerical integration. The only Galerkin discretization is due to H. Andrä (Karlsruhe), who realized the symmetric formulation in three–dimensional linear elasticity.

4 Conclusions

The problems and results shown here present a first attempt for defining a set of benchmark problems for boundary element methods. Up to now we still discuss the choice of benchmarks as well as the presentation of the results. In this first set of problems we asked mostly for the solution of mixed boundary value problems, where the results are given either in tables or by graphics. Hence, still a fair and detailed comparison of the results and of the methods used, is not possible. Therefore we startet a second attempt stating the problems and the results required more detailed in corresponding tables [1]. Everybody is invited to join this forum, to propose benchmark problems and to describe specified results to be listed in those tables.

The main goal hereby should be the comparison of mathematical methods, i.e. the numerical amount of work with respect to the accuracy of the solution. Moreover, we are also interested in the solution of real application problems.

Acknowledgements

The authors thank all participants in the discussions and all contributors to this first set of benchmark problems, whose results made these first comparisons possible.

References

1. M. Kuhn, O. Steinbach. Benchmarking for Boundary Element Methods. Technical Report No. 96–5, Universität Linz and Preprint Universität Stuttgart, 1996.

Index

Printing: Mercedesdruck, Berlin
Binding: Buchbinderei Lüderitz & Bauer, Berlin